Thermodynamics

Thermodynamics

Dwight C. Look, Jr.

Harry J. Sauer, Jr.

University of Missouri, Rolla

Brooks/Cole Engineering Division

Monterey, California

the whole art
of teaching
is only the art
of awakening
the natural
curiosity
of young minds . . .
Anatole France

Sponsoring Editor: Ray Kingman
Production Services Coordinator: Stacey C. Sawyer
Production Service: Mary Forkner, Publication Alternatives
Manuscript Editor: Don Yoder
Interior Design: Michael Rogondino
Cover Design: Lenn Chrestman
Illustrations: Pat Rogondino
Typesetting: Interactive Composition Corporation

Brooks/Cole Engineering Division
A Division of Wadsworth, Inc.

Printed in the United States of America

10 9 8 7 6 5 4 3

Library of Congress Cataloging in Publication Data

Look, Dwight C., 1938–
    Thermodynamics.

    Bibliography: p.
    Includes index.
    1. Thermodynamics. I. Sauer, Harry J.,
1935–   . II. Title
QC311.L584   536'.7   81-17097
ISBN 0-8185-0491-9   AACR2

Preface

Energy is of great importance to everyone, especially to the engineers of today and tomorrow. Thermodynamics, the science of energy, is an integral element in the education of *all* types of engineers. However, the needs of the various engineering disciplines often differ considerably regarding thermodynamics and its applications. Some branches require a thorough and rigorous treatment. For this approach, many excellent thermodynamic textbooks are currently available.

The goal of this book is to provide a basic introduction to the *art and science* of engineering thermodynamics along with many demonstrations of the significance of thermodynamics. This text attempts to provide, under a single cover, a firm physical understanding of thermodynamics and the basic tools necessary to obtain quantitative solutions to common engineering applications involving energy and its conversion, conservation, and transfer.

Even though this book is directed toward junior or senior students who are not majoring in mechanical engineering, it may serve as a handy reference for mechanical engineers as well. Part 1 of the book is devoted to basic thermodynamics, which is presented in essentially the classic way; Part 2 presents the basic uses of these principles. Chapters 1 through 3 discuss the fundamentals and basic concepts of thermodynamics. Chapter 4 discusses the first law of thermodynamics in its various operational forms. This is one of the most important chapters of the book and should be studied carefully. Chapters 5 and 6 deal with the elusive second law of thermodynamics and its restricting nature.

Part 2 of the book begins with Chapter 7. This chapter presents examples of simple thermal systems using thermodynamic principles. Chapter 8 introduces some of the complications of systems in use today. The quality of energy is the subject of Chapter 9. In Chapter 10 we discuss mixtures and psychrometrics and their relationship to environmental control. Reacting systems (combustion) are briefly covered in Chapter 11.

The last chapter of this textbook is devoted to heat transfer. At first glance, it may seem somewhat unorthodox to present the modes of heat transfer in a thermodynamics book. Nevertheless, we believe that this brief coverage should be made available to those who wish to teach it since heat transfer does deal with thermal energy and the results must conform to the rules, laws, and definitions presented in the basic thermodynamics course. Chapter 12 starts with a brief introduction, deals directly with the subject of conduction, and finishes with a discussion of convection and radiation. Included in the conventional presentation of radiation is a discussion of solar energy.

The Appendices are divided into two sections: A and B. Appendices A-1 to A-6 present tables of physical constants and properties; of particular importance are the abbreviated steam tables (both SI and English). Appendices B-1 to B-10 present special discussions that may be covered in detail at the discretion of the instructor.

We should point out that SI units are used in conjunction with English units in the text. Our intent is to allow the student to become comfortable with both systems.

Since thermodynamics and its application may be viewed as an art as well as a science, experience with problems must be obtained as a prelude to true understanding. Thus the serious student should try to solve the homework problems at the end of each chapter. In addition to these problems, a large number of examples are distributed throughout the text. Our approach in this regard is based on what Confucius is reputed to have said:

> *I hear, and I forget . . .*
> *I see, and I remember . . .*
> *I do, and I understand*

Acknowledgments

It is impossible to acknowledge all the people who have, in one way or another, contributed to this book. Through writing, discussion, and lecturing, many people have helped. Unfortunately, however, the sources of many good ideas, examples, problems, approaches, and techniques have long been forgotten. In the following pages we trust that adequate recognition is given throughout to sources of information. Nevertheless, there are obvious notions that may be perceived, and for these backgrounds we are indebted.

Moreover, we appreciate the efforts of the teachers and many students who suffered through the development of this text and its classroom testing. Their suggestions and their encouragement contributed greatly to the completion of the book. Thank you. Also, we especially thank the following manuscript reviewers for their many helpful suggestions: Dale Beckstead; C. Birnie, Pennsylvania State University; Marty Cerza, Rutgers University; George T. Craig, San Diego State University; George P. Mulholland, New Mexico State University; Thornton W. Price, Arizona State University; C. M. Simmang, Texas A & M University; and Edward O. Stoffel, California Polytechnic University at San Luis Obispo.

D. C. Look, Jr.
H. J. Sauer, Jr.

Contents

Appendices

Part

Principles of Thermodynamics

Chapter

1

Fundamental Concepts and Definitions

Welcome to the world of thermodynamics! To make your way in this world, you must tread along the path with your eyes wide open, for it is beset with difficulties. Thus, the first steps along the way will be to prepare you to understand this world. To accomplish this, Chapter 1 presents the basic units, definitions, and concepts. In particular, we will discuss the ideas of heat and work.

1–1 The Nature of Thermodynamics

Thermodynamics is the science of energy, the transformation of energy, and the accompanying change in the state of matter. Because every engineering operation involves an interaction between energy and matter, the principles of thermodynamics can be found, in whole or in part, in all engineering activities. Thus, the professional engineer must have more than a nodding acquaintance with thermodynamics—occupational advancement may depend upon an understanding of the science.

The word "thermodynamics" brings to mind a picture of thermal energy in motion. This picture is misleading, however, because the classic methods of thermodynamics—the methods that we will study—deal with systems in equilibrium only. That is, classical thermodynamic methods are dependent upon the end conditions and require that all of the "acting power" be balanced. In addition, the rates of everyday processes cannot be inferred by this study (unfortunately for the engineer). Thus "thermostatics" or "equilibrium thermodynamics" might be more accurate names for this science and this course.

Thermodynamics may be studied from either a microscopic (Lagrangian) or macroscopic (Eulerian) point of view. The microscopic view considers matter to be composed of molecules and concerns itself with the actions of these individual molecules (an individual-particle approach). The macroscopic view is concerned with effects of the action of *many* molecules. This approach then considers the average properties of a large group of molecules (a state-of-the-system approach). When the study is macroscopic, the field is called *classic thermodynamics* or just *thermodynamics*. When the study is microscopic, the science is called *statistical thermodynamics* and includes *kinetic theory*, *statistical mechanics*, *quantum mechanics*, and *wave mechanics*.

Thermodynamics is a physical theory of great generality impinging on practically every phase of human experience. It may be called the description of the behavior of matter in equilibrium and its changes from one equilibrium state to another. Thermodynamics is based on two master concepts and two great principles. The concepts are energy and entropy; the principles are the first and second laws of thermodynamics. These are not really laws in the strict physical sense, since they do not describe regularities in experience directly; rather they are hypotheses whose use is justified by the agreement of their consequences with experience. The idea of energy is the embodiment of the attempt to find in the physical universe an invariant—something that remains constant in the midst of obvious flux. It is in the transformation process that nature appears to exact a penalty, and this is where the second principle makes its appearance. Every naturally occurring transformation of energy is accompanied, somewhere, by a loss in the availability of energy for the future performance of work.

Some History

The subject of thermodynamics has evolved since the beginning of the eighteenth century. G. W. Leibnitz laid the groundwork for a formal statement of the first law of thermodynamics in 1695. He showed that the sum of kinetic and potential energies remains constant in an isolated system. However, the real birth of this science came about by the operation of the first working steam engines (T. Savery in 1685 and T. Newcomen in 1712). Statements of the first and second laws of thermodynamics were made in the nineteenth century. They were preceded, however, by late-eighteenth-century developments of Joseph Black (specific heat, latent heats of transformation and "the caloric theory"—ca., 1720). The unique caloric theory was used in explanations of calorimetry and other processes involving heat. We hasten to say that this caloric theory was used to explain these physical phenomena . . . erroneously. It was erroneous because in the explanation energy was created from friction. Substantial experimental proof of the error in the caloric theory was presented by James Joule between 1843 and 1849. (This work is also the basis for the first law of thermodynamics.) Works of H. Helmholtz in 1847, Lord Kelvin in 1848, and R. Clausius in 1850, along with that of Joule, finally laid the caloric theory to rest.

The German physicist Rudolf Clausius (1822–1888) invented the concept of entropy to describe quantitatively the loss in available energy in all naturally occurring transformations. Thus, although the natural tendency is for heat to flow from a hot to a colder body with which it is placed in contact, corresponding to an increase in entropy, it is perfectly possible to make heat flow from the colder body to the hot body—as is done every day in a refrigerator. But everyone knows that this costs money. It does not take place naturally or without some extra effort exerted somewhere.

Clausius epitomized the fundamental principles of thermodynamics: *The energy of the world stays constant; the entropy of the world increases without limit.* If the essence of the first principle in everyday life is that we cannot get something for nothing, the second principle emphasizes that every time we do get something we reduce by a measurable amount the opportunity to get that something in the future, until ultimately the time will come when there will be no more getting. This is the "heat death" envisioned by Clausius. The whole universe will have reached a dead level of temperature; and though the total amount of energy will be the same as ever, there will be no means of making it available—the entropy will have reached its maximum value.

By the beginning of the twentieth century, thermodynamics had been structured by such great scientists as R. Clausius, J. C. Maxwell, M. Planck, J. H. Poincaré, and J. W. Gibbs into a sound science. Other approaches by such scientists as C. Carathéodory and H. B. Callen have surfaced recently. These approaches are an axiomatic structure—that is, a listing of abstract statements that serve as a foundation for the science. More thumbnail historical sketches of many important persons of thermodynamics may be found in Appendix B-10.

System and Surroundings

Most applications of thermodynamics require the definition of a **system** and its **surroundings**. A system can be any object, any quantity of matter, any region of space, selected for study and set apart (mentally) from everything else; the everything else becomes the surroundings. The systems of interest in thermodynamics are finite, and the point of view taken is macroscopic rather than microscopic. That is to say, no account is taken of the detailed structure of matter; only the general characteristics of the system, such as its temperature and pressure, are regarded as thermodynamic coordinates. These characteristics are advantageously dealt with because they have a direct relation to our sense perceptions and are measurable.

A **thermodynamic system** *is thus a region in space or quantity of matter within a prescribed volume, set apart for this purpose of analysis.* This defining volume may be either moveable or fixed and either real or imaginary. What is not system is referred to as the surroundings (i.e., the universe will consist of the system plus the surroundings).

An *isolated* **system** can exchange neither mass nor energy with its surroundings. If a system is not isolated, its boundaries may permit either mass or energy or both to be exchanged with its surroundings. If the exchange of mass is allowed, the system is said to be open; if only energy and not mass may be exchanged, the system is closed (but not isolated) and its mass is constant. *Thus the* **closed system** *is a fixed mass that remains unchanged in amount and identity. The* **open system** *is indicated as a region in space.* In general, empty space does not have thermodynamic properties; only matter does. Hence the open system specification implies that the system consists of all the matter that is within the volume (the space) at the instant in which an analysis is made. The volume (frequently called a control volume) can expand, contract, or move, and it need not be contiguous.

Closure

Thermodynamics, like all sciences, is built upon a logical sequence of basic laws. These laws were, of course, deduced from experimental observation. In the sections that follow, we present these laws and the related thermodynamic properties and apply them to a number of representative examples. The student should try to gain a thorough understanding of the fundamentals and learn to apply them to thermodynamics problems. The purpose of the examples and problems is to further this twofold objective. It is not necessary to memorize numerous equations, for problems are best solved by applying the definitions and laws of thermodynamics.

Thermodynamic reasoning is deductive rather than inductive. That is, the reasoning is always from the general law to the specific case. To illustrate the elements of thermodynamic reasoning that are similar to other ways of reasoning and those that are different, we may divide the analytic processes arbitrarily into two steps:

1. The first step is the idealization or substitution of an analytic model for a real system—a step that is taken in all engineering sciences.

These idealizations are fairly easy to make after a little experience. Skill in making them is an essential part of the engineering art.

2. The second step, unique to thermodynamics, is the deductive reasoning from the first and second laws of thermodynamics.

These steps will involve an energy balance, a suitable properties relation, and an accounting of entropy changes.

1–2 Definition of Units

This book presents thermodynamics from a macroscopic point of view. As such, the subject is presented in terms of quantities that are measurable. This means that mastery of the units involved is essential. Acquiring the habit of including units with every property discussed and every number calculated will result in fewer embarrassing errors.

To present the basic time unit, one must choose between the mean solar day and the sideral day. Both are measures of one complete revolution of the earth relative to a fiducial point. In the first case this point is our own sun, and in the second case it is some other easily defined but fixed star. For our purposes the mean solar day is used. The resulting standard time unit is the mean solar second. (There are 86,400 mean solar seconds [(24 hours/day) (60 minutes/hour) (60 seconds/minutes)] in one mean solar day.) In the English system "seconds" is abbreviated "sec" whereas in SI the abbreviation is "s." Because both systems are used in this book, both "sec" and "s" will be found.

The basic unit of length is the meter. The currently defined length of a meter is 1,650,763.73 wavelengths of the orange line of krypton 86. Though somewhat obscure to you, this standard is an invariant, accessible standard, easily reproduced in the laboratory. The yard has been defined (at least in the English-speaking countries) as 0.9144 meters. Thus the inch is defined as 1/36 of this length, or

$$1 \text{ in.} = 0.0254 \text{ meter} = 2.54 \text{ cm}$$

Mass and system are intimately related in this book. In fact, a system identity depends upon the characteristics of the mass. We will use as a basic unit of mass, the pound mass (designated lbm)*. This unit is defined from the standard kilogram as

$$1 \text{ lbm} = 0.45359237 \text{ kg} \quad \text{or} \quad 1 \text{ kg} = 2.2046 \text{ lbm}$$

The relationship of force and mass is a very important one and is sometimes a difficult matter for students. Force and mass are related by

---

*Another unit sometimes seen is the slug. In terms of equivalence,

$$1 \text{ slug} = 32.174 \text{ lbm} = 14.594 \text{ kg}$$
$$g_c = 1 \text{ slug-ft/lbf-sec}^2$$

Newton's second law of motion, which states that the force acting on a body is proportional to the rate of change of momentum. For our purposes, this rate reduces to the product of the mass and the acceleration in the direction of the force:

$$F \propto ma$$

Notice that this equation indicates that

$$\text{lbf} \; [=] \; \text{lbm-ft/sec}^2$$

(The sign [=] implies "has units of.") The two sides of this unit equation do not look alike. Obviously, the proportionality constant indicated by Newton's second law must correct this problem. Thus:

$$F = \frac{ma}{g_c} \tag{1-1}$$

where g_c is a constant that relates force and mass (and also length and time).

To define g_c, we must recall that in the English engineering system of units, force is a fundamental quantity while mass (as discussed above) is referred to as a derived quantity. The standard pound force is defined as the gravitational pull of the earth of a standard mass at a particular place on the earth. When this mass is suspended at sea level, one pound force is defined to be numerically equal to one pound mass. Also, at sea level the standard acceleration of gravity is 32.1740 ft/sec². Substitution of this information into Newton's second law yields

$$1 \; \text{lbf} = \frac{1 \; \text{lbm}(32.174 \; \text{ft/sec}^2)}{g_c}$$

or

$$g_c = 32.174 \; \text{lbm-ft/lbf-sec}^2$$

Note that at a location where the acceleration of gravity is 29 ft/sec² a mass of 10 lbm exerts a force of 9 lbf. Perhaps a better way to present this information is

$$1 \; \text{lbf} = 32.174 \; \text{lbm-ft/sec}^2$$

The term pound and the symbol lb will not be used in this book. The reason for this is to avoid confusion between lbm and lbf.

In the International System (SI), the unit of force may be deduced from Newton's second law. In this system, the proportionality constant g_c takes the form

$$g_c = 1 \; \text{kg m/(N} \cdot \text{s}^2)$$

In order to keep the physical quantities of mass and force separate, you will need to remember that weight is equivalent to force and not mass. Also, a good clue is that because force is defined with respect to gravitational pull, the weight (force) varies with altitude and mass does not.

The units of pound-mole and gram-mole will not be used in this book—although they are very common units, particularly to the chemist. These quantities are defined as the molecular weight of any substance, element or compound, in pounds or grams (respectively). If these units are encountered, division by the appropriate molecular weight (e.g., lbm/lb-mole) will change the units to those of this book.

EXAMPLE 1-1

A 1.25-kg mass is accelerated by a force of 11.5 lbf. What is the acceleration in ft/sec² and cm/s²?

SOLUTION:
$$F = \frac{ma}{g_c}$$

or

$$a = \frac{Fg_c}{m} = \frac{(11.5 \text{ lbf})(32.174 \text{ lbm-ft/lbf-sec}^2)}{(1.25 \text{ kg})2.2046 \text{ lbm/kg}}$$

$$= 134.26 \text{ ft/sec}^2$$

$$= 134.26 \text{ ft/sec}^2 \ (2.54 \text{ cm/in.})(12 \text{ in./ft})$$

$$= 4092.40 \text{ cm/s}^2$$

1–3 Properties and States

A **property** of a system is any characteristic of the system. A listing of a sufficient number of independent properties constitutes a complete definition of the state of a system. The common thermodynamic properties are specific volume or density, temperature, pressure, internal energy, enthalpy, and entropy. Other names for thermodynamic properties include state variables and thermodynamic coordinates.

The **state** of a system is its condition or configuration described in sufficient detail that one state may be distinguished from all other states. The state is described by macroscopic properties. On the other hand, a property is any measurement or quantity derived from a measurement used to describe the state of the system. The word macroscopic is implied in the word property because our ability to measure microscopic quantities is very limited. Thus the unique description of the state of the system by properties requires that unambiguous values be used for the characterization. No hysteresis effects are allowed. (Each property has exactly the same value for a given state regardless of the method used to arrive at that state.) In addition to defining a property, the minimum number of properties needed for this unique description is of great importance.

Two types of properties are encountered in the study of thermodynamics: intensive and extensive. An **intensive property** is independent of the mass enclosed by the boundaries of the system; an **extensive property** is directly and linearly proportional to the mass of the system. Therefore, such properties as temperature and pressure are intensive while properties such as kinetic energy and volume are extensive. Because the mass dependence in extensive properties is linear, division of the extensive property by the mass of the system yields an intensive property of the system. The usual example here is the specific volume $(v = V/m)$.

Specific Volume or Density

Density (usually denoted by the symbol ρ) is an intensive property we have met before. It is the ratio of the volume and its corresponding mass. It is a perfectly good property, but (strictly as a matter of convenience) we will deal with the reciprocal of the density and call it the specific volume (denoted by the symbol v).

During the course of this book, we may refer to the density (specific volume) at a point. Obviously, the concept of a point may be misleading—if the point is at the nucleus of an atom, the density will be quite large. Conversely, if the point is not at a nucleus, the density will be quite small. For this reason, the density is defined as

$$\rho = \lim_{\Delta V \to A} \left(\frac{\Delta m}{\Delta V} \right)$$

where A is a very small volume, but large enough to contain enough atoms or molecules to be statistically significant. (That is, it is a continuum.) Then $v = 1/\rho$.

EXAMPLE 1-2

A box having a volume of 2.1 ft³ contains 19.4 lbm of gas. What is the specific volume of the gas? If 8.6 lbm of gas escapes, determine the specific volume and the final density of the system.

SOLUTION:
$$v = \frac{V}{m} = \frac{2.1 \text{ ft}^3}{19.4 \text{ lbm}} = 0.1083 \text{ ft}^3/\text{lbm}$$

After 8.6 lbm escapes, 10.8 lbm of the gas remains in the same volume. Thus

$$v = \frac{V}{m} = \frac{2.1 \text{ ft}^3}{10.8 \text{ lbm}} = 0.1944 \text{ ft}^3/\text{lbm}$$

$$\rho = \frac{1}{v} = 5.143 \text{ lbm/ft}^3$$

Pressure

Pressure is defined as the normal component of force per unit area. Assuming a fluid (liquid or gas) is not flowing, the pressure at any point in the fluid is isotropic. Pressures (i.e. stresses) exist in solids but cannot be measured except at the surface. Note the phrase "pressure at a point" may be misleading. So, for the same reasons presented in the discussion of specific volume, we must consider pressure to be defined as

$$p = \lim_{\Delta A \to B} \left(\frac{\Delta F_n}{\Delta A} \right)$$

where ΔF_n is the normal component of the force acting on ΔA and B is a very small area but sufficiently large to encompass enough atoms and molecules to be statistically significant (a continuum).

The gauges (both vacuum and pressure) we meet every day usually measure the difference between the absolute pressure and the local ambient pressure. (This is sometimes the atmospheric pressure.) In our study of thermodynamics we must use the absolute pressure. The relation between these two pressures is illustrated in Figure 1-1.

Perhaps a better way to present this information is by the equation

$$p(\text{absolute}) = p(\text{ambient}) + p(\text{pressure gauge})$$

or

$$p(\text{absolute}) = p(\text{ambient}) - p(\text{vacuum gauge}) \tag{1-2}$$

Many ingenious devices are used to measure pressure: the deadweight piston gauge, manometer, barometer, McLeod gauge, Bourdon gauge, strain gauge, and more. The Bourdon gauge, the type most often used in industry, is in a class called *pressure transducers*. To make the measurement, an elastic element is used to convert fluid energy to mechanical energy. Figure 1-2 presents the details of this device.

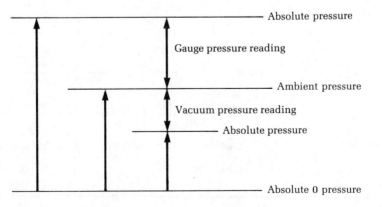

Figure 1-1. Terms used in pressure measurement.

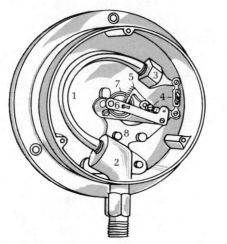

1. Bourdon tube
2. Tube socket
3. Tip
4. Adjustable linkage
5. Geared sector
6. Pointer shaft
7. Hair spring
8. Support for mechanism

Figure 1-2. Details of Bourdon Pressure Gauge. *(Reprinted by permission from Crosby Valve & Gage Co.)*

EXAMPLE 1-3

What is the pressure of the fluid in the chamber indicated in the accompanying figure if atmospheric pressure p_a is 14.7 lbf/in.^{2}, $y_2 = 1$ m, $y_1 = 0.51$m, $\rho = 13.6$ gm/cm^3, and $g = 29.4$ ft/sec^2?

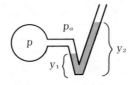

SOLUTION: $p_a + \dfrac{\rho g y_2}{g_c} = p + \dfrac{\rho g y_1}{g_c}$

$$p = p_a + \frac{\rho g}{g_c}(y_2 - y_1) \qquad \text{(absolute pressure)}$$

$$p - p_a = +\frac{\rho g}{g_c}(y_2 - y_1) \qquad \text{(gauge pressure)}$$

$$p = 14.7\frac{\text{lbf}}{\text{in.}^2} + 13.6\frac{\text{gm}}{\text{cm}^3}\,29.4\frac{\text{ft}}{\text{sec}^2}\frac{\text{lbf-sec}^2}{32.174\ \text{lbm-ft}}(0.49\text{m})$$

$$\times \left(\frac{2.2046\ \text{lbm}}{1000\ \text{gm}}\right)\left(\frac{2.54\ \text{cm}}{\text{in.}}\right)^3\frac{39.4\ \text{in.}}{\text{m}}$$

$$= 14.7\ \text{lbf/in.}^2 + 8.7\ \text{lbf/in.}^2 = 23.4\ \text{lbf/in.}^2$$

$$= 0.161\ \text{MPa}$$

EXAMPLE 1-4

A mercury barometer used to measure pressure in a chamber reads 27.5 in. of mercury. The local ambient pressure is 29.5 in. of mercury. What is the gauge pressure in $lbf/in.^2$?

SOLUTION:

$$p(gauge) = p(abs) - p(amb)$$

$$= (27.5 - 29.5) \text{ in. Hg}$$

$$= -2 \text{ in. Hg} \left(\frac{0.49 \text{ lbf/in.}^2}{\text{in. Hg}} \right)$$

$$= -0.98 \text{ lbf/in.}^2$$

$$= -0.00676 \text{ MPa}$$

Temperature and Temperature Scales

When we touch an object and sense heat or cold, we associate this sensation with "the temperature." Often this familiar property measure fools us. As engineers we know that this definition is of no use to us. We need another approach.

To understand this idea of temperature, let us postulate the following. All parts of a system completely isolated will eventually come to and remain at the same temperature. We will refer to this temperature as *thermodynamic equilibrium*. Now consider two blocks (A and B) of a material that have been isolated such that they are in thermodynamic equilibrium with themselves. Now suppose we remove the isolation requirement and bring blocks A and B into direct physical contact. If we detect no change in *any* observable property of either block, we say that

$$T_A = T_B$$

That is to say, block A and block B are in thermal equilibrium with each other.

Let us carry this procedure one step further by considering another set of blocks, A, B, and C. Now conduct the same experiment with blocks A and C and with blocks B and C. Of course the outcomes are

$$T_A = T_C \quad \text{and} \quad T_B = T_C$$

In fact we would conclude that

$$T_A = T_B$$

Why?

The answer to this question is the zeroth law of thermodynamics. The **zeroth law of thermodynamics** states that *when two bodies are in thermal equilibrium with a third body, they in turn are in thermal* *equilibrium with each other.* As self-evident as this law seems, it really is not. That is, try as you may, you cannot derive this law from other laws. It must be accepted and used because it is our basic measurement of temperature. It is called the zeroth law because it logically precedes the first and second laws of thermodynamics (although it was not developed in that order).

Since it may be unsafe to bring two substances into direct physical contact (water and sulfuric acid or water and potassium, for example), we need a yardstick of thermodynamic equilibrium. Marks may be placed on a column of mercury, and this device can then be used to make thermodynamic equilibrium comparisons with other substances. However, with this procedure a problem remains—how to analytically compare the thermal equilibrium of two substances when they are not the same. Thus, we need a standard temperature scale. This scale must be flexible enough to be used with many different temperature measuring devices regardless of their makeup.

To create such a scale, let us assume that the temperature T depends linearly on an observable property x. (The linear scale is not unique.) Thus,

$$T = ax + b$$

We must also define some standard points. The familiar standard points are the ice and steam points of water. The **ice point** occurs for thermodynamic equilibium of ice (solid) and air-saturated water (liquid) at a pressure of 1 atm (0 C, 32 F, 273 K, 491.7 R). The **steam point** occurs for the thermodynamic equilibrium of water (liquid) in contact with water vapor (gas) at a pressure of 1 atm (100 C, 212 F, 373 K, 672.7 R). At this point, of course, the word saturated and the designations C, F, K, and R are supposedly unknown to you. They will be defined later. Now let us say the temperature is T_i when the observable property has a value of x_i at the ice point. Similarly, we will use T_s for x_s at the steam point. A little algebraic manipulation to determine a and b results in an expression for the temperature in terms of observable properties and defined temperatures:

$$T = (T_s - T_i)\left(\frac{x - x_i}{x_s - x_i}\right) + T_i \tag{1-3}$$

By defining the ice point temperature and the number of units between the ice and steam points, we now have a scale.

Notice that two scales (say A and B) may be compared immediately. That is, for the linear case

$$\frac{T_A - T_{A_i}}{T_{A_s} - T_{A_i}} = \frac{T_B - T_{B_i}}{T_{B_s} - T_{B_i}} = \frac{x - x_i}{x_s - x_i} \tag{1-4}$$

The two commonly used scales for temperature measurement are the Fahrenheit scale and the Celsius (formerly the centigrade) scale.

Note that the ice and steam points are not the only fiducial points in use today, nor are the Fahrenheit and Celsius the only scales. In terms of nomenclature, the Fahrenheit temperature will be denoted by F and the Celsius temperature will be denoted by C. The commonly used symbol for degree (°) will not be used to represent temperature. (It will be used to represent angular measurement.) The symbol T will be used to denote temperature regardless of scale.

It should be noted that the Tenth Conference on Weights and Measures in 1954 alternately defined the Fahrenheit and Celsius scales. In doing so they eliminated the two fixed-point approach in favor of a single fixed point and the magnitude of the degree. The single fixed point is the **triple point of water** *which is defined as 0.01 C. The triple point is the state of a substance in which solid, liquid, and vapor coexist in equilibrium.* For most purposes there is essential agreement between the two fixed-point and the one fixed-point scales.

In our study of thermodynamics, absolute temperatures must be used (just like absolute pressures). Absolute scales exist which correspond to both the Fahrenheit and the Celsius scales. The absolute Fahrenheit scale is called Rankine (denoted by the symbol R) while the absolute Celsius scale is called Kelvin (denoted by the symbol K). Later efforts will be made to show the basis for these relations, but for now they are given as

$$K - C = 273.2 \tag{1-5}$$

$$R - F = 459.7 \tag{1-6}$$

EXAMPLE 1-5

A new temperature scale is being introduced. This scale has a defined ice point at 0 and a steam point at 86. What is the temperature on this scale of absolute zero?

SOLUTION: Assuming a linear scale and comparing it with the Fahrenheit scale, we get

$$T = \frac{T_s - T_i}{T_{F_s} - T_{F_i}}(T_F - T_{F_i})$$

$$= \frac{86}{180}(T_F - 32) = 0.4778(-459.7 - 32)$$

$$= -234.92$$

EXAMPLE 1-6

Instead of a linear temperature scale, let us assume that the temperature T depends in the following fashion on an observable property x:

$$T = a \ln x + b$$

If $x_i = 5$ in. and $x_s = 25$ in. while $T_i = 0$ and $T_s = 100$, what is the distance in inches between $T = 0$ and $T = 25$ and between $T = 75$ and $T = 100$?

SOLUTION:
$$\left. \begin{array}{r} 0 = a \ln 5 + b \\ 100 = a \ln 25 + b \end{array} \right\} \quad \begin{array}{l} a = 62.1335 \\ b = -100 \end{array}$$

Thus $T = 62.13 \ln x - 100$ or

$$x = \exp\left(\frac{T + 100}{62.13}\right)$$

$$\Delta x (T = 0 \text{ and } 25) = 7.4767 - 5 = 2.4767 \text{ in.}$$

$$\Delta x (T = 75 \text{ and } 100) = 25 - 16.7185 = 8.285 \text{ in.}$$

Internal Energy

Internal energy relates to the energy *possessed* by a material due to the motion of the molecules, their position, or both. This form of energy may be divided into two parts. **Kinetic internal energy** *is due to the velocity of the molecules.* **Potential internal energy** *is due to the attractive forces existing between molecules.* Changes in the velocity of molecules are indicated by temperature changes of the system whereas variations in position are denoted by changes in phase of the system.

The internal energy of a system under consideration will be denoted by the symbol U. Because this energy is directly proportional to the mass of the system, it is, therefore, an extensive property. To change this quantity to an intensive property we only need to divide by the mass. This quantity would then be the specific internal energy, denoted by the symbol, u. Oddly enough the "specific" part is usually dropped (the exception is the specific volume), because the overall situation dictates whether u or U is the concern. Thus, the term internal energy will refer to both cases.

Enthalpy

It is not uncommon that during the thermodynamic analysis of a process (or a series of processes) a combination of properties may be presented in some convenient form. One such combination is $U + pV$. Therefore, we find it convenient to define a new extensive property called **enthalpy.**

$$H \equiv U + pV \tag{1-7}$$

Or per unit mass:

$$h \equiv u + pv \tag{1-8}$$

Notice that like internal energy, we have enthalpy and specific enthalpy. As has been previously stated the overall situation will dictate which form of the enthalpy is meant.

Entropy

Unlike such terms as pressure, temperature, and energy, the word *entropy* is essentially unheard in everyday conversation. Nevertheless, this unfamiliar term is helpful in making thermodynamic decisions. Later, when we define entropy by a mathematical expression, you may obtain an intuitive feeling for its usefulness. You may not be able to associate it with an easily understood physical picture or model, however. This, of course, adds to its elusiveness. Thus you should concentrate on learning how to *use* this property rather than trying to determine what it is.

To help develop your intuition, think of entropy S as a measure of the chaotic nature (the "mixed-upness") of a system or state. As the system becomes more disordered, its entropy increases. On the other hand, if a system is completely ordered the entropy should have a minimum value (maybe zero.) Boltzmann formed this idea into an operational expression. He hypothesized that $S = k \ln (\Omega) + S_0$, where k is called the Boltzmann constant and Ω is called the thermodynamic probability.* Later Planck suggested that S_0 be zero.

This was the beginning of statistical thermodynamics. In applying this theory you must be careful about the definition of the thermodynamic probability. You must also be aware that this statistical theory is most applicable to systems consisting of a very large number of "particles." When this procedure is applied to a molecular system, the results can be shown to agree with quantities that can be measured. A detailed discussion of this relationship and the calculation of absolute entropies is beyond the scope of this book.

1–4 States

The **state** *of a macroscopic system is the condition of the system characterized by the values of its properties.* We will direct our attention toward what are known as equilibrium states. The word *equilibrium* is being used in its generally accepted context—the state of balance (when forces are equal). In future discussion, the term *state* will refer to an

---

* The thermodynamic probability Ω of a macrostate is the number of corresponding microstates. For example, a macrostate of 7 on two dice may be obtained in six ways (microstates). Note that W is always greater than (or equal to) 1; it is never less than 1.

equilibrium state unless otherwise noted. The concept of equilibrium is an important one, since it is only in an equilibrium state that thermodynamic properties have any real meaning. By definition: *A system is in* **thermodynamic equilibrium** *if it is not capable of a finite, spontaneous change to another state without a finite change in the state of the surroundings.* This definition implies that all thermodynamic properties have the same value at all points of the system. There are many types of equilibrium, all of which must be met to fulfill the condition of thermodynamic equilibrium. If a system is in *thermal* equilibrium, the system is at the same temperature as the surroundings and the temperature is the same throughout the whole system. If a system is in *mechanical* equilibrium, no part of the system is accelerating ($\Sigma F = 0$) and the pressure within the system is the same as in the surroundings. If a system is in *chemical* equilibrium, the system does not tend to undergo a chemical reaction; the matter in the system is said to be inert.

When a system is isolated, it is not affected by its surroundings. Nevertheless, changes may occur in the system that can be detected with measuring devices such as thermometers and pressure gauges. Such changes are observed to cease after a period of time, however, and the system is said to have reached a condition of **internal equilibrium** such that it has no further tendency to change. For a closed system that may exchange energy with its surroundings, a final static condition may also eventually be reached such that the system is not only internally at equilibrium but also in **external equilibrium** with its surroundings.

An **equilibrium state** represents a particularly simple condition of a system. This state is subject to precise mathematical description because the system exhibits a set of identifiable, reproducible properties. Indeed, the word *state* represents the totality of macroscopic properties associated with the system. Certain properties are readily measurable (T and p) whereas other properties, such as internal energy, are recognized only indirectly. The number of properties that may be arbitrarily set at given values in order to fix the state of a system (that is, to fix *all* properties of the system) depends on the nature of the system. This number, which is generally small, is the number of properties that may be selected as independent variables for a system. These properties then represent one set of thermodynamic coordinates for the system.

To the extent that a system exhibits a set of identifiable properties, it has a thermodynamic state whether or not the system is at equilibrium. Moreover, the laws of thermodynamics have general validity, and their application is not limited to equilibrium states. The importance of equilibrium states in thermodynamics derives from the fact that a system at equilibrium exhibits a set of *fixed* properties that are independent of time and may, therefore, be measured with precision. Furthermore, such states are readily reproduced from time to time and from place to place.

Any property of a thermodynamic system has a fixed value in a given equilibrium state, regardless of how the system arrives at that state. Therefore, the change that occurs in the value of a property when a system is altered from one equilibrium state to another is always the same. This is true regardless of the method used to bring about a change between the two end states. The converse of this statement is equally true. If a measured quantity always has the same value between two given states, that quantity is a measure of the change in a property. This latter assertion will be useful to us in connection with the conservation of energy principle.

The uniqueness of a property value for a given state can be described mathematically in the following manner. The integral of an **exact differential** dY is given by

$$\int_1^2 dY = Y_2 - Y_1 = \Delta Y$$

Thus the value of the integral depends solely on the initial and final states. But the change in the value of a property likewise depends only on the end states. Hence the differential change dY in a property Y is an exact differential. Throughout this text, the infinitesimal variation of a property will be signified by the differential symbol d preceding the property symbol. For example, the infinitesimal change in the pressure p of a system is given by dp. The finite change in a property is denoted by the symbol Δ (capital delta)—for example, Δp. The change in a property value ΔY always represents the final value minus the initial value. This convention must be kept in mind.

Use of the symbol δ, instead of the usual differential operator d, is intended as a reminder that some quantities depend on the process and are not a property of the system. Thus both δ and d represent a small quantity.

1–5 Processes

A **process** is a change in state that can be stated as any change in the properties of a system. A process is described in part by the series of states the system passes through. Often, but not always, some sort of interaction between the system and surroundings occurs during a process; the specification of this interaction completes the description of the process.

A description of a process typically involves specification of the initial and final equilibrium states, the path (if identifiable), and the interactions that take place across the boundaries of the system during the process. **Path** in thermodynamics refers to the specification of a series of states through which the system passes. Of special significance in thermodynamics is a **quasi-static process** or path. During such a process the system internally must be infinitesimally close to a state of equilibrium at all times. That is, the path of a quasi-static process is a series of equilibrium steps. Although a quasi-static process is an idealization, many actual processes approximate quasi-static conditions closely. It is extremely helpful in engineering analysis that the initial and final states of a nonequilibrium process must be equilibrium states. This is essential because certain intermediate information during the nonequilibrium process is missing. Nevertheless, we are still able to predict various overall effects even though a detailed description is not possible.

Some processes have special names:

1. If the pressure does not change, the process is an **isobaric** or **constant-pressure** process.

2. If the temperature does not change, the process is an **isothermal** process.

3. If the volume does not change, the process is **isometric**.

4. If no heat is transferred to or from the system, the process is **adiabatic**.

5. If there is no change in entropy, the process is **isentropic**.

6. The whole series of processes for which pV^n is constant is referred to as **polytropic**.

A cycle is a process or, more frequently, a series of processes in which the initial and final states of the system are identical. Therefore, when all the processes of the cycle have been completed, all the properties assume these initial values.

Reversible Process

All natural occurring changes or processes are irreversible. Like a clock, they tend to run down and cannot rewind themselves. Familiar examples are the transfer of heat with a finite temperature difference, the mixing of two gases, a waterfall, a chemical reaction. All of these changes *can* be reversed—we can transfer heat from a region of low temperature to one of higher temperature, we can separate a gas into its components, we can cause water to flow uphill. The important point is that we can do these things *only at the expense of some other system*, which itself becomes run down.

A process is said to be **reversible** *if its direction can be reversed at any stage by an infinitesimal change in external conditions.* If we consider a connected series of equilibrium states, each representing only an infinitesimal displacement from the adjacent one, but with the overall results a finite change, then we have a reversible process.

All actual processes can be made to approach more or less closely a reversible process by suitable choice of conditions; but like the absolute zero of temperature, the strictly reversible process is purely a concept that aids in the analysis of certain problems. Nevertheless, the approach of actual processes to this ideal limit can be made almost as close as we please. The closeness of approach is generally limited by economic factors rather than purely physical ones. The truly reversible process would require an infinite time for its completion, but we are generally in more of a hurry than that.

The sole reason for the invention of the concept of the reversible process is to establish a standard for the comparison of actual processes. The reversible process is one that gives the maximum accomplishment—that is, it yields the greatest amount of work or requires the least amount of work to bring about a given change. It tells us the maximum efficiency toward which we may strive but which we never expect to equal. Without such an absolute standard, the attempts of engineers to improve processes would be but shots in the dark. With the reversible process as our standard, we know at once whether a process is highly efficient or whether it is very inefficient and, therefore, capable of considerable improvement.

Another aspect of the reversible process will be found useful in certain arguments. Since the reversible process represents a succession of equilibrium states, each only a differential step from its neighbor, it can be represented as a continuous line on a process indicator. The irreversible process cannot be so represented. We can note the terminal states and indicate the general direction of change, but it is inherent in the nature of the irreversible process that a complete path of the change is indeterminate and therefore cannot be drawn as a line on a thermodynamic diagram.

Irreversibilities always lower the efficiency of processes. Their effect in this respect is identical with that of friction, which is one cause of irreversibility. Conversely, no process more efficient than a reversible process can even be imagined. The reversible process is an abstraction, an idealization, which is never achieved in practice. It is, however, of enormous utility because it allows calculation of work from knowledge of the system's properties alone. Moreover, it represents a standard of perfection that cannot be exceeded because:

1. It places an upper limit on the work that may be *obtained* from a work-producing process.

2. It places a lower limit on the work *put into* a work-requiring process.

Process Indicators

In the course of our study, it will be convenient to represent a reversible process graphically. The coordinates of these graphic representations

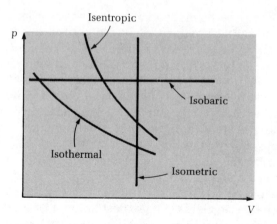

Figure 1-3. (p, V) Diagram indicating various processes.

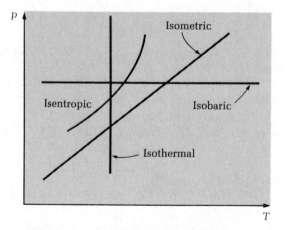

Figure 1-4. (p, T) Diagram indicating various processes.

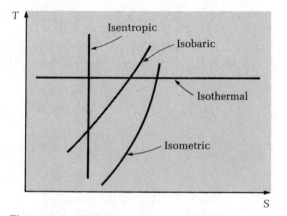

Figure 1-5. (T, S) Diagram indicating various processes.

may be any two thermodynamic properties. The most common coordinates are pressure and volume (p, V). Other coordinates used a great deal are pressure and temperature (p, T) and temperature and entropy (T, S). Figures 1-3, 1-4, and 1-5 present some of the special processes mentioned earlier. Note that the special processes represented by lines other than those parallel to a coordinate axis are only general trends and not necessarily the same for all possible physical conditions and processes.

Irreversible Process

All natural processes are irreversible. Everyday experience yields the following list of factors that make processes irreversible:

1. Friction
2. Free expansion (unrestrained expansion)
3. Inelastic deformation
4. Heat transferred across a finite temperature difference
5. Mixing of substances
6. All chemical reactions
7. Sudden change of phase
8. I^2R loss in electrical resistors
9. Hysteresis effects

Therefore, it may be concluded that an irreversible process is one in which dissipative effects occur or one which is not executed quasi-statically. (During a quasi-static process the system is at all times infinitesimally near a state of thermodynamic equilibrium.) This process is sometimes referred to as a quasi-equilibrium process as well. Regardless of the name, whenever a finite unbalanced force is encountered (mechanical, thermal, chemical, and so forth) an irreversible process occurs.

Polytropic Process

There are many different paths that a process may take. However, a number of these paths can be described by the equation

$$pV^n = \text{constant}$$

where p = pressure, V = volume, and n is a constant referred to as the **polytropic exponent or index**. Figure 1-6 illustrates the (p, V) diagram for such a process and the area bounded by such a curve. Although not all processes are in this category, some of the common ones are included. The convenience of this flexible description is seen immedi-

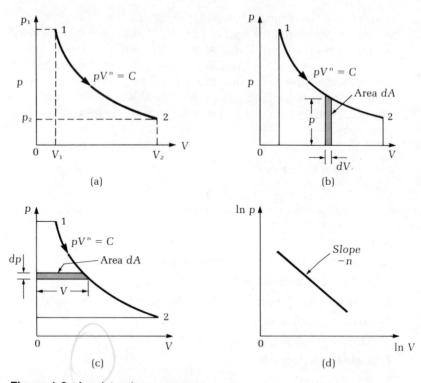

(a)   (b)

(c)   (d)

Figure 1-6. A polytropic process.

ately when we plot ln p versus ln V. As presented in Figure 1-6, the slope of the curve is $-n$. If a (p, V) diagram is viewed (Figure 1-7), we notice that a large family of processes results. The easiest to see is $n = 0$, which represents an isobaric process. It will be shown later that when we are dealing with an ideal gas, $n = 1$ represents an isothermal process. We will also see that when $n = k$, we are dealing with an isentropic process. (The index k represents the ratio of specific heats at

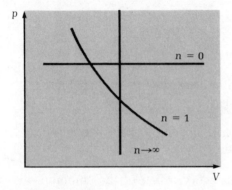

Figure 1-7. (p, V) Diagram of polytropic processes.

constant pressure and constant volume.) And as $n \to \infty$, we are dealing with an isometric process. This may be easily seen by taking the derivative of the definition of polytropic processes and rearranging to the form

$$\frac{dp}{dV} = -n\frac{p}{V}$$

Since both p and V are finite numbers, the right side of this equation increases as n increases. Viewing the left side of this equation, we see that since p is finite, dp is also finite and this side of the equation will approach infinity only as dV approaches zero (is isometric).

EXAMPLE 1-7

Consider the five ideal-gas processes (ab, bc, cd, da, and ac) sketched on the (p, V) diagram. Sketch these same processes on the (p, T) and (T, V) diagram.

SOLUTION:

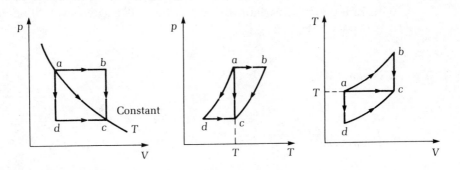

EXAMPLE 1-8

Suppose that 3 lbm of a gas are compressed from 14.7 psia and 70 F to 60 psia in a process such that $pv^n = $ constant, where $n = 1.4$. If the initial volume of this gas is 45 ft³, find the final volume and temperature. Assume that p, V, and T are interrelated by the equation of state

$$pv = RT \qquad (R \text{ constant})$$

SOLUTION:
$$V_2 = V_1\left(\frac{p_1}{p_2}\right)^{1/n} = 45 \text{ ft}^3\left(\frac{14.7}{60}\right)^{1/1.4} = 16.478 \text{ ft}^3$$

Eliminating V by using the equation of state, we get

$$T_2 = T_1\left(\frac{p_2}{p_1}\right)^{(n-1)/n} = (70 + 460)\left(\frac{60}{14.7}\right)^{0.4/1.4} = 792.14 \text{ R}$$

EXAMPLE 1-9

A gas expands from 0.850 MPa to 0.5 MPa. If the initial volume of the gas is 10^{-2} m³ and the expansion takes place such that pV^n = constant, what is the final volume if $n = 1.67$?

SOLUTION:

$$V_2 = V_1 \left(\frac{p_1}{p_2}\right)^{1/n}$$

$$= 10^{-2} \text{ m}^3 \left(\frac{0.85}{0.5}\right)^{1/1.67} = 10^{-2}(1.7)^{0.6} \text{ m}^3$$

$$= 1.375(10^{-2}) \text{ m}^3$$

1–6 Point and Path Functions

Let us digress a short while to present the mathematical background for some important thermodynamic quantities. Recall from your elementary calculus that when we are dealing with a differential of a function f with independent variables x and y, there is a quick way to determine if it is exact. If

$$df = Mdx + Ndy$$

it is exact if

$$\frac{\partial M}{\partial y} \equiv \frac{\partial N}{\partial x} \tag{1-9}$$

Furthermore,

$$\int_1^2 df = f(x_2, y_2) - f(x_1, y_1)$$

and

$$\int_1^2 df + \int_2^1 df = 0$$

Thus the integral of this differential depends only on the end points. In particular, it does not depend on the path from point 1 to point 2; and upon returning to the initial point 1 it results in zero. Therefore, it is a **point function.** One may immediately note that the various thermodynamic properties already discussed fall into this category. For example, temperature and pressure do not depend on the means (path) of obtaining a particular magnitude.

On the other hand, what is the situation if the following is true?

$$\frac{\partial M}{\partial y} \neq \frac{\partial N}{\partial x}$$

Mathematically, we would refer to this situation as one in which df is an inexact differential. We would also conclude that, in general,

$$\int_{1}^{2} df \neq f(x_2, y_2) - f(x_1, y_1)$$

and

$$\int_{1}^{2} df + \int_{2}^{1} df \neq 0$$

That is, going from point 1 to point 2 by one path and then returning to point 1 by another path will not yield zero. Therefore f would be called a **path function.** We have not yet met a thermodynamic quantity that falls into this category. We will, however, in the next sections on work and heat.

EXAMPLE 1-10

Determine whether R is a point function or path function of the two paths $y = 2x^2$ and $y = 8x$ connecting points $(0, 0)$ and $(4, 32)$ if:

1. $dR = xdy + ydx$

2. $dR = xdy + 2ydx$

SOLUTION: 1. Check by definition that $\partial M/\partial x = \partial N/\partial y$ or $1 = 1$. This implies that this is an exact differential; that is, R is a point function. On path 1, $y = 2x^2$, we get

$$R = \int_{0,0}^{4,32} (xdy + ydx) = \int_{0}^{4} (4x^2\, dx + 2x^2\, dx) = \int_{0}^{4} 6x^2\, dx$$

$$= 2x^3 \bigg|_{0}^{4} = 128$$

On path 2, $y = 8x$, we get

$$R = \int_{0}^{4} (8x\, dx + 8x\, dx) = 16 \int_{0}^{4} x\, dx$$

$$= 8x^2 \bigg|_{0}^{4} = 128$$

Therefore it appears that $\int_1 dR + \int_2 dR = 0$ (exact).

2. As in part 1, check the definition—this time, $\partial M/\partial x = 1$ and $\partial N/\partial y = 2$. Therefore this is inexact. So on path 1,

$$R = \int_{0,0}^{4,32} (x\, dy + 2y\, dx) = \int_{0}^{4} (4x^2\, dx + 4x^2\, dx) = 8 \int_{0}^{4} x^2\, dx$$

$$= \tfrac{8}{3}x^3 \Big|_0^4 = 170.7$$

On path 2,

$$R = \int_0^4 (8x\ dx + 16x\ dx) = 24 \int_0^4 x\ dx$$

$$= 12\,x^2 \Big|_0^4 = 192$$

Therefore $\int_1 dR + \int_2 dR \neq 0$ (inexact).

1–7 Work

Work, denoted W, is a form of energy in transit. **Work** is the mechanism by which energy is transferred across the boundary between systems by reason of the difference in pressure (or force of any kind) of the two systems. The transfer is always in the direction of the lower pressure. If the total effect produced in the system can be reduced to the raising of a weight, then nothing but work has crossed the boundary. Notice the conspicuous absences of temperature or temperature difference in this definition. Also, work is not *possessed* by the system; it occurs only when energy is transferred.

Work is, by a general definition, the energy resulting from a force having moved through a distance and excludes energy transfer resulting from a temperature difference. If the force varies with distance l, work may be expressed as $\delta W = F\,dl$ or

$$W = \int_0^x F\ dl \tag{1-10}$$

In thermodynamics, one often finds work done by a force distributed over an area—for example, by a pressure p acting through a volume change dV, as in the case of a fluid pressure exerted on a piston (a closed system). In this event,

$$\delta W = F\ dl\,(A/A) = \left(\frac{F}{A}\right) A\ dl$$

$$= p\ dV \tag{1-11}$$

where p is an external pressure exerted on the system. Note that work has been designated as a path function. This may be easily seen by sketching the process on a (p, V) diagram (see Figure 1-8). Depending on which path you consider, $p_1(V)$ or $p_2(V)$, the magnitude of the work, represented by the area under the curve, will be different. In fact,

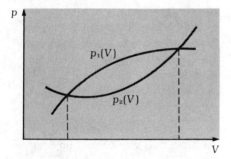

Figure 1-8. Work calculations.

$$\int p_1 \, dV > \int p_2 \, dV$$

Notice that since we have sketched these processes we have tactfully assumed that they are reversible. Thus Equation (1-11) is valid only for reversible closed-system processes.

Work done *by* a system is considered positive; work done *on* a system is considered negative. From the preceding discussion we note that positive work implies an increase in volume and negative work implies a decrease in volume. This is not to say that no work is done if there is no volume change. In fact, there are examples of net negative work (on the system) with no volume change.

The units of work (force times distance) include ft-lbf, Btu, kW-hr and kJ. Power is the *rate* of doing work and involves units such as ft-lbf/hr, Btu/hr, horsepower, and kW. There are a variety of ways in which work may be done on a system or by a system. **Mechanical or shaft work,** *W, is the energy delivered or absorbed by a mechanism such as a turbine, air compressor, or internal combustion engine.* Shaft work can always be evaluated from the basic relation for work.

For the open system, in addition to the work done at a moving boundary there is always flow work to be considered. **Flow work** *consists of the energy carried into or transmitted across the system boundary as a result of the fact that a pumping process occurs somewhere outside the system, causing the fluid to enter the system.* It might be more easily conceived as the work done by the fluid just outside the system on the adjacent fluid entering the system to push it into the system. Flow work also occurs as fluid leaves the system. In this case the fluid in the system does work on the fluid just leaving the system. As an analogy, consider two people as particles of fluid, one in the doorway and one just outside. Flow work would be done by the person outside if he shoved the person in the doorway into the room (system). As stated previously, all of this energy was really provided by a pump or other mechanical means outside the system. But since it is outside and only indirectly affects the system, we consider the direct effect—

that is, the work of fluid on fluid. Hence

$$\text{Flow work (per unit mass)} = \frac{1}{m} \int F \, dx = \frac{1}{m} \int pA \, dx \quad (p = \text{constant})$$

$$= \frac{p}{m} \int_0^v dV \tag{1-12}$$

where v is the specific volume, or the volume displaced per unit mass.

We can see from this discussion that work must be done in causing fluid to flow into or out of a system. This work is called flow work. Other names such as flow energy and displacement energy are sometimes used. Disagreement on the name used for this quantity results from the fact that the pV term is generally derived as a work quantity; yet it is unlike other work quantities since it is expressed in terms of a point function. Because it is so expressed, some engineers prefer to group it with stored energy quantities and sometimes speak of it as "transported energy" or "convected energy" instead of work. But it must be remembered that pV can be treated as energy only when a fluid is crossing a system boundary. For a closed system, pV does not represent any form of energy. Incidentally, arguments as to whether pV really is flow work or flow energy are fruitless. Both terms are used. In addition to mechanical work and flow work, the types most frequently encountered in thermodynamics, work may be done due to surface tension, the flow of electricity, magnetic fields, and in many other ways.

For nonflow processes (a closed system), the form of mechanical work most frequently encountered is that done at the moving boundary of a system, such as the work done in moving the piston in a cylinder, and may be expressed in equation form for reversible processes as $W = \int p \, dV$. For the nonflow process, we can generally express work as follows:

$$\text{CLOSED SYSTEM} \quad W = \int p \, dV + \cdots \tag{1-13}$$

where the dots indicate other ways in which work can be done by the system or on the system (including the work lost because of irreversibility).

A useful expression for the work done in a particular type of open system, the frictionless **steady-flow process,** will now be derived by following these steps:

1. Sketch a differential volume element including indications of the forces acting on the element.

2. Apply Newton's second law to this differential volume element, and solve the resulting expression for the force. The force is the driving force causing the fluid to flow.

3. Use the definition ($W = \int F \, dL$) to obtain an expression for the work done by this force on the system.

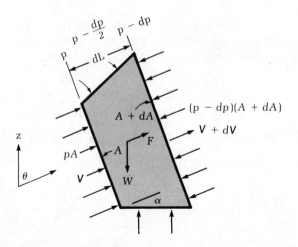

Figure 1-9. Differential volume of fluid in frictionless steady flow.

Figure 1-9 illustrates a free-body diagram using a differential volume.
From the geometry in Figure 1-9, the mass of the system is $\rho(A + dA/2)dL$ whereas the acceleration is dV/dt. The product of these two quantities must be equal to the sum of forces acting on the system. Thus,

$$pA - (p - dp)(A + dA) - \frac{mg}{g_c} \cos \theta + \left(p - \frac{dp}{2}\right) dA + F$$

GRAV. FORCE

$$= Adp + \frac{dp\, dA}{2} - \frac{mg}{g_c} \cos \theta + F$$

$$= \frac{\rho}{g_c}\left(A + \frac{dA}{2}\right) dL \frac{dV}{dt}$$ *X SEE, ½ WAY THROUGH*

2c ONLY. FACTOR

By applying the definition of work, noting that $dz = dL \cos \theta$ and $V = dL/dt$, the following expression is obtained:

$$\delta(\text{work}) = F\, dL = -V\, dp + \frac{mV\, dV}{g_c} + \frac{mg}{g_c} dz$$

For convenience, terms of the magnitude $dA\, dL$ and $dp\, dA$ are assumed negligible. Thus the work per unit mass is

OPEN SYSTEM

$$\delta(\text{work/mass}) = -v\, dp + \frac{V\, dV}{g_c} + \frac{g}{g_c} dz$$

Or, for flow between sections a finite distance apart,

$$w = -\int v\, dp + \Delta\left(\frac{V^2}{2g_c}\right) + \frac{g}{g_c} \Delta z \tag{1-14}$$

Equation (1-14) represents the work done in a frictionless steady-flow process. Notice that it was deduced from forces (Newton's second law). Later we shall obtain the same expression using the conservation of energy.

EXAMPLE 1-11

For the polytropic process discussed as Example 1-8, compute the work done.

SOLUTION: By definition (Equation 1-13),

$$W = \int p \, dV = \int \frac{C}{V^n} \, dV = \frac{C}{1-n} V^{1-n} \Big|_{V_1}^{V_2}$$

$$= \frac{1}{1-n}\left(CV_2^{1-n} - CV_1^{1-n}\right)$$

$$= \frac{1}{1-n}\left(p_2 V_2^n V_2^{1-n} - p_1 V_1^n V_1^{1-n}\right)$$

$$= \frac{p_2 V_2 - p_1 V_1}{1-n}$$

$$= \frac{(60 \text{ lbf/in.}^2)(16.478 \text{ ft}^3) - (14.7 \text{ lbf/in.}^2)45 \text{ ft}^3}{1-1.4}$$

$$= -817.95 \frac{\text{lbf-ft}^3}{\text{in.}^2}\left(\frac{144 \text{ in.}^2}{\text{ft}^2}\right) = -1.178(10^5) \text{ ft-lbf}$$

(handwritten annotations: $pv^n = $ constant, $p = \dfrac{C}{v^n}$, CLOSED SYSTEM)

EXAMPLE 1-12

Determine the mechanical work done (per mass) in a frictionless, steady-flow, polytropic process. Assume negligible changes in kinetic and potential energies.

SOLUTION: By definition (Equation 1-14),

$$w = -\int v \, dp = -c' \int p^{1/-n} \, dp = \frac{-c'}{1-(1/n)} p^{1-(1/n)} \Big|_{p_1}^{p_2}$$

$$= \frac{-nc'}{n-1}\left(p_2^{1-(1/n)} - p_1^{1-(1/n)}\right)$$

$$= \frac{-n}{n-1}\left(p_2^{1/n} v_2 \, p_2^{1-(1/n)} - p_1^{1/n} v_1 \, p_1^{1-(1/n)}\right)$$

$$= \frac{-n}{n-1}(p_2 v_2 - p_1 v_1) = n\left(\frac{p_2 v_2 - p_1 v_1}{1-n}\right) = \frac{nR}{1-n}\left(T_2 - T_1\right)$$

(handwritten annotations: CLOSED, OPEN)

If, as in Example 1-11,

$$p_1 = 14.7 \text{ psia} \qquad p_2 = 60 \text{ psia}$$

$$T_1 = 70 \text{ F} \qquad n = 1.4$$
$$V_1 = 45 \text{ ft}^3/\text{lbm}$$

then $w = -1.649(10^5)$ft-lbf/lbm.

EXAMPLE 1-13

The gas in a closed system experiences a volume change of 0.15 to 0.05 m³. During this process the pressure was 0.35 MPa. Determine the work.

SOLUTION: By definition,

$$W = \int p \, dV$$
$$= p(V_2 - V_1) \qquad \text{since } p \text{ is constant}$$
$$= -0.35 \text{ MPa } (0.15 - 0.05) \text{ m}^3$$
$$= -35,000 \text{ m}^3 \text{ Pa}$$
$$= -35,000 \text{ N} \cdot \text{m} = -35 \text{ kJ}$$

1-8 Heat

Heat Q *is the energy that is transferred across the boundary between systems by reason of the difference in temperature of the two systems.* The transfer is always in the direction of the lower temperature. Being transitory, heat is not a property. It is redundant to speak of heat being transferred, for the term *heat* itself signifies energy in transit. Nevertheless, in keeping with common usage, we will refer to heat as being transferred.

Although a body or system cannot "contain" heat, it will be useful in discussing many processes to speak of *heat received* or *heat rejected* so that the direction of heat transfer relative to the system is immediately obvious. This usage should not be construed as meaning that heat is substance. In terms of a sign convention, if in an energy interaction the system receives heat, Q is said to be positive; if it loses heat, Q is negative. Conversely if Q is positive the system gains energy, and if Q is negative the system loses energy. If a system neither loses nor gains heat in an energy interaction, the corresponding process is called **adiabatic.**

Because heat is not a characteristic of a system, it is not a property. This, of course, means that it is a path function. Therefore, it is like work in that it is an inexact differential and must be denoted by the symbols δQ. To compute the heat transferred in a process we must know the path (i.e., a functional relationship of state variables—a process equation). Once this is known, the following integration will yield the heat transfer:

$$\int_1^2 \delta Q = {}_1Q_2$$

Unfortunately, this process equation is usually not known and the quantity, ${}_1Q_2$ (the heat transferred during the process from state 1 to state 2) cannot be independently determined. The units of this quantity are Btu, kW-hr, kJ, and the like. The time rate of change of heat is denoted by the symbol $\dot{Q}$, or

$$\dot{Q} \equiv \frac{\delta Q}{dt}$$

and has units of Btu/hr, kW, and the like. As with the other variables discussed, the intensive form of the heat, Q, is denoted by the symbol q (and has units of Btu/lbm or kW-hr/kg)

$$q \equiv \frac{Q}{m} \tag{1-15}$$

The phase "specific" heat is *never* used to represent the quantity q.

1-9 Conservation of Mass

To make any scientific analysis, rules or laws that we know to be true must be applied. In particular, some of our basic definitions of system types put restrictions on the mass of a system (e.g., closed systems require no mass changes). To test whether system energy changes produce significant mass changes, recall (from the theory constructed by Einstein) the well-known mass-energy equivalence relation:

$$E \equiv \frac{mc^2}{g_c} \tag{1-16}$$

Because we are dealing with energy changes, we must determine the equivalent mass change to state a conservation principle accurately. To make this approximate calculation, consider the mass change resulting from a one kilowatt-hour (or 3413 Btu) energy change. Note that nothing has been said concerning what form of energy is changed (heat, work, kinetic energy, etc.) just that it is changing. Differentiating Equation (1-16) yields upon rearrangement

$$\Delta m = \Delta E \frac{g_c}{c^2}$$

$$= \frac{(1000 \text{ W-hr})(\text{kg-m}/(\text{N} \cdot \text{s}^2))}{(2.9979(10^8))^2 \text{m}^2/\text{s}^2} \left(\frac{3600\text{N} \cdot \text{m}}{\text{W-hr}}\right)$$

$$= 4.0056(10^{-11}) \text{ kg}$$

For all intents and purposes, there will be no change in mass when the energy of the system changes. Therefore, as far as we are concerned we may assume that mass is conserved with no significant error.

We may deduce a word expression for a general conservation of mass:

$$\begin{pmatrix} \text{Net rate of mass} \\ \text{flow out of a volume} \\ \text{of interest} \end{pmatrix} + \begin{pmatrix} \text{rate of accumulation of} \\ \text{mass in this volume of} \\ \text{interest} \end{pmatrix} = 0 \qquad (1\text{-}17)$$

The result of describing each of these terms mathematically is

$$\nabla \cdot \rho V + \frac{\partial \rho}{\partial t} = 0^* \qquad (1\text{-}18)$$

where V is a velocity vector and the dot operation represents the divergence. This gruesome-looking expression may be simplified for our purposes—that is, steady flow (no time dependence). The mass rate of flow of a fluid passing through a cross-sectional area A in this case is

$$\dot{m} = \frac{AV}{v} = \rho AV \qquad (1\text{-}19)$$

where V is the average velocity of the fluid in a direction normal to the plane of the area A, and v is the specific volume of the fluid. For steady flow (open system) with fluid entering a system at section 1 and leaving at section 2,

$$\dot{m}_1 = \dot{m}_2 = \frac{A_1 V_1}{v_1} = \frac{A_2 V_2}{v_2} \qquad (1\text{-}20)$$

This formula is the **continuity equation of steady flow.** It is an important relation and is frequently used, since it can readily be extended to any number of system inlets and outlets.

EXAMPLE 1-14

A liquid enters a constant cross-sectional-area pipe at a pressure of 60 psia, a specific volume of 0.01610 ft³/lbm, and a velocity of 12 ft/sec. What would be the velocity of this liquid at the pipe exit if the pressure were 53 psia and the specific volume 0.01663 ft³/lbm?

SOLUTION:

$$\dot{m} = \rho VA = \frac{VA}{v}$$

Hence

$$V_{\text{exit}} = V_{\text{ent}} \left(\frac{v_{\text{exit}}}{v_{\text{ent}}} \right)$$

$$= 12 \left(\frac{0.01663}{0.01610} \right) \text{ ft/sec}$$

$$^*\nabla = i \frac{\partial}{\partial x} + j \frac{\partial}{\partial y} + k \frac{\partial}{\partial z}.$$

$$= 12.40 \text{ ft/sec}$$

$$= 3.78 \text{ m/s}$$

PROBLEMS

1-1 In an environmental test chamber, an artificial gravity of 5.5 ft/sec² is produced. How much would a 205-lbm man weigh inside the chamber?

1-2 Determine the specific volume of a gas at 500 kPa and 20 C. Assume that $v = RT/p$ and $R = 287 \text{ N} \cdot \text{m}/(\text{kg} \cdot \text{K})$.

1-3 Assume a pressure gauge and a barometer reads 33 lbf/in.² and 26.27 in. Hg. Calculate the absolute pressure in psia, psfa, and atm.

1-4 The pressure of a partially evacuated enclosure is determined to be 26.8 in. Hg when the local barometer reads 29.5 in. Hg. Determine the absolute pressure in in. Hg, psia, atm, and microns of Hg.

1-5 A vertical cylinder containing air is fitted with a piston of 68 lbm and cross-sectional area of 35 in². The ambient pressure outside the cylinder is 14.6 psi and the local acceleration due to gravity is 31.1 ft/sec². What is the air pressure inside the cylinder in psia and in psig?

1-6 The pressure in the air space above an enclosed 76 ft (vertical measurement) column of water is 34 psia (i.e., water plus the air space height is 82 ft) if you assume the average density of the water is 62.4 lbm/ft³, what is the pressure of the water at ground level in psig and in psia?

1-7 A water manometer used to measure the pressure rise across a fan reads 1.1 in. H₂O when the density of the water is 62.1 lbm/ft³. Determine the pressure difference in psi.

1-8 A thermometer reads 72 F. Specify the temperature in C, K, and R.

1-9 One gallon of fuel oil having a heating value of 139,000 Btu/gal burns in a home furnace. Determine the mass loss (converted to energy) per gallon of fuel burned.

1-10 Air with a density of 0.075 lbm/ft³ enters a steady-flow system through a 12-in.-diameter duct with a velocity of 10 ft/sec. It leaves with a specific volume of 5.0 ft³/lbm through a 4-in.-diameter duct. Determine: (a) the mass flow rate (in lbm/hr) and (b) the outlet velocity (in ft/sec).

1-11 Air at 60 psia, 100 F is trapped in a cylinder/piston arrangement. The following data represents the compression of this air.

Pressure, lbf/in.²	Volume, in.³
60	80.0
80	60
100	45
120	35
140	30
160	25
180	20

Determine the work required for compression of the air assuming a reversible process.

1-12 Convert the following Celsius temperatures to Fahrenheit temperatures: (a) -30 C, (b) -10 C, (c) 0 C, (d) 200 C, and (e) 1050 C.

1-13 On the Réaumur temperature scale, the ice point is zero and the steam point is 80. What is the Réaumur temperature of absolute zero?

1-14 Under conditions of thermal equilibrium, assume that the thermometric function to establish a scale is

$$T(x) = b + a \ln x$$

Determine the constants a and b in general form in terms of an ice-point temperature T_i at x_i and a steam-point temperature T_s at x_s.

1-15 Someone has proposed the use of three new thermodynamic quantities: x, y, and z. Assuming the following definitions, are they properties?

$$x = \int (p \, dv - v \, dp)$$
$$y = \int (p \, dv + v \, dp)$$
$$z = \int (R \, dT + p \, dv) \qquad R = pv/T, \text{ a constant}$$

1-16 Gas is trapped in a cylinder/piston arrangement (see the sketch). If p (initial) $= 13{,}789.5$ Pa and V (initial) $= 0.02832$ m³, determine the work assuming that the volume is increased to 0.08496 m³ in a constant-pressure process.

Gas

1-17 For the same conditions as Problem 1-16, determine the work done if the process is polytropic with $n = 1$ ($pV = $ constant).

1-18 For the same conditions as Problem 1-16, determine the work done if the process is polytropic with $n = 1.4$ ($pV^{1.4} = $ constant).

1-19 The following figure shows two processes a-c and a-b sketched on a $(p-V)$ plane. Sketch these processes on $(p-T)$ and $(T-V)$ planes.

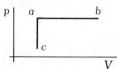

1-20 On the Jovian temperature scale the freezing and boiling points of water are 100 Z and 1000 Z. Set up relations between this scale and both the Fahrenheit and Celsius scales. Assume linear scaling in all cases. What is absolute zero on the Z scale?

1-21 The accompanying sketch indicates a complicated compartment arrangement: a and b. The ambient pressure, p_{amb} is 30.0 in. Hg. If gauge C reads 620,528 Pa and gauge B reads 275,790.3 Pa, determine the reading of gauge A and convert it to an absolute value.

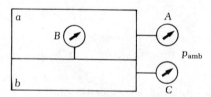

1-22 Determine the work done by 1 kg of fluid as it expands slowly inside a cylinder/piston arrangement from an initial pressure of 80 psia and 1 ft^3 to a final volume of 4 ft^3 if the process relations are:

1. $p = -20V + 100$. (The units of this result will be psia if V is in ft^3.)

2. $pV^2 =$ constant.

1-23 A very unusual quantity $\bar{f}$ is defined as

$$d\bar{f} = \left(3V^2p^6 + 2Vp + \sqrt{\frac{p}{V}}\right)dV + \left(6V^3p^5 + V^2 + \sqrt{\frac{V}{p}}\right)dp$$

where p [=] psia and V [=] ft^3. Is $\bar{f}$ a thermodynamic property?

1-24 Sketch, on a (p, V) diagram, a process in which $pV =$ constant from (p_1, V_1) to (p_2, V_2) $(p_1 > p_2)$. Also indicate this same process as it would appear on a (p, T) and (T, V) diagram.

1-25 A cylinder containing a gas is fitted with a piston having a cross-sectional area of 0.029 m^2. Atmospheric pressure is 0.1035 MPa and the acceleration due to gravity is 30.1 ft/sec^2. To produce an absolute pressure on the gas of 0.1517 MPa, what mass (kg) of piston is required?

1-26 A cylinder containing air at 29.4 C is fitted with a piston having a cross-sectional area of 0.029 in². The mass of the piston, which is above the air, is 160.6 kg and the acceleration due to gravity is 9.144 m/s. The cylinder has a volume of 9.63 m³. Determine the mass of air trapped beneath the cylinder. Atmospheric pressure is 0.10135 MPa. For air, pv = RT, where R = 0.29 kN·m/kg·K.

1-27 Water (density of 990 kg/m³) is discharged by a pump at a rate of 3 (10⁵) cm³/min from a pipe. (a) Find the mass flow rate (kg/min). (b) Convert this rate to lbm/hr.

1-28 The water level in a sealed tank is 26 m above the ground. The pressure in the air space above the water is 0.250 MPa (gauge). The average density of the water is 1000 kg/m³. What is the pressure of the water at ground level?

1-29 A gas (ρ = 1.20 kg/m³) enters a steady-flow system through a 5-cm-diameter tapering duct with a velocity of 3.5 m/s. It leaves the duct with a specific volume of 0.31 m³/kg through a 1.6-cm-diameter constriction. Determine: (a) the mass flow rate (kg/hr) and (b) the outlet velocity (m/s).

1-30 A person with a barometer is driving up a mountain in Colorado. In the foothills of the mountain the barometer reads 75 cm Hg absolute. Several hours later it reads 70 cm Hg absolute. Assuming the average density of the atmospheric air is 1.2 kg/m³, estimate the altitude change experienced on this trip.

1-31 A thermocouple is a device that generates a voltage E (in mV) in a circuit when one end is maintained at 0 C (the cold junction) and the other junction is used as a probe to measure temperature T (C). The voltage–temperature relationship is $E = aT + bT^2$, where the constants a and b have magnitudes 0.26 and 5 (10⁻⁴). After stating the units of a and b, determine the temperatures for each of the following millivolt readings: (a) 10 mV, (b) 20 mV, (c) 50 mV.

1-32 A pump is used to remove water from a very large cave. To estimate the work done by this pump, let us model the process as a frictionless, steady-flow process. Assume the water enters the pump at 0.0689 MPa and leaves it at 3.516 MPa with an average density of 995 kg/m³. What is your estimate of the work of the pump if there is no change in kinetic or potential energies?

1-33 A compressor is a device used to increase the pressure of a gas. Estimate the work of a compressor in changing the properties of the gas from 0.1724 MPa, 9.4625 kg/m³ to 1.241 MPa, 35.286 kg/m³. Assume that the process is a frictionless, steady-flow process and the compression is carried out such that $pV^{1.5}$ = constant is valid.

Chapter

2

Physical Properties

The physical world is part of our study of thermodynamics. In particular, we must use the physical properties of the substances of this world if we wish to describe a change in any part of it. Therefore, the concepts of phase diagram, states of a system, specific heats, and property tables (in particular, steam tables) are presented in this chapter.

2–1 Phases of a Pure Substance

All substances may exist in various forms. Water, for example, may exist as a vapor, a liquid, and a solid. Thus we speak of phases. A **phase** *of a substance is any homogeneous part of a system that is physically distinct and separated by definite boundaries (phase boundaries).* In fact, some substances exist in more phases than we may realize. Water has several distinct solid phases, as do sulfur and carbon. Solid, liquid, and vapor mixtures may constitute a single multicomponent phase, no matter how many substances are included, as long as the mix is homogeneous. A solution of sugar and water is one phase of two constituents, for example. Of course, it is possible to add sugar until the water is unable to dissolve it. (Part of the sugar remains in the solid phase.) A layered solution of oil and water is a two-phase situation—each layer is homogeneous and the two are separated by a phase boundary.

Certain characteristics of molecular structure may aid your intuitive grasp of thermodynamics. Solids, for example, whether crystalline or noncrystalline, are a tightly bound three-dimensional array. The molecules of the solid are essentially fixed in position and the molecular density is in the order of 10^{36} molecules per cubic centimeter. A variety of extremely powerful short-range cohesive forces hold the solid together. In the case of liquids, the molecular spacing is of the same order of magnitude as the solid, but slightly larger. The intermolecular forces are such that there is no rigid three-dimensional structure, although small numbers of molecules do adhere to each other in a fashion similar to a solid. Because of this relaxed structure, the liquid will not withstand shear. As for gases, the molecular density is of the order of 10^{19} molecules per cubic centimeter. Thus the molecules in a gas are very far apart. There is no position restriction since the gas molecules are in continuous, chaotic motion. The various intermolecular forces are overcome (as far as we are concerned) by adding thermal energy to the substances.

In this book we will consider only pure substances—that is, ones that are homogeneous with an invariant chemical composition. Different phases may exist, but the chemical composition remains the same for all phases. Thus a mixture of solid carbon dioxide and CO_2 gas is a pure substance in two phases. In reality, liquid air or gaseous air is not a pure substance. Nevertheless, we will consider a mixture of gases to be a pure substance as long as there is no change in phase.

2-2 Equilibrium of a Pure Substance

How does one describe the state of a pure substance? Obviously, a number of independent properties (or variables) are needed. *A property (variable) is* **independent** *if it may assume many values without affecting any other independent variable.* Conveniently, only two independent properties are required to specify the state of a pure substance. Literally, once the two properties are known, *any* other property may be found (in theory, at least). For example, given the two independent properties of pressure and specific volume (p, v), such properties as temperature, entropy, and the like can be found. The important point is the unique specification of the two independent properties. That is, care must be taken that the two properties selected are *truly* independent.

In the study of thermodynamics you will find that pressure and temperature are the variables of concern. That is, if pressure and temperature are independent, the state is one of single phase (solid, liquid, gas, or vapor). If the pressure and temperature are not independent, there is a fixed expression, $p = p(T)$, relating the two properties, and the system consists of two phases (e.g., liquid and vapor) coexisting in equilibrium. In this case, pressure and temperature do not uniquely define the state of the system (i.e., they are the same single bit of information). Recall from your experience that only one temperature exists for a given pressure where a change of phase occurs, and, similarly, only one pressure exists for a given temperature when a change of phase occurs (water -212 F and 14.69 psia). Thus, some other property must be used to uniquely define the state (e.g., specific volume, enthalpy, entropy, or the like) or two other properties (not including p or T). That is (p, v) and (T, v) are essentially the same information pair when a change of phase occurs. So you might use an (h, s) or (p, h) pair for the explicit identification of the state.

2–3 Equilibrium Thermodynamic Properties: An Example

To make this point clear, consider a system consisting of a liquid sealed in a cylinder/piston arrangement (see the first piston in Figure 2-1). This magic piston maintains a constant pressure of p lbf/in² within the cylinder at all times regardless of the interaction. Assume that the initial temperature of the liquid is T, which is less than the boiling temperature, T_{sat} at the pressure p. Thus, as heat is added to the liquid, its temperature increases toward T_{sat}. The change in volume of a liquid (with temperature) is small until $T = T_{sat}$. At this point the liquid will begin to boil, producing a vapor and a drastic increase in volume (p is still constant). Eventually all of the liquid will boil away, filling the container with vapor at $T = T_{sat}$. Further addition of heat to the system increases T (larger than T_{sat}), the volume will increase greatly, and the pressure is still constant. Finally, if this substance is highly superheated it may be called a gas.

The saturation condition described here is vaporization (it could be condensation as well) and is characterized by a **saturation temperature—saturation pressure pair** *(T_{sat} at a corresponding p_{sat}). That is, the*

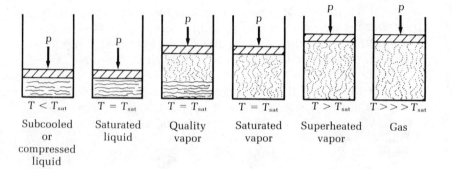

$T < T_{sat}$	$T = T_{sat}$	$T = T_{sat}$	$T = T_{sat}$	$T > T_{sat}$	$T >>> T_{sat}$
Subcooled or compressed liquid	Saturated liquid	Quality vapor	Saturated vapor	Superheated vapor	Gas

Figure 2-1. Thermodynamic fluid states.

liquid will always boil (or condense) at $T = T_{sat}$ if $p = p_{sat}$ and vice versa. (From your experience you know that for water this saturation pair is 212 F, 14.7 psia.)

In the preceding example, note that the substance existed as a liquid and as a vapor at $T = T_{sat}$ (p was p_{sat}). When the container is filled only with liquid at saturation, it is referred to as a **saturated liquid.** Similarly, if the container is filled only with vapor for saturation conditions, it is called a **saturated vapor.** A **compressed liquid** is encountered when the $p > p_{sat}$ and $T = T_{sat}$. A **subcooled liquid** occurs when $T < T_{sat}$ and $p = p_{sat}$. Notice that a liquid exists in either case, and it may be compressed or subcooled depending on the method used to arrive at that state. On the other hand, a **superheated vapor** occurs if $T > T_{sat}$ and $p = p_{sat}$. (This is an expanded vapor.) In this superheated vapor condition as in the subcooled (or compressed) liquid state, p and T are independent properties. They are not independent at saturation.

The region of most interest is when some vapor and liquid coexist for a given p_{sat} and T_{sat}. Because the pressure and temperature are dependent in this region, a new variable, the quality, is defined. **Quality (x) is the ratio of the mass of the vapor and the mass of the liquid plus the vapor.** This variable exists only within the saturation region, and since it is independent of mass, it is an intensive property.

Figure 2-2 may be used to clarify the important points of the preceding discussion. It is a (p, T) diagram illustrating the general relationships of the various phases (solid, liquid, and vapor). The preceding example may be represented by the portion of the line (AB) from the liquid phase to the vapor phase. Notice the process crosses the vaporization line where the liquid and vapor phases coexist in equilibrium. This line extends from the triple point (line) to the critical

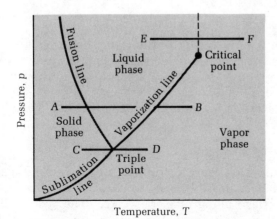

Figure 2-2. The pure substance.

point. The **triple point** (line) is where all three phases exist in equilibrium whereas the **critical point** is where the specific volume of the liquid and the specific volume of the vapor are identical. Included in this figure are the fusion and sublimation lines as well as the region representing the solid phase. The fusion line represents the conditions where the solid phase and the liquid phase coexist in equilibrium. Similarly the sublimation line represents the coexistence of the solid and vapor phases. In both of these regions a saturated solid state exists.

Also presented in this figure are a series of constant-pressure transitions. The line (AB) begins in the solid region. As the temperature is increased at constant pressure, the solid first melts at the fusion temperature. It then proceeds across the liquid phase, eventually boiling at a temperature higher than the fusion temperature. Upon further temperature increase it becomes superheated. The constant pressure line (CD) executes the same transition as the (AB) line, except it goes through the triple point. For transitions at pressures lower than the triple-point pressure, no liquid phase is encountered. The line (EF) is also unique in that because it is above the critical point one cannot identify where the transition from liquid to vapor occurs. Sometimes the general term "fluid" is used in this region.

Two final points need emphasis. The first point has to do with the variables to be used when describing a substance in the saturation region. Because pressure and temperature remain constant for the saturated-liquid to saturated-vapor transition, variables such as p and v or T and x must be used. The second point is that air will be treated only as a pure substance as long as any process using air is well away from the saturation region (or boundary)—only in its gas phase.

2–4 Thermodynamic Surfaces

If we could plot every value of equilibrium pressure, specific volume, and temperature of a substance on a three-dimensional space, we would have a complete thermodynamic description of that substance. Figure 2-3 depicts a substance that contracts on freezing (as most do) such as CO_2; Figure 2-4 represents a substance that expands on freezing (water).

Phase Diagrams

Figures 2-3 and 2-4, which represent a thermodynamic surface, are a conceptual convenience. Possibly a more usable form would be the projection of the surface along one of its axes. The (p, T) diagram showing a saturation line is probably the most commonly used of these

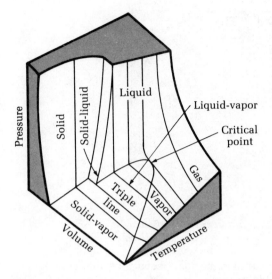

Figure 2-3. Thermodynamic surface for a substance that contracts on freezing.

projections—it is called a **phase diagram.** Every point (p, T) on this diagram represents an equilibrium state of a pure substance. If the pressure and temperature values place the point on a curve of this diagram, the condition represents a two-phase situation (i.e., solid-vapor, solid-liquid, or liquid-vapor) coexisting in equilibrium. Otherwise, only single phases of the pure substance are represented. In our

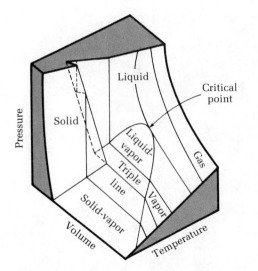

Figure 2-4. Thermodynamic surface for a substance that expands on freezing.

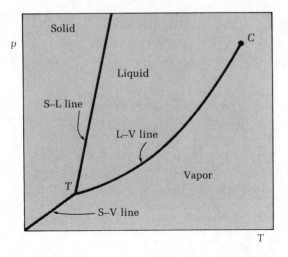

Figure 2-5. Typical phase diagram for a substance that contracts on freezing (*C* = critical point, *T* = triple point).

study of thermodynamics we will be particularly interested in the region in and about the liquid-vapor (vaporization) curve. Figures 2-5 and 2-6 illustrate the (p, T) projections of Figures 2-3 and 2-4. You may easily pick out the regions of supercooled liquid, superheated vapor, compressed solid, and so forth.

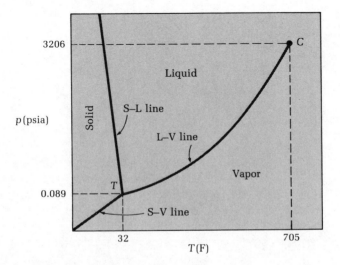

Figure 2-6. Approximate phase diagram for water (not to scale).

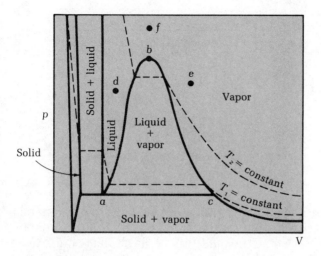

Figure 2-7. (p, V) diagram of a substance that contracts on freezing, ($T_2 > T_1$).

Other Useful Diagrams

Apart from the phase diagrams, other projections are useful in analyzing thermodynamic processes. The most popular is the (p, V) diagram. Figure 2-7 illustrates a (p, V) diagram of a substance that contracts on freezing. Curve ab is the saturated liquid line, curve bc is the saturated vapor line, and ac is the triple-point line. Point b is the critical point, point d is a liquid, point e is a vapor, and point f is a fluid (neither liquid nor vapor). Among the various saturation regions exhibited, the liquid + vapor region is most important to us. The broken lines in Figure 2-7 are lines of constant temperature. Notice that in all two-phase regions constant-temperature lines coincide with constant-pressure lines.

Typical Values of Characteristic Points

As we have seen, the intersection of the vaporization line, fusion line, and sublimation line on a (p, T) diagram represents the triple point or triple-point line. When a substance can exist in more than three phases, there will be more than one triple point (lines) for that substance. Table 2-1 lists triple-point values for several substances.

The critical point, where ρ(vapor) = ρ(liquid), represents the extreme condition where the identification of liquid and vapor phases is possible. For pressures and temperatures higher than those of the critical point, the liquid and vapor phases are indistinguishable. Below these values, the transition from liquid phase to vapor phase is easily seen. (A phase boundary exists during the change of phase.)

Table 2-1 Triple-Point Data

Substance	p, psia	p, MPa	T, F	T, C
Ammonia	0.88	0.0061	−108	−78
Carbon dioxide	75	0.517	−71	−57
Helium	0.731	0.0050	−456	−271
Hydrogen	1.021	0.0070	−434	−259
Nitrogen	1.817	0.0125	−346	−210
Oxygen	0.022	0.0002	−361	−218
Water	0.0886	0.00061	32.02	+0.01

The conditions of the critical point are hard to understand because these characteristic values of pressure and temperature are not experienced every day. To understand this point of phase indistinguishability, consider a pure substance characterized by point a of Figure 2-8. The two phases will be separated by the phase boundary (a meniscus). Upon heating at constant specific volume, the pressure and temperature will increase (along line a–c). Thus, the phase boundary rises because the fraction of liquid increases. Note that from point b, the saturated liquid point of the given specific volume, to point c, the increasing pressure (and temperature) is just compressing the liquid. Thus we started in a saturation state and ended up with only liquid—even though the temperature increased. Consider another situation where the initial state of the substance is characterized by point d of Figure 2-8. Again the phase boundary separates the two phases. If we increase the pressure and temperature (at constant volume) along line d–f, we note that the phase boundary eventually falls because the fraction of vapor increases. From point e, the saturated vapor point, to point f, the increasing pressure (and temperature) is superheating the vapor. Thus, we started in a saturation state and ended up only with vapor. Finally,

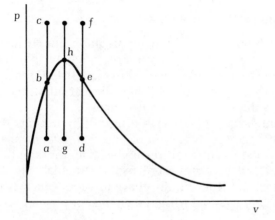

Figure 2-8. Critical-point experiments (not to scale).

Table 2-2   Critical-Point Data

Substance	p, psia	p, MPa	T, F	T, C
Ammonia	1639	11.3	270	132
Carbon dioxide	1071	7.38	88	31
Helium	34	0.234	−450	−268
Hydrogen	188	1.296	−400	−240
Mercury	> 2939	> 20	> 2820	> 1550
Nitrogen	493	3.399	−233	−147
Oxygen	731	5.040	−182	−119
Water	3206	22.10	705	374

consider the substance in a condition characterized by point g. Note that the specific volume of point g is identical to that of the critical point h. If we again consider the constant specific volume process, the phase boundary will vanish as the pressure approaches the critical value. This indicates that no boiling or condensation took place—phase identity is lost. Continued pressure increase compresses the single phase that remains. Typical values of the critical-point pressure and temperature are presented in Table 2.2.

Tables of Properties

An equation of state relates three properties of a substance. We are familiar with several simple p, V, T equations of state. Others exist and are not so simple. A few will be presented later. It is possible to express the internal energy, enthalpy, and entropy as functions of any two of the state properties p, V, and T. Unfortunately, this relationship cannot be expressed by simple equations for most substances of engineering importance. Therefore, the properties of these substances must be determined by measurements and then supplemented by interpolation. The results of these measurements and calculations are presented in tables or charts that, in general, all have the same form. In the case of steam, the tables of the American Society of Mechanical Engineers (ASME) are very popular. An abbreviated form appears as Appendix A-1 and their use will be discussed in detail later.

Generally speaking, the properties usually tabulated are p and T (directly measurable and usually controllable); v (generally useful all around); h and u (useful in applications of the first law); and s (to be discussed later). Specific tabulations are usually presented for the following states:

1. *Saturated liquid—vapor*: In this case, either p or T is used as the independent property. As a result, the saturation tables include both a whole number temperature and a whole number pressure listing. These two tables present the same data because p and T are not

independent and are presented as a convenience. Note that another property is required to make an exact determination of the state of the system.

2. *Superheated vapor*: In this case p and T may be used as the defining independent properties. Of course other properties (i.e., v, h, and s) may be used. Regardless, two properties are required to specify the state of the substance.

3. *Compressed liquid*: In this case p and T may be used as the defining independent properties (as well as others). Unfortunately, these listings are few and far between—and when available are difficult to handle. As a result, an approximation procedure with the saturation tables is used. The basis for the approximation is the fact that liquids are essentially incompressible (pressure effects are small), and therefore the primary property is the temperature. Thus, the saturated liquid value at the given temperature is used.

4. *Saturated solid (solid–vapor equilibrium)*: In this case, the properties are seldom tabulated for classic thermodynamics.

A complete description of a point in the saturation region depends not only on temperature (pressure) but on the proportions of liquid and vapor (that is, the quality). The use of the quality is consistent with the requirement that two independent variables are needed to define a state uniquely. As we have seen, quality (x) is defined as the fraction by mass of vapor in a mixture of liquid and vapor. The term *quality* has no meaning outside the saturation region. The limiting values of quality are zero for the saturated liquid and 1 for saturated vapor alone.

Consider a liquid–vapor mixture (saturated) at a given pressure. The liquid in the mixture has a specific volume of v_f. The vapor in the mixture has a specific volume of v_g. Recall that v_f is the specific volume of the saturated liquid $(x = 0)$ whereas v_g is the specific volume of the saturated vapor $(x = 1)$. The specific volume of the mixture may be related to the specific volumes of the saturated liquid and vapor, v_f and v_g, at the same pressure (temperature). Noting (1) that the specific volume of the mixture, v, must have a value between v_f and v_g and (2) that the total volume of the mixture is the sum of the volumes of the liquid and vapor that are present, we may write

$$v = \frac{V}{m} = \frac{V_f + V_g}{m_f + m_g} = \frac{m_f v_f + m_g v_g}{m_f + m_g} \qquad (2\text{-}1)$$

where m_f and m_g represent the mass of the liquid and vapor, respectively. Note that the definition of specific volume has also been used. By using the definition of quality

$$x \equiv \frac{m_g}{m_f + m_g} \qquad (2\text{-}2)$$

Equation (2-1) reduces to

$$v_x = (1 - x)v_f + xv_g \tag{2-3}$$

or

$$v_x = v_f + x(v_g - v_f) \tag{2-4}$$

or

$$v_x = v_g - (1 - x)(v_g - v_f)$$

The difference $(v_g - v_f)$ is denoted by the symbol v_{fg}, so that

$$v_x = v_f + xv_{fg} \tag{2-5}$$

and

$$v_x = v_g - (1 - x)v_{fg}$$

The internal energy, enthalpy, and entropy are extensive properties, since they depend on the mass of the system. Therefore in the saturated region the internal energy, enthalpy, and entropy may be computed directly from the saturated liquid value, the saturated vapor value, and the quality. Thus

$$u_x = \frac{U_x}{m} = \frac{U_f + U_g}{m_f + m_g} = \frac{m_f}{m_f + m_g}u_f + \frac{m_g}{m_f + m_g}u_g$$

$$= (1 - x)u_f + xu_g \tag{2-6}$$

$$= u_f + xu_{fg} \tag{2-7}$$

$$= u_g - (1 - x)u_{fg}$$

Exactly similar expressions are obtained in the same fashion for the enthalpy and entropy in the saturated region.

Steam

Water is a common working substance in thermodynamics. Steam engines for pumping water, steam boilers, power plants—all have been an integral part of life since the industrial revolution. Certainly steam's high latent heat, moderate density, and reasonable vapor pressure make it an economical fluid.

Since steam is a prime mover in the development of mechanical work, interest in its physical properties has been great. As might be expected, no simple equation of state exists. Therefore, to present property information in an accurate but convenient form for hand or graphic calculation, various property charts and tabulations were developed. Numerical calculations are greatly aided by the use of these tables and diagrams giving internal energy, enthalpy, entropy, and the like over a range of pressures and temperatures.

Table 2-3 presents an abbreviated version of the thermodynamic properties of water at saturation. (See the Appendix for more complete

Table 2-3   Thermodynamic Properties of Water at Saturation

T, F	Absolute Pressure, P lb/in²	Absolute Pressure, P in. Hg	Specific Volume, ft³/lbm Sat. Liquid v_l	Specific Volume, ft³/lbm Evap. v_{lg}	Specific Volume, ft³/lbm Sat. Vapor v_g	Enthalpy, Btu/lbm Sat. Liquid h_l	Enthalpy, Btu/lbm Evap. h_{lg}	Enthalpy, Btu/lbm Sat. Vapor h_g	Entropy, Btu/lbm (F) Sat. Liquid s_l	Entropy, Btu/lbm (F) Evap. s_{lg}	Entropy, Btu/lbm (F) Sat Vapor s_g	T, F
34	0.095999	0.19546	0.01602	3061.7	3061.7	2.01	1074.03	1076.04	0.00409	2.1755	2.1796	34
35	0.099908	0.20342	0.01602	2947.8	2947.8	3.02	1073.46	1076.48	0.00612	2.1700	2.1761	35
36	0.10396	0.21166	0.01602	2838.7	2838.7	4.02	1072.90	1076.92	0.00815	2.1644	2.1726	36
37	0.10815	0.22020	0.01602	2734.1	2734.1	5.03	1072.33	1077.36	0.01018	2.1589	2.1691	37
38	0.11249	0.22904	0.01602	2633.8	2633.8	6.03	1071.77	1077.80	0.01220	2.1535	2.1657	38
39	0.11699	0.23819	0.01602	2537.6	2537.6	7.04	1071.20	1078.24	0.01422	2.1480	2.1622	39
40	0.12164	0.24767	0.01602	2445.4	2445.4	8.04	1070.64	1078.68	0.01623	2.1426	2.1588	40
41	0.12646	0.25748	0.01602	2356.9	2356.9	9.05	1070.06	1079.11	0.01824	2.1372	2.1554	41
42	0.13145	0.26763	0.01602	2272.0	2272.0	10.05	1069.50	1079.55	0.02024	2.1318	2.1520	42
43	0.13660	0.27813	0.01602	2190.5	2190.5	11.05	1068.94	1079.99	0.02224	2.1265	2.1487	43
44	0.14194	0.28899	0.01602	2112.3	2112.3	12.06	1068.37	1080.43	0.02423	2.1211	2.1453	44
45	0.14746	0.30023	0.01602	2037.3	2037.3	13.06	1067.81	1080.87	0.02622	2.1158	2.1420	45
46	0.15317	0.31185	0.01602	1965.2	1965.2	14.06	1067.24	1081.30	0.02820	2.1105	2.1387	46
47	0.15907	0.32387	0.01602	1896.0	1896.0	15.06	1066.68	1081.74	0.03018	2.1052	2.1354	47
48	0.16517	0.33629	0.01602	1829.5	1829.5	16.07	1066.11	1082.18	0.03216	2.0999	2.1321	48
49	0.17148	0.34913	0.01602	1765.7	1765.7	17.07	1065.55	1082.62	0.03413	2.0947	2.1288	49
50	0.17799	0.36240	0.01602	1704.3	1704.3	18.07	1064.99	1083.06	0.03610	2.0895	2.1256	50
51	0.18473	0.37611	0.01602	1645.4	1645.4	19.07	1064.42	1083.49	0.03806	2.0842	2.1223	51
52	0.19169	0.39028	0.01602	1588.7	1588.7	20.07	1063.86	1083.93	0.04002	2.0791	2.1191	52
53	0.19888	0.40492	0.01603	1534.3	1534.3	21.07	1063.30	1084.37	0.04197	2.0739	2.1159	53
54	0.20630	0.42003	0.01603	1481.9	1481.9	22.08	1062.72	1084.80	0.04392	2.0688	2.1127	54
55	0.21397	0.43564	0.01603	1431.5	1431.5	23.08	1062.16	1085.24	0.04587	2.0637	2.1096	55
56	0.22188	0.45176	0.01603	1383.1	1383.1	24.08	1061.60	1085.68	0.04781	2.0586	2.1064	56
57	0.23006	0.46840	0.01603	1336.5	1336.5	25.08	1061.04	1086.12	0.04975	2.0535	2.1033	57
58	0.23849	0.48558	0.01603	1291.7	1291.7	26.08	1060.47	1086.55	0.05168	2.0485	2.1002	58
59	0.24720	0.50330	0.01603	1248.6	1248.6	27.08	1059.91	1086.99	0.05361	2.0434	2.0970	59
60	0.25618	0.52160	0.01603	1207.1	1207.1	28.08	1059.34	1087.42	0.05553	2.0385	2.0940	60
61	0.26545	0.54047	0.01604	1167.2	1167.2	29.08	1058.78	1087.86	0.05746	2.0334	2.0909	61
62	0.27502	0.55994	0.01604	1128.7	1128.7	30.08	1058.22	1088.30	0.05937	2.0284	2.0878	62
63	0.28488	0.58002	0.01604	1091.7	1091.7	31.08	1057.65	1088.73	0.06129	2.0235	2.0848	63
64	0.29505	0.60073	0.01604	1056.1	1056.1	32.08	1057.09	1089.17	0.06320	2.0186	2.0818	64
65	0.30554	0.62209	0.01604	1021.7	1021.7	33.08	1056.52	1089.60	0.06510	2.0136	2.0787	65
66	0.31636	0.64411	0.01604	988.63	988.65	34.07	1055.97	1090.04	0.06700	2.0087	2.0757	66
67	0.32750	0.66681	0.01605	956.76	956.78	35.07	1055.40	1090.47	0.06890	2.0039	2.0728	67

Temp.											
68	2.0698	1.9990	0.07080	1090.91	1054.84	36.07	926.08	926.06	0.01605	0.69021	0.33900
69	2.0668	1.9941	0.07269	1091.34	1054.27	37.07	896.97	896.95	0.01605	0.71432	0.35084
70	2.0639	1.9893	0.07458	1091.78	1053.71	38.07	867.97	867.95	0.01605	0.73916	0.36304
71	2.0610	1.9845	0.07646	1092.21	1053.14	39.07	840.47	840.45	0.01605	0.76476	0.37561
72	2.0580	1.9797	0.07834	1092.65	1052.58	40.07	813.97	813.95	0.01606	0.79113	0.38856
73	2.0551	1.9749	0.08022	1093.08	1052.01	41.07	788.40	788.38	0.01606	0.81829	0.40190
74	2.0522	1.9701	0.08209	1093.52	1051.46	42.06	763.75	763.73	0.01606	0.84626	0.41564
75	2.0494	1.9654	0.08396	1093.95	1050.89	43.06	739.97	739.95	0.01606	0.87506	0.42979
76	2.0465	1.9607	0.08582	1094.38	1050.32	44.06	717.03	717.01	0.01606	0.90472	0.44435
77	2.0437	1.9560	0.08769	1094.82	1049.76	45.06	694.90	694.88	0.01607	0.93524	0.45935
78	2.0408	1.9513	0.08954	1095.25	1049.19	46.06	673.51	673.52	0.01607	0.96666	0.47478
79	2.0380	1.9466	0.09140	1095.68	1048.62	47.06	652.93	652.91	0.01607	0.99900	0.49066
80	2.0352	1.9419	0.09325	1096.12	1048.07	48.05	633.03	633.01	0.01607	1.0323	0.50701
81	2.0324	1.9373	0.09510	1096.55	1047.50	49.05	613.82	613.80	0.01608	1.0665	0.52382
82	2.0297	1.9328	0.09694	1096.98	1046.93	50.05	595.27	595.25	0.01608	1.1017	0.54112
83	2.0269	1.9281	0.09878	1097.42	1046.37	51.05	577.36	577.34	0.01608	1.1380	0.55892
84	2.0242	1.9236	0.10062	1097.85	1045.80	52.05	560.06	560.04	0.01608	1.1752	0.57722
85	2.0214	1.9189	0.10246	1098.28	1045.23	53.05	543.35	543.33	0.01609	1.2136	0.59604
86	2.0187	1.9144	0.10430	1098.71	1044.67	54.04	527.21	527.19	0.01609	1.2530	0.61540
87	2.0160	1.9099	0.10611	1099.14	1044.10	55.04	511.62	511.60	0.01609	1.2935	0.63530
88	2.0133	1.9054	0.10794	1099.58	1043.54	56.04	496.54	496.52	0.01610	1.3351	0.65575
89	2.0106	1.9009	0.10976	1100.01	1042.97	57.04	481.98	481.96	0.01610	1.3779	0.67678
90	2.0079	1.8963	0.11158	1100.44	1042.40	58.03	467.90	467.88	0.01610	1.4219	0.69838
91	2.0053	1.8919	0.11339	1100.87	1041.84	59.03	454.28	454.26	0.01610	1.4671	0.72059
92	2.0026	1.8874	0.11520	1101.30	1041.27	60.03	441.12	441.10	0.01611	1.5136	0.74340
93	2.0000	1.8830	0.11701	1101.73	1040.70	61.03	428.40	428.38	0.01611	1.5613	0.76684
94	1.9974	1.8786	0.11881	1102.16	1040.13	62.03	416.09	416.07	0.01611	1.6103	0.79091
95	1.9947	1.8741	0.12061	1102.59	1039.56	63.03	404.19	404.17	0.01612	1.6607	0.81564
96	1.9922	1.8698	0.12241	1103.02	1039.00	64.02	392.67	392.65	0.01612	1.7124	0.84103
97	1.9996	1.8654	0.12420	1103.45	1038.43	65.02	381.53	381.51	0.01612	1.7655	0.86711
98	1.9870	1.8610	0.12600	1103.88	1037.86	66.02	370.75	370.73	0.01612	1.8200	0.89388
99	1.9844	1.8566	0.12778	1104.31	1037.29	67.02	360.32	360.30	0.01613	1.8759	0.92137
100	1.9819	1.8523	0.12957	1104.74	1036.72	68.02	350.22	350.20	0.01613	1.9334	0.94959
101	1.9793	1.8480	0.13135	1105.17	1036.16	69.01	340.44	340.42	0.01614	1.9923	0.97854
102	1.9768	1.8437	0.13313	1105.59	1035.58	70.01	330.98	330.96	0.01614	2.0529	1.0083
103	1.9743	1.8394	0.13490	1106.02	1035.01	71.01	321.82	321.80	0.01614	2.1149	1.0388
104	1.9718	1.8351	0.13667	1106.45	1034.44	72.01	312.95	312.93	0.01614	2.1786	1.0700
105	1.9693	1.8309	0.13844	1106.88	1033.87	73.01	304.36	304.34	0.01615	2.2440	1.1021
106	1.9668	1.8266	0.14021	1107.30	1033.29	74.01	296.04	296.02	0.01615	2.3110	1.1351

Source: Reprinted by permission from ASHRAE, 1977, Fundamentals.

versions of both English and SI tables). In Table 2-3, the first two columns after the saturation temperature give the corresponding saturation pressure in pounds force per square inch (lbf/in.^{2}) and inches of mercury (in. Hg). The next three columns give specific volume in cubic feet per pound mass (ft^3/lbm). The first of these lists the saturated liquid specific volume, v_f whereas the third column lists the saturated vapor specific volume, v_g. The second column lists the difference, $(v_g - v_f)$

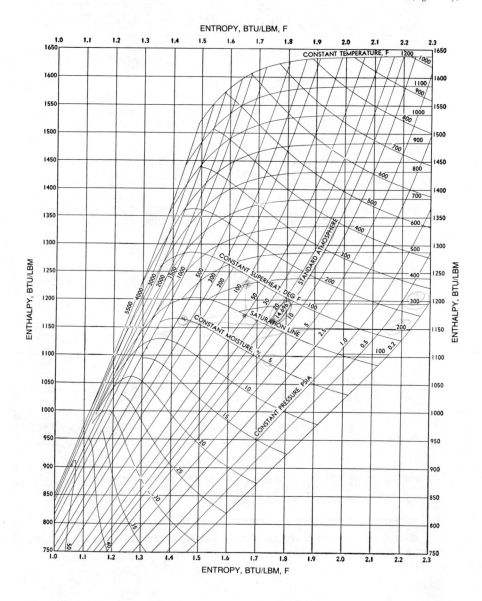

Figure 2-9. Plot of the properties of steam (Mollier diagram). (*Courtesy of Babcock & Wilcox, a McDermott Company.*)

and is designated v_{fg}. This quantity represents the change in specific volume of steam in a constant-pressure phase change. To calculate the specific volume of a substance within the saturation region (the saturation dome) one must remember that the total volume consists of the sum of the volumes of the liquid and the vapor that are present. The magnitude of these components depends upon the quality (the ratio of the mass of vapor to the mass of the liquid plus vapor in the saturation state). So, for a substance of mass m and quality x, the liquid volume is $m(1 - x)v_f$ and the vapor volume is mxv_g. Therefore, the total specific volume is

$$v = xv_g + (1 - x)v_f \tag{2-8}$$

Alternate forms of Eq. (2-8) may be obtained by using $v_{fg} = v_g - v_f$:

$$v = v_f + xv_{fg} \tag{2-9a}$$

$$= vg - (1 - x)v_{fg} \tag{2-9b}$$

The same procedure is followed for determining the enthalpy and the entropy for quality conditions:

$$h = xh_g + (1 - x)h_f \tag{2-10}$$

$$s = xs_g + (1 - x)s_f \tag{2-11}$$

Internal energy can then be obtained from the definition of enthalpy as $u = h - p\bar{v}$.

If the substance is a compressed or subcooled liquid, the thermodynamic properties of specific volume, enthalpy, internal energy, and entropy are strongly temperature-dependent (rather than pressure-dependent) and thus may be approximated, if compressed liquid tables are not available, by the corresponding values for saturated liquid (v_f, h_f, u_f, s_f) at the existing temperature.

In the superheated region, thermodynamic properties must be obtained from superheat tables or a plot of the thermodynamic properties, commonly called a **Mollier diagram**. The Mollier diagram is an enthalpy–entropy (h, s) plot—an example is shown in Figure 2-9.

EXAMPLE 2-1

Find the specific volume, the internal energy, and the enthalpy of steam at 500 F and a quality of 0.7.

SOLUTION: $\quad v = xv_g + (1-x)v_f$

$\qquad\qquad = [0.7(0.6749) + 0.3(0.0204)] \text{ ft}^3/\text{lbm} \qquad$ (from Table A-1-1)

$\qquad\qquad = 0.4786 \text{ ft}^3/\text{lbm}$

$\qquad h = xh_g + (1 - x)h_f$

$\qquad\qquad = [0.7(1202.2) + 0.3(487.9)] \text{ Btu/lbm} \qquad$ (from Table A-1-1)

$\qquad\qquad = 987.53 \text{ Btu/lbm}$

From the definition, we get $u = h - pv$. Hence

$$u = xu_g + (1 - x)u_f$$

$$= x(h_g - pv_g) + (1 - x)(h_f - pv_f)$$

$$= [xh_g + (1 - x)h_f] - p[xv_g + (1 - x)v_f]$$

$$= 987.8 \text{ Btu/lbm}$$

$$- 680.8 \text{ lbf/in.}^2 (0.4786 \text{ ft}^3/\text{lbm})\left(\frac{144 \text{ in.}^2}{\text{ft}^2}\right)\left(\frac{\text{Btu}}{778.3 \text{ ft-lbf}}\right)$$

$$= 927.49 \text{ Btu/lbm}$$

EXAMPLE 2-2

A vessel having a 14-ft³ volume contains 4 lbm of water vapor and liquid in equilibrium at 85 psia. Find the volume and mass of each component.

SOLUTION: $v = \dfrac{14 \text{ ft}^3}{4 \text{ lbm}} = 3.5 \text{ ft}^3/\text{lbm}$

$$= xv_g + (1 - x)v_f$$

$$= [5.168x + 0.01761(1 - x)] \text{ ft}^3/\text{lbm} \qquad \text{(from Table A-1-1)}$$

$$x = 0.6761$$

$$m_g = xm_{total} = 0.6761(4) \text{ lbm} = 2.704 \text{ lbm}$$

Thus

$$m_f = 1.296 \text{ lbm}$$

TOTAL
MIXTURE
VOLUME $\leftarrow$ $V_g = m_g v_g = 2.704(5.168) \text{ ft}^3 = 13.98 \text{ ft}^3$

Thus $\leftarrow$ SPECIFIC VOLUME

$$V_f = 0.02 \text{ ft}^3$$

EXAMPLE 2-3

Determine the temperature (if superheated) or quality (if saturated) of:

1. Water: 80 F and 20 ft³/lbm

2. Water: 100 lbf/in.² and 5.27 ft³/lbm

SOLUTION: 1. From the steam table (Table A-1-1) if 80 F is the saturation temperature, $v_g = 633.3$ ft³/lbm and $v_f = 0.01607$ ft³. Since the specific volume of the system is between v_f and then v_g, it is in the saturated region. Thus

$$x = \frac{v - v_f}{v_{fg}} = \frac{20 - 0.01607}{633.3} = 0.0316$$

2. If 100 psia is the saturation pressure, then the specific volume of the system should be between v_g (4.432 ft³/lbm) and v_f (0.01774 ft³/lbm). It is not. Thus it is superheated; $T = 450$ F.

EXAMPLE 2-4

What must be the quality of the steam at 300 psia such that it will reach the critical state if heated at constant volume?

SOLUTION: The critical state of steam (from Table A-1-1) is

$$p = 3208 \text{ lbf/in.}^2$$

$$v = 0.0508 \text{ ft}^3/\text{lbm}$$

$$T = 705 \text{ F}$$

The specific volume of the system remains constant since neither the volume nor the mass changes. Hence

$$v \text{(crit. state)} = xv_g \quad \text{(at 300 psia)}$$
$$+ (1 - x)v_f \quad \text{(at 300 psia)}$$
$$0.0508 = x(1.5427) + (1 - x)0.0189$$
$$x = 0.0209$$

EXAMPLE 2-5

A 270-ft³ rigid vessel contains 2.5 lbm of water (both liquid and vapor in thermal equilibrium) at a pressure of 1 psia. Calculate the volume and mass of both the liquid and the vapor.

SOLUTION: $v \text{(system)} = \dfrac{270 \text{ ft}^3}{2.5 \text{ lbm}} = 108 \text{ ft}^3/\text{lbm}$

Using Table 2-3 instead of the more comprehensive Table A-1-1, we get

$$v = xv_g + (1 - x)v_f$$
$$108 = x\,331 + (1 - x)0.01614$$
$$\text{(saturation temperature} = 102 \text{ F)}$$
$$x = 0.3263$$

Using the definition of quality, we get the following masses of the vapor and liquid:

$$m_g = xm_{total} = 0.326(2.5 \text{ lbm}) = 0.815 \text{ lbm}$$

and

$$m_f = (1 - x)m_{total} = 0.674(2.5 \text{ lbm}) = 1.685 \text{ lbm}$$

Using the definition of specific volume, we get the volumes of the vapor and liquid:

$$V_g = m_g v_g = 0.815 \text{ lbm } (331 \text{ ft}^3/\text{lbm}) = 269.77 \text{ ft}^3$$

$$V_f = m_f v_f = 1.685 \text{ lbm } (0.01614 \text{ ft}^3/\text{lbm}) = 0.0272 \text{ ft}^3$$

EXAMPLE 2-6

A mixture of water and steam occupies a volume of 1m^3. The mass of this combination is 50 kg. Determine the quality at 300 C.

SOLUTION: From Table A-1-5 we get (at 300 C)

$$v_f = 0.001404 \text{ m}^3/\text{kg}$$

$$v_g = 0.02165 \text{ m}^3/\text{kg}$$

The specific volume of the system is, by definition,

$$v = \frac{V}{m} = \frac{1 \text{ m}^3}{50 \text{ kg}} = 0.02 \text{ m}^3/\text{kg}$$

$$= x v_g + (1 - x) v_f$$

or

$$x = \frac{v - v_f}{v_g - v_f} = \frac{0.02 - 0.001404}{0.02165 - 0.001404}$$

$$= 0.9185$$

EXAMPLE 2-7

A mixture of steam and water at 0.100 MPa has a quality of 0.8. What is the specific enthalpy of the system?

SOLUTION: From Table A-1-5 at 0.1 MPa, we get

$$h_f = 417.5 \text{ kJ/kg} \quad \text{and} \quad h_{fg} = 2258 \text{ kJ/kg}$$

So

$$h = h_f + x h_{fg}$$

$$= [417.5 + 0.8 (2258)] \text{ kJ/kg}$$

$$= 2223.9 \text{ kJ/kg}$$

Refrigerant-12

The thermodynamic properties of the refrigerants used in vapor-compression systems are found in similar tables (see Appendix A-2). However, for these refrigerants the common Mollier plot is the pressure–enthalpy diagram illustrated in Figure 2-10.

2–5 Specific Heats and Latent Heat of Transformation

Two functions which are useful in our study of thermodynamics are the specific heats at constant volume and at constant pressure. The **constant-pressure specific heat** (denoted by the symbol c_p) is defined as

$$c_p \equiv \left(\frac{\partial h}{\partial T}\right)_p \tag{2-12}$$

The **constant-volume specific heat** (denoted by the symbol c_v) is defined as

$$c_v \equiv \left(\frac{\partial u}{\partial T}\right)_v \tag{2-13}$$

Note that the constant-pressure and constant-volume specific heats are thermodynamic properties since the only terms appearing in the definitions (h, u, p, v, and T) are properties.

The term **heat capacity,** though often used instead of specific heat, is not exactly correct.* The amount of heat that must be added to a closed system in order to accomplish a given change of state depends on how the process is carried out. Only for a reversible process in which the path is fully specified is it possible to relate the heat to a property of the system. On this basis we define heat capacity in general by

$$C_x = \left(\frac{\delta Q}{dT}\right)_x \tag{2-14}$$

where x indicates that the process is reversible and the path is fully specified. We could define a number of heat capacities according to this prescription, but only two are in common use. These are C_v, heat capacity at constant volume, and C_p, heat capacity at constant pressure. In both cases the system is presumed to be closed and of constant composition. By definition,

$$C_v = \left(\frac{\delta Q}{dT}\right)_v \tag{2-15}$$

---

* When the caloric theory was in vogue, heat was assumed to be "possessed" by a substance—thus, a substance had the capacity to possess heat. Do not let this unfortunate carryover phrase confuse you. These specific "heats" are properties of the systems if carefully defined.

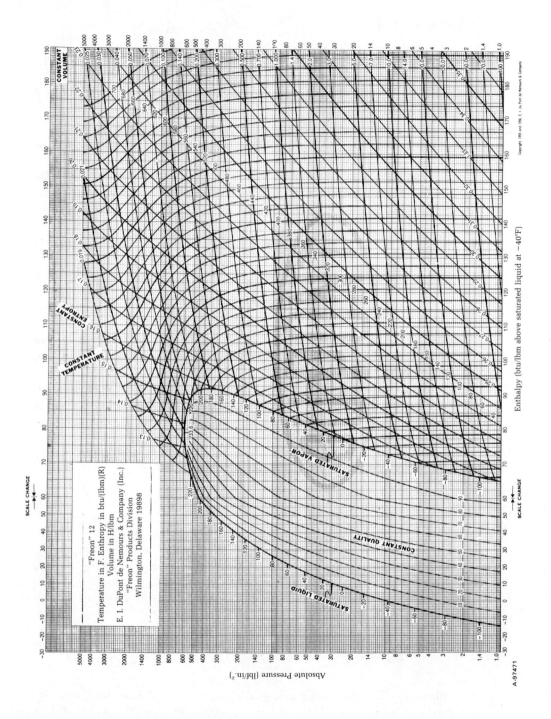

Figure 2-10. Pressure–enthalpy diagram for refrigerant-12. (© 1955–1956 by DuPont Company. Used by permission.)

which represents the amount of heat required to increase the temperature by dT when the system is held at constant volume.

Similarly,

$$C_p = \left(\frac{\delta Q}{dT}\right)_p \tag{2-16}$$

which represents the amount of heat required to increase the temperature by dT when the system is heated in a reversible process at constant pressure. After we have discussed the first law of thermodynamics, you will see the relationship between specific heat and heat capacity (see Appendix B-2).

One other specific heat is used. The **polytropic specific heat** is defined as

$$c_n = \frac{c_p - n c_v}{1 - n} \tag{2-17}$$

where n is the index of a polytropic process ($pv^n = $ constant).

The **latent heat of transformation*** *is defined as the ratio of the heat supplied, Q, to the mass m undergoing a change in phase.* Later we will see that the latent heat of transformation in any change of phase is equal to the difference in the two possible saturated states involved (at the same pressure). Another way of saying this is that the latent heat of a pure substance is the amount of heat that must be added to a unit mass of a substance at a given pressure to change its phase. Therefore:

Latent heat of vaporization $\quad h_{fg} = h_g - h_f$

Latent heat of fusion $\quad h_{if} = h_f - h_i$

Latent heat of sublimation $\quad h_{ig} = h_g - h_i$

Notice that these latent heats are also thermodynamic properties just like the specific heats we defined. In fact, it has been suggested that names like enthalpy of vaporization be used instead. Engineers and scientists have not accepted this suggestion, however.

EXAMPLE 2-8

For steam at 450 psia and 550 F, use the steam tables to estimate values of

$$\left(\frac{\partial h}{\partial T}\right)_p \quad \text{and} \quad \left(\frac{\partial v}{\partial T}\right)_p$$

SOLUTION: Using the superheated steam tables, we must form the ratio of

---

* Another unfortunate carryover name from the days of the caloric theory.

$\Delta h / \Delta T$ and $\Delta v / \Delta T$ with p held constant. From the tables, we get

$$\frac{\Delta h}{\Delta T} = \frac{1303.4 - 1238.3}{100} \frac{Btu}{lbm\text{-}F} = 0.651 \text{ Btu/lbm-F}$$

and

$$\frac{\Delta v}{\Delta T} = \frac{1.3005 - 1.1228}{100 \text{ F}} \text{ ft}^3/lbm = 0.00177 \text{ ft}^3/lbm\text{-}F$$

EXAMPLE 2-9

The experimentally determined specific heat of a substance has been determined to be

$$c_p = 0.338 - \frac{123.86}{T} + \frac{4.14\,(10^4)}{T^2} \; [=] \; \frac{Btu}{lbm\text{-}R} \quad \substack{\text{FROM CURVE} \\ \text{FITTING}}$$

in the temperature range of 540–9000 R (to within 1.7%). Determine the average (mean) specific heat in the ranges of 1000–3000 R and 7000–9000 R.

SOLUTION: $c_p(\text{mean}) = \dfrac{\displaystyle\int_1^2 c_p \, dT \; \Delta h}{T_2 - T_1}$

$$= 0.338 - \frac{123.86}{T_2 - T_1} \ln\left(\frac{T_2}{T_1}\right) - \frac{4.14\,(10^4)}{(T_2 - T_1)} \left(\frac{1}{T_2} - \frac{1}{T_1}\right)$$

1000–3000 R: $c_p(\text{mean}) = 0.284$ Btu/lbm-R

7000–9000 R: $c_p(\text{mean}) = 0.323$ Btu/lbm-R

PROBLEMS

2-1   Complete the following table:

Substance	T, F	p, psia	v, ft³/lbm	u, Btu/lbm
	20			
Freon-12		50	0.6	
		50		
	100	1000		
Water	80		20	
		1000		

(Table continued on next page)

2-1 continued (side to side)

Substance	ENTHALPY h, Btu/lbm	s, Btu/lbm-R	Condition x, SH or SC
Freon-12	22.83		
			75 SH
Water			
	1123		

2-2 Complete the following table:

Substance	T, F	p, psia	v, ft^3/lbm	Condition (x, SH, or SC)
H_2O	600	140		
H_2O		2000	0.018439	
Freon-12	120	35		
Freon-12	120			$x = 0.62$

2-3 For H_2O complete the following:

a. $p = 1000$ psia	$T = 150$ F	$v = $ ___ ft^3/lbm	$h = $ ___ Btu/lbm	$s = $ ___ Btu/lbm-R
b. $p = $ 30 psia	$T = 150$ F	$v = $ ___ ft^3/lbm	$h = $ ___ Btu/lbm	$s = $ ___ Btu/lbm-R
c. $p = $ ___ psia	$T = 250$ F	$v = $ ___ ft^3/lbm	$h = $ ___ Btu/lbm	$s = 1.21$ Btu/lbm-R
d. $p = $ 30 psia	$T = $ ___ F	$v = 1.4$ ft^3/lbm	$h = $ ___ Btu/lbm	$s = $ ___ Btu/lbm-R
e. $p = $ 200 psia	$T = 600$ F	$v = $ ___ ft^3/lbm	$h = $ ___ Btu/lbm	$s = $ ___ Btu/lbm-R

2-4 Steam in a boiler has been determined to have an enthalpy of 1100 Btu/lbm and an entropy of 1.56 Btu/lbm-R. What is its internal energy in Btu/lbm?

2-5 Water at 30 psig is heated from 62 to 115 F. Determine the change in enthalpy per pound.

2-6 A hot water heater has 2.0 gal/min entering at 50 F and 40 psig. The water leaves the heater at 160 F and 39 psig. Determine: (a) the change in enthalpy per pound and (b) the amount of water leaving (in gal/min) if the heater is operating under steady-flow conditions.

2-7   Water at 1000 psia and 200 F enters the steam-generating unit of a power plant and leaves the unit as steam at 1000 psia and 1600 F. Determine the following properties:

Inlet	Outlet
$v =$	$v =$
$h =$	$h =$
$u =$	$u =$
$s =$	$s =$
Condition:	Condition:

2-8   In a proposed automotive steam engine, the steam after expansion would reach a state at which the pressure is 20 psig and the volume occupied per pound mass is 4.8 ft³/lbm. Atmospheric pressure is 15 psi. Determine the following properties of the steam at this state:

$$T =$$

$$u =$$

Condition:

2-9   As the pressure in a steam line reaches 100 psia, the safety valve opens and releases steam to the atmosphere in a constant-enthalpy process across the valve. The temperature of the escaping steam (after the valve) was measured at 250 F. Determine the temperature of the steam in the line as well as its specific volume and condition.

2-10   Freon-12 enters the evaporator of a freezer at −20 F with a quality of 85%. The refrigerant leaves the evaporator at 15 psia with an entropy of 0.1835 Btu/lbm-F. Determine the following properties at each state:

Inlet	Outlet
$p =$	$T =$
$s =$	$v =$
	$h =$
	$u =$
	Condition:

2-11   Refrigerant-12 is compressed in a piston/cylinder system having an initial volume of 80 in.³. Initial pressure and temperature are 20 psia and 140 F. The process is *isentropic* to a final pressure of 175 psia. Determine:

    a. The final temperature (F)

    b. The mass of R-12 (lbm)

    c. The change in enthalpy (Btu/lbm)

    d. The change in internal energy (Btu)

2-12 In an ideal low-temperature refrigeration unit, Freon-12 is compressed isentropically from saturated vapor at 15.3 psia to a pressure of 200 psia. Determine the change in internal energy across the compressor per pound of Freon-12.

2-13 Refrigerant-12 vapor enters a compressor at 25 psia and 40 F; the mass rate of flow is 5 lbm/min. What is the smallest-diameter tubing that can be used if the velocity of refrigerant must not exceed 20 ft/sec?

2-14 Freon-12 is compressed in a residential air conditioner from saturated vapor at 40 F to superheated vapor at 100 psia having an entropy of 0.170 Btu/lbm-F. Determine the change in enthalpy for this compression process.

2-15 In a household refrigerator, Freon-12 enters the compressor as saturated vapor at 30 F. If the process across the compressor is isentropic and the discharge pressure is 150 psia, determine the refrigerant temperature at the compressor outlet.

2-16 Water is pumped through pipes embedded in the concrete of a large dam. The water in picking up the heat of hydration of the curing increases in temperature from 50 to 100 F. Water pressure is 500 psia. Determine: (a) the change in enthalpy per pound of water and (b) the change in entropy per pound of water.

2-17 Steam enters the condenser of a modern power plant with a temperature of 90 F and a quality of 0.98 (98% by mass vapor). The condensate (water) leaves at 1 psia and 80 F. Determine the change in specific volume between inlet and outlet of the condenser.

2-18 Determine the average heat capacity of a substance that receives 250 kJ of heat and experiences an 85 C temperature change. The mass of the substance is 4 lbm.

2-19 In the condenser of an air conditioning unit, refrigerant-12 is cooled at constant pressure from a superheated vapor at 125 psia and 140 F to a liquid that is subcooled by 6 F. Determine the change in internal energy per pound of refrigerant-12.

2-20 A 7.57-m^3 rigid tank contains 0.546 kg of H_2O at 37.8 C. The H_2O is then heated to 204.4 C. Determine: (a) the initial and final pressures of the H_2O in the tank (in MPa) and (b) the change in internal energy (in kJ).

2-21 A cylinder fitted with a piston contains steam initially at 0.965 MPa and 315.6 C. The steam then expands in an isentropic process (the entropy s remains constant) to a final pressure of 0.138 MPa. Determine the change in internal energy per pound of steam.

2-22 What is the condition (T and x) of H_2O in the following states?

    a. 10 $lbf/in.^2$; 1100 Btu/lbm (h)

    b. 1000 $lbf/in.^2$; 0.4 ft^3/lbm

    c. 80 F; 0.05 lbm/ft^3

    d. 225 F; 0.00245 lbm/ft^3

    e. 30 $lbf/in.^2$; 0.0168 ft^3/lbm

2-23 What is the condition (T and x) of Freon-12 in the following states?

    a. 50 psia; 0.96 ft^3/lbm

    b. 40 psia; 0.96 ft^3/lbm

    c. 70 psia; 0.77 ft^3/lbm

    d. 70 psia; 0.55 ft^3/lbm

    e. 102 F; 5 lbm/ft^3

2-24 A rigid vessel of 0.25 ft^3 volume contains 1 lbm of liquid and vapor H_2O in equilibrium at 100 F. The vessel is slowly heated. Will the liquid level inside the vessel eventually rise to the top of the container or drop toward the bottom? Why? What would happen if the vesssel contained 10 lbm instead of 1 lbm? Why?

2-25 Six lbm of H_2O is contained in a 20-ft^3 container (liquid and vapor at equilibrium) at 100 $lbf/in.^2$. Calculate: (a) the volume and mass of liquid and (b) the volume and mass of vapor.

2-26 Suppose that 0.3 lbm of H_2O (liquid and vapor in equilibrium) is contained in a vertical cylinder/piston arrangement (see sketch) at 120 F. Initially, the volume beneath the 250-lbm piston (area of

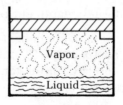

120 in.^{2}) is 1.054 ft^3. With the atmospheric pressure of 14.7 lbf/in.^{2} (g = 30.0 ft/sec^2), the piston is resting on the stops. Heat is applied to the arrangement until there is only saturated vapor inside.

a. Show this process on a (T, V) diagram.

b. What is the temperature of the H$_2$O when the piston first rises from the stops?

c. Determine the work done.

2-27 Determine the work done by a 2-lbm steam system as it expands slowly in a cylinder/piston arrangement from the initial conditions of 324 psia and 12.44 ft^3 to the final conditions of 25.256 ft^3 in accordance with the following relations:

a. $p = 20V + 75.12$, where if V [=] ft^3, then p [=] psia

b. pV = constant

2-28 The accompanying sketch is a general (p, v) diagram of a substance that expands on freezing.

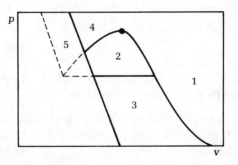

a. Using s for solid, l for liquid, and v for vapor, indicate the phases existing at each location from 1 to 6.

b. Label the critical point and the triple point on a (p, v) diagram.

c. Make a representative (p, T) diagram.

2-29 A mass m_1 of a liquid at temperature T_1 is mixed with a mass m_2 of the same liquid but at temperature T_2. If $m_2 = 0.833m_1$ and $c_{p2} = 1.5c_{p1}$, what is the final temperature of the mixture in terms of T_1 and T_2? The system is thermally insulated. (*Hint:* Recall the definition of heat capacity and the definition of thermally insulated.)

2-30 A rigid vessel contains saturated Freon-12 at 15.6 C. Determine: (a) the volume and mass of liquid and (b) the volume and mass of vapor at the point necessary to make the Freon pass through the critical state (or point) when heated.

2-31 What is the condition (T and x) of H_2O in the following states?

    a. 20 C; 2000 kJ/kg (u)

    b. 2 MPa; 0.1 m³/kg

    c. 140 C; 0.5089 m³/kg

    d. 4 MPa; 25 kg/m³

    e. 2MPa; 0.111 m³/kg

2-32 Compute the enthalpy of vaporization of water at a pressure of 10 MPa.

2-33 Determine the specific enthalpy of superheated ammonia vapor at 1.3 MPa and 65 C given the following data for specific enthalpy:

T	h, 180 psia	h, 220 psia
140 F	668 Btu/lbm	662 Btu/lbm
160 F	681 Btu/lbm	675.8 Btu/lbm

2-34 Determine the specific entropy of evaporation of steam at standard atmospheric pressure (in kJ/(kg·K)).

2-35 The specific heat of a gas at constant pressure is given as 0.24 Btu/lbm-R (room temperature). What is this specific heat in units of kJ/(kg·K)? ~1.0

2-36 Show how it is possible to change a vapor into a liquid without condensation.

2-37 Considering a pump to be a frictionless, steady-flow device, estimate the work done (no kinetic or potential energy changes) per kilogram of water entering the pump at 0.01 MPa and 40 C and leaving at 0.35 MPa.

2-38 What must be the quality of the steam at 2 MPa such that, if heated in an isometric process, it will pass through the critical point?

2-39 An 85-m³ rigid vessel contains 10 kg of water (both liquid and vapor in thermal equilibrium at a pressure of 0.01 MPa). Calculate the volume and mass of both the liquid and vapor.

2-40 For steam at 3 MPa and 300 C, estimate the values of

$$\left.\frac{\partial h}{\partial T}\right)_p \quad \text{and} \quad \left.\frac{\partial v}{\partial T}\right)_p$$

2-41 An experimentally determined specific heat relation for a substance is

$$c_p = 0.2e^{0.0015T}$$

where the T is in degrees Celsius. Estimate the mean specific heat in the range from 1000 to 1500 K.

Chapter

3

Gases

An equation of state of a substance is a relationship among any three state variables. For convenience, we use pressure, specific volume, and temperature. This p-v-T relationship exists for every substance: solid, liquid, and vapor. Unfortunately, most equations of state are not known since they are extremely complicated. As a result, accurate equations of state for wide pressure and temperature ranges are few and far between. This chapter presents some of these approximate equations of state, beginning with that for an ideal gas.

3–1 Ideal Gas

Experimental evidence indicates that for gases at "low" pressure and "high" temperature, the equation of state can be represented in an extremely simple form:

$$p\bar{v} = RT \quad \longrightarrow \quad R = \bar{R}\,\frac{1}{mol.wt.}$$

where R is the gas constant for that gas. Even though this ideal-gas equation will be used a great deal in this text, it must be remembered that it is at best only an approximation. Its use is one of convenience since it is easily understood, presents the appropriate trends, aids in

$$\bar{v} = (vol.)(mol.wt.)$$

$$p\bar{v}_m = \bar{R}T$$

developing the correct intuition, and may be used to present the computational procedure. Steam is not an ideal gas in this text unless explicitly indicated.

Equation of State

Consider some experimental measurements of the pressure, volume, temperature, and mass of a certain gas over wide ranges of these variables. Let us correlate these data at a given absolute temperature and display this information on a plot of $p\bar{v}/T$ versus p, where the actual volume V is divided by the number of moles of the gas used ($V/\bar{n} = \bar{v}$, the molar specific volume). *It is found experimentally that these ratios all lie on a smooth curve, whatever the temperature, but the ratios at different temperatures lie on different curves.* Figure 3-1 shows a typical set of curves for a number of different temperatures. Note that these curves converge to exactly the same point on the vertical axis, whatever the temperature (and the curves for all other gases converge to exactly the same point). This limit of the ratio $p\bar{v}/T$, common to all gases, is called the **universal gas constant** and is denoted by $\bar{R}$.

$$\bar{R} = 1545 \text{ ft-lbf/lb-mole R} \quad \Leftarrow \quad 10.73 \frac{\text{FT}^3 \text{ PSIA}}{\text{LB MOLE }^\circ\text{R}}$$

$$= 8.3143 \times 10^3 \text{ J/kg-mole K}$$

$$= 1.986 \text{ Btu/lb-mole R}$$

$$= 1.986 \text{ cal/gm-mole K}$$

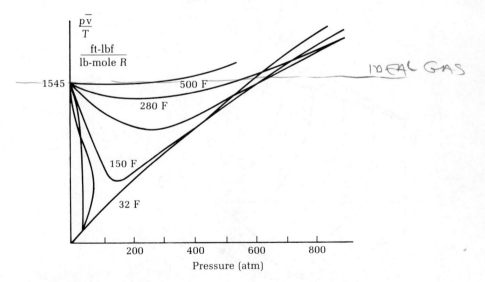

Figure 3-1. *$p\bar{v}/T$ versus p.*

Thus

$$\lim_{p \to 0} \left(\frac{p\bar{v}}{T} \right) = \bar{R} \tag{3-1}$$

For low pressure, this ideal-gas equation of state is

$$p\bar{v} = \bar{R}T \tag{3-2}$$

or $\nu = \dfrac{J}{M}$

$$pV = n\bar{R}T \qquad n = \dfrac{m}{M} \tag{3-3}$$

Recall that the number of moles is the mass divided by the molecular weight ($n = m/M$). Thus

$$p\frac{V}{m} = \frac{n}{m} \bar{R}T \qquad \checkmark \; \text{o}^\circ\text{k}$$

or

$$R = \frac{\bar{R}}{M}$$

$$p\hat{v} = \frac{\bar{R}}{M} T$$

$$p\bar{v} = RT \tag{3-4}$$

Notice that where $\bar{R}$ is a universal gas constant (a number), R is a gas constant that depends on the molecular weight of the gas. It may also be seen that

$$\frac{pV}{T} = n\bar{R} \qquad \text{(Boyle's law)} \tag{3-5}$$

$$\rho = \frac{1}{\nu} = \frac{p(\text{Mol.Wt.})}{\bar{R}\,T}$$

$$p(\text{Vol}) = (\text{mass})\,RT$$

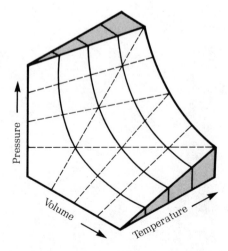

Figure 3-2. *p-v-T* surface for an ideal gas.

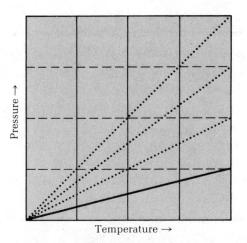

Figure 3-3. Phase diagram for an ideal gas.

is valid from this ideal-gas equation. Other information now becomes clear with the aid of the ideal-gas expression. For example, at standard conditions (273 K and 1 atm)

$$\overline{v} = \frac{\overline{R}T}{p} = 22.4146 \text{ m}^3/\text{mole}$$

Figures 3-2 to 3-4 present a portion of the p-v-T surface, the phase diagram, and a (p, v) projection for an ideal gas. The main point to note is that the equation of state is an experimental addition, not a theoretical deduction of thermodynamics.

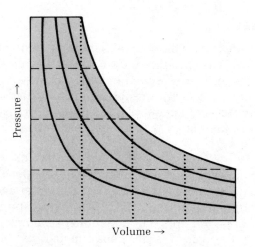

Figure 3-4. (p, v) diagram for an ideal gas.

EXAMPLE 3-1

A room contains 10,000 ft³ of air at 80 F and 29.0 in. of mercury. What is the mass of the air?

SOLUTION: $pV = mRT$ or $m = \dfrac{pV}{RT}$

$$m = \frac{(29 \text{ in. Hg})(0.4912 \text{ lbf/in.}^2/\text{in. Hg})(10^4 \text{ ft}^3)(144 \text{ in.}^2/\text{ft}^2)}{(53.34 \text{ ft-lbf/lbm-R})(540 \text{ R})}$$

$$= 712.1 \text{ lbm}$$

$80\,°F + 460\,°R$

EXAMPLE 3-2

The density of ammonia at 32 F and 1 atm is 0.04813 lbm/ft³. If we assume that ammonia is an ideal gas, what is the gas constant?

SOLUTION: $pv = RT$ or $R = \dfrac{pv}{T} = \dfrac{p}{\rho T}$

$$R = \frac{(14.7 \text{ lbf/in.}^2)(144 \text{ in.}^2/\text{ft}^2)}{(0.04813 \text{ lbm/ft}^3)(492 \text{ R})}$$

$$= 89.39 \text{ lbf-ft/lbm-R}$$

$NH_3 = 17$

As a check, recall that $M \doteq 17$ lbm/lb-mole. Thus

$$\bar{R} = MR = 1520 \text{ ft-lbf/lb-mole-R}$$

which is not quite 1545 ft-lbf/lb-mole R for ideal gas behavior.

Properties of Ideal Gases

Stating that a gas is ideal says much more than that the equation of state is $pv = RT$. The effect of this assumption on the properties of internal energy, enthalpy, constant-pressure specific heat, constant-volume specific heat, and entropy are very noticeable. Appendix B-1 shows that, for an ideal gas, the internal energy is a function of temperature only. From the definition of enthalpy it is easily seen that

$$h = u + pv = u + RT \qquad\qquad (3\text{-}6)$$

Therefore, the enthalpy is also a function of temperature only. This is a great simplification because most substances have internal energies and enthalpies that are functions of more than just temperature (e.g., specific volume or pressure—$u(T, p)$). The functional relationship for this internal energy of an ideal gas, $u = u(T)$, may be deduced in principle from the definition of the constant-volume specific heat:

$$c_v = \left(\frac{\partial u}{\partial T}\right)_v \tag{2-12}$$

Because of the ideal gas assumption, this definition reduces for an ideal gas to

$$c_v = \frac{du}{dT}$$

$$du = c_v \, dT \tag{3-7}$$

Similarly, the functional relationship for the enthalpy for an ideal gas, $h = h(T)$, may be deduced in principle from the definition of the constant-pressure specific heat:

$$c_p = \left(\frac{dh}{dT}\right)_p \tag{2-13}$$

Using the same reasoning as before, this relation reduces to

$$c_p = \frac{du}{dT}$$

$$dh = c_p \, dT \tag{3-8}$$

Notice that Equations (3-7) and (3-8) imply that not only are the internal energy and the enthalpy functions of temperature only, but the constant-pressure and the constant-volume specific heats are also functions of temperature only. That is, these two equations are valid regardless of the process and how the pressure, specific volume, and the like vary.

Using this information and beginning with Equation (3-6), we may obtain an interesting relation.

$$dh = du + R \, dT$$

$$c_p \, dT = c_v \, dT + R \, dT$$

or

$$c_p - c_v = R \tag{3-9}$$

The unique feature of Equation (3-9) is that while c_p and c_v are functions of temperature (only), the difference is a constant.

To emphasize the preceding information concerning *all* processes involving an ideal gas, we consider Figure 3-5. In accordance with the equation of state, constant-temperature lines on a (p, v) diagram are hyperbolas. According to the preceding information,

$$u(T_2) = u(T_1) + \int_{T_1}^{T_2} c_v \, dT \tag{3-10}$$

$$u(T_2) - u(T_1) = c_v(T_2 - T_1) \qquad c_v \text{ considered constant over the temperature range } T_1 \rightarrow T_2$$

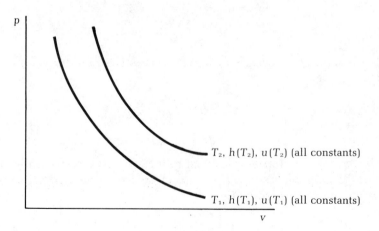

Figure 3-5. (p, v) diagram emphasizing unique results for an ideal gas.

and

$$h(T_2) = h(T_1) + \int_{T_1}^{T_2} c_p \, dT \tag{3-11}$$

$$h(T_2) - h(T_1) = c_p(T_2 - T_1) \qquad c_p \text{ considered constant over the temperature range } T_1 \rightarrow T_2$$

Thus lines of constant temperature are also lines of constant internal energy and also lines of constant enthalpy. Transition from one constant-temperature line to another, no matter what path is followed, results in the same change of these energy forms. Hence the equation for u that involves c_v (constant-volume specific heat) is not limited to constant-volume processes. The relationship $\delta q = c_v \, dT$ is restricted to constant-volume processes.

Entropy remains a function of both temperature and pressure, but a fairly simple one is given by the equation

$$ds = c_p \frac{dT}{T} - R \frac{dP}{P} \tag{3-12}$$

$$s(T_2, p_2) - s(T_1, p_1) = c_p \ln\left(\frac{T_2}{T_1}\right) - R \ln\left(\frac{p_2}{p_1}\right) \qquad c_p \text{ constant}$$

This expression will be derived later when the second law of thermodynamics and its uses are understood.

The ratio of specific heats is often denoted by

$$k = \frac{c_p}{c_v} \tag{3-13}$$

This is a useful quantity in calculations for ideal gases. Appendix A-3-4 gives the ideal-gas values for some common gases. It is not uncommon to find c_p and c_v as functions of k.

Notice that by using Equation (3-9) we obtain

$$\frac{c_p}{c_v} - 1 = \frac{R}{c_v} = k - 1$$

or

$$c_v = \frac{R}{k - 1} \tag{3-14}$$

and in a similar fashion

$$c_p = \frac{Rk}{k - 1} \tag{3-15}$$

The ideal gas is, of course, an idealization. No real gas exactly satisfies these equations over any finite range of temperature and pressure. All real gases approach ideal behavior at low pressures, however, and in the limit as $p \to 0$ they do in fact meet these requirements. Thus the equations for an ideal gas provide good approximations to real-gas behavior at low pressures. Moreover, because of their simplicity they are very useful.

To use the ideal gas approximation in an engineering calculation, one must be concerned with the resulting accuracy. So far the phases "low pressure" and "high temperature" have been stated as the conditions for this approximation. About the only "rule of thumb" that may be used is: you may expect only a low percentage error if the calculation is concerned with temperatures well above the critical temperature and pressures well below the critical pressure of the substance used. Thus nitrogen, $T(\text{crit}) = -233$ F, $p(\text{crit}) = 493$ psia, is very nearly an ideal gas at 80 F and 14.7 psia whereas carbon dioxide, $T(\text{crit}) = 88$ F, $p(\text{crit}) = 1071$ psia, is not. Since the critical temperature of steam is 705 F, it is rarely considered an ideal gas; thus we must use the steam tables.

EXAMPLE 3-3

For low-pressure oxygen in the temperature range of 80 to 4500 F, the experimentally detetermined specific heat at consant pressure is

$$c_p = 0.35 - \frac{5.375}{\sqrt{T}} + \frac{47.813}{T} \; [=] \; \frac{\text{Btu}}{\text{lbm-R}} \quad \big\} \; \text{FROM DATA}$$

where T is in degrees Rankine. Compute the mean specific heat c_v (mean) between 100 and 1200 F.

SOLUTION: Assume that the low-pressure oxygen is an ideal gas and recall the following:

1. $c_p - c_v = R$

2. $c_p(\text{mean}) = \dfrac{1}{T_2 - T_1} \displaystyle\int_{T_1}^{T_2} c_p \, dT \quad = \; h_2 - h_1$

and $c_v(\text{mean}) = \dfrac{1}{T_2 - T_1} \displaystyle\int_{T_1}^{T_2} c_v \, dT$

Thus

$$c_v = c_p - R \quad \text{and} \quad c_v(\text{mean}) = c_p(\text{mean}) - R$$

$$c_v(\text{mean}) = \frac{1}{1200 - 100} \int_{560}^{1660} \left(0.35 - \frac{5.375}{\sqrt{T}} + \frac{47.813}{T} \right) dT - R$$

$$= \left\{ \frac{1}{1100} \left[0.35(1100) - 10.75(\sqrt{1660} - \sqrt{560}) \right. \right.$$

$$\left. \left. + 47.813 \ln\left(\frac{1660}{560}\right) \right] - \frac{48.28}{778} \right\} \text{Btu/lbm-R}$$

conversion

$$= 0.1683 \text{ Btu/lbm-R}$$

EXAMPLE 3-4

Using the formulas of Example 3-3, calculate the change of enthalpy per lbm of oxygen when heated from 100 to 1200 F.

SOLUTION: $\quad h_2 - h_1 = \int c_p \, dT$

$$h(1200) - h(100) = \int_{560}^{1660} \left(0.35 - \frac{5.375}{\sqrt{T}} + \frac{47.813}{T} \right) dT$$

$$= 253.39 \text{ Btu/lbm}$$

Reversible Adiabatic Process

The **frictionless adiabatic process** is often encountered in thermo-dynamic analyses. For this reason it is necessary to establish the corre-sponding process equation. The process equation relates two properties (as opposed to an equation of state which relates three properties) and describes a sequence of states characterizing the process. Having this two-variable equation, the equation of state may be used to eliminate one of the two properties in favor of the third property. Appendix B-2 presents the derivation of the (p, v) process equation for a frictionless adiabatic process of an ideal gas with a constant specific heats ratio in a closed system. The result is

$$pv^k = \text{constant} \tag{3-16}$$

Be careful at this point. Although Equation (3-13) was derived under many specific constraints (e.g., ideal gas—closed system), it would be possible in the laboratory to set up such a process with any gas. The resulting progression of states would not necessarily represent an adia-batic, much less frictionless, process for that gas. In addition, Equation (3-16) is valid for an open system. This is true because it is an equation of properties (exact differentials) and depends only on the end points for a frictionless adiabatic process.

To relate any two states suffering a frictionless adiabatic process with Equation (3-16) and the ideal gas equation of state yields two expressions

$$p_1 v_1^k = p_2 v_2^k \tag{3-17}$$

and

$$\frac{p_1 v_1}{T_1} = \frac{p_2 v_2}{T_2} \tag{3-18}$$

Solving these two expressions simultaneously yields three expressions involving (p, v), (p, T) or (T, v). They are

$$\frac{p_1}{p_2} = \left(\frac{v_2}{v_1}\right)^k = \left(\frac{T_1}{T_2}\right)^{k/(k-1)} \tag{3-19}$$

$$1.40 = n_{AIR}$$

It may be easily seen that Equation (3-16) is a special case of the more general polytropic process discussed in Chapter 1. Therefore, for many but not all frictionless processes of an ideal gas, the polytropic process relation is

$$pv^n = \text{constant} \tag{3-20}$$

where n is a constant. Notice that Equations (3-19) may be rewritten with the index n:

$$\frac{p_1}{p_2} = \left(\frac{v_2}{v_1}\right)^n = \left(\frac{T_1}{T_2}\right)^{n/(n-1)} \tag{3-21}$$

EXAMPLE 3-5

Helium expands isentropically from 85 to 50 psia. If the initial volume is 10^4 cm^3, what is the final volume?

SOLUTION:
$$v_2 = v_1 \left(\frac{p_1}{p_2}\right)^{1/k}$$

$$K = \frac{C_P}{C_V}$$

Since k for helium is 1.667, we get

$$v_2 = 10^4 \text{ cm}^3 \left(\frac{85}{50}\right)^{1/1.667} = 1.375(10^4) \text{ cm}^3$$

3–2 Approximate Equations of State

The equation of state for an ideal gas holds exactly for a real gas only at near-zero pressure and approximately at moderate temperatures and somewhat higher pressures. Near the critical point, for example, the deviation of real gas from ideal-gas behavior is great. Many equations have been proposed for real gases. Some are empirical; others are deduced from assumptions regarding molecular properties.

Clausius Gas

Realistically, one cannot accept all the restrictions imposed on an ideal gas. As a result, Clausius reasoned that the next step was to account for the finite volume occupied by the molecules. The result is the Clausius equation of state:

$$p(\bar{v} - b) = \bar{R}T \tag{3-22}$$

Though it is a convenience pedagogically, this equation of state is no real improvement.

Van der Waals Gas

Van der Waals (1873) included a second correction term to account for the intermolecular forces. (Molecules do not actually have to collide to exert forces on one another.) This semitheoretical improvement over the ideal-gas equation is

$$\left(p + \frac{a}{\bar{v}^2}\right)(\bar{v} - b) = \bar{R}T \tag{3-23}$$

Obviously b, called the covolume, accounts for the finite molecular volume (as was suggested by Clausius as well). That is, if the volume of the molecule is b, then the space between molecules is $(v - b)$. The a term is the intermolecular force of attraction term. Thus, according to van der Waals, the pressure of a real gas is less than that of an ideal gas

Table 3-1 Approximate Values for the van der Waals Constants

Gas	a		b	a	b
	$\dfrac{\text{atm-ft}^6}{\text{mole}^2}$	$\dfrac{\text{psia-ft}^6}{\text{mole}^2}$	$\dfrac{\text{ft}^3}{\text{mole}}$	$\dfrac{\text{Nm}^4}{(\text{kg-mole})^2}$	$\dfrac{\text{m}^3}{\text{kg-mole}}$
Air	344	5,052	0.587		
Ammonia	1,070	15,720	0.596		
Carbon dioxide	926	13,600	0.686	366	0.0429
Carbon monoxide	381	5,598	0.639		
Freon-12	2,718	39,950	1.595		
Helium	8.57	126	0.372	3,440	0.0234
Hydrogen	62.8	922	0.427	24.8	0.0266
Nitrogen	346	5,082	0.618		
Oxygen	350	5,140	0.510	1.38	0.0318
Water vapor	1,400	20,580	0.488	5.80	0.0319

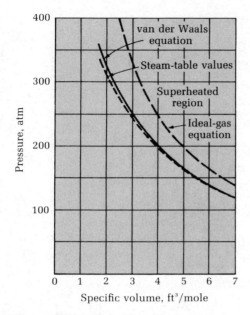

Figure 3-6. $(p, \overline{v})$ diagram for water.

for the same temperature and specific volume due to these forces. He thus postulated that this pressure reduction is proportional to $(1/\overline{v})^2$. The constants a and b are evaluated from experimental data. In partic- ular, the experimental data relative to the isotherm through the critical point is used. Some typical results are listed in Table 3-1. To emphasize that this equation of state is still an approximation, Figure 3-6 presents a quick comparison of steam-table values, the ideal-gas equation, and the van der Waals equation.

Other Forms

Many other forms of equations of state have been proposed. The follow- ing is just a partial list of the most popular ones.

[handwritten annotation:] $R = 10.73 \ \frac{FT^3 \ PSI}{LB \ MOLE \ °R}$

$(FOR \ H_2O) \ 18.02 \ \frac{LB}{LB \ MOLE}$

1. Dieterici (two unknowns):

$$p(\overline{v} - b) = \overline{R}Te^{-a/\overline{v}RT} \qquad \textbf{(3-24)}$$

[handwritten annotation:] $p = 0$ $-\partial\beta = 0$

$Z^3 - Z^2 + (\alpha - \beta^2 - \beta)Z - \alpha\beta = 0$

$\alpha = .4275 \ \frac{P_r}{T_r^{2.5}}$

2. Redlich–Kwong (two unknowns):

$$p = \frac{\overline{R}T}{\overline{v} - b} - \frac{a}{\sqrt{T}\,(\overline{v}^2 + \overline{v}b)} \qquad \textbf{(3-25)}$$

[handwritten annotation:] $\beta = .09664 \ \frac{P_r}{T_r}$

3. Callendar (two unknowns):

$$p(\overline{v} - b) = \overline{R}T - \frac{ap}{T^n} \qquad (n = 3.333) \qquad \textbf{(3-26)}$$

4. Saha–Bose (two unknowns):

$$p = -\frac{\bar{R}T}{2b} e^{-a/\bar{R}T\bar{v}} \ln\left(\frac{\bar{v} - 2b}{\bar{v}}\right) \tag{3-27}$$

5. Berthelot (two unknowns):

$$p(\bar{v} - b) = \bar{R}T - \frac{a(\bar{v} - b)}{T\bar{v}^2} \tag{3-28}$$

6. Clausius (II) (three unknowns):

$$\left[p + \frac{a}{T(\bar{v} + c)^2}\right](\bar{v} - b) = \bar{R}T \tag{3-29}$$

7. Beattie–Bridgman (five unknowns):

$$p\bar{v} = \bar{R}T\left[1 + \frac{B_0}{\bar{v}}\left(1 - \frac{b}{\bar{v}}\right)\right]\left(1 - \frac{c}{\bar{v}T^3}\right) - \frac{A_0}{\bar{v}}\left(1 - \frac{a}{\bar{v}}\right) \tag{3-30}$$

8. Benedict–Webb–Rubin (eight unknowns):

$$p\bar{v} = \bar{R}T + \frac{\bar{R}TB_0 - A_0 - C_0/T^2}{\bar{v}} + \frac{\bar{R}Tb - a}{\bar{v}^2}$$

$$+ \frac{\alpha a}{\bar{v}^5} + \frac{e}{\bar{v}^2 T^2}\left(1 + \frac{\gamma}{\bar{v}^2}\right)e^{-\gamma/\bar{v}^2} \tag{3-31}$$

9. Martin–Hou (nine unknowns):

$$p = \frac{\bar{R}T}{\bar{v} - b} + \frac{A_1 + A_2T + A_3 e^{-5.475\,T/T_c}}{(\bar{v} - b)^2} \tag{3-32}$$

$$+ \frac{A_4 + A_5T + A_6 e^{-5.475\,T/T_c}}{(\bar{v} - b)^3} + \frac{A_7}{(\bar{v} - b)^4} + \frac{A_8T}{(\bar{v} - b)^5}$$

Another useful form, called the **virial form** of the equation of state of a real gas from a theoretical point of view, is

$$p\bar{v} = A + \frac{B}{\bar{v}} + \frac{C}{\bar{v}^2} + \cdots \tag{3-33}$$

or

$$p\bar{v} = A^1 + \frac{B^1}{p} + \frac{C^1}{p^2} + \cdots \tag{3-34}$$

where A, B, C, A^1, B^1, and so forth are functions of temperature called **virial coefficients.** Thus, for an ideal gas, it is evident that $A = A^1 = RT$ and that all other virial coefficients are zero. The van der Waals equation can be put in virial form by first rearranging to the form

$$p\bar{v} = \bar{R}T\left(1 - \frac{b}{\bar{v}}\right)^{-1} - \frac{a}{\bar{v}}$$

The binomial theorem may be used to expand the term

$$\left(1 - \frac{b}{\bar{v}}\right)^{-1} = 1 + \frac{b}{\bar{v}} + \frac{b^2}{\bar{v}^2} + \cdots$$

Hence

$$p\bar{v} = \bar{R}T + \frac{(\bar{R}Tb - a)}{\bar{v}} + \frac{\bar{R}Tb^2}{\bar{v}^2} + \cdots$$

3–3 Real Gases

Compressibility Factor

The compressibility factor is defined as

$$p\bar{v} = z\bar{R}T$$

$$Z = \frac{p\bar{v}}{\bar{R}T} \tag{3-35}$$

For an ideal gas $Z \equiv 1$ for all p. Thus Equation (3-35) is a modification of the ideal gas equation where the value of the quantity $(Z - 1)/Z$ represents the relative deviation from ideal gas behavior. As you recall, the equation of state is an experimental addition to our study of thermodynamics—so must be the values of Z.

Figure 3-7 presents the diagram of the compressibility factor for nitrogen. The solid lines of the figure are isotherms (except for the saturation dome). As is easily seen from this figure, all isotherms approach one as the pressure decreases from the critical pressure. Note also that only the high temperature (approximately 300 K and slightly less) isotherms have values of $Z \simeq 1$ over a wide pressure range $(dZ/dp \simeq 0)$.

The (Z, p) diagram of the compressibility factor is not unique. Another view, a (T, p) diagram for superheated steam is presented in Figure 3-8. Though not quite as obvious as from Figure 3-7, the error that would result in using the ideal gas equation of state is shown.

EXAMPLE 3-6

Using the van der Waals equation of state, find the indicated limits:

1. $\lim\limits_{p \to 0} (Z)_T$

2. $\lim\limits_{p \to 0} (\alpha)_T$, where $\alpha = (\bar{R}T/p) - \bar{v}$ and is denoted as the residual volume. The subscript T implies constant-temperature operations.

SOLUTION: $\qquad p = \dfrac{\bar{R}T}{\bar{v} - b} - \dfrac{a}{\bar{v}^2} \qquad$ (van der Waals equation of state)

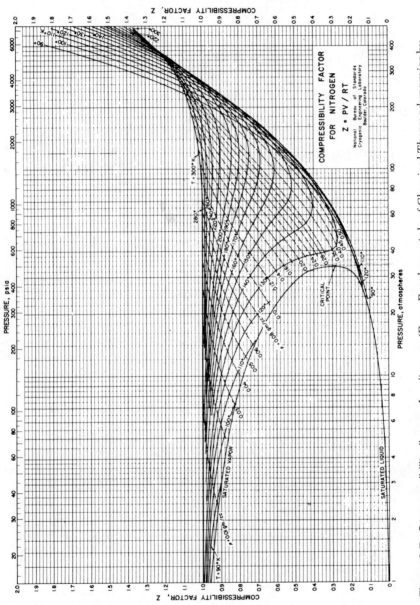

Figure 3-7. Compressibility diagram for nitrogen. (From *Fundamentals of Classical Thermodynamics* by Van Wylen/Sonntag, Copyright © 1965 by John Wiley & Sons, Inc. Reprinted by permission of John Wiley & Sons, Inc.)

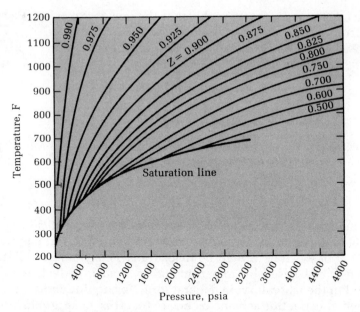

Figure 3-8. Compressibility factors for superheated steam. (Data from G. A. Hawkins and J. T. Agnew, *Combustion*, vol. 16, 1944, p. 46. From *Engineering Thermodynamics* by Jones/Hawkins, Copyright © 1960 by John Wiley & Sons, Inc. Reprinted by permission of John Wiley & Sons, Inc.)

1. $Z \equiv \dfrac{p\bar{v}}{\bar{R}T} = \dfrac{\bar{v}}{\bar{v} - b} - \dfrac{a}{\bar{v}\bar{R}T} = \dfrac{1}{1 - b/\bar{v}} - \dfrac{a}{\bar{v}\bar{R}T}$

Thus

$$\lim_{p \to 0}(Z)_T = \lim_{p \to 0}\left(\frac{1}{1 - b/\bar{v}} - \frac{a}{\bar{v}\bar{R}T}\right)$$

$$= \frac{1}{1 - 0} - 0 = 1 \qquad (\text{since } p \to 0,\ \bar{v} \to \infty)$$

2. $\alpha = \dfrac{\bar{R}T}{p} - \bar{v} = -b + \dfrac{a(\bar{v} - b)}{\bar{v}^2 p} = -b + \dfrac{a}{p\bar{v}} - \dfrac{ab}{p\bar{v}^2}$

Thus

$$\lim_{p \to 0}(\alpha)_T = -b + a\,\lim_{p \to 0}\left(\frac{1}{p\bar{v}}\right)_T - ab\,\lim_{p \to 0}\left(\frac{1}{p\bar{v}^2}\right) = -b + \frac{a}{\bar{R}T}$$

Reduced Coordinates

The compressibility factor scheme has one main disadvantage—one chart is needed for each gas. Because it is extremely convenient to have one chart for all gases, a generalized compressibility figure has been produced. The basis for this scheme is the **law of corresponding states.** This principle may be stated as follows: If two or more substances have the same reduced pressure, p_R, and reduced temperature, T_R, then the reduced volume, v_R, should also be equal. These reduced quantities are defined as

$$p_R = \frac{p}{p_c}, \ T_R = \frac{T}{T_c}, \text{ and } v_R = \frac{v}{v_c} \tag{3-36}$$

where (p_c, T_c, v_c) are the critical pressure, temperature, and specific volume, respectively, of the substance. Thus, the equation of state for all gases is

$$Z = Z(p_R, T_R)$$

Even though this procedure is most useful near the critical point, it is not 100% accurate even there. The main source of error is the measurement of v_c—thus v_R will be inaccurate. This is not to say that this scheme is not based on experimental evidence. In fact, it has been verified in the laboratory, in general form, for liquids and gases. Therefore, when using this scheme be aware that it is an approximation that is better than using ideal gas relations even though it is not completely correct.

EXAMPLE 3-7

Careful study of experimental data indicates that the critical point is a point of inflection for an isotherm. Thus

$$\left.\frac{\partial p}{\partial v}\right)_T = 0 \quad \text{and} \quad \left.\frac{\partial^2 p}{\partial v^2}\right)_T = 0$$

at the critical point. Using the Berthelot equation of state, show that the constants a and b are

$$a = 27\bar{R}^2 \frac{T_c^3}{64p_c}$$

and

$$b = \frac{\bar{v}_c}{3}$$

SOLUTION: The Berthelot equation of state is

$$p = \frac{\bar{R}T}{\bar{v} - b} - \frac{a}{T\bar{v}^2}$$

At the critical point,

$$\left.\frac{\partial p}{\partial \bar{v}}\right)_T = \frac{-\bar{R}T_c}{(\bar{v}_c - b)^2} + \frac{2a}{T_c\bar{v}_c^3} = 0 \quad \text{and}$$

$$\left(\frac{\partial^2 p}{\partial \bar{v}^2}\right)_T = \frac{2\bar{R}T_c}{(\bar{v}_c - b)^3} - \frac{6a}{T_c\bar{v}_c^4} = 0$$

Solving these two equations simultaneously yields

$$b = \frac{\bar{v}_c}{3} \quad \text{and} \quad a = \frac{27\,\bar{R}^2 T_c^3}{64 p_c} = 3\,p_c\,\bar{v}_c^2 T_c$$

EXAMPLE 3-8

Air at 260 C has a specific volume of $7.822(10^{-3})$ m³/kg. Determine the pressure of this air by using the van der Waals equation of state with $T_c = 132.6$ K and $p_c = 3.769$ MPa.

SOLUTION: The van der Waals equation is

$$\left(p + \frac{a}{v^2}\right)(v - b) = RT$$

Thus a, b, and R must be determined.

$$\bar{R} = 8.3143 \text{ kJ/kg-mole-K}$$

and

$$R = \frac{\bar{R}}{M}$$

where $M \doteq 28.97$ kg/kg-mole. Hence $R = 0.287$ kJ/(kg·K).

From the procedure introduced in Example 3-7, we can determine a. The result is

$$a = \frac{27}{64} \frac{R^2 T_c^2}{p_c}$$

$$= \frac{27}{64}(0.287 \text{ kJ/kg·K})^2 \frac{(132.6 \text{ K})^2}{3.769 \text{ MPa}}$$

$$= 162.1 \text{ (m}^3\text{/kg)}^2 \text{ Pa}$$

From the same sources,

$$b = \frac{RT_c}{8p_c}$$

$$= (0.287 \text{ kJ/kg·K}) \frac{(132.6 \text{ K})}{8\,(3.769 \text{ MPa})}$$

$$= 1.262\,(10^{-3}) \text{ m}^3\text{/kg}$$

So,

$$p = \frac{RT}{v - b} - \frac{a}{v^2}$$

Direct substitution (keep track of the units) yields

$$p = 20.68 \text{ MPa}$$

EXAMPLE 3-9

For an ideal gas, $Z \equiv 1$ at any state. Determine the compressibility factor Z_c for a Berthelot gas at the critical point.

SOLUTION: By definition,

$$Z \equiv \frac{p\bar{v}}{\bar{R}T}$$

and

$$p = \frac{\bar{R}T}{\bar{v} - b} - \frac{a}{T\bar{v}^2}$$

Thus

$$Z = \frac{\bar{v}}{\bar{v} - b} - \frac{a}{\bar{v}\bar{R}T^2}$$

Now at the critical point

$$Z_c = \frac{\bar{v}_c}{\bar{v}_c - b} - \frac{a}{\bar{v}_c \bar{R}T_c^2}$$

The values of a and b may be obtained from Example 3-7. Thus

$$Z_c = \frac{\bar{v}_c}{\bar{v}_c - \bar{v}_c/3} - \frac{3p_c\bar{v}_c^2 T_c}{\bar{v}_c \bar{R}T_c^2}$$

$$= \frac{3}{2} - 3\frac{p_c\bar{v}_c}{\bar{R}T_c} = \frac{3}{8}$$

EXAMPLE 3-10

7 OCT 85 Homework

Calculate the specific volume of nitrogen at a pressure of 80 atm (1176 psia) and a temperature of 150 K. Compare this value with that obtained using the ideal-gas equation.

SOLUTION: For N_2:

$$T_c = 126 \text{ K}$$

$$p_c = 33.5 \text{ atm}$$

$$R = 55.1 \text{ ft-lbf/lbm-R}$$

Thus

$$T_r = \frac{T}{T_c} = \frac{150 \text{ K}}{126 \text{ K}} = 1.190$$

$$p_r = \frac{p}{p_c} = \frac{80}{33.5} = 2.388$$

From the compressibility chart (Figure 3-7), we get

$$Z = 0.54$$

$$pv = ZRT$$

[handwritten: $p\bar{v} = ZRT_cT_R$]

or

$$v = \frac{ZRT}{p} = \frac{0.54 \,(55.15 \text{ ft-lbf/lbm-R}) \,(150 \text{ K})}{(1176 \text{ lbf/in.}^2) \,(144 \text{ in.}^2/\text{ft}^2)}$$

$$= 0.02638 \text{ ft}^3/\text{lbm (K/R) 1.8 R/K}$$

$$= 0.04748 \text{ ft}^3/\text{lbm}$$

For the ideal-gas case,

[handwritten: $\bar{v} = \frac{RT}{p} = \frac{55.1(266)}{1176(144)}$]

$$v = 0.08793 \text{ ft}^3/\text{lbm}$$

A generalized compressibility chart may be deduced from the law of corresponding states. That is:

$$Z = Z_c \frac{p_R}{T_R} v_R, \quad \text{but } v_R = v_R(p_R, T_R)$$

$$= Z_c F(p_R, T_R) \tag{3-37}$$

Figures 3-9 and 3-10 are examples of the generalized compressibility chart. When using these charts one must always remember that the results are an approximation. Thus the generalized compressibility chart should be used only when the data are not available for the gas in question.

EXAMPLE 3-11

What is the specific volume of nitrous oxide ($T_c = 557$ R; $p_c = 1054$ lbf/in.^{2}) at a pressure of 2108 lbf/in.^{2} and a temperature of 208 F?

SOLUTION:
$$T_R = \frac{668}{557} = 1.2 \qquad p_R = \frac{2108}{1054} = 2$$

$$R = \frac{1545 \text{ ft-lbf/mole-K}}{44 \text{ lbm/mole}} = 35.1 \text{ ft-lbf/lbm-K}$$

From Figure 3-9, we get $Z = 0.58$. Thus

$$v = \frac{ZRT}{p} = \frac{0.58 \,(35.1 \text{ ft-lbm/lbm-R}) \, 668 \text{ R} \,(\text{ft}^2/144 \text{ in.}^2)}{2108 \text{ lbf/in.}^2}$$

$$= 0.0448 \text{ ft}^3/\text{lbm}$$

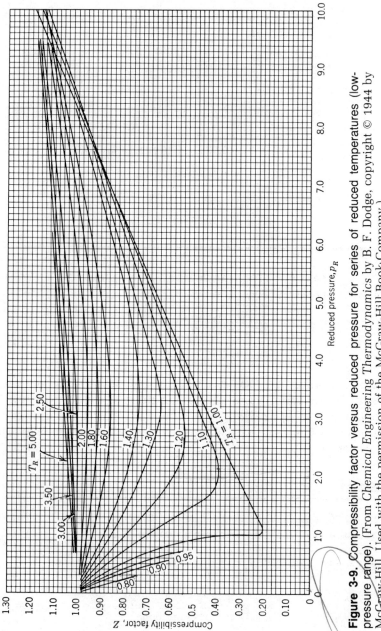

Figure 3-9. Compressibility factor versus reduced pressure for series of reduced temperatures (low-pressure range). (From *Chemical Engineering Thermodynamics* by B. F. Dodge, copyright © 1944 by McGraw-Hill. Used with the permission of the McGraw-Hill Book Company.)

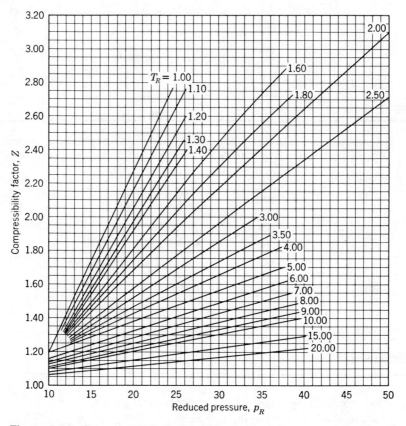

Figure 3-10. Compressibility factor versus reduced pressure for series of reduced temperatures (high-pressure range). (From *Chemical Engineering Thermodynamics* by B. F. Dodge, copyright © 1944 by McGraw-Hill. Used with the permission of the McGraw-Hill Book Company.)

Notice that if the ideal-gas approximation had been used,

$$v = \frac{RT}{p} = 0.0772 \text{ ft}^3/\text{lbm}$$

EXAMPLE 3-12

If the specific volume and pressure of nitrous oxide are 0.0448 ft³/lbm and 2104 lbf/in.² respectively, what is the temperature? (Recall that $T_c = 557$ R and $p_c = 1054$ lb/in.²)

SOLUTION: Recall that $pv = ZRT = ZRT_c T_R$. Thus

$$T_R = \frac{pv}{ZRT_c} = \frac{(2108 \text{ lbf/in.}^2)(0.0448 \text{ ft}^3/\text{lbm})}{Z\, 35.1 \text{ ft-lbf/lbm-K}\, 557 \text{ R}} \left(\frac{144 \text{ in.}^2}{\text{ft}^2} \right)$$

$$= \frac{0.6956}{Z}$$

By plotting $T_R Z = 0.6956$ on Figure 3-9, we get $Z = 1.2$. Thus

$$T_R = 0.5797$$

$$T = T_R T_c = 322.9 \ R$$

EXAMPLE 3-13

Steam at a pressure of 0.015 MPa and temperature of 650 C has a specific volume of 0.0268 m³/kg. If $R = 0.4615$ kJ/(kg·K), what is the compressibility factor?

SOLUTION: By definition,

$$Z = \frac{pv}{RT}$$

$$= \frac{(0.015 \ \text{MPa}) \ (0.0268 \ \text{m}^3/\text{kg})}{(0.4615 \ \text{kJ/kg·K})(923 \ \text{K})}$$

$$= 0.9436$$

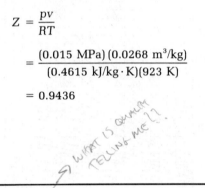

PROBLEMS

3-1 Find the missing value in each case:
a. H₂O: $p = 80$ psia; $x = 0.85$; $T =$ _____ F
b. Air: $p = 80$ psia; $T = 260$ F; $v =$ _____ ft³/lbm

3-2 A jet engine operates with an air/fuel ratio of 0.017 lbm fuel per pound of air. The fuel flow is 5280 lbm/hr. Air enters the engine at 1.8 psia and −30 F with a velocity of 550 ft/sec. Determine: (a) the air flow in ft³/min at the inlet and (b) the inlet area in ft².

3-3 Air is drawn into the compressor of a jet engine at 8 psia and +10 F. It is compressed isentropically to 40 psia. Determine: (a) the temperature after compression; (b) the specific volume before compression; and (c) the change in specific enthalpy for the process.
a. $T_2 =$ _____
b. $v_1 =$ _____
c. $h_2 - h_1 =$ _____

3-4 An automobile engine has a compression ratio (V_1/V_2) of 8.0. If the compression is isentropic and the initial temperature and pressure are 85 F and 14.7 psia, respectively, determine (a) the temperature and pressure after compression and (b) the change in enthalpy for the process.

3-5 As air flows across the cooling coil of an air conditioner at the rate of 8500 lbm/hr, its temperature drops from 79 to 55 F. Determine the internal energy change in Btu/hr.

3-6 Carbon monoxide is discharged from an exhaust pipe at 120 F and 0.12 psia. Determine its specific volume.

3-7 Twenty ft^3/min of carbon monoxide (CO) at 640 F is cooled to 80 F at atmospheric pressure. Determine: (a) the specific change in entropy of the CO for the process and (b) the enthalpy change per hour.

3-8 Oxygen is compressed from 14.7 psia and -260 F to 2000 psia and 40 F. Determine: (a) the range in specific volume and (b) the final internal energy.

3-9 Carbon dioxide (CO_2) is heated in a constant-pressure process from 60 F and 14.7 psia to 300 F. Determine, per unit mass, the changes in (a) enthalpy, (b) internal energy, (c) entropy, and (d) volume.

3-10 Air is used for cooling an electronics compartment. The air enters at 60 F and leaves at 105 F. Pressure remains essentially atmospheric (14.7 psia). Determine: (a) the change in internal energy of the air as it flows through the compartment and (b) the change in specific volume.

3-11 Air flows through a 3-in-diameter pipe at the rate of 1 lbm/sec. At section 1, the air has a velocity of 18 ft/sec, a temperature of 100 F, and an enthalpy of 134 Btu/lbm. Downstream at section 2, the air reaches a temperature of 240 F and a pressure of 19 psia. Determine the velocity at section 2.

3-12 Nitrogen is compressed in a cylinder having an initial volume of 0.25 ft^3. Initial pressure and temperature are 20 psia and 100 F. If the process is adiabatic and the final volume is 0.1 ft^3 at a pressure of 100 psia, determine:
a. The final temperature (F)
b. The mass of nitrogen (lbm)
c. The change in enthalpy (Btu)
d. The change in entropy (Btu/R)
e. The change in internal energy (Btu)

3-13 Air undergoes a steady-flow, reversible process. The initial state is 200 psia and 180 F and the final state is 20 psia and 785 F. Determine the final specific volume, the change in specific volume, the change in specific enthalpy, the change in specific internal energy, and the change in specific entropy. If the initial or inlet air flow rate is 155 ft^3/min, what is the exit or final flow rate (in lbm/hr and in ft^3/min)?

3-14 Air is compressed in a piston/cylinder system having an initial volume of 80 in³. Initial pressure and temperature are 20 psia and 140 F. The final volume is one-eighth of the initial volume at a pressure of 175 psia. Determine:

a. The final temperature (F)
b. The mass of air (lbm)
c. The change in internal energy (Btu)
d. The change in enthalpy (Btu)
e. The change in entropy (Btu/lbm-R)

3-15 Air, behaving as a perfect gas with $pv = RT$, is compressed *reversibly* in a cylinder by a piston. The 0.12 lbm of air in the cylinder is initially at 15 psia and 80 F, and the compression process takes place *isothermally* to 120 psia. Determine the work required to compress the air (in Btu).

3-16 Air is heated as it flows through a constant-diameter tube in steady flow. The air enters the tube at 50 psia and 80 F and has a velocity of 10 ft/sec at entrance. The air leaves at 45 psia and 255 F.

a. Determine the velocity of the air (ft/sec) at the exit.
b. If 23 lbm/min of air is to be heated, what diameter (in.) tube must be used?

3-17 One m³ of an ideal gas expands in an isothermal process from 760 to 350 kPa. Determine the work done by this gas.

3-18 An ideal gas expands in a reversible adiabatic process ($k = 1.4$) from 850 to 500 kPa. Determine the final volume if the initial volume is 100 m³.

3-19 Fifty kg of water and steam in equilibrium and at 300 C occupies a volume of 1 m³. What is the percentage of water (that is, the moisture content, $1 - x$)?

3-20 In a closed system, an ideal gas undergoes a process from 75 psia and 5 ft³ to 25 psia and 9.68 ft³. If c_p is constant, $\Delta H = -62$ Btu, and $c_v = 0.754$ Btu/lbm-R, determine: (a) ΔU, (b) c_p, and (c) R.

3-21 The temperature of an ideal gas remains constant while the pressure changes from 15 to 120 psia. If the initial volume is 2.8 ft³, what is the final volume?

3-22 Determine, by means of the ideal-gas equation, the pressure of 7 lbm of nitrogen at 752 F contained in a vessel having a volume of 0.5 ft³.

3-23 In the accompanying sketch, an ideal gas is taken from point a to point b along two paths; ab and adb. If $p_2 = 2p_1$ and $v_2 = 2v_1$, compute the work done in each process in terms of R and T.

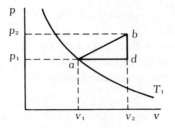

3-24 Sketch, on a (p, v) diagram, a process in which pv = constant is satisfied from (p_1, v_1) to (p_2, v_2) for an ideal gas $(p_1 > p_2)$. Also indicate this same process as it would appear on (p, T) and (T, v) diagrams.

3-25 The five processes $(a-b, b-c, c-d, d-a, \text{and } a-c)$ sketched on the adjacent (p, v) plane are for an ideal gas. Indicate the same processes on the (p, T) and (T, v) planes.

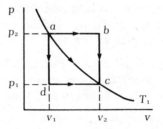

3-26 Three lbm of an ideal gas in a closed system is compressed frictionlessly and adiabatically from 14.7 psia and 70 F to 60 psia. For this gas $c_p = 0.238$ Btu/lbm-F, $c_v = 0.169$ Btu/lbm-F, and $R = 53.7$ ft-lbf/lbm-R. Compute: (a) the final volume if the initial volume is 40.3 ft^3 and (b) the final temperature.

3-27 Air expands in an adiabatic fashion from 25 psia and 140 F to 15 psia and 40 F. (Assume that the specific heats are constant.) What is the change in entropy?

3-28 In a closed system, 4 lbm of air $(\bar{C}_v = 4.96$ Btu/mole-R; $k = 1.4)$ is heated at constant pressure from 30 psia and 40 F to 140 F. What is the change in internal energy?

3-29 Prove that for an ideal gas

$$du = \frac{1}{k-1} d(pv) \quad \text{and} \quad dh = \frac{k}{k-1} d(pv)$$

3-30 Air at the rate of 5.52 m^3/min at 21 C and 0.1035 MPa enters the turbo-supercharger on an automobile engine. The air is compressed by the supercharger to 0.2413 MPa in an isentropic process. Determine:

a. The mass flow rate through the supercharger (kg/hr)

b. The air temperature after compression (C)

c. The change in enthalpy of the air across the supercharger (kJ/hr)

3-31 A cylinder fitted with a piston contains oxygen initially at 0.965 MPa and 315.5 C. The oxygen then expands in an isentropic process. (The entropy s remains constant to a final pressure of 0.1379 MPa.) Determine the change in internal energy per kg of oxygen.

3-32 Air enters an air conditioning duct at a rate of 56.62 m³/min at 4.44 C and 0.1035 MPa. The air discharges from the duct at 15.5 C and 0.1035 MPa. Determine

a. Mass flow rate of air (kg/hr)

b. Volume flow rate at discharge (m³/min)

c. Change in enthalpy of air between inlet and outlet (kJ/hr)

3-33 Water vapor at 10 MPa and 400 C has a specific volume of 0.026 m³/kg. Compute Z, p_R, and T_R if $p_c = 22.12$ MPa, $T_c = 647.3$ K, and $R = 0.4618$ kJ/(kg·K).

3-34 Carbon dioxide at 0.1 MPa and 0.5 m³/kg may be approximated by the van der Waals equation of state. If $p_c = 7.386$ MPa and $T_c = 304.2$ K, determine the temperature.

3-35 Nitrogen is cooled at constant pressure from 3000 to 300 K. Determine the heat transferred in this constant-pressure process (per unit mass). Assume that the empirical expression for c_p is

$$c_p = 1.3953 - 0.1832 \, (10^5) \, T^{-1.5} + 0.3832 \, (10^6) \, T^{-2}$$

$$- \, 0.2931 \, (10^8) \, T^{-3}$$

where c_p has units of kJ/(kg·K) if T is in degrees Kelvin.

3-36 Calculate the heat rejected (per unit mass) of an ideal gas whose molecular weight is 26 kg/kg-mole. Assume that $k = 1.26$, $c_v = 6$ kJ/(kg-mole·K), and the heat reject occurs at constant pressure from 100 to 1100 K.

3-37 For propane $T_c = 370$ K, $p_c = 4.26$ MPa, and $R = 0.18855$ kJ/(kg·K). Estimate the specific volume of propane at 6.8 MPa, 171 C. What is the percent difference if propane is assumed to be an ideal gas?

3-38 The coefficient of thermal expansion is defined as

$$\alpha = \frac{1}{v} \left(\frac{\partial v}{\partial T} \right)_p$$

and the coefficient of isothermal compressibility is defined as

$$\kappa = -\frac{1}{v}\left(\frac{\partial v}{\partial p}\right)_T$$

Determine these coefficients for the following equations of state:
a. Ideal gas: $p = RT/v$
b. Dieterici: $p = [RT/(v - b)]e^{-a/RTv}$
c. Van der Waals: $p = [RT/(v - b)] - a/v^2$

3-39 Transform the Clausius equation of state into its virial forms of density $(1/v)$ (at least four coefficients).

3-40 Recalculate as directed in Problem 3-39 but use the van der Waals equation of state.

3-41 Recalculate as directed in Problem 3-39 but use the Dieterici equation of state.

3-42 Air is compressed in a frictionless, steady-flow process from 0.1 MPa and 27 C to 0.9 MPa. What is the work of compression and the change in entropy per pound (mass) of air, assuming the process is (a) isothermal and (b) polytropic with $n = 1.4$?

3-43 A pump in a power plant handles 950,000 L/min of water at 150 C and 0.05 MPa and increases the water pressure to 10 MPa $(v = 0.063 \text{ m}^3/\text{kg})$. Determine:
a. Mass flow rate (kg/hr)
b. Volume flow rate at discharge (L/min)
c. The temperature change across the pump (C)
d. The enthalpy change across the pump (kJ/kg)

3-44 Determine the specific volume of propane at 6.9 MPa and 150 C. Compare this with the specific volume obtained by using the ideal-gas equation. For propane $T_c \doteq 370$ K, $p_c \doteq 4.25$ MPa, and $R = 35.1$ ft-lbf/lbm-R.

3-45 For a gas that obeys the van der Waals equation of state, determine the equation that describes a reversible adiabatic process, (that is, (p, v) and (T, v) equations). Assume that c_p and c_v are constant. (*Hint:* See Equation (9) of Appendix B-1 and follow the procedure of Appendix B-2-1.)

Chapter

4

The First Law of Thermodynamics

Now we come to one of the most important aspects of thermo-dynamics—in fact, possibly the most important single law of the phys-ical world. Because of its significance, this may be the chapter that requires the greatest amount of your effort. This first law brings together heat and work, first for closed systems and then for open systems.

4–1 Forms of Energy

Thermodynamics is the science founded on a law of the conservation of energy. This law says in effect that energy can neither be created nor destroyed. Heat and work are transitory forms of energy; they lose their identity as soon as they are absorbed by the body or region to which they are delivered. From our previous study, recall that work and heat are not possessed by a system and, therefore, are not properties. Thus, if there is a net transfer of energy across the boundary from a system (such as heat, work, or both), where did this energy come from? The only answer is that it must have come from energy stored in the system. This stored energy may be assumed to reside within the bodies or regions with which it is associated. In thermodynamics, accent is placed on the *changes* of this stored energy rather than on absolute quantities.

Stored Forms of Energy

Energy is stored in many forms. Some examples that quickly come to mind are thermal (internal) energy, mechanical energy, chemical energy, and atomic (nuclear) energy. One may easily see that stored energy is concerned with:

The molecules of the system (internal energy)

The system as a unit (kinetic and potential energy)

The arrangement of the atoms (chemical energy)

Cohesive forces within the nucleus (nuclear energy)

Molecular stored energy is associated with the relative position and velocity of the molecules; the total effect is called **internal** or **thermal energy.** It is called thermal energy because it cannot be readily converted into work. The stored energy associated with the velocity of the system is called **kinetic energy;** the stored energy associated with the position of the system is called **potential energy.** These are both forms of mechanical energy since they can be converted readily and completely into work. Although chemical and atomic energy would be included in any accounting of stored energy, engineering thermodynamics frequently confines itself to systems that do not undergo changes in these forms of energy due both to time limitations and to avoid duplication of effort. If the basic principles are understood, it is a simple matter to include these effects. These stored-energy forms require some thought before we go on. To emphasize their use, let us consider each form in a little more detail.

Thermal (Internal) Energy, U. As discussed in Chapter 1, internal energy relates to the energy *possessed* by a material due to the motion of molecules, their position, or both. This form of energy may be divided into two parts:

1. Kinetic internal energy is due to the velocity of the molecules.

2. Potential internal energy is due to the attractive forces existing between molecules.

Changes in the velocity of molecules are indicated by temperature changes of the system; variations in position are denoted by changes in phase of the system.

Potential Energy, PE. Potential energy is the energy *possessed* by the system due to its elevation or position. This potential energy is equivalent to the work required to lift the medium from an arbitrary zero elevation to its elevation z in the absence of friction. Using Newton's second law,

$$F = \frac{ma}{g_c} = \frac{mg}{g_c}$$

and the definition of work:

$$PE = W = \int_0^z F \, dx = \int_0^z m \frac{g}{g_c} \, dx = \frac{mg}{g_c} Z \qquad \textbf{(4-1)}$$

Kinetic Energy, KE. Kinetic energy is the energy *possessed* by the system as a result of its velocity. It is equal to the work that could be done in bringing to rest a medium that is in motion, with a velocity V, in the absence of gravity. Again Newton's second law is used in the following form:

$$F = \frac{ma}{g_c} = -\frac{m}{g_c} \frac{dV}{dt}$$

Thus

$$KE = W = \int_0^z F \, dx = -\int_v^0 \frac{m}{g_c} \frac{dV}{dt} \, dx = -\int_v^0 \frac{mV \, dV}{g_c} = \frac{mV^2}{2g_c} \qquad \textbf{(4-2)}$$

Chemical Energy, E_c. Chemical energy is *possessed* by the system because of the arrangement of the atoms comprising the molecules. Reactions that liberate energy are termed **exothermic**; those that absorb energy are termed **endothermic.**

Nuclear Energy, E_a. Nuclear energy is *possessed* by the system due to the cohesive forces holding the protons and neutrons together as the nucleus of the atom.

Transient Forms of Energy

The transient forms of energy with which we will be dealing are *work* and *heat*. Since we have already discussed these forms in Chapter 1, there is no need to discuss them further—except to suggest that you review the derivation of work for the special case of a frictionless steady-flow process. (This process will be presented in detail later.)

Closure

Table 4-1 summarizes the energy forms we have just discussed. This table divides the forms of energy into two categories: transient and possessed. Notice that the transient forms are path functions and thus can only be identified as they cross the system boundary. On the other hand, the possessed forms are point functions.

Table 4-1 Forms of Energy

Transient Forms of Energy	
Work	Potential: force
Heat	Potential: temperature

Energy Possessed by Substances and Systems	
By substances as entities:	
Potential	Manifested by elevation
Kinetic	Manifested by velocity
Internal or intrinsic:	
Thermal	
Molecular kinetic	Manifested by temperature
Molecular potential	Manifested by phase
Chemical	Manifested by changes in molecular composition
Nuclear	Manifested by changes in atomic composition

4–2 The First Law of Thermodynamics

From the previous discussion of the first law of thermodynamics—the law of conservation of energy—we can conclude that for any system, open or closed, there is an "energy balance" such as

$$\begin{pmatrix} \text{Net amount of energy} \\ \text{added to system} \end{pmatrix} = \begin{pmatrix} \text{net increase in stored} \\ \text{energy of system} \end{pmatrix}$$

or

$$\text{Energy in} - \text{energy out} = \text{increase in energy in system} \quad \textbf{(4-3)}$$

With both open and closed systems, energy can be added to the system or taken from it by means of heat and work.

First Law for Closed Systems

As stated above, the first law of thermodynamics is a statement of the principle of conservation of energy. It asserts that the net flow of energy across the boundary of a system is equal to the change in energy of the system. Since we are discussing only transients, we need consider only two types of energy flow across a boundary—work done on or by the system and heat. Therefore, the first law for closed systems executing a cyclic process is as follows: *During any cycle a system undergoes, the cyclic integral of the heat is proportional to the cyclic integral of work.* *

---

* Another way of looking at this definition is that the net change of energy in a closed-system cyclic process is zero.

Hence

$$J \oint \delta Q = \oint \delta W \qquad \textbf{(4-4)}$$

where $\oint \delta Q$ = net heat transfer during cycle

$\oint \delta W$ = net work during cycle

J = proportionality factor = 778.2 ft-lbf/Btu $(1\ N \cdot m/J)$

= mechanical equivalent of heat

The conclusion from the experimentally verified equation (4-4) is a quite accurate form of the first law, but it is not the most convenient. Notice that this principle is called a law. That is, it is guaranteed to be correct for all situations. If you doubt the universal accuracy of this statement, the only rebuttal one can make is that in all its forms and applications, it has not been found to be in error.

Equation (4-4) applies only to closed-system cyclic processes. Let us extend this formula to noncyclic processes (the change of state of a system). The noncyclic form may be deduced from the cyclic form directly. To do this, consider a change in state of the system from state 1 to state 2 by path A and then back to state 1 by path B (see Figure 4-1).

Applying the cyclic form of the first law to this cycle we get

$$0 \equiv \oint_{1-A-2-B-1} (\delta Q - \delta W) = \int_1^2 (\delta Q - \delta W)_A + \int_2^1 (\delta Q - \delta W)_B \qquad \textbf{(4-5)}$$

The mechanical equivalent of work has been omitted merely as a matter of convenience. Thus we assume that work and heat will be cast in the same units. Now we repeat the operation, but this time we substitute path C for path B and write out the cyclic form of the first law:

$$0 \equiv \oint_{1-A-2-C-1} (\delta Q - \delta W) = \int_1^2 (\delta Q - \delta W)_A + \int_2^1 (\delta Q - \delta W)_C \qquad \textbf{(4-6)}$$

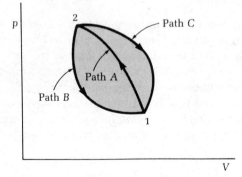

Figure 4-1. (p, V) diagram of a cyclic process.

Now subtract Equation (4-6) from (4-5). The result is

$$\int_2^1 (\delta Q - \delta W)_B - \int_2^1 (\delta Q - \delta W)_C = 0 \qquad (4\text{-}7)$$

Rearrange Equation (4-7) to the following form:

$$\int_2^1 (\delta Q - \delta W)_B = \int_2^1 (\delta Q - \delta W)_C \qquad (4\text{-}8)$$

Notice that since paths B and C are arbitrary, the quantity $(\delta Q - \delta W)$ depends only on the initial and final equilibrium states of the system. Thus we have a definition: The quantity $(\delta Q - \delta W)$, the difference in two path functions, is a point function (an exact differential). This quantity is an energy that has been stored in the system. Thus

$$dE = \delta Q - \delta W \qquad \boxed{\text{CLOSED SYSTEM}} \qquad (4\text{-}9)$$

A better form for Equation (4-9) might be

$$E_2 - E_1 = {_1}Q_2 - {_1}W_2 \qquad (4\text{-}10)$$

where ${_1}Q_2$ and ${_1}W_2$ are the heat transferred and work done on or by the system in going from state 1 to state 2. Of course, E_1 and E_2 are the values of the stored energy in the system at the beginning and end of the process.

For our purposes, the stored energy consists of the system's kinetic, potential, and internal energies. Thus

$$\delta Q = \delta W + dU + d(KE) + d(PE) \qquad (4\text{-}11)$$

This division of the stored energy is merely one of convenience. In fact, a possible interpretation of this arrangement is that the kinetic and potential energies have been separated out of the stored-energy term, leaving all energies that are not kinetic or potential. This remainder, regardless of the cause of the energy, is called internal energy. *Regardless of the interpretation, the existence of internal energy has been demonstrated directly from the conservation of energy (the first law).*

As far as the engineer is concerned, these results imply directly that it is impossible to construct a machine operating in cycles which, in any number of cycles, will put out more energy in the form of work than is absorbed in the form of heat. A machine that could do this would be called a "perpetual motion machine of the first kind." Thus the first law might be rephrased as: "A perpetual motion machine of the first kind is impossible."

Integration of Equation (4-11) using Equations (4-1) and (4-2) yields

$${_1}Q_2 = U_2 - U_1 + \frac{m}{2g_c}(V_2^2 - V_1^2) + \frac{mg}{g_c}(Z_2 - Z_1) + {_1}W_2 \qquad (4\text{-}12)$$

$\Delta U = q - W$

$\rightarrow$ QUALITY "x" $u_i = (1-x)u_f + x u_g$

$u_f = h_f - p\bar{v}_f$

$u_g = h_g - p\bar{v}_g$

where V designates velocity (not volume). It can be seen that by dividing by the mass of the system, the specific version is obtained:

$$_1q_2 = u_2 - u_1 + \frac{1}{2g_c}(V_2^2 - V_1^2) + \frac{g}{g_c}(Z_2 - Z_1) + _1w_2 \qquad \textbf{(4-13)}$$

By dividing Equation (4-11) by dt and taking the limit, we obtain the rate form of the first law:

$$_1\dot{Q}_2 = \frac{dU}{dt} + \frac{d}{dt}\left(\frac{mV^2}{2g_c}\right) + \frac{d}{dt}\left(\frac{mgZ}{g_c}\right) + _1\dot{W}_2 \qquad \textbf{(4-14)}$$

EXAMPLE 4-1

Suppose that 180 Btu of heat is added to a closed system executing a process from state 1 to state 2 where the internal energy is increased by 100 Btu. To restore the closed system to its initial state (state 2 to state 1), 95 Btu of work is done on the system. What is $_2Q_1$?

SOLUTION:

$$_1Q_2 = 180 \text{ Btu}$$

$$_2W_1 = -95 \text{ Btu}$$

$$E_2 - E_1 = 100 \text{ Btu}$$

The general statement of the first law is

$$\oint(\delta Q - \delta W) = 0$$

$$_1Q_2 + _2Q_1 - _1W_2 - _2W_1 = 0$$

or

$$_2Q_1 = _1W_2 + _2W_1 - _1Q_2$$

We know $_1Q_2$ and $_2W_1$, but not $_1W_2$. But

$$_1Q_2 - _1W_2 = E_2 - E_1$$

$$_1W_2 = _1Q_2 - (E_2 - E_1) = [180 - (100)] \text{ Btu}$$

$$= 80 \text{ Btu}$$

Therefore

$$_2Q_1 = [80 + (-95) - 180] \text{ Btu} = -195 \text{ Btu} = -2.057(10^5) \text{ J}$$

EXAMPLE 4-2

Suppose that 0.4 lbm of an ideal gas expands frictionlessly in a closed system from 25 psia, 165 F, until its volume triples. The expansion follows the path of pV = constant from an initial volume of 2.5 ft^3. Also, 31.77 Btu/lbm of heat is added to this ideal gas during this process. What is Δu?

SOLUTION: $_1q_2 - _1w_2 = \Delta u$ and pV = constant = p_1V_1

$$_1W_2 = \frac{1}{m} \int_1^2 p\, dV = \frac{p_1V_1}{m} \int_1^2 \frac{dV}{V} = \frac{p_1V_1}{m} \ln\left(\frac{V_2}{V_1}\right)$$

$$= \frac{25\dfrac{\text{lbf}}{\text{in.}^2}\, 2.5\ \text{ft}^3\, 144\dfrac{\text{in.}^2}{\text{ft}^2}\, \ln(3)}{0.4\ \text{lbm}} \left(\frac{\text{Btu}}{778\ \text{ft-lbf}}\right)$$

$$= 31.77\ \text{Btu/lbm}$$

Therefore

$$\Delta u = (31.77 - 31.77)\ \text{Btu/lbm} = 0$$

EXAMPLE 4-3

The input work to a paddle wheel used to stir a bowl of water is 6000 Btu. At the same time, 2400 Btu of heat is rejected by the bowl of water. Determine the internal energy change of the system.

SOLUTION:
$$U_2 - U_1 = {}_1Q_2 - {}_1W_2$$

WHY IS THE WORK NEGATIVE?

$$= [-2400 - (-6000)]\ \text{Btu}$$

BECAUSE WORK IS DONE ON SYSTEM!!

$$= 3600\ \text{Btu}$$

Note that work is done on the bowl of water, but there is essentially no volume change ($\int p\, dV = 0$).

EXAMPLE 4-4

A rigid vessel whose volume is 4 ft³ contains steam at 250 F and 45% quality. Determine the heat transferred if the vessel is cooled to 50 F.

SOLUTION: The first law:

$$\Delta u = {}_1q_2 - {}_1w_2$$

The rigid vessel does not change volume; thus ${}_1w_2 = 0$ and

$$_1q_2 = u_2 - u_1$$

Condition 1 is 250 F and $x = 0.45$ (that is, it is saturated). Using Appendix A-1-1 for a saturation temperature of 250 F we get

CALC.
u = h - pv ??

$$u_1 = xu_{g_1} + (1 - x)u_{f_1}$$

$$= [0.45\,(1087.9) + 0.55\,(218.5)]\ \text{Btu/lbm}$$

$$= 609.7\ \text{Btu/lbm}$$

$$u_2 = ?\quad \text{(Is it saturated?)}$$

To check this, let us look at the specific volume since it remains constant (that is, the system is closed—$dm = 0$ and the vessel is rigid).

$$v_1 = xv_{g_1} + (1 - x)v_{f_1}$$

$$= [0.45\,(13.83) + 0.55\,(0.01700)]\ ft^3/lbm = 6.2329\ ft^3/lbm$$

$$\equiv v_2$$

If state 2 is saturated, 50 F is the saturated temperature and $v_{g_2} = 1704.8\ ft^3/lbm$ and $v_{f_2} = 0.016023\ ft^3/lbm$. Since $v_{g_2} > v_2 > v_{f_2}$, it is saturated in state 2 and we must get the quality:

$$v_2 = x_2 v_{g_2} + (1 - x_2)v_{f_2}$$

$$x_2 = \frac{v_2 - v_{f_2}}{v_{g_2} - v_{f_2}} = \frac{6.2329 - 0.01602}{1704.8 - 0.01602} = 0.00366$$

and

$$u_2 = x_2 u_{g_2} + (1 - x)u_{f_2}$$

$$= [0.00366\,(1027.2) + 0.99634\,(18.06)]\ Btu/lbm$$

$$= 21.78\ Btu/lbm$$

Therefore

$$_1q_2 = 587.9\ Btu/lbm$$

EXAMPLE 4-5

Liquid water at 60 F is trapped (no vapor is present) in a cylindrical piston arrangement (see sketch a). The piston's weight is such that the pressure of the liquid is 100 psia. A fire under the arrangement (see sketch b) causes the piston to rise frictionlessly until it lodges at a point where the volume is 12.88 ft³. More, heat (nobody put out the fire) is transferred to the water until it exists as a saturated vapor. Determine the heat transferred to the water and the work done by the water. The mass of the liquid is 4 lbm.

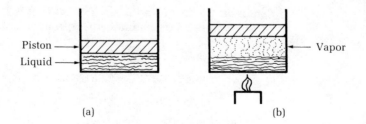

Piston

Liquid

Vapor

(a)   (b)

SOLUTION:

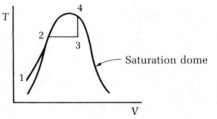

The accompanying sketch shows this process on a (T, V) diagram (not to scale). To approximate V_1 ($= mv_1$), assume $v_1 = v_f$ ($T = 60$ F). Thus $V_1 = 4$ lbm (0.016033 ft^3/lbm) = 0.06414 ft^3. Note from the description of the processes that process 1–2–3 is at constant pressure while process 3–4 is at constant volume. The liquid (in state 1) first heats up, expanding slightly to the saturation temperature (state 2) where boiling occurs and continues to point 3. The piston lodges at this point. The specific volume here is

$$v_3 = \frac{12.88 \text{ ft}^3}{4 \text{ lbm}} = 3.22 \text{ ft}^3/\text{lbm}$$

Boiling continues from state 3 to state 4, but at constant volume. Note that $v_3 = v_4 = v_{g_4}$. From Appendix A-1-1, the saturation temperature is 353.08 F and the saturation pressure is 140 psia. The work done may be computed directly from definition:

$$W_{\text{total}} = {}_1W_2 + {}_2W_3 + {}_3W_4$$

$$= p_1 (V_2 - V_1) + p_1 (V_3 - V_2) + 0$$

$$= 100 \frac{\text{lbf}}{\text{in.}^2} (12.88 - 0.06414) \text{ ft}^3 \frac{144 \text{ in.}^2}{\text{ft}} \frac{\text{Btu}}{778 \text{ ft-lbf}}$$

$$= 237.2 \text{ Btu}$$

$${}_1Q_4 = m (u_4 - u_1) + W_{\text{total}}$$

$$= 4 \text{ lbm} (1110.3 - 28) \text{ Btu/lbm} + 237.2 \text{ Btu}$$

$$= 4565 \text{ Btu}$$

EXAMPLE 4-6

In a closed system, a gas undergoes a reversible, constant-pressure volume change (0.15 to 0.05 m^3). During this process 25 kJ of heat is rejected. What is the internal energy change? The pressure is 0.35 MPa.

SOLUTION: From the first law:

$${}_1Q_2 = {}_1W_2 + \Delta U$$

Rearranging (note that it is a reversible process), we get

$$\Delta U = {_1}Q_2 - \int p \, dV$$

$$= -25 \text{ kJ} - p(V_2 - V_1) \qquad \text{(Remember: heat rejection)}$$

$$= -25 \text{ kJ} - 0.35 \text{ MPa } (0.05 - 0.15) \text{ m}^3$$

$$= 10 \text{ kJ}$$

EXAMPLE 4-7

Suppose that 7 kg of a substance receives 250 kJ of heat in an isometric change of temperature of 85 C. Estimate the average specific heat of this substance during this process. Assume it is a reversible process.

SOLUTION: Using the first law,

$$\Delta U = {_1}Q_2 - {_1}W_2$$

Note that the work is zero, so

$${_1}Q_2 = \Delta U = mc_v \, \Delta T \qquad \text{if } c_v \text{ is constant}$$

Thus

$$c_v = \frac{Q}{m \, \Delta T}$$

$$= \frac{250 \text{ kJ}}{7 \text{ kg } 85 \text{ C}}$$

Note that an increment of C is equal to an increment of K. Therefore,

$$c_v = 0.420 \text{ kJ/(kg} \cdot \text{K)}$$

First Law for Open Systems

Recall that an open system is a region of space surrounded by a boundary or surface (imaginary) through which mass, as well as energy, may propagate. Often this open system is called a **control volume** and the bounding (imaginary) surface is termed the **control surface.** Notice that for this situation heat, work, and mass flow across the control surface. An example of this type of device would be an air compressor or a turbine. There is no hard and fast rule in selecting the control volume and surface; their size and shape are arbitrarily selected to facilitate the analysis.

There is an additional mechanism for increasing or decreasing the stored energy of an open system. When mass enters the system, the stored energy of the system is increased by the stored energy of the entering mass. The stored energy of a system is decreased whenever mass leaves the system because the mass takes stored energy with it. If we distinguish this transfer of stored energy of mass crossing the system boundary from heat and work, then

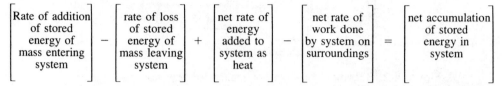

$$\begin{bmatrix} \text{Rate of addition} \\ \text{of stored} \\ \text{energy of} \\ \text{mass entering} \\ \text{system} \end{bmatrix} - \begin{bmatrix} \text{rate of loss} \\ \text{of stored} \\ \text{energy of} \\ \text{mass leaving} \\ \text{system} \end{bmatrix} + \begin{bmatrix} \text{net rate of} \\ \text{energy} \\ \text{added to} \\ \text{system as} \\ \text{heat} \end{bmatrix} - \begin{bmatrix} \text{net rate of} \\ \text{work done} \\ \text{by system on} \\ \text{surroundings} \end{bmatrix} = \begin{bmatrix} \text{net accumulation} \\ \text{of stored} \\ \text{energy in} \\ \text{system} \end{bmatrix}$$

The net *exchange* of energy between the system and its surroundings must be balanced by the *change* in the system's energy. "Exchange of energy" includes our definition of energy in transition being either work or heat. However, we must describe what is meant by the energy of the system and the energy associated with mass entering or leaving the system.

The energy E of the system is a property of the system and consists of all the various forms in which energy is characteristic of a system. These forms include potential energy (due to position) and kinetic energy (due to motion). Note that since work and heat are *energy in transition* and are not characteristic of the system, they are not included here. As with the closed system, all the energy of an open system— exclusive of kinetic and potential energy—is called internal energy. The symbol for internal energy per unit mass is u; $U = mu$.

We must now describe precisely what is meant by the energy associated with mass entering or leaving the system. Each pound of mass that flows into or out of the system carries with it the energy characteristic of that pound of mass. This energy includes the internal energy u plus the kinetic and potential energies.

If we investigate the flow of mass across the boundary of a system, we find that work is always done on or by a system where fluid flows across the system boundary. Therefore, the work term in an energy balance for an open system is usually separated into two parts:

1. The work required to push a fluid into or out of the system (flow work)

2. All other forms of work

To understand the first type of work, flow work, consider Figure 4-2. The mass entering from the left does so as the result of force F_1. This force, acting through a distance L_1, does work:

$$\text{Work}_\text{in} = F_1 \times L_1 = p_1 A_1 L_1 = p_1 V_1 \qquad \textbf{(4-15)}$$

Similarly, a force F_2 is required to remove mass from the control volume:

$$\text{Work}_\text{out} = F_2 \times L_2 = p_2 A_2 L_2 = p_2 V_2 \qquad \textbf{(4-16)}$$

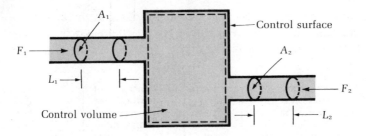

Figure 4-2. Schematic of a control volume for flow work calculations.

Therefore, the flow work per unit mass crossing the boundary of a system is pv. If the pressure or the specific volume or both vary as a fluid flows across a system boundary, the flow work is calculated by integrating $\int pv \, \delta m$, where δm is an infinitesimal mass crossing the boundary. The symbol δm is used instead of dm because the amount of mass crossing the boundary is not a property. Since the mass *within* the system is a property, the infinitesimal change in mass in the system is properly represented by dm.

The work term in an energy balance for an open system is, as we have seen, usually separated into two parts: flow work and all other forms of work.

The term *work* (W), without modifiers, is conventionally understood to stand for all other forms of work except flow work, and the complete two-word name is always used when referring to flow work.

An equation representing the generalized first law (for a one-inlet and one-outlet situation) can now be written with the symbols we have defined. As in Figure 4-3, we will let δm_1 be the mass entering the system and δm_2 be the mass leaving in a time interval dt. The first law in differential or incremental form directly from the previously stated

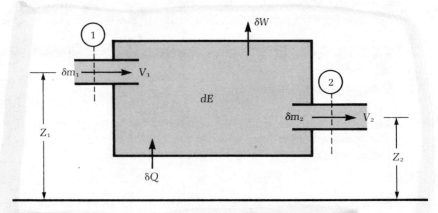

Figure 4-3. Energy flows in a general thermodynamic system.

word equation becomes

$$[\delta m\,(e + pv)]_{\text{in}} - [\delta m\,(e + pv)]_{\text{out}} + \delta Q - \delta W = dE$$

By substituting the exact expression for the stored energy, we get

$$\delta m_1 \left(u_1 + p_1 v_1 + \frac{V_1^2}{2g_c} + Z_1 \frac{g}{g_c} \right)$$

$$- \delta m_2 \left(u_2 + p_2 v_2 + \frac{V_2^2}{2g_c} + Z_2 \frac{g}{g_c} \right) + \delta Q - \delta W = dE \qquad \textbf{(4-17)}$$

where δQ and δW are the increments of work and heat and dE is the differential change in the energy of the system. Recall that E or U (or e or u) are properties of the system. As such, they are treated like any other property such as temperature, pressure, density, or viscosity. The combination of properties $u + pv$ is also a property, which we previously defined as enthalpy. There is nothing magic about enthalpy. It is merely used for simplicity and speed in obtaining numerical values for this combination.

In terms of enthalpy, the generalized first law equation becomes

$$\delta m_1 \left(h_1 + \frac{V_1^2}{2g_c} + \frac{g}{g_c} Z_1 \right)$$

$$- \delta m_2 \left(h_2 + \frac{V_2^2}{2g_c} + \frac{g}{g_c} Z_2 \right) + \delta Q - \delta W = dE \qquad \textbf{(4-18)}$$

or, in integrated form,

$$\int_0^{m_1} \delta m_1 \left(h_1 + \frac{V_1^2}{2g_c} + \frac{g}{g_c} Z_1 \right) - \int_0^{m_2} \delta m_2 \left(h_2 + \frac{V_2^2}{2g_c} + \frac{g}{g_c} Z_2 \right)$$

$$+ Q - W = E_{\text{final}} - E_{\text{initial}}$$

or, if divided by the time interval Δt,

$$\frac{\delta m_1}{\Delta t} \left(h_1 + \frac{V_1^2}{2g_c} + Z_1 \frac{g}{g_c} \right) - \frac{\delta m_2}{\Delta t} \left(h_2 + \frac{V_2^2}{2g_c} + Z_2 \frac{g}{g_c} \right) + \frac{\delta Q}{\Delta t} - \frac{\delta W}{\Delta t} = \frac{dE}{\Delta t}$$

as

$$\Delta t \to 0 \qquad \frac{\delta Q}{\Delta t} \to \dot{Q} \qquad \frac{\delta W}{\Delta t} \to \dot{W} \qquad \frac{\delta m_1}{\Delta t} \to \dot{m}_1 \qquad \frac{\delta m_2}{\Delta t} \to \dot{m}_2 \qquad \frac{dE}{\Delta t} \to \frac{dE}{dt}$$

$$\dot{m}_1 \left(h_1 + \frac{V_1^2}{2g_c} + Z_1 \frac{g}{g_c} \right) - \dot{m}_2 \left(h_2 + \frac{V_2^2}{2g_c} + Z_2 \frac{g}{g_c} \right) + \dot{Q} - \dot{W} = \frac{dE}{dt} \qquad \textbf{(4-19)}$$

where $\dot{Q}$ and $\dot{W}$ are the heat flow and work rates; $\dot{W}$ is recognized as power.

Equation (4-19) and the general form of the conservation of mass equation must now be solved simultaneously. Because of the obvious

GENERAL EQUATION

FOR AN INSTANCE IN TIME.

$mu_2 - mu_1$

difficulties, even for the apparently simple (one inlet, one outlet) situation, special cases are considered. These special cases are selected because they best model the physical situation encountered in many engineering problems. These special cases are the so-called **steady state/steady flow** and **uniform state/uniform flow.** The steady-state/ steady-flow case represents a condition such that at each point in space there is no variation of any property with respect to time. Moreover, the following assumptions are made:

1. The properties of the fluids crossing the boundary remain constant at each point on the boundary.

2. The flow rate at each section where mass crosses the boundary is constant. (The flow rate cannot change as long as all properties, including velocity, at each point remain constant.)

3. All interactions with the surroundings occur at a steady rate. (Thus the total mass within the control volume remains constant with respect to time.)

A simple example of this condition is the nozzle of a low-speed wind tunnel (Figure 4-4). Note that the velocity at section 1 is much less than the velocity at 2; but in sections 1 and 2 these velocities will maintain their respective magnitudes forever. Thus the velocity may change with position, but not with time.

The uniform-state/uniform-flow case represents a condition such that at each point of space any property has the same value at any instant of time. However, any property of the substance in the control volume may change with time, but it will have a uniform value throughout the control volume. Moreover, the following assumption is made: The properties of the fluids crossing the boundary remain constant with respect to time over the areas of the control surface where flow occurs. A simple example of this condition is the filling of an aerosol can (Figure 4-5). When the valve is opened, deodorant fills the can until the pressure, which is initially at 0 psia, has increased to the supply line pressure. Notice that the mass flow rate, initially very high, would drop as the can is filled.

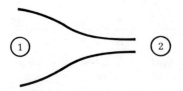

Figure 4-4. Schematic of a low-speed wind tunnel nozzle.

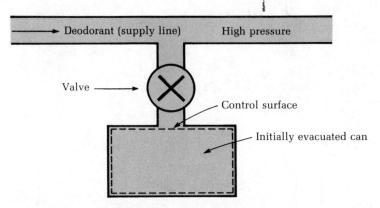

Figure 4-5. Schematic of aerosol can being filled.

For the *steady-state/steady-flow case* and for one inlet and one outlet, Equation (4-19) reduces to

$$\dot{m}_2\left(h_2 + \frac{V_2^2}{2g_c} + Z_2\frac{g}{g_c}\right) - \dot{m}_1\left(h_1 + \frac{V_1^2}{2g_c} + Z_1\frac{g}{g_c}\right) = \dot{Q} - \dot{W} \qquad \textbf{(4-20)}$$

The subscript 1 implies inlet and 2 implies outlet. Recall that the conservation of mass relation reduces to the following simple form for these conditions:

$$\dot{m}_1 = \dot{m}_2 \qquad \textbf{(4-21)}$$

Therefore, Equation (4-20) becomes

$$q - w = (h_2 - h_1) + \frac{1}{2g_c}(V_2^2 - V_1^2) + \frac{g}{g_c}(Z_2 - Z_1) \qquad \textbf{(4-22)}$$

Even though Equation (4-22) looks quite similar to Equation (4-13) (closed system), they are vastly different. *Each* term of both expressions represents a different physical quantity.

As a matter of completeness, the following expressions are given for the steady-state/steady-flow case for multiple inlets and outlets:

$$\sum_{\text{out}} \dot{m}\left(h + \frac{V^2}{2g_c} + \frac{g}{g_c}Z\right) - \sum_{\text{in}} \dot{m}\left(h + \frac{V^2}{2g_c} + \frac{g}{g_c}Z\right)$$
$$= \dot{Q} - \dot{W} \qquad \textbf{(4-23)}$$

$$\sum_{\text{out}} \dot{m} = \sum_{\text{in}} \dot{m} \qquad \textbf{(4-24)}$$

$$\dot{m} = \rho V A$$

For the uniform-state/uniform-flow case, Equation (4-19) has the form

$$\sum \dot{m}_{in}\left(u + pv + \frac{V^2}{2g_c} + \frac{g}{g_c}Z\right)_{in}$$

$$-\sum \dot{m}_{out}\left(u + pv + \frac{V^2}{2g_c} + \frac{g}{g_c}Z\right)_{out} + {}_cQ_f - {}_cW_f$$

$$= \left[m_f\left(u + \frac{V^2}{2g_c} + \frac{g}{g_c}Z\right)_f - m_i\left(u + \frac{V^2}{2g_c} + \frac{g}{g_c}Z\right)_i\right]_{system} \quad \textbf{(4-25)}$$

where the subscript f indicates condition of the control volume at the end of a time interval. At the beginning of this time interval, the control volume conditions are indicated by i. Note also that continuity is

$$\sum_{out} \dot{m} = \sum_{in} \dot{m} - \left(\frac{dm}{dt}\right)_{cv} \quad \textbf{(4-26)}$$

EXAMPLE 4-8

A system receives 0.756 kg/s of a fluid at a velocity of 36.58 m/s at an elevation of 30.48 m. At the exit, at an elevation of 54.86 m, the fluid leaves at a velocity of 12.19 m/s. The enthalpies of entering and exiting fluid are 2791.2 kJ/kg and 2795.9 kJ/kg, respectively. If the work done by the system is 4.101 kW, determine the heat supplied.

SOLUTION: Continuity:

$$\dot{m}_{in} = \dot{m}_{out} \quad (\text{or } \dot{m}_1 = \dot{m}_2 = 0.756 \text{ kg/s})$$

First law:

$$q - w = h_2 - h_1 + \frac{1}{2g_c}(V_2^2 - V_1^2) + \frac{g(Z_2 - Z_1)}{g_c}$$

Rearranging the first law yields

$$\dot{Q} = \dot{m}q = \dot{W} + \dot{m}(h_2 - h_1) + \frac{\dot{m}}{2g_c}(V_2^2 - V_1^2) + \frac{g\dot{m}}{g_c}(Z_2 - Z_1)$$

$$= 4.101 \text{ kJ/s} + 0.756 \text{ kg/s} \{(2795.9 - 2791.2) \text{ kJ/kg}$$

$$\left(+ \frac{[(12.19 \text{ m/s})^2 - (36.58 \text{ m/s})^2]}{2 \text{ kg m/(N} \cdot \text{s}^2)}\right.$$

$$+ 9.80 \text{ m/s} (54.86 - 30.48) \text{ m/kg m/(N} \cdot \text{s}^2)\}$$

$$= 4.101 \text{ kJ/s} + 0.756 \text{ kg/s} (4.7 \text{ kJ/kg} - 0.59524 \text{ kJ/kg}$$

$$+ 0.2394 \text{ kJ/kg})$$

$$= (4.101 + 3.553 - 0.45 + 0.181) \text{ kJ/s}$$

$$= 7.385 \text{ kJ/s} = 443.1 \text{ kJ/min} = 26.586 \text{ mJ/hr}$$

Note that the total energy available for heat transfer

$$|W| + |\Delta h| + |\Delta KE| + |\Delta PE|$$

is 8.285 kJ/s. Thus the fractional contribution of each energy form is

$$W \Rightarrow \frac{4.101}{8.285} = 0.495$$

$$\Delta h \Rightarrow \frac{3.553}{8.285} = 0.429$$

$$\Delta KE \Rightarrow \frac{0.45}{8.285} = 0.054$$

$$\Delta PE \Rightarrow \frac{0.181}{8.285} = 0.022$$

These relative percentages indicate that both the potential and the kinetic energy contribute only a small fraction to the energy transfer in this particular thermodynamic system.

EXAMPLE 4-9

A nozzle in a steam system passes 10^4 lbm/min. The initial and final pressures and velocities are respectively 250 psia, 1 psia, 400 ft/sec and 4000 ft/sec. Assuming the process is adiabatic, what is Δh?

SOLUTION:

This is a one-inlet/one-outlet situation. Thus

$$\dot{m}_{in} = \dot{m}_{out}$$

and

$$q - w + h_1 - h_2 + \frac{1}{2g_c} (V_1^2 - V_2^2) + \frac{g}{g_c} (Z_1 - Z_2) = 0$$

or

$$h_2 - h_1 = \frac{1}{2g_c} (V_1^2 - V_2^2)$$

$$= \frac{1}{64.4 \text{ lbm-ft/lbf-sec}^2} \frac{[(400)^2 - (4000)^2] \text{ ft}^2/\text{sec}^2}{778 \text{ ft-lbf/Btu}}$$

$$= -316 \text{ Btu/lbm}$$

EXAMPLE 4-10

A high-speed turbine produces 1 hp while operating on compressed air. The inlet and outlet conditions are 70 psia, 85 F, and 14.7 psia, −50 F, respectively. Assume the kinetic and potential energy differences to be very small ($\Delta KE = 0$ and $\Delta PE = 0$). What mass flow rate is required?

SOLUTION:

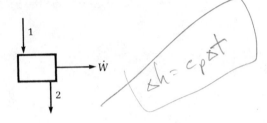

Continuity:

$$\dot{m}_1 = \dot{m}_2$$

First law:

$$\dot{m}_1\, h_1 = \dot{m}_2\, h_2 + \dot{W} \Rightarrow \dot{m} = \frac{\dot{W}}{(h_1 - h_2)}$$

Assuming air to be an ideal gas, we get

$$\dot{m} = \frac{\dot{W}}{c_p\, \Delta T} = \frac{1 \text{ hp (2545 Btu/hr hp)}}{0.24 \text{ Btu/lbm-R } [85 - (-50)] \text{ R}}$$

$$= 78 \text{ lbm/hr}$$

EXAMPLE 4-11

To obtain dry saturated steam, superheated steam and water are mixed. Assume the following data:

Superheated	Water	Dry Steam
400 psia	420 psia	300 psia
600 F	100 F	
2000 lbm/hr	?	

Find the mass rate of water to the mixture.

SOLUTION: Continuity:

$$\dot{m}_w + 2000 \text{ lbm/hr} = \dot{m}_v$$

First law: Assume $\Delta(KE) = \Delta(PE) = \dot{W} = \dot{Q} = 0$. Therefore

$$\sum_{\text{in}} \dot{m}h = \sum_{\text{out}} \dot{m}h$$

So

$$\dot{m}_w\, h_w + \dot{m}_s\, h_s = \dot{m}_v\, h_v$$

From Appendix A-1:

$$h_w \doteq 68 \text{ Btu/lbm} \qquad h_s = 1307 \text{ Btu/lbm} \qquad h_v = 1203 \text{ Btu/lbm}$$

Thus

$$\dot{m}_w\, 68 + 2000\,(1307)\text{ lbm/hr} = (2000 \text{ lbm/in.} + \dot{m}_w)\, 1203$$

$$\dot{m}_w = \frac{2000\,(1307 - 1203)}{1203 - 68}\text{ lbm/hr}$$

$$= 183.3 \text{ lbm/hr}$$

Guidelines for Energy Analysis

At this point, let us review a few of the major points to be considered in a thermodynamic analysis. The following procedures will be helpful in solving problems:

1. Sketch the system in general and select a control volume for analysis.

2. Determine the important energy interactions. (Remember the conventions.)

3. Write the first law for the system.

4. Determine the nature of the process between the initial and final states. Sketch a diagram of the process.

5. Make the idealizations or assumptions necessary to complete the solution.

6. Obtain physical data for the substance under study (equation of state, graphs, tabular data).

7. Complete the solution, taking great care to check the units in each equation.

Sketching the system and the process, though difficult habits to develop, are important steps in the solution that are often neglected. On the system sketch, indicate all the relevant energy terms. These sketches will help you approach a problem in a straightforward manner. Equally important is the process diagram on thermodynamic coordinates, such as a (p, v) plane. The value of this diagram will become apparent throughout the book as more properties are introduced and problems become more complex.

Most open-system processes are dominated by only a few effects. Generally the principal effect is easily identified. Heat exchangers, for example, are dominated by heat interaction and the accompanying enthalpy change. Thus we may assume in this case that kinetic and potential energy changes are negligible. Similarly, the same conclusion could be obtained for the work term in such devices as pumps, compressors, and turbines. But it should be clear that nozzles, diffusers, jets, and rocket engines have kinetic energy change as a principal effect. In the analysis of these systems ΔKE must be retained. Similarly, liquid flows with great elevation differences such as liquid flow from a tank or a hydroelectric station have ΔPE as a principal term. Analysis of open systems is greatly facilitated by the engineer's perception.

One final note: The subscripts "in" and "out" of the steady-flow equations are explicit but cumbersome. To simplify the notation it is common to use 1 for in and 2 for out in examples and exercises.

4–3 Enthalpy, the Joule–Thomson Coefficient, and the Porous Plug

As we have seen, enthalpy is a property, not an energy. It is a combination of variables that are used for convenience. Certainly enthalpy has its use. To emphasize this point, let us consider the classic porous plug experiment. This porous plug is just a constriction in a pipe (Figure 4-6). Think of it as a partial obstruction of the pipe that will slow down, but not block off, the flow of a fluid. In engineering systems, this plug may be a valve, denoted by valve symbol $\otimes$. In a time interval Δt, a mass m of the fluid will go from the left side (at temperature T_1 and pressure p_1) through the plug and into the right side (at temperature T_2 and pressure p_2). If we restrict our attention to this mass and consider it as our system, we may apply the first law for a closed system (Equation 4-13). If in addition we assume $V_1 \simeq V_2$, the pipe to be insulated, and $Z_1 = Z_2$, we get directly from the first law

$$0 = u_2 - u_1 + {}_1w_2 \qquad (4\text{-}27)$$

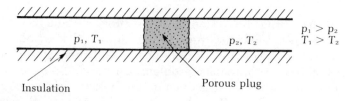

Figure 4-6. Porous plug experimental setup.

As in the discussion of flow work, the work done on the system to push mass m through the plug is (volume goes from V_1 to zero)

$$w_{in} = -p_1 V_1 \tag{4-28}$$

The work done by the fluid passing through the plug is (volume goes from zero to V_2)

$$w_{out} = p_2 V_2 \tag{4-29}$$

Thus Equation (4-27) becomes

$$0 = u_2 - u_1 + p_2 V_2 - p_1 V_1$$

or

$$u_2 + p_2 V_2 = u_1 + p_1 V_1 \tag{4-30}$$

or

$$h_2 = h_1$$

Therefore, the enthalpy is constant from one side of a valve to the other. Figure 4-7 illustrates a valve used in refrigeration systems. The same relation could be derived from an open-system analysis.

Now let us carry this idea one step further by adjusting the back pressure (formerly p_2) to p_3. We could execute the previous argument and end up with $h_1 = h_3$. Of course, we could change this back pressure many times; the result is that

$$h_1 = h_2 = h_3 = h_4 = \cdots$$

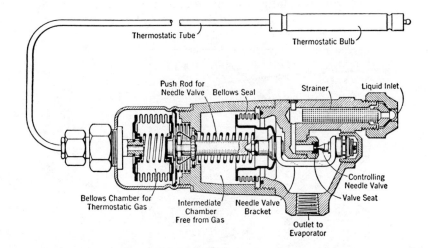

Figure 4-7. Thermostatic expansion valve. (Fig. 16-5, p. 558, in *Environmental Engineering: Analysis and Practice* by Burgess H. Jennings (intext). Copyright 1939, 1944, 1956, 1958, © 1970 by Harper & Row, Publishers, Inc. Reprinted by permission of Harper & Row Publishers, Inc.)

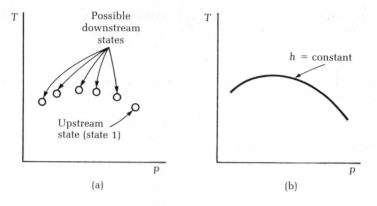

Figure 4-8. (a) Raw (T, p) data for flow through a porous plug. (b) The limiting case (sketch of a smooth curve through the data points) of a constant h line.

If each (T, p) pair were then plotted, Figure 4-8 would result. The slope of the curve of Figure 4-8b at any point is the **Joule–Thomson coefficient**, $\mu = (\partial T / \partial p)_h$. If, in fact, we executed this experiment for many constant h values, Figure 4-9 would result. Notice in this figure that a dashed line intersects each constant h line at its peak. This is called the inversion line and is defined as the locus of points for which the Joule–Thomson coefficient, μ, is equal to zero. It is a dividing line. On the left side of the inversion curve, the Joule–Thomson coefficient is greater than zero. Therefore, if both sides of the valve are represented by states on the left of the inversion line, the pressure drop through the valve (e.g., $a \rightarrow b$) results in a temperature drop as well. On the other hand, on the right side of this inversion curve the Joule–Thomson

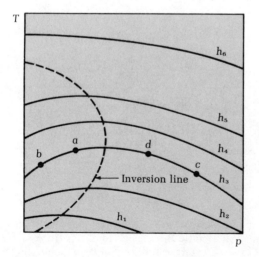

Figure 4-9. Constant-enthalpy lines ($h_1 < h_2 < h_3 < \ldots$) for a pure substance

coefficient is less than zero. Thus, if both sides of the valve are represented by states on the right side of the inversion curve, the pressure drop through the valve (e.g., $c{\rightarrow}d$) results in a temperature rise. It is possible, of course, to have a transition across the inversion curve. In that event positive slopes ($d{\rightarrow}b$) or negative slopes ($c{\rightarrow}a$) may occur.

This effect has been used as a means of refrigeration. When it is, care must be taken to be sure that the high-pressure side exhibits the higher temperature (e.g., near the inversion point).

EXAMPLE 4-12

Steam at 250 psia and 650 F is flowing in a pipe. A valve connecting the pipe to an evacuated container is opened, allowing steam to fill the container. Assume that $W = Q = \Delta KE = \Delta PE = 0$. Determine the final temperature of the steam in the container.

SOLUTION: Notice that this is a uniform-state example. Continuity:

$$m_{out} - m_{in} = m_{init} - m_{final}$$

$h_1 = h_2$

or

$$m_{in} = m_{final} \qquad (m_{out} = m_{init} = 0)$$

First law:

$$m_{in} h_{in} = m_{final} u_{final}$$

Therefore

$$u_{final} = h_{in}$$
$$= 1344.9 \text{ Btu/lbm}$$

The final conditions of the container are $p = 250$ psia and $u = 1344.9$ Btu/lbm. Thus $T_{final} \doteq 942$ F.

EXAMPLE 4-13

Steam is throttled from a saturated liquid at 212 F to a temperature of 50 F. What is the quality of the steam after passing through the expansion valve? Assume $\Delta PE = \Delta KE = Q = W = 0$.

SOLUTION:

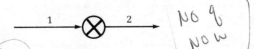

The throttling process requires that $\Delta h = 0$ through the valve. Thus $h_1 = h_2$. Continuity:

$$\dot{m}_1 = \dot{m}_2$$

First law:

$$\dot{m}_1 h_1 = \dot{m}_2 h_2$$

Thus

$$h_1 = h_2 = x h_{g_2} + (1 - x) h_{f_2}$$

$$x = \frac{h_1 - h_{f_2}}{h_{g_2} - h_{f_2}} = \frac{180.17 - 18.05}{1083.4 - 18.05}$$

$$= 0.152$$

Incidentally, the average Joule–Thomson coefficient is

$$(\mu_i)_{\text{avg}} \equiv \frac{\Delta T}{\Delta p}\bigg)_h = \frac{(50 - 212) \text{ F}}{(0.178 - 14.7) \text{ psia}} = 11.16 \text{ in}^2 \text{ F/lbf}$$

PROBLEMS

4-1 Complete the following table based on 1 *lbm* of matter.

| | Problem A | | Problem B | |
| | Steam Turbine | | Electrically Heated Wire | |
	Inlet	Outlet	Initial	Final
Pressure (psia)	1000	1.0	14.7	14.7
Temperature (F)	1000	101.74	80	380
Specific volume (ft³/lbm)		264.71	0.01	0.02
Enthalpy (Btu/lbm)		923	32.0	152.0
Entropy (Btu/lbm-R)	1.652	1.652	1.07	1.94
Kinetic energy (Btu/lbm)	2.5	0	0	0
Potential energy (Btu/lbm)	0.6	0.6	0	0
Internal energy (Btu/lbm)	1350.9	874.0	31.97	151.95
Flow work (Btu/lbm)	153.8	49.0		
Heat (Btu/lbm)	−40			
Work (Btu/lbm)			0	

(Table continued on next page.)

	Problem C		Problem D	
	Nozzle		Missile Nose Cone	
	Inlet	Outlet	Initial	Final
Pressure (psia)	100	40	14.0	3.0
Temperature (F)	500	300	75	2300
Specific volume (ft³/lbm)	5.59	11.04	2.0	2.0
Enthalpy (Btu/lbm)	1278.6	1193.8	63	
Entropy (Btu/lbm-R)	1.7085	1.7085	0.65	2.75
Kinetic energy (Btu/lbm)	0.3		0.2	0
Potential energy (Btu/lbm)	1.2		0.3	95
Internal energy (Btu/lbm)		1111.8		1717
Flow work (Btu/lbm)	103.5			0
Heat (Btu/lbm)	0			
Work (Btu/lbm)			0	

USE EQ. 4-20, 4-22

4-2 A system has a mass flow rate of 1 lbm/sec. The enthalpy, velocity, and elevation at entrance are, respectively, 100 Btu/lbm, 100 ft/sec, and 300 ft. At exit, these quantities are 99 Btu/lbm, 1 ft/sec, and −10 ft. Heat is transferred to the system at 5 Btu/sec. How much work is done by this system (a) per pound of fluid, (b) per minute, and (c) in kilowatts? $\dot{Q} - \dot{W} = \dot{m}_2 \left(h_2 + \frac{v_2^2}{2g_c} + z_2 \frac{g}{g_c} \right) - (\dot{m})(h_1 + \dots$

4-3 Consider 10 lbm of air that is initially at 14.7 psia and 100 F. Heat is transferred to the air until the temperature reaches 500 F. Determine the change of internal energy, the change in enthalpy, the heat transfer, and the work done for (a) a constant-volume process and (b) a constant-pressure process.

$pv = mRT$

4-4 Suppose that 30,000 lbm/hr of water at 500 psia and 200 F enters the steam-generating unit of a power plant and leaves the unit as steam at 500 psia and 1500 F. Determine the size of the unit in Btu/hr. RATE OF HEAT ADDITION. $\dot{Q}$

4-5 Suppose that 5 gpm of water at 30 psig is heated from 62 to 164 F. If electrical heating elements are used, determine: (a) the wattage required and (b) the current (amps) if a single-phase 220-V circuit is used.

4-6 While trapped in a cylinder, 5 lbm of air is compressed isothermally (a water jacket is used around the cylinder to maintain constant temperature) from initial conditions of 14.7 psia and 60 F to a final pressure of 115 psia. Determine for the process (a) the work required (Btu) and (b) the heat removed (Btu).

4-7 Air at the rate of 40 lbm/sec is drawn into the compressor of a jet engine at 8 psia and -10 F and is compressed reversibly and adiabatically (which is also isentropically) to 40 psia. Determine the size of compressor required (hp).

4-8 In a conventional power plant, 1,500,000 lbm/hr of steam enters a turbine at 1000 F and 500 psia. The steam expands isentropically to 15 psia. Determine the ideal turbine rating (kW).

4-9 After being heated in the combustion chamber of a jet engine, air at low velocity and 1600 F enters the jet nozzle. Determine the maximum velocity (ft/sec) that can be obtained from the nozzle if the air discharges from it at 700 F.

4-10 An air conditioning coil cools 2000 ft³/min of air at 14.7 psia and 80 F to 45 F. Determine the rating of the air conditioner (Btu/hr).

4-11 Steam enters the condenser of a modern power plant at a pressure of 1 psia and a quality of 0.98. The condensate leaves at 1 psia and 80 F. Determine: (a) the heat rejected per pound and (b) the change in specific volume between inlet and outlet.

4-12 In a diesel engine, air expands reversibly against a piston during combustion, which occurs at a constant pressure of 650 psia. At the end of combustion the temperature is 4868 R. For a 30-in.³ displacement of the piston during this combustion process, determine the power obtained (hp) if combustion occurs 600 times per minute.

4-13 Air is used for cooling an electronics compartment. Atmospheric air enters at 60 F and the maximum allowable air temperature is 100 F. If the equipment in the compartment dissipates 3600 W of energy to the air, determine the necessary air flow rate in (a) lbm/hr and (b) ft³/min at inlet conditions.

4-14 A refrigeration unit employing refrigerant-12 is shown in the adja-

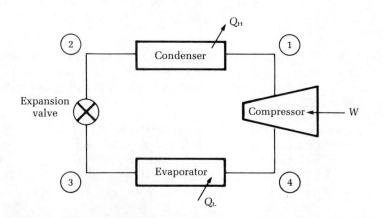

cent sketch. Condensing pressure is 216 psia. Evaporator temperature is -10 F. The unit is rated at 66,000 Btu/hr for cooling. If the velocity in the line between evaporator and compressor is not to exceed 5 ft/sec, determine the inside diameter (in.) of the tubing to be used.

4-15 What *minimum* size of motor (hp) would be necessary for a pump that handles 85 gpm of city water while increasing the water pressure from 15 to 90 psia?

4-16 For a normal city water supply having a pressure of 50 psig at negligible velocity, use the first law of thermodynamics to determine: (a) the *maximum* velocity (ft/sec) that can be obtained from a fire or garden hose nozzle and (b) the *maximum* elevation (ft) to which this water will flow without additional pumps.

4-17 Freon-12 at 180 psia and 100 F flows through the expansion valve in a vapor-compression refrigeration system. The pressure leaving the valve is 20 psia. Determine the quality of the Freon leaving the expansion valve. State clearly all assumptions you made in solving this problem.

4-18 Hot gases enter a row of blades of a gas turbine with a velocity of 1800 ft/sec and leave with a velocity of 400 ft/sec. There is an increase in the specific enthalpy of 2.2 Btu in the blade passage. If the mass flow of gases is 520 lbm/min, determine the blade horsepower.

4-19 A simple coal-fired steam power plant has a rated capacity of 500,000 kW when operating between limiting conditions of 500 psia and 600 F at the steam-generating unit outlet (also turbine inlet) and 1 psia at the turbine outlet. Steam leaves the turbine with a quality of 90% and enters the condenser. Neglecting pump work, determine:
a. The heat required at the steam-generating unit (in Btu/hr)
b. The amount of coal burnt if the coal has a heating value of 11,000 Btu/lbm (in lbm/hr)
c. The thermal pollution (Btu/hr)

4-20 A tank having a volume of 200 ft³ contains saturated vapor (steam) at a pressure of 20 psia. Attached to this tank is a line in which vapor at 100 psia and 400 F flows. Steam from this line enters the vessel until the pressure is 100 psia. If there is no heat transfer from the tank and the heat capacity of the tank is neglected, calculate the mass of steam that enters the tank.

4-21 During the operation of a steam power plant, the heat input at the steam-generating unit was 6.4×10^8 Btu/hr and the net output of the plant was 75,000 kW. Determine the heat rejected at the condenser (thermal pollution) in Btu/hr.

4-22 During the operation of a steam power plant, the steam flow rate was 500,000 lbm/hr with turbine inlet conditions of 500 psia and 1000 F and turbine exhaust (condenser inlet) conditions of 1.0 psia (90% quality). Determine: (a) turbine output in kW and (b) condenser heat rejection rate in Btu/hr.

4-23 A closed system rejects 25 kJ while experiencing a volume change of 0.1 m³ (0.15 to 0.05 m³). Assuming a reversible constant-pressure process at 350 kPa, determine the change in internal energy.

4-24 An open system is described by the following information:

	Input	Output
Velocity	36.58 m/s	12.19 m/s
Elevation	30.48 m	54.86 m
Enthalpy	2791.2 kJ/kg	2795.9 kJ/kg
Mass rate	0.756 kg/s	0.756 kg/s

If the work rate is 4.101 kW, what is the heat rate?

4-25 In the standard home freezer refrigeration unit, a capillary tube is often used to produce a throttling (constant enthalpy) process. In one such system, R-12 is throttled from saturated liquid at 151 psia to a pressure of 12 psia. Determine:

a. Initial temperature (F)
b. Final temperature (F)
c. Final condition (SC, x, or SH)
d. Change in specific volume (ft³/lbm)

4-26 Steam is throttled from a saturated liquid at 212 F to a temperature of 50 F. What is the quality of the steam after passing through this expansion valve? Assume $\Delta PE = \Delta KE = Q = W = 0$. Also estimate the Joule–Thomson coefficient.

4-27 A high-speed turbine produces 1 hp while operating on compressed air. The inlet and outlet conditions are 70 psia and 85 F, 14.7 psia and −50 F, respectively. Assume $\Delta KE = \Delta PE = 0$. What mass flow rate is required?

4-28 An inventor claims to have a closed system that operates continuously and produces the following energy effects during the cycle: net $Q = 3$ Btu and net $W = 2430$ ft-lbf. Prove or refute the claim.

4-29 A tank contains a fluid that is stirred by a paddle wheel. The work input to the paddle wheel is 4085 Btu. The heat transferred from the tank is 1300 Btu. Considering the tank and the fluid as a closed system, determine the change in the internal energy of the system.

4-30 Steam at 100 lbf/in.² and 400 F enters a rigid, insulated nozzle with a velocity of 200 ft/sec. It leaves at a pressure of 20 lbf/in.². Assuming the enthalpy at the entrance is 1227.6 Btu/lbm and that at the exit is 1148.4 Btu/lbm, determine the exit velocity.

4-31 A nozzle in a steam system passes 10^5 lbm/min. The initial and final pressure and velocities are 250 psia, 1 psia, 400 ft/sec, and 4000 ft/sec, respectively. Assuming the process is adiabatic, what is the change in enthalpy per pound mass?

4-32 Suppose that 6 lbm of steam of 200 psia and 80% quality is heated in a closed-system frictionless process until the temperature is 500 F. Calculate the heat transferred if the process is carried out at (a) constant pressure and (b) constant volume.

4-33 Consider the accompanying control volume schematic in which $\dot{m}$ is in lbm/hr and h is in Btu/lbm. If $\dot{W} = 845$ Btu/hr and $\Delta PE = \Delta KE = 0$, what is $\dot{Q}$?

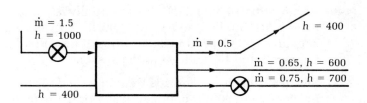

4-34 Suppose that 40,000 Btu/hr is transferred from a steam turbine while the mass flow rate is 10,000 lbm/hr. Assuming the following data are known for the steam entering and leaving the turbine, find the work rate if g = 32.17 ft/sec.

	Conditions	
	Inlet	Outlet
Pressure	200 psia	15 psia
Temperature	700 F	—
Velocity	200 ft/sec	600 ft/sec
Elevation	16 ft	10 ft
h	1361.2 Btu/lbm	1150.8 Btu/lbm

Also determine the fraction of the power contributed by each term of the first law as compared to the change in enthalpy.

4-35 An ideal gas, for which $c_p = 7R/2$, is taken from point a to point b (see sketch) along two paths: ab and adb. Let $p_2 = 2p_1$ and $v_2 = 2v_1$. Compute the heat supplied to the gas in each process (in terms of R and T_1 only). This is a closed system and all processes are assumed reversible.

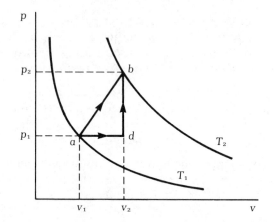

4-36 Assume that 0.5 lbm of steam initially at 20 psia and 228 F ($u_1 = 1081$ Btu/lbm and $v_1 = 20.09$ ft³/lbm) is in a closed rigid container. It is heated to 440 F ($u_2 = 1158.9$ Btu/lbm). Determine the amount of heat added to the system.

4-37 Steam at 100 lbf/in.² and 400 F enters a rigid insulated nozzle with a velocity of 200 ft/sec. It leaves at a pressure of 20 lbf/in.² and a velocity of 2000 ft/sec. Assuming that the enthalpy at the entrance (h_i) is 1227.6 Btu/lbm, determine the magnitude of the enthalpy at the exit (h_e).

4-38 In a closed system, 2 lbm of air ($\overline{Cv} = 4.96$ Btu/mole-R; $k = 1.4$) is heated at constant pressure from 30 psia and 40 F to 140 F. Because of friction, 10 Btu of work is done. Calculate the amount of heat added to the air.

4-39 Suppose that 3 lbm of an ideal gas in a closed system is compressed frictionlessly and adiabatically from 14.7 psia and 70 F to 60 psia. For this gas $c_p = 0.238$ Btu/lbm-F, $c_v = 0.169$ Btu/lbm-F and $R = 53.7$ ft-lbf/lbm-R. Compute: (a) the initial volume, (b) the final volume, (c) the final temperature, and (d) the work.

4-40 For the series of processes indicated in the sketch (a cycle), show:
a. $q_{ab} = h_b - h_a$
   $w_{ab} = h_b - h_a - u_b + u_a$
b. $q_{bc} = 0$
   $w_{bc} = u_b - u_c$

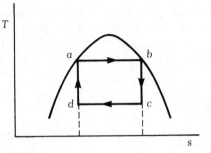

4-41 The heat from students, from lights, through the walls, and so forth, to the air moving through a classroom is 22,156 kJ/hr. Air is supplied to the room from the air conditioner at 12.8 C. The air leaves the room at 25.5 C. Specify:
a. Air flow rate (kg/hr)
b. Air flow rate (m³/hr) at inlet conditions
c. Duct diameter (m) for air velocity of 183 m/min.

4-42 Water is throttled (constant-enthalpy process) across a valve from 20.0 MPa and 260 C to 0.143 MPa. Determine the temperature and condition (SH or SC) after the valve.

4-43 Air is throttled (constant-enthalpy process) across a valve from 20.0 MPa and 260 C to 0.143 MPa. Determine the temperature and specific volume after the valve and the change in entropy across the valve (kJ/kg · R).

4-44 Saturated steam at 0.276 MPa flows through a 5.08-cm inside diameter pipe at the rate of 7818 kg/hr. Determine the kinetic energy of the steam in kJ/kg.

4-45 The mass rate of air flow into a nozzle is 100 kg/s. If the discharge pressure and temperature are 0.1 MPa and 270 C and the inlet conditions are 1.4 MPa and 800 C, determine the outlet diameter of the nozzle in meters.

4-46 A 216-in³ tank (6 in. on a side) contains saturated vapor (steam) at 0.143 MPa. This tank is attached to a line carrying a vapor at 0.7 MPa and 200 C. The steam from the line enters the tank until the pressure is 0.7 MPa. Assuming the tank is insulated and that we may neglect the heat capacity of the tank, calculate the mass of the steam that entered the tank.

4-47 During the operation of a steam power plant, the steam flow rate was 230,000 kg/hr for turbine inlet conditions of 3.5 MPa and 550 C; turbine exhaust (condenser inlet) conditions were 0.01 MPa (85% quality). Determine: (a) the turbine output in kW and (b) the condenser heat rejection rate in kJ/hr.

4-48 Steam at 0.7 MPa and 205 C enters a rigid, insulated nozzle with a velocity of 60 m/s. It leaves at a pressure of 0.14 MPa. Assuming the enthalpy at the entrance is 0.793 kJ/kg and that at the exit is 0.742 kJ/kg, determine the exit velocity.

Chapter

5

Thermodynamic Systems and Cyclic Processes

The first law of thermodynamics, as stated in the previous chapter, relates heat and work. In addition, it makes possible the definition of stored energy. According to this law, as long as the energy is conserved any process is possible; there are no restrictions as to which way a process will go. Thus, if restricted only by the first law, a power plant could be operated by taking energy out of the air and a ship could cross the ocean by extracting energy from the water. Experience suggests that this is not the way the world is, however. Though it is not impossible to get energy from these sources, it requires an energy input. In general, nature is such that you cannot get something for nothing. (In fact, you do not break even.)

Because the first law is incomplete in this regard, another law will be introduced later. This second law of thermodynamics in fact restricts the direction of possible processes. And, like the first law, it conforms to our intuition regarding the way nature must execute a process. A hot cup of coffee loses heat to the surroundings, for example, and therefore cools. None of us has seen a hot cup of coffee get hotter by being exposed to cool surroundings.

Before we can get to an exact statement of the second law, we must lay the foundation. This chapter presents the definitions and concept of cyclic processes as well as the Carnot cycle.

5–1 Heat Engines and Thermal Efficiency

The **heat engine** is a device that does net *positive work* as a result of heat transfer from a high-temperature reservoir with some heat transfer to a low-temperature reservoir while operating in a thermodynamic cycle—the steam engine is a device that satisfies these criteria. Figure 5-1 shows a schematic of a heat engine. Heat Q_H leaves the high-temperature reservoir at temperature T_H and goes into the engine. The heat engine does work W and rejects heat Q_L into the low-temperature reservoir at temperature T_L. This device is operating in a thermodynamic cycle. Thus, by applying the first law,

$$W = Q_H - Q_L > 0 \tag{5-1}$$

Notice that both Q_H and Q_L are positive;* the sign designating the heat rejected from the engine, Q_L, has been included; and most important, not all of the heat added to the system is converted to work.

The concept of **thermal efficiency** is quite convenient for engineers—particularly since it is apparent from the preceding discussion (as well as from our intuition) that not all of the energy put into a device goes directly into work. Thermal efficiency, η is usually defined in general terms as output over input:

$$\eta = \frac{\text{output}}{\text{input}} \tag{5-2}$$

This quantity has meaning only for cyclic processes and is sometimes

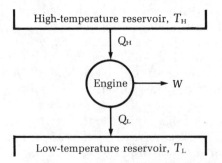

Figure 5-1. Schematic of a heat engine.

---

* Important note: This is a departure from our sign convention in that Q_L is negative when the working fluid is considered to be the system. In this chapter we will use Q_H to represent the magnitude of heat transfer to or from the high temperature body and Q_L the magnitude of the heat transfer to or from the low temperature body. The direction (sign) will be evident.

expressed in percent. The output and input quantities may be deduced in various ways—money or energy, for instance. For most engineering purposes, energy (or power) is the subject of this definition. Thus for a heat engine

$$\eta = \frac{W}{Q_H}$$ (5-3)

Notice that work W is what is desired when the heat engine is used whereas Q_H is the input—the thing that costs to get W.

EXAMPLE 5-1

The thermal efficiency of a particular engine is 33%. Determine:
1. The heat supplied per 1800 W-hr of work developed
2. The ratio of heat supplied to heat rejected
3. The ratio of the work developed to heat rejected

SOLUTION: 1. $\eta = \dfrac{W}{Q_H} \Rightarrow Q_H = \dfrac{W}{\eta} = \dfrac{(1800/0.293)\ \text{Btu}}{0.33} = 18{,}616\ \text{Btu}$

2. $\dfrac{Q_H}{Q_L} = \dfrac{-Q_H}{W - Q_H} = \dfrac{18{,}616}{12{,}472} = 1.4925$

3. $\dfrac{W}{Q_L} = \dfrac{1800/0.2928}{12{,}472} = 0.4925$

EXAMPLE 5-2

An inventor is trying to persuade you to invest in his new heat engine. His claim is that for a heat input of 10^4 Btu, the engine rejects only 1.7568 kW-hr and is 46% efficient. Would you invest in this device?

SOLUTION:

GOES TO Q_L

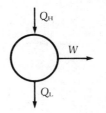

$$Q_H = 10^4\ \text{Btu}$$

$$\eta = 0.46$$

Thus

$$W = \eta Q_H = 4600\ \text{Btu}$$

and

$$Q_L = Q_H - W = 5400 \text{ Btu}$$

According to the inventor,

$$Q_L = 1.7568 \text{ kW-hr}$$

$$= \frac{1.7568 \text{ kW-hr}}{0.000293 \text{ kW-hr/Btu}}$$

$$= 5995 \text{ Btu}$$

Thus the actual heat rejection is larger than that obtained by the stated efficiency. The real efficiency is 40% . . . less than claimed. Do not invest!

5–2 Heat Pumps and Refrigerators

The **heat pump** or the **refrigerator** is *a device that requires net negative work in order to have heat transfer from a low-temperature reservoir to a high-temperature reservoir while operating in a thermodynamic cycle.* Figure 5-2 illustrates the schematic for both heat pumps and refrigerators. In both cases, Q_L is extracted from the low-temperature reservoir at temperature T_L at the expense of work W being done on the system. Heat Q_H is then rejected into the high-temperature reservoir at temperature T_H. Again, applying the first law for cyclic processes,

$$-W = -Q_H + Q_L \tag{5-4}$$

or

$$Q_H = W + Q_L > 0 \tag{5-5}$$

Notice that Q_H, the heat rejected into the high-temperature reservoir, is

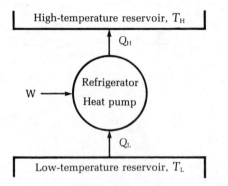

Figure 5-2. Schematic of heat pumps and refrigerators.

always greater than Q_L, the heat extracted from the low-temperature reservoir.

To compute thermal efficiency, we must first decide whether the device of interest is a heat pump or a refrigerator. In fact, efficiency is an awkward word to use while discussing these devices. Hence we will keep the definition but use instead the term **coefficient of performance.**

A **refrigerator** is a device whose purpose is to remove heat from a low-temperature reservoir (to make ice, for example). Thus the coefficient of performance for cooling (COP_c) is

$$COP_c \ (= \eta_R) = \frac{Q_L}{W} = \frac{Q_L}{Q_H - Q_L} \tag{5-6}$$

Similarly, a **heat pump** is a device whose purpose is to put heat into the high-temperature reservoir (to heat your home in the winter, for example). Thus the coefficient of performance for heating (COP_h) is

$$COP_h \ (= \eta_{HP}) = \frac{Q_H}{W} = \frac{Q_H}{Q_H - Q_L} \tag{5-7}$$

Notice in Equations (5-6) and (5-7) that if Q_H is only slightly greater than Q_L, both η_R and η_{HP} may be very much greater than 1—thus the awkwardness of the word *efficiency* and the switch to *coefficient of performance*.

5–3 Reservoirs

At this point we must discuss the reservoirs that are essential to the preceding definitions. A **reservoir** is an imaginary device that does not change temperature when heat is added to or taken away from it. One gets the feeling from the name that great quantities of heat are stored in these reservoirs. Since heat cannot be stored, this is not the case; they merely act as a source or sink for energy. To understand the concept, recall the definition of heat capacity:

$$C_x = \left. \frac{\delta Q}{dT} \right)_x \tag{2-14}$$

Thus, from the word definition, a reservoir has an infinite heat capacity (δQ is finite and $dT = 0$). Since this is impossible, however, let us define a heat reservoir in this way: a body with a mass so large that the heat absorbed or rejected does not cause an appreciable change in any of the thermodynamic coordinates. This definition implies the following:

1. Its temperature essentially remains constant (isothermal) during the absorption or rejection of heat.

2. In being subjected to absorption or rejection of heat, it does so in a reversible fashion. (It is in equilibrium at all times.)

3. Only the transient form of energy called heat is allowed to cross the boundaries.

Examples of heat reservoirs are easy to imagine. Dropping a 50-lbm block of ice into the center of Lake Michigan will not appreciably affect the temperature of the water in Chicago. Lighting a match will not appreciably affect the temperature in the Astrodome.

5–4 Processes and Cycles—Reversible and Irreversible

The preceding discussion implies that it is impossible to have a heat engine with 100% efficiency. In fact, certain processes that may be conceived will not occur and others may change the system and surroundings so that neither may be returned reversibly to their initial state. From the discussion of Chapter 1, we know these as irreversible processes.

The questions that should occur to you now are: If I cannot get 100% efficiency, what is the largest efficiency I can expect? And how do I determine whether a process is impossible? The answers to both these questions involve the second law of thermodynamics and will be discussed in detail in Chapter 6. Moreover, we must understand the ideal process that we have called reversible. Engineers are, of course, interested in reversible processes and cycles because more work is delivered than by a corresponding irreversible process. Similarly, refrigerators and heat pumps whose operation may be described as reversible require less work input than their irreversible counterparts.

Reversible Processes

As we have seen, reversible processes are involved with equilibrium states. That is, the reversible process is the result of an infinitesimal deviation from equilibrium—thus requiring an infinite execution time. Thus in addition to the earlier definition—a process whose direction can be reversed at any stage by an infinitesimal change in external conditions—it might help you to think of this ideal process as a succession of equilibrium states. Processes and cycles that do not fulfill this concept are irreversible processes.

Causes of Irreversibility

Irreversible processes are the result of everyday events. The following list presents only four of the most common:

1. *Friction:* This involves the interaction of solids, liquids, and gases with each other and the other phases. The work done to overcome this friction is lost as useful energy.

2. *Heat transfer across a finite temperature difference:* This is by definition a nonequilibrium situation since the work required to restore the system to its initial state is lost. (A refrigerator and heat pump are required.) Notice that an isothermal heat transfer must occur (between system and surroundings) in order to have a reversible heat transfer process.

3. *Unrestrained or free expansion:* This is the classic example of a gas and a vacuum separated by a partition. When the partition is removed, the gas expands into the vacuum. This process is irreversible because of the loss of ability to do work (unrestrained expansion).

4. *Mixing:* Work must be done to separate the components that were mixed.

Inelastic deformations, I^2R losses, and combustion are other causes of irreversibility. It must be remembered that although a system may be experiencing a process that is irreversible, the system itself may be restored to its initial state at the expense of energy.

5–5 The Carnot Cycle

The Cycle

In order to determine the maximum possible efficiency for a physical situation, let us consider a particular type of heat engine operating between high-temperature and low-temperature reservoirs. This cycle operates such that every process is reversible. Thus we may conclude that if every process is reversible, this cycle is reversible. French engineer, N. L. S. Carnot devised this cycle in a treatise on the second law of thermodynamics published in 1824.* Appendix B-4 presents a demonstration that this is the most efficient cycle that can operate between two constant-temperature reservoirs.

The Carnot cycle consists of alternate reversible, isothermal and

---

* In 1824, Sadi Carnot published *Reflections on the Motive Power of Fire.* In this book Carnot made three important contributions: the concept of reversibility, the concept of a cycle, and the specification of the heat engine producing maximum work when operating in a cycle between two fixed temperature reservoirs. Carnot's book is available as a paperback from Dover Publications under the title *Reflection on the Motive Power of Heat and on Machines Fitted to Develop This Power.* In 1943, ASME published a translation of Carnot's work made by Robert H. Thurston in 1890.

adiabatic processes that may occur in either a closed or an open system (Figure 5-3). The heat source and sink are placed in contact with the device to accomplish the required isothermal heat addition (a–b) and rejection (c–d). The insulation replaces the heat reservoirs for executing the reversible adiabatic processes involving expansion (b–c) and compression (d–a). Notice that the process characteristics for good heat transfer and good work transfer are not the same—in fact, they can be in conflict. Figure 5-3b illustrates an open system executing the Carnot cycle; in this case the work and heat transfer processes are assigned to separate devices. For both the open and closed systems the changes in state of the working fluid are shown on the (p, V) diagram (Figure 5-3c).

If we attempt to use the Carnot cycle, we will encounter problems (irreversibilities in the form of finite temperature differences during the heat transfer processes and fluid friction during work transfer processes). Moreover, the compression process (d–a) is difficult to perform and requires an input of work from the turbine output. As we move from the ideal Carnot cycle to actual systems (plants), the size and cost of equipment would be very high, and consequently other cycles appear more attractive as models for study. We will discuss them later.

A Carnot cycle may be executed by any thermodynamic system whether it is mechanical, electrical, or magnetic. Operation of a Carnot cycle involves (1) a system, (2) a high-temperature reservoir at temperature T_H, (3) a low-temperature reservoir at temperature T_L, (4) a means of insulating the system from one or both of the reservoirs when needed, and (5) a surrounding that periodically absorbs heat and does work on the system. Thus the Carnot cycle, is defined by two adiabatic and two isothermal processes. The order of processes for a heat engine, using any working fluid, are:

1. A reversible isothermal process at T_H—heat, Q_H, is transferred from the high-temperature reservoir at T_H (process a–b)

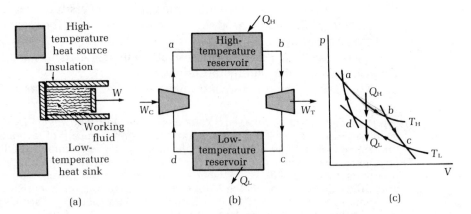

Figure 5-3. Carnot cycle heat engine: (a) closed system: (b) open system; (c) (p, V) diagram.

2. A reversible adiabatic process—the temperature of the working fluid is reduced from T_H to T_L (process b–c)

3. A reversible isothermal process—heat, Q_L, is transferred to the low-temperature reservoir at T_L (process c–d)

4. A reversible adiabatic process—the temperature of the working fluid is increased from T_L to T_H (process d–a)

Figure 5-3c illustrates the Carnot cycle on a (p, V) diagram. Notice that in process a–b heat is absorbed by the system and positive work is done; as for process b–c, no heat is transferred but positive work is done; in process c–d, heat is rejected by the system and negative work is done (that is, work done on the system—$dV < 0$); in process d–a, no heat is transferred and more negative work is done. We must note that in going through the cycle we end up with net positive work. Note also that the heat absorption and rejection processes of the Carnot cycle are *always* isothermal. An isothermal process may also be a constant-pressure process or a constant-internal-energy process, but this characteristic is merely incidental to the execution of a Carnot cycle. The Carnot cycle requires that the heat absorption and rejection processes be reversible and isothermal—nothing else.

To relate this cycle to a physical situation, consider Figure 5-4. This schematic represents a heat engine that is operating as a steam power plant. Process 1 is a reversible isothermal process where heat from the high-temperature reservoir is transferred to the working fluid in a SGU (steam-generating unit or boiler). Keeping the pressure constant during the boiling will require the change of phase to occur at constant temperature, T_H. (Note that this is a saturation temperature–pressure pair.) Process 2 is a reversible adiabatic process where energy is removed from the working fluid in a turbine. Because of this energy loss, the temperature of the working fluid is reduced from the high reservoir

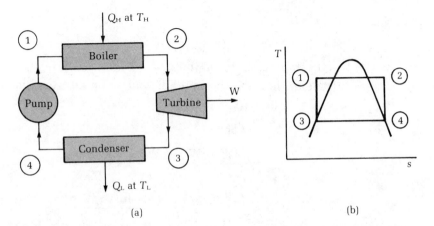

(a)   (b)

Figure 5-4. Schematic of a heat engine that operates on a Carnot cycle.

temperature, T_H, to the low reservoir temperature, T_L. (Note that this energy loss cannot be in the form of heat.) Process 3 is a reversible isothermal process where heat from the working fluid is transferred to the low-temperature reservoir in a condenser. Again constant pressure condensation requires a constant temperature process at T_L. In completing the cycle, a reversible adiabatic process (process 4) is used to increase the working fluid temperature from the low reservoir temperature, T_L, to the high reservoir temperature, T_H. Note that we have assumed that on process 3 the working fluid leaves the condenser as a saturated liquid and, in process 4, work is the energy expended to raise the temperature of the working fluid. Finally, it should be noted that a refrigerator or heat pump could be obtained by reversing every process of this reversible Carnot cycle.

Carnot Cycle Efficiency

The efficiency of a Carnot cycle will be derived here, but under the assumption that the working substance is an ideal gas with constant specific heats. To determine this quantity, we begin by computing the work for each process of the Carnot cycle, assuming a closed system.

The work done in process a–b (see Figure 5-3) is positive:

$$_aW_b = \int_a^b p \, dV = RT_H \ln\left(\frac{V_b}{V_a}\right) \qquad \text{(isothermal, positive work)} \qquad \textbf{(5-8)}$$

Similarly, in process b–c the work is positive:

$$_bW_c = \int_b^c p \, dV = -\int du = c_v(T_H - T_L)$$

$$\text{(adiabatic, positive work)} \qquad \textbf{(5-9)}$$

The work done in processes c–d and d–a are negative:

$$_cW_d = \int_c^d p \, dV = RT_L \ln\left(\frac{V_d}{V_c}\right) \qquad \textbf{(5-10)}$$

and

$$_dW_a = c_v(T_L - T_H) \qquad \textbf{(5-11)}$$

We also note that $Q_H = {}_aW_b$ and $_cW_d = Q_L$. (Both are isothermal processes.) By definition, the thermal efficiency of a heat engine is

$$\eta = \frac{W_{net}}{Q_H}$$

$$= \frac{_aW_b + {}_bW_c + {}_cW_d + {}_dW_a}{_aW_b}$$

$$= \frac{T_H \ln(V_b/V_a) + T_L \ln(V_d/V_c)}{T_H \ln(V_b/V_a)} \qquad \textbf{(5-12)}$$

Recall that for reversible adiabatic processes

$$TV^{k-1} = \text{constant}$$

Thus for our system

$$\frac{V_b}{V_a} = \frac{V_c}{V_d}$$

Therefore

$$\eta = \frac{T_H - T_L}{T_H} \tag{5-13}$$

It may be easily shown that the coefficients of performance of the refrigerator and heat pump are respectively

$$COP_c(=\eta_R) = \frac{T_L}{T_H - T_L} \tag{5-14}$$

and

$$COP_h(=\eta_{HP}) = \frac{T_H}{T_H - T_L} \tag{5-15}$$

In closing, we will make one final observation. If we compare Equations (5-3), (5-6), and (5-7) with Equations (5-13), (5-14), and (5-15), respectively, a special relation may be deduced:

$$\frac{Q_L}{Q_H} = \frac{T_L}{T_H} \qquad \text{RANKINE or KELVIN} \tag{5-16}$$

This functional relationship was proposed by Lord Kelvin during his studies of thermodynamic scales of temperature. With this formula he was able to deduce an absolute temperature scale merely by stating the magnitude of the degree. To demonstrate this let us assume a Carnot heat engine operating between the steam and ice points of water. Thus, this reversible engine receives heat, Q_H, at temperature T_s (vaporization temperature) and rejects heat, Q_L, at temperature T_i (the fusion temperature). Measurement of the ratio Q_H/Q_L yields

$$\frac{Q_H}{Q_L} = 1.3661$$

Using Equation (5-16) indicates that

$$\frac{T_s}{T_i} = \frac{Q_H}{Q_L} = 1.3661 \tag{5-17}$$

Notice that this is also consistent with Equation (5-13) and that Equation (5-17) involves the two unknown T_s, vaporization or steam-point temperature, and T_i, fusion or ice-point temperature, (i.e., one equation and two unknowns). To obtain another equation relating these two unknowns, you must decide on the number of degrees you want between the ice and steam points. Recall for the Fahrenheit scale

$$T_s - T_i = 180 \qquad\qquad (5\text{-}18)$$

Solving Equations (5-17) and (5-18) simultaneously yields

$$T_s = 671.7 \text{ R} \quad \text{and} \quad T_i = 491.7 \text{ R}$$

Thus, the absolute Fahrenheit (Rankine) and the Fahrenheit scales are related by

$$R = F + 459.7 \text{ R} \qquad\qquad (1\text{-}6)$$

Similarly, if we select 100 degrees between the ice point and the steam point we get the relations

$$T_s - T_i = 100$$

Solving this equation simultaneously with Equation (5-17) yields

$$T_s = 373.2 \text{ K} \quad \text{and} \quad T_i = 273.2 \text{ K}$$

The absolute Celsius (Kelvin) and the Celsius scales are related by

$$K = C + 273.2 \qquad\qquad (1\text{-}5)$$

EXAMPLE 5-3

1. Calculate the thermal efficiency of a Carnot cycle heat engine operating between 1051 and 246 F.

2. What would be the coefficient of performance of this device if it were reversed and run as a heat pump? As a refrigerator?

SOLUTION: 1.

$$1 - \frac{246 + 460}{1051 + 460} = .5328$$

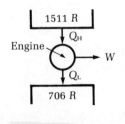

$$\eta = 1 - \frac{706}{1511} = 0.533$$

2.

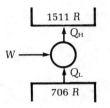

$$\text{COP}_h\,(=\eta_{\text{HP}}) = \frac{1511}{1511 - 706} = 1.877$$

$$\text{COP}_c\,(=\eta_{\text{R}}) = \frac{706}{1511 - 706} = 0.877$$

Note: $\eta_{\text{HP}} - \eta_{\text{R}} = 1$.

EXAMPLE 5-4

A Carnot refrigerator is working between reservoirs of -30 and 32 C. What is the coefficient of performance? If an actual refrigerator has a coefficient of performance that is 75% of this Carnot value, calculate the refrigerating effect.

SOLUTION: $T_L = -30 + 273 = 243$ K   and $T_H = 32 + 273 = 305$ K

$$\text{COP}_h\,(=\eta_{\text{R}}) = \frac{T_L}{T_H - T_L} = 3.92$$

$$\text{COP}\,(=\eta_{\text{act}}) = 0.75\,\eta_{\text{R}} = 2.94$$

$$= \frac{Q_L}{W}$$

Thus for 1 kW of work (power) put into the refrigerator, the refrigerating effect is 2.94 kW.

PROBLEMS

5-1 Determine the applicable efficiency or coefficient of performance for each of the following:

a. A refrigerator with EER (Energy Efficiency Ratio) = 6.25 Btu/hr/W

b. A 600-MW steam power plant with a thermal pollution rate of 3.07×10^9 Btu/hr

5-2 Determine the applicable efficiency or coefficient of performance for each of the following:

a. An ideal heat pump using Freon-12 and operating between pressures of 35.7 and 172.4 psia.

b. A refrigerator providing 4500 Btu/hr of cooling while drawing 585 W.

c. A heat engine to recover the thermal energy in the ocean by operating between warmer surface waters of 82 F and the colder 45 F water at 1200 ft.

5-3 A heat pump is used in place of a furnace for heating a house. In winter, when the outside air temperature is 10 F, the heat loss from the house is 60,000 Btu/hr if the inside is maintained at 70 F. Determine the minimum electric power required to operate the heat pump (in kW).

5-4 A gas turbine has an efficiency of 18%. Heat in the amount of 18,000 Btu is released for every pound of fuel consumed. The horsepower developed is 8000. What is the rate of fuel consumption?

5-5 A heat pump is used in place of a furnace for heating a house. In winter, when the outside air temperature is −10 C, the heat loss from the house is 200 kW if the inside is maintained at 21 C. Determine the minimum electric power required to operate the heat pump.

5-6 Assuming that the temperature of the surroundings remains at 60 F, determine the minimum increase in operating temperature needed to effect an increase in thermal efficiency from 30 to 40%.

5-7 Solar energy is to be used to warm a large collector plate. This energy will, in turn, be transferred as heat to a fluid in a heat engine, and the engine will reject energy as heat to the atmosphere. Experiments indicate that about 200 Btu/hr/ft^2 of energy can be collected when the plate is operating at 190 F. Estimate the minimum collector area required for a plant producing 1 kW of useful shaft power when the atmospheric temperature is 70 F.

5-8 A Carnot engine operates between a heat source at 1200 F and a heat sink at 70 F. If the output of the engine is 200 hp, compute the heat supplied, the heat rejected, and the thermal efficiency of the heat engine.

5-9 The efficiency of an ideal engine discharging heat to a cooling pond at 80 F is 30%. If the cooling pond receives 800 Btu/min, what is the power output of the engine? What is the source temperature?

5-10 A Carnot refrigerator is used for making ice. Water freezing at 32 F is the cold body, and the heat is rejected to a river at 72 F. How much work is required to freeze 2000 lbm of ice? (The latent heat of fusion of ice is 144 Btu/lbm.)

5-11 A Carnot engine operating between 750 and 300 K produces 100 kJ of work. Determine: (a) the thermal efficiency and (b) the heat supplied.

5-12 A reversed Carnot cycle operating between −20 and 30 C receives 126.575 kJ of heat. If this cycle is operating as a refrigerator, determine: (a) the thermal efficiency and (b) the heat rejected.

5-13 Reconsider Problem 5-12 but assume the device is a heat pump this time.

5-14 (a) Calculate the thermal efficiency of a Carnot cycle heat engine operating between 1051 and 246 F. (b) What would be the coefficient of performance of this device if it were reversed to run as a heat pump? As a refrigerator?

5-15 A Carnot refrigerator is used to remove 300 Btu/hr from a region at −160 F and to discharge this heat to the atmosphere at 40 F. The Carnot refrigerator is to be driven by a Carnot engine operating between a reservoir at 1140 F and the atmosphere (40 F). How much heat must be supplied (in Btu/hr) to the Carnot engine from the 1140 F reservoir? What is the work done by the Carnot engine? What is the work done on the Carnot refrigerator? What are the efficiencies of both the Carnot refrigerator and the Carnot engine?

5-16 What are the differences and similarities of a heat pump and a refrigerator?

5-17 A Carnot engine operates between a source at 800 F and a sink of 100 F. If 200 Btu is rejected each minute to the sink, compute the power output.

5-18 A Carnot engine receives 15 Btu/sec from a source at 900 F and delivers 6000 ft-lbf/sec of power. Determine the efficiency and the temperature of the receiver (sink).

5-19 Two Carnot heat engines operate in series between a source at 527 C and a sink at 17 C. The first engine rejects 400 kJ to the second engine. If both engines have the same efficiency, calculate:
a. The temperature of the source for the second engine (that is, the first engine's output)
b. The heat taken by the first engine from the 527 C source
c. The work done by each engine
d. The efficiencies of each engine

5-20 A refrigerator is operating on a Carnot cycle between reservoirs of −6 and 22 C. Calculate the coefficient of performance, the refrigeration effect, and the heat rejected to the high-temperature reservoirs per kJ of work supplied.

5-21 The load on a residential air conditioner is 10.55 kW when the outdoor air temperature is 35 C and the indoor temperature is maintained at 23.9 C. Determine the minimum power requirement (kW) to operate the air conditioner.

5-22 Air is compressed at the steady rate of 12,727 kg/hr from 0.1035 MPa, and 18.3 C to 0.621 MPa, isothermally. Determine the minimum size of compressor required, (in kW) using *two* different methods.

5-23 Determine the thermal efficiency of a Carnot cycle heat engine in terms of the isentropic compression ratio ($r_k = V_{large}/V_{small}$).

5-24 What is the expression for the efficiency of a Carnot cycle heat engine if the working fluid is a gas obeying the Clausius equation of state: $p(v - b) = RT$? Can you prove it?

5-25 Determine the boiling point and freezing point of water on a temperature scale where $T_{bp} - T_{fp} = 80$.

5-26 The low-temperature reservoir of a Carnot heat engine is at 10 C. If you wish to increase the efficiency of this heat engine from 40 to 55%, by how many degrees must you increase the temperature of the high-temperature reservoir?

Chapter

6

The Second Law of Thermodynamics

This chapter presents the second law of thermodynamics. This law will conform to your intuition—once you understand its significance and can apply it properly to both closed and open systems.

6–1 The Second Law from Classic Thermodynamics

The second law involves the fact that processes proceed in a certain direction and not in the opposite direction. A cup of coffee cools as heat is transferred to the surroundings, but heat will not flow from the surroundings to the hotter cup of coffee. This example and a host of others are matters of common experience—so common, in fact, that it seems almost absurd to make such obvious statements. Nevertheless, the second law of thermodynamics is nothing more or less than a generalized statement of such common observations.

A system that undergoes a series of processes and always returns to its initial state is said to have gone through a cycle. For the closed system undergoing a cycle, the first law of thermodynamics is

$$\oint \delta Q = \oint \delta W \tag{6-1}$$

The symbol $\oint$ stands for the cyclic integral of the increment of heat or work. Any heat supplied to a cycling system must be balanced by an equivalent amount of work done by the system. Or vice versa: Any work done on the cycling system results in an equivalent amount of heat given off.

Many examples exist of work that is completely converted into heat. Although a cycling system that completely converts heat into work has never been observed, such complete conversion would not be a violation of the first law. The fact that heat cannot be completely converted into work is the basis for the second law of thermodynamics. The justification for the second law is empirical.

The second law has been stated in different ways,* all of which are equivalent. We will discuss two of them: the Kelvin–Planck statement and the Clausius statement.

The Kelvin–Planck statement of the second law is: *It is impossible for any cycling device to exchange heat with only a single reservoir and produce positive work.* In other words, the Kelvin–Planck statement says that heat cannot be continuously and completely converted into work; a fraction of the heat must be rejected to another reservoir at a lower temperature. The second law thus places a restriction on the first law in relation to the way energy is transferred. Work can be continuously and completely converted into heat, but not vice versa.

If the Kelvin–Planck statement were not true and heat could be completely converted into work, the heat might be obtained from a low-temperature source, converted into work, and the work converted back into heat in a region of higher temperature. The net result of this series of events would be the flow of heat from a low-temperature region to a high-temperature region with no other effect. This phenomenon has never been observed and is contrary to all our experience.

The Clausius statement of the second law is: *No process is possible whose sole result is the removal of heat from a reservoir at one temperature and the absorption of an equal quantity of heat by a reservoir at a higher temperature.* This statement does not say that it is impossible to transfer heat from a lower-temperature body to a higher-temperature body. This is exactly what a refrigerator does when it receives an energy input, usually in the form of work. This transfer of energy from the surroundings constitutes an effect other than the transfer of heat from the lower-temperature body to the higher-temperature body; thus the "sole result" of the Clausius statement includes effects within the refrigerating device itself.

The consequences of the Clausius and Kelvin–Planck statements of the second law are equivalent. This equivalence is demonstrated in Appendix B-3 by showing that the violation of either statement can always be made to result in a violation of the other.

---

* For example: "There exist arbitrarily close to any given state of a system other states which cannot be reached from it by reversible adiabatic processes"—Carathéodory's statement of the second law.

6–2 Corollaries to the Second Law

Many useful concepts result from the second law, all helpful in one way or another. Some corollaries of the second law are given below:

Corollary A: No engine operating between two given reservoirs can have a greater efficiency than a reversible engine operating between the same two reservoirs (see Appendix B-4).

Corollary B: All reversible engines operating between the same temperature limits have the same efficiency.

Corollary C: The efficiency of any reversible engine operating between two reservoirs is independent of the nature of the working fluid and depends only on the temperature of the reservoirs.

Corollary D: It is theoretically impossible to reduce the temperature of a system to absolute zero by a series of finite processes.

Corollary E (the thermodynamic temperature scale): Define the ratio of two temperatures as the ratio of the heat absorbed by a Carnot engine to the heat rejected when the engine is operated between reservoirs at these temperatures. Thus the equality $Q_L/Q_H = T_L/T_H$ becomes a matter of definition, and the fundamental problem of thermometry, that of establishing a temperature scale, reduces to a problem in calorimetry.

Corollary F (the inequality of Clausius): When a system is carried around a cycle and the heat δQ added to it at every point is divided by its temperature at that point, the sum of all such quotients is less than zero for irreversible cycles and in the limit is equal to zero for reversible cycles (see Appendix B-5):

$$\oint \frac{\delta Q}{T} \leq 0 \qquad \text{(6-2)}$$

PROPERTY = ENTROPY

Corollary G: There exists a property (denoted by S) of a system such that a change in its value is equal to

$$S_2 - S_1 = \int_1^2 \frac{\delta Q}{T} \qquad \text{(6-3)}$$

for any reversible process undergone by the system between states 1 and 2. This property is called entropy. There will be more on this subject in Section 6-6.

Corollary H (principle of the increase of entropy): In any process whatever between two equilibrium states of a system, the increase in entropy of the system plus the increase in entropy of its surroundings is equal to or greater than zero.

Corollaries *A* and *B* are sometimes referred to as the Carnot principle, the Carnot theorem, or the Carnot theorem and corollary.

6–3 The Second Law from Statistical Thermodynamics

The significance of the statistical interpretation of the second law of thermodynamics may be implied by considering a gas trapped in a light-weight container. The particles of this gas are in continuous and chaotic or random motion. The motion of these molecules varies from very slow speeds to very fast speeds. This distribution of speeds is due to interparticle and particle-wall collisions and is not uniform. The average speed of these particles is approximately that of a pressure wave (sound). In the case of air at room temperature the speed is about 1100 ft/sec. In addition to magnitude changes, the particles will experience many direction changes. However, because of the enormous number of particles in this container, we would expect the number of particles traveling in a given direction to remain essentially constant. In addition, because the container is not moving, the velocities are distributed essentially equally in all directions. Experience has shown that all of these particles will not acquire suddenly and simultaneously a velocity in the same direction (i.e., the container might jump). Although, from the standpoint of the first law of thermodynamics, such a happening is possible. Thus, although it is highly unlikely, this possibility cannot be excluded. Therefore, the dogmatic statement of its non-occurrence cannot be made. Experience has shown that these processes do not occur often enough that it is of any real practical value to an engineer. So a more accurate statement of the second law is: "It is highly improbable that a process occurs with a cycling device whose sole result is the removal of heat from a high-temperature (heat) reservoir and the production of positive work." The word *impossible* has been replaced with *improbable*. Thus, the second law is a statement of the improbability of the spontaneous transition of a system from a highly probable state to one of lower probability.

As an example of the enormous numbers involved, consider the everyday situation of one deck of cards arranged in four hands of thirteen cards each. The number of different hands possible is 635,013,559,600. Of this number there are only four possibilities of getting a hand with only one suit in it. A 4–3–3–3 suit arrangement can occur 100,358,782,000 ways. Thus, the chance of getting thirteen cards of one suit (a very orderly arrangement) compared to the 4–3–3–3 arrangement (a not-so-orderly arrangement) is 1 to 100,358,782,000/4. In principle, the statistics of particles of a gas are the same as for the deck of cards—except that you are considering much larger numbers. Thus the chances of an orderly arrangement essentially is nonexistent.

EXAMPLE 6-1

You are to buy a refrigerator unit that maintains a volume of 15,000 ft^3 at 20 F while operating in a warehouse where the temperature is 80 F. A salesman tells you he has just such a device with a coefficient of performance of 9. Would you buy this device from him?

SOLUTION: If the refrigerator ran on a Carnot cycle,

$$\text{COP} \ (= \ \eta_R) = \frac{T_L}{T_H - T_L} = \frac{480}{60} = 8$$

The salesman claims COP($= \eta_R$) = 9 for these temperatures. It is impossible. Do not buy it! Even if he had claimed that COP ($= \eta_R$) = 8, you should not buy it since this figure represents ideal performance. (Though not impossible, it is highly improbable.)

EXAMPLE 6-2

Consider temperature reservoirs at 1000 and 500 R. To understand the Clausius inequality, let us consider three cases:

1. Heat conduction between the reservoirs

2. A heat engine between these reservoirs with an efficiency of 25%

3. The same as case 2, but with $\eta = 50\%$

SOLUTION: 1. We know that conduction is an irreversible process. Thus $\oint \delta Q/T < 0$. Let us be sure. If $Q_H = 2000$ Btu, then $Q_L = -2000$ Btu. Therefore

$$\frac{Q_H}{T_H} + \frac{Q_L}{T_L} = \left(\frac{2000}{1000} - \frac{2000}{500}\right)\frac{\text{Btu}}{\text{R}} = -2 \ \text{Btu/R}$$

$$T_H = 1000 \ R$$
$$\downarrow Q_H$$
$$\downarrow Q_L$$
$$T_L = 500 \ R$$

2.

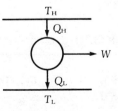

$$W = \eta Q_H = 500 \ \text{Btu} \quad \text{and} \quad Q_L = Q_H - W = 1500 \ \text{Btu}$$

Thus

$$\frac{Q_H}{T_H} + \frac{Q_L}{T_L} = \left(\frac{2000}{1000} - \frac{1500}{500}\right)\frac{Btu}{R} = -1\ Btu/R$$

3. This time $\eta = 50\%$, $W = \eta Q_H = 1000$ Btu, and $-Q_L = Q_H - W = 1000$ Btu. Thus

$$\frac{Q_H}{T_H} + \frac{Q_L}{T_L} = \left(\frac{2000}{1000} - \frac{1000}{500}\right)\frac{Btu}{R} = 0$$

Therefore, we have a reversible engine. We could have determined this by noting that

$$\eta = 1 - \frac{T_L}{T_H} = 1 - \frac{500}{1000} = 0.5$$

EXAMPLE 6-3

The machine represented by the schematic is to be analyzed. Is it possible?

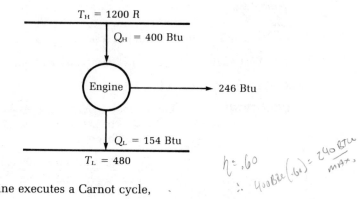

$T_H = 1200$ R

$Q_H = 400$ Btu

Engine → 246 Btu

$Q_L = 154$ Btu

$T_L = 480$

$h = .60$

$\therefore 400 Btu (.60) = 240\frac{BTU}{MAX}$

SOLUTION: If the machine executes a Carnot cycle,

$$\eta_{Carnot} = 1 - \frac{T_L}{T_H} = 1 - \frac{480}{1200} = 0.6$$

The efficiency of this machine is

$$\eta = \frac{W}{Q_H} = \frac{246}{400} = 0.61 > \eta_{Carnot}$$

It is impossible.

6–4 The Physical Meaning of Entropy

What has entropy to do with this? Entropy describes the chaotic nature (the "mixed-upness") of a system or state. As the system becomes more disordered, its entropy increases. On the other hand, if a system is completely ordered the entropy should have a minimum value (maybe zero). Boltzmann formed this idea into an operational expression. He hypothesized that $S = k \ln (\Omega) + S_0$, where k is called the Boltzmann constant and Ω is called the thermodynamic probability. Later Planck suggested that S_0 be zero.

This was the beginning of statistical thermodynamics. In applying this theory you must be careful as to the definition of the thermodynamic probability. (It is the number of microstates per corresponding macrostate and is always greater than or equal to one.) You must also be aware that this statistical theory is most applicable to systems consisting of a very large number of "particles." When this procedure is applied to a molecular system the results can be shown to agree with measurable quantities.

From the macroscopic point of view, Rankine (in 1851) demonstrated analytically that the ratio of the heat exchanged in a reversible process to the temperature of the interaction defined a thermodynamic function that was not consumed in a reversible cycle. The following year, Clausius independently derived the same result but identified the function as a property of a system and designated it as the entropy (Greek: evolution).

Borrowing directly from Carnot, Clausius theorized zero net entropy change for a system plus its surroundings in a reversible cycle; for any other cyclic process, he hypothesized an entropy increase. Because the efficiency of a reversible cycle is the maximum possible, this increase signified a loss in work-performing capability. The concept of energy served as a measure of the quantity of heat, but entropy served as a measure of its quality. Clausius also concluded that, although the energy of the world is constant, the entropy will increase indefinitely due to the irreversible nature of real processes. This extreme represents a condition whereby there are no thermal potential differences in the universe, a state described by Boltzmann as heat death.

At this point you may be saying, "Who cares whether or not a process is irreversible?", "Who cares if the entropy of the universe is increased by this irreversible process?", and "Who cares about disorder? The first law of thermodynamics is still valid, and as an engineer I am concerned with energy." The answer to each of these questions is, "You have lost something!" This loss is particularly important for engineers because you have lost the opportunity to do work, and once it is lost it can never be recovered. Though somewhat abstract, the classic example of this loss is the mixing of a very hot and a very cold fluid reservoir, say water. In principle at least, you could operate a mechanical device between these two reservoirs to obtain net positive work. If these two reservoirs are mixed with each other in an adiabatic process, the total energy of the separate two fluids and the final mixture is the same. (The first law is valid.) Now you have only one reservoir, and (according to the Kelvin–Planck statement) a device cannot produce positive work while receiving energy from only one reservoir.

From the concept of entropy, this irreversible process created an entropy increase. That is, even though the first law is not contradicted, the net entropy of the process increased. The term "net" is used because the entropy change of the hot fluid is negative (i.e., S decreased, $\Delta S_{hot} < 0$), the entropy change of the cold fluid is positive (i.e., S increased, $\Delta S_{cold} > 0$), and the increase is greater than the decrease ($|\Delta S_{cold}| > |\Delta S_{hot}|$). Because there is no entropy reservoir, this entropy increase was created by the process, and once it is created it can never be destroyed. This point describes the aura of mystery around this whole concept—net entropy is created in irreversible processes. Therefore a concluding synopsis is

> First law: Energy cannot be created or destroyed.
> Second law: Entropy can be ~~created~~ but not destroyed.
>   ACCELERATED

6–5 More on Corollary *D*

In the last half of the seventeenth century, a scientist, Guilaume Amontons, was very interested in thermometry. He was concerned with defining the lower limit of temperature—how cold can matter get? Using a volume of air at 0 C, he discovered that if you heated this air it expanded and, in cooling down, it contracted. In fact, by his estimate the volume changed by 1/240 of the 0 C volume per degree C regardless of the heating or cooling process. (We assume the pressure was constant.) His logic dictated that with this estimate, at −240 C, the volume of the air would be zero. Thus he concluded that the absolute zero would be −240 C. This is an amazing discovery and surprisingly accurate for that time in history. (It is actually −273 C.)

The question of whether you could actually cool matter to absolute zero appeared to be of no concern until very late in the nineteenth century (1898). An experimentalist and educator by the name of Dewar reduced the temperature of some matter to within 11 C of absolute zero.

The next important event in the history of this lower limit was due to Nernst. In 1906 he announced this third law of thermodynamics (corollary D)—while absolute zero can be approached to an arbitrary degree, it can never be reached.

With this pronouncement by Nernst, two other ideas were put forward by the scientific community:

1. It is the entropy and not the energy which tends to zero as the temperature approaches zero.
2. Even at absolute zero, there is some energy left in matter.

Thus, if the energy in matter were associated with thermal motion (a function of temperature), how did this "zero-degree" energy manifest itself?

It turns out that the crucial point is the entropy and its disappearance as the temperature approaches zero. Unfortunately because of its definition $(dS = \delta Q/T)$ the concept of entropy is not as often used as energy, even by very technically proficient scientists and engineers. The lack of a convenient physical picture is probably the reason. The statistical approach to the concept of entropy proposed by Boltzmann is helpful. The thermodynamic probability of Boltzmann's statement is a measure of disorder in a system. With this idea, the third law would imply absolute order at absolute zero. Experiment has, in fact, indicated increasing order as the temperature goes down. The result is another statement of the third law (called Nernst's theorem): The absolute entropy of a pure crystalline substance in complete internal equilibrium is zero at zero degrees absolute. Experiments indicate that Nernst's theorem is valid—there is some question of nuclear spin energies at absolute zero which has not been completely resolved.

The question now is, what about substances that are not pure crystalline? Do they possess zero entropy at absolute zero temperature? Boltzmann's and later Planck's, interpretation of entropy $(S = k \ln \Omega)$ does yield a clue to the explanation of this problem. For a pure crystalline substance, there is only one molecular configuration, thus the thermodynamic probability is one and the entropy is zero. For other than pure crystalline substances, more than one molecular configuration exists; the thermodynamic probability is greater than one and the entropy is greater than zero. Therefore, the final conclusion is that the entropy of a substance will not be zero unless the molecular configuration of the substance has been arranged to its highest possible ordered configuration.

At this point you probably have noticed a problem. In order to reach absolute zero temperature, you must decrease the entropy of a substance to zero (or nearly). Yet the second law of thermodynamics states that the entropy of the universe must increase (the universe must become more mixed up). Everyday experience gives proof of the validity of the second law. For example, a new deck of cards is very ordered; once the order is disturbed by shuffling it may not be reorder by further shuffling. To reorder the cards you must expend considerably more energy than it takes to shuffle. Similarly, a jar filled with red and white sand (red on one side and white on the other) turns pink when stirred. It will never go back to all red on one side and all white on the other regardless of how you stir the sand. To reorder you must expend a lot of energy. It seems to follow that to obtain absolute zero temperature you would have to expend a lot of energy.

6–6 Entropy—the Working Definition

The second law of thermodynamics is the basis of the definition of entropy. Since any attempt to generally apply the physical picture of entropy (mixed-upness) is extremely difficult, we will define it mathematically.

Second Law for Closed Systems

As we did with the first law, let us consider a change from state 1 to state 2 by path A, then back to state 1 by path B, and then apply the equality of Clausius (reversible process). Figure 6-1 illustrates the situation. Now apply the equality of Clausius to the $(A–B)$ cycle:

$$0 \equiv \oint_{1A-2-B-1} \frac{\delta Q}{T} = \int_1^2 \left(\frac{\delta Q}{T}\right)_A + \int_2^1 \left(\frac{\delta Q}{T}\right)_B \tag{6-4}$$

Now repeat this operation, but over the (A–C) cycle:

$$0 \equiv \oint_{1-A-2-B-1} \frac{\delta Q}{T} = \int_1^2 \left(\frac{\delta Q}{T}\right)_A + \int_2^1 \left(\frac{\delta Q}{T}\right)_C \tag{6-5}$$

If we subtract Equation (6-5) from Equation (6-4) and rearrange, the result is

$$\int_2^1 \left(\frac{\delta Q}{T}\right)_B = \int_2^1 \left(\frac{\delta Q}{T}\right)_C \tag{6-6}$$

The quantity $\delta Q/T$ depends only on the initial and final equilibrium states and not on the paths (since B and C are arbitrary). Thus we have a definition: *The quantity $\delta Q/T$, a path function divided by a property, is a point function (an exact differential):*

$$dS \equiv \frac{\delta Q}{T}\Big)_{rev} \tag{6-7}$$

The integral of Equation (6-7) represents the change in entropy of a system during a reversible change of state (and is corollary G):

$$S_2 - S_1 = \int_1^2 \frac{\delta Q}{T}\Big)_{rev} \tag{6-8}$$

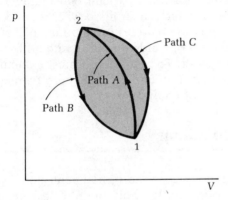

Figure 6-1. (p, V) diagram of a cyclic process.

Of course, the actual integration of Equation (6-8) depends on the relationship of the heat and the temperature. Regardless, the equation represents the procedure to compute the entropy change along any arbitrary reversible path. In fact, it is more than that. It dictates the procedure for entropy change calculation whether it is a reversible or irreversible path. That is, entropy is an exact differential and only the end points are important. Unfortunately, very much like the stored energy deduced from the first law, Equation (6-8) yields only entropy change information and nothing about absolute values.

Now let us pause and reiterate the important conclusions obtained so far:

1. Entropy is defined for an equilibrium state *only*.

2. Only entropy *differences* may be computed.

3. Entropy is independent of the history of the system. (It is a property.)

4. Entropy changes may be computed from the heat transfer for reversible processes *only*.

5. Entropy changes for an irreversible process from one equilibrium state to another may be determined by:
   a. Devising a reversible process connecting the same two end states
   b. Using tables (subtract the two end-state values)
   c. Using $S = S(T, p)$ if this functional relationship is known

Entropy Used as a Coordinate

Let us note that for reversible processes

$$\delta Q \equiv T\,dS \tag{6-9}$$

Thus the heat transfer may be computed directly from properties and is just the area under the process curve represented on a (T, S) diagram (see Figure 6-2). Thus

$$_1Q_2 = \int_1^2 T\,dS \tag{6-10}$$

Now we will reconsider the Carnot cycle in light of the concept of entropy. Figure 6-3 is a (T, S) diagram of a Carnot cycle. Notice that the reversible adiabatic processes of the cycle require

$$S_2 - S_3 = S_4 - S_1 = 0 \tag{6-11}$$

That is, they are constant-entropy or isentropic processes. Let us compare the heat transferred in processes 1–2 and 3–4. Using Equation (6-10)—that is, $_2Q_3 = {}_4Q_1 = 0$—we get

$$_1Q_2 = T_H(S_2 - S_1) \tag{6-12}$$

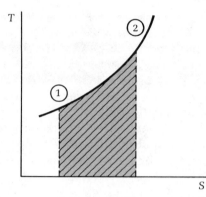

Figure 6-2. (T, S) diagram.

and

$$_3Q_4 = T_L(S_4 - S_3) \tag{6-13}$$

Notice that $_3Q_4$ in Equation (6-13) is inherently a negative number (it is the heat rejected) and that $(S_2 - S_1)$ and $(S_3 - S_4)$ are of equal magnitude. Dividing Equation (6-12) by Equation (6-13) yields

$$\frac{_1Q_2}{_3Q_4} = \frac{Q_H}{Q_L} = \frac{T_H}{T_L} \tag{5-16}$$

In Figure 6-3, we note that the area within the rectangle represents the net work done. The (T, S) diagram also graphically illustrates that as T_H increases, the efficiency increases; moreover, as T_L decreases, the efficiency also increases (that is, the W area becomes larger). Figure 6-4 may help to strengthen your intuition of the representation of various processes on a (T, S) diagram. Note that the irreversible process is represented by a dashed line in Figure 6-4. This is done because the exact path is unknown. Therefore, the area beneath this dashed line (the irreversible process) is meaningless because there is no real boundary.

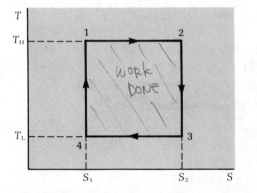

Figure 6-3. (T, S) diagram of a Carnot cycle.

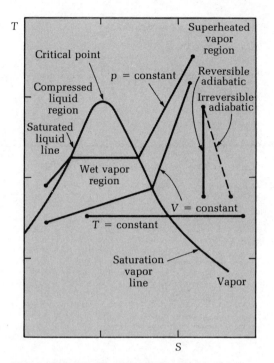

Figure 6-4. (T, S) diagram for liquid and vapor.

EXAMPLE 6-4

A cylinder fitted with a piston contains 1 lbm of steam at 14.7 psia and 400 F. The piston is moving so that the steam is compressed in a reversible isothermal process until the steam is a saturated vapor. Determine the work done on the system and the heat transfer for the process.

SOLUTION: The process on a (T, S) diagram would look like the sketch.

$Q = T ds$

Using Appendix A-1:

$$_1Q_2 = T(s_2 - s_1)\, m$$

ISOTHERMAL

$$= 860\ \text{R}(1.5274 - 1.8743)\frac{\text{Btu}}{\text{lbm}}\frac{\text{lbm}}{\text{R}} = -298.3\ \text{Btu}$$

Using the first law and Appendix A-1:

$$_1W_2 = {}_1Q_2 - (\Delta u)m$$

$$= -298.3 \text{ Btu} - 1 \text{ lbm} (1116.6 - 1145.6) \text{ Btu/lbm}$$

$$= -269.3 \text{ Btu}$$

EXAMPLE 6-5

Resketch the process indicated on the (p, V) diagram on a (T, S) diagram if the substance is an ideal gas.

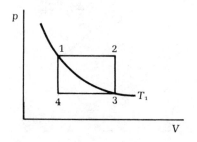

SOLUTION: 1. On a (T, S) diagram, the isothermal process 1–3 will be a horizontal straight line.

2. Noting that the temperature at point 2 is greater than T_1, there is an entropy increase from (1–2).

3. Similarly, $T_4 < T_1$ so we have a temperature decrease.

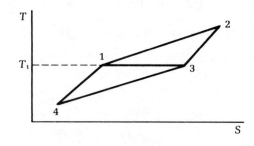

Relevant Thermodynamic Relations

At this point, it is convenient to obtain relations involving the properties deduced from the first and second laws of thermodynamics. To begin, let us consider the first law for a closed system with $\Delta PE = \Delta KE = 0$:

$$\delta Q = dU + \delta W \tag{4-11}$$

If we restrict our point of view to reversible processes, then

$$\delta Q = T \, dS \tag{6-9}$$

and

$$\delta W = p \, dV \tag{1-11}$$

Substitution of these two expressions into Equation (4-11) yields

$$T \, dS = dU + p \, dV \tag{6-14}$$

Another expression may be deduced by using the differential of enthalpy $(H = U + pV)$:

$$dH = dU + p \, dV + V \, dp$$

Using Equation (6-14) yields

$$T \, dS = dH - V \, dp \tag{6-15}$$

In their mass-independent forms, Equations (6-14) and (6-15) are

$$T \, ds = du + p \, dv \tag{6-16}$$

and

$$T \, ds = dh - v \, dp \tag{6-17}$$

Equations (6-16) and (6-17) are valid for *any* process of a pure substance as long as the resultant integration is performed between equilibrium states. The reason they are true for any process is that both equations deal *only* with properties.

Now let us consider these $T \, ds$ equations and assume that the working substance is an ideal gas. Thus Equations (6-16) and (6-17) become

$$ds = c_v \frac{dT}{T} + R \frac{dv}{v} \tag{6-18}$$

and

$$ds = c_p \frac{dT}{T} - R \frac{dp}{p} \tag{3-12}$$

With only a little effort (and remembering that $c_p - c_v = R$ for an ideal gas), it can be shown that there is a third $T \, ds$ relation:

$$ds = c_p \frac{dv}{v} + c_v \frac{dp}{p} \tag{6-19}$$

And if we assume that c_p and c_v are constants, the three equations above become (integrating from some reference point T_o, v_o, p_o, and s_o)

$$s - s_o = c_v \ln\left(\frac{T_2}{T_o}\right) + R \ln\left(\frac{V_2}{V_o}\right) \tag{6-20}$$

IDEAL GAS ONLY

IDEAL
GAS
ONLY

$$s - s_o = c_p \ln\left(\frac{T_2}{T_o}\right) - R \ln\left(\frac{p_2}{p_o}\right) \qquad \text{(6-21)}$$

$$s - s_o = c_p \ln\left(\frac{V_2}{V_o}\right) + c_v \ln\left(\frac{p_2}{p_o}\right) \qquad \text{(6-22)}$$

For a pictorial view of the significance of each term, consider Figure 6-5 and the various transitions from point o to point a. In the case of path $o–b–a$, use Equation (6-20). The convenience of the use of this equation can be seen by noting that for the path $o–b$, a constant-volume process, Equation (6-20) reduces to

$$s_b - s_o = c_v \ln\left(\frac{T_b}{T_o}\right) \qquad \text{CONSTANT VOLUME}$$

while in the case of the path $b–a$, Equation (6-20) reduces to

$$s_a - s_b = R \ln\left(\frac{V_a}{V_o}\right) \qquad \text{CONSTANT TEMPERATURE}$$

Thus for the complete transition ($o–b–a$),

$$s_a - s_o = c_v \ln\left(\frac{T_a}{T_o}\right) + R \ln\left(\frac{V_a}{V_o}\right)$$

Similar correspondences may be made between path $o–c–a$ and Equation (6-21) and path $o–d–a$ and Equation (6-22).

A Word About Irreversible Processes

Earlier in this chapter we used the equality portion of the Clausius inequality (that is, for reversible processes) to deduce a property called entropy. The defining equation is Equation (6-3).

REVERSIBLE ADIABATIC
IS ISENTROPIC

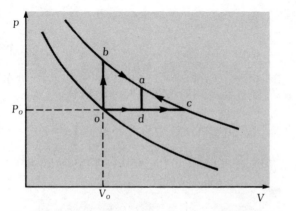

Figure 6-5. (p, V) diagram of various transitions $o–a$.

If $\int \delta Q/T$ is considered along several irreversible paths, the resultant value will be different for each path. That is, $\int_{\text{irrev}} \delta Q/T$ is *not* a property. In fact, it can be shown that $\oint \delta Q/T < 0$. The area beneath the path of an irreversible process on a (T, S) diagram has no significance. It does not represent the heat transfer because

$$Q_{\text{irrev}} \neq \int T\, dS$$

In fact, it can be shown that

$$Q_{\text{irrev}} < \int T\, dS \qquad\qquad (6\text{-}23)$$

Therefore, the important fact to notice is that the definition should be

$$dS \geq \frac{\delta Q}{T} \qquad\qquad (6\text{-}24)$$

where the equal sign is for reversible processes and the unequal sign is for irreversible processes.

Principle of the Increase of Entropy

Simply stated, the principle of the increase of entropy is: *Net entropy never decreases.* To stimulate your intuition, Figure 6-6 illustrates a cycle with an irreversible process. Starting at point a, let us execute an irreversible adiabatic process to point b. The corresponding entropy change is

$$\Delta S = S_b - S_a \qquad\qquad (6\text{-}25)$$

In going from point b to point c, we execute a reversible adiabatic

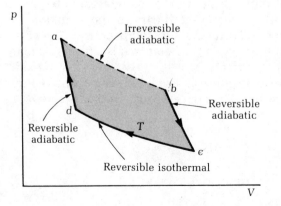

Figure 6-6. (p, V) diagram of a cycle with an irreversible process.

(isentropic) process such that $S_c = S_b$. Thus the net entropy change so far is

$$\Delta S = S_c - S_a \tag{6-26}$$

Now go to point d by an isothermal process at the temperature T. Point d is selected such that a reversible adiabatic (isentropic) process will return the system to point a. Thus $S_a = S_d$ and the total change in entropy in the process a–b is

$$\Delta S = S_c - S_d \tag{6-27}$$

Since this is a cycle, the change in stored energy is zero, net positive work is done, and the only heat transfer occurs during process c–d. (Note that $_cQ_d$ is an inherently negative number.) Therefore

$$W = -_cQ_d \tag{6-28}$$
$$= T(S_c - S_d)$$

This implies that

$$S_c - S_d > 0 \tag{6-29}$$

Thus, from Equation (6-27),

$$\Delta S > 0 \tag{6-30}$$

Of course, if process a–b had been reversible, the change in entropy would have been zero. Thus the mathematical version of the increase in entropy principle is

$$\Delta S_{\text{universe}} \geq 0 \tag{6-31}$$

Hence the engineer has a criterion for the permissible direction of a process: If the entropy does not increase (or remain constant for the ideal case), the process will not go. To illustrate this point, suppose you are asked to make a quick determination of the financial feasibility of a new cold-start heat pump system to be manufactured by the company for which you work. Figure 6-7 illustrates the situation. The temperature of the home is T_1 (the outside temperature). A reversible heat pump is used to put heat Q into the home to raise the inside temperature to T_2. In doing so, it extracts heat $(Q - W)$ from the outside air. Of course, a reversible heat pump is used because it would indicate the minimum work required. Thus the entropy changes of the three parts of this "universe" are

$$\Delta S_{\text{reservoir}} = -\frac{Q - W}{T_1}$$

$$\Delta S_{\text{rev. heat pump}} = 0$$

$$\Delta S_{\text{home}} = S_2 - S_1$$

Body whose temperature is to be
raised from T_1 to T_2

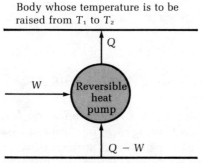

Reservoir at T_1 (outside air)

Figure 6.7. Feasibility model of a cold-start heat pump system.

By applying our principle, we get

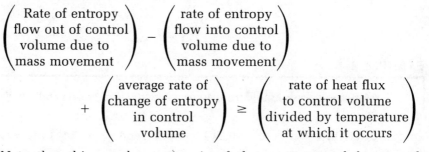

$$S_2 - S_1 - \frac{Q - W}{T_1} \geq 0$$

or

$$W \geq Q - T_1(S_2 - S_1) \tag{6-32}$$

Of course, the minimum work required is

$$W_{min} = Q - T_1(S_2 - S_1)$$

Thus W_{min} will give you a lower limit on the cost to run this new system. Therefore, you may make the decision dependent upon the relative cost of energy.

Open System

The formulation of a general entropy equation is as follows

$$\begin{pmatrix} \text{Rate of entropy} \\ \text{flow out of control} \\ \text{volume due to} \\ \text{mass movement} \end{pmatrix} - \begin{pmatrix} \text{rate of entropy} \\ \text{flow into control} \\ \text{volume due to} \\ \text{mass movement} \end{pmatrix}$$

$$+ \begin{pmatrix} \text{average rate of} \\ \text{change of entropy} \\ \text{in control} \\ \text{volume} \end{pmatrix} \geq \begin{pmatrix} \text{rate of heat flux} \\ \text{to control volume} \\ \text{divided by temperature} \\ \text{at which it occurs} \end{pmatrix}$$

Note that this word equation is of the same general form as the definition deduced for closed systems but includes terms that account for the net entropy change due to the mass crossing the control surface.

Thus for the general case of an open system, the second law can be written

$$dS_{\text{system}} \geq \left(\frac{\delta Q}{T}\right)_{\text{rev}} + \delta m_i s_i - \delta m_e s_e \qquad \text{(6-33)}$$

or

$$dS_{\text{system}} = \left(\frac{\delta Q}{T}\right)_{\text{rev}} + \delta m_i s_i - \delta m_e s_e + dS_{\text{irr}} \qquad \text{(6-34)}$$

where $\delta m_i s_i$ is the entropy increase due to the mass entering, $\delta m_e s_e$ is the entropy decrease due to the mass leaving, $\delta Q/T$ is the entropy change due to heat transfer alone between system and surroundings, and dS_{irr} is the entropy created or produced due to irreversibilities.

For the special case of steady state/uniform properties, this equation translates to

$$\sum_{\text{out}} \dot{m}s - \sum_{\text{in}} \dot{m}s \geq \oint \frac{\delta \dot{Q}}{T} \qquad \text{(6-35)}$$

where A is the area through which the heat is transferred.

For a one-inlet/one-outlet situation,

$$s_{\text{out}} - s_{\text{in}} \geq \oint \frac{\delta Q}{mT} \qquad \text{(6-36)}$$

And if, in addition, the process is adiabatic, then

$$s_{\text{out}} \geq s_{\text{in}} \qquad \text{(6-37)}$$

For completeness, the uniform-flow/uniform-properties case is included:

$$\sum_{\text{out}} ms - \sum_{\text{in}} ms + (m_f s_f - m_i s_i) \geq \int \oint \frac{\delta \dot{Q}}{T} \, dt \qquad \text{(6-38)}$$

or

$$(m_f s_f - m_i s_i)_{\text{system}} = \int_{\text{rev}} \frac{\delta Q}{T} - \sum_{\text{out}} ms + \sum_{\text{in}} ms + \Delta S_{\text{irr}} \qquad \text{(6-39)}$$

EXAMPLE 6–6

Air expands irreversibly from 40 psia and 360 F to 20 psia and 220 F. Calculate Δs.

SOLUTION: $c_p = 0.24$ Btu/lbm-R   and $c_v = 0.171$ Btu/lbm-R pg. 546

Assuming that air is an ideal gas, we get

$$s_2 - s_1 = c_p \ln\left(\frac{T_2}{T_1}\right) - R \ln\left(\frac{p_2}{p_1}\right)$$

$$= \left[(0.24 \ln\left(\frac{680}{820}\right) - (0.24 - 0.171) \ln\left(\frac{20}{40}\right)\right] \text{ Btu/lbm-R}$$

[IF AIR EXPANDED REVERSIBLY, $\Delta S = 0$.]

$$= (-0.0449 + 0.0478)\ \text{Btu/lbm-R}$$

$$= 0.002896\ \text{Btu/lbm-R}$$

∴ NET INCREASE IN ENTROPY.

EXAMPLE 6–7

Is the adiabatic expansion of superheated steam from 1500 F and 300 psia to 1200 F and 180 psia possible?

SOLUTION:

$$\Delta s = (1.9227 - 1.9572)\ \text{Btu/lbm-R}$$

$$= -0.0345\ \text{Btu/lbm-R}$$

Since this is an adiabatic expansion, Δs should be greater than zero (or equal to zero if it is a reversible expansion). It is not possible.

EXAMPLE 6–8

A Carnot cycle heat engine uses steam as the working fluid. Heat is absorbed by the steam at 212 F while it changes from a saturated liquid to a saturated vapor. Heat is rejected by this engine at 100 F. Determine the beginning and ending qualities of the heat rejection portion of the cycle.

SOLUTION:

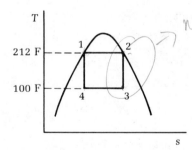

NOTE: ISENTROPIC

We seek x_3 and x_4. Notice that $s_3 = s_2$ and $s_1 = s_4$. Thus

$$s_1 = 0.3121\ \text{Btu/lbm-R} = s_4$$

$$= x_4 s_{g_4} + (1 - x_4)s_{f_4} = 1.9825 x_4 + 0.1295(1 - x_4)$$

$$x_4 = \frac{0.3121 - 0.1295}{1.9825 - 0.1295} = 0.0985$$

Similarly,

$$s_2 = 1.7568\ \text{Btu/lbm-R} = s_3$$

$$= x_3 s_{g_3} + (1 - x_3)s_{f_3} = x_3 s_{g_4} + (1 - x_3)s_{f_4}$$

$$x_3 = \frac{1.7568 - 0.1295}{1.9825 - 0.1295} = 0.8783$$

EXAMPLE 6–9

Air expands through a nozzle at a rate of 2 lbm/sec. The outlet and inlet conditions are indicated in the sketch. Assuming the process to be reversible and adiabatic, what is the exit velocity?

SOLUTION:

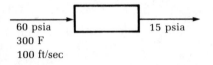

60 psia
300 F
100 ft/sec

15 psia

Continuity:

$$\dot{m}_1 = \dot{m}_2$$

First law:

$$h_1 + \frac{V_1^2}{2g_c} = h_2 + \frac{V_2^2}{2g_c}$$

Second law:

$$s_1 = s_2$$

Thus

$$V_2^2 = V_1^2 + 2g_c(h_1 - h_2) \qquad h = c_p \Delta t$$

$$= V_1^2 + 2g_c c_p(T_1 - T_2) \qquad (\text{Air} \doteq \text{ideal gas})$$

For a reversible adiabatic process,

$$T_2 = T_1\left(\frac{p_2}{p_1}\right)^{\frac{k-1}{k}}$$

For air $k = 1.4$. Thus

$$T_2 = 760\left(\frac{15}{60}\right)^{0.4/1.4} = 511.4 \text{ R}$$

$$V_2^2 = (100 \text{ ft/sec})^2$$

$$+ 64.34 \frac{\text{lbm-ft}}{\text{lbf-sec}^2} 0.248 \frac{\text{Btu}}{\text{lbm-R}} (248.6 \text{ R})\left(\frac{\text{ft}}{\text{ft}}\right) 778 \frac{\text{ft-lbf}}{\text{Btu}}$$

$$V_2 = 1760 \text{ ft/sec}$$

EXAMPLE 6–10

A cylinder contains 10 lbm of superheated steam at 400 F and 140 psia. The steam is compressed isothermally to a saturated vapor requiring 800 Btu of work done on the cylinder. During the process, the heat transfer takes place with the surroundings at 400 F. Is this process possible?

Q = du+W

SOLUTION: Using Appendix A-1:

$$u_1(400\ F,\ 140\ \text{psia}) = 1131.4\ \text{Btu/lbm}$$

$$u_2 = u_g(400\ F) = 1116.6\ \text{Btu/lbm}$$

$$_1Q_2 = m(u_2 - u_1) + _1W_2$$

$$= 10\ \text{lbm}\ (1116.6 - 1131.4)\ \text{Btu/lbm} - 800\ \text{Btu}$$

↗ *WORK DONE ON THE SYSTEM.*

$$= -948\ \text{Btu} \qquad \text{(a heat loss)}$$

$$\Delta S_{sys} = m(s_2 - s_1) = 10\ \text{lbm}\ (1.5274 - 1.6085)\ \text{Btu/lbm-R}$$

$$= -0.811\ \text{Btu/R}$$

↗ *+ NUMBER, BECAUSE SURR. GAINS HEAT.*
∴ S_{SURR.} = +

$$\Delta S_{surr} = -\frac{_1Q_2}{T_0} = \frac{948\ \text{Btu}}{860} = 1.1023\ \text{Btu/R}$$

$$\Delta S_{univ} = \Delta S_{sys} + \Delta S_{surr} = 0.2913\ \text{Btu/R}$$

Thus this process is possible.

IRREVERSIBLE

EXAMPLE 6–11

A turbine receives steam at 100 psia and 500 F. The steam expands in a reversible and adiabatic process and leaves the turbine at 14.7 psia and 240 F. Does this process violate the second law?

SOLUTION: In the case of a reversible adiabatic process, the second law requires $s_{out} \geq s_{in}$. From the tables,

$$s_{in}(100\ \text{psia},\ 500\ F) = 1.7088$$

$$s_{out}(14.7\ \text{psia},\ 240\ F) = 1.7764$$

∴ ADIABATIC / IRREVERSIBLE COMPRESSOR.

This process does not violate the second law.

EXAMPLE 6–12

A cylinder contains 1 kg of steam at a pressure of 0.7 MPa and entropy of 6.5 kJ/(kg · K). This steam is heated reversibly at constant pressure until the temperature is 250 C. Determine the heat supplied and sketch this process on a (T, S) diagram.

SOLUTION: Using Appendix Table A-1-5 we can see that the initial condition of the steam is one of saturation. The quality is determined by using entropy; thus

$$x_i = \frac{6.5 - 1.992}{4.713} = 0.957$$

To determine the heat, we must apply the first law with $\Delta PE = \Delta KE = 0$:

$$q - w = \Delta u = u_f - u_i$$

and

$$w = \int_i^f p \, dv = p(v_f - v_i)$$

So

$$q = h_f - h_i$$

Thus

$$h_i = h_f + x h_{fg}$$

$$= 697 + 0.957(2066) = 2672 \text{ kJ/kg}$$

The final state is 0.7 MPa and 250 C. Interpolating in Appendix Table A-1-7 yields an h_f for this superheated state:

$$h_f = 2954 \text{ kJ/kg}$$

Thus

$$q = (2954 - 2672) \text{ kJ/kg}$$

$$= 282 \text{ kJ/kg}$$

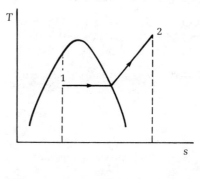

EXAMPLE 6–13

Air at a temperature of 15 C and a pressure of 0.1 MPa is contained in a cylinder of 0.02 m³ volume. From this initial condition the following cycle is executed: constant-volume heating to a pressure of 0.42 MPa; constant-pressure cooling to the original temperature; and finally a constant-temperature pressure decrease to the original conditions. Sketch this series of processes on a (T, s) diagram and determine the entropy changes.

SOLUTION:

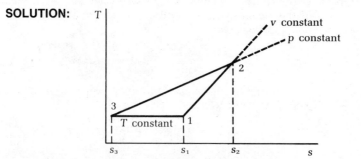

Let us assume that air is an ideal gas. We know that $T_1 = (15 + 273)$ K $= 288$ K and $p_1 = 0.1$ MPa and that $v_1 = v_2$ while $p_2 = 0.42$ MPa. Thus

$$\frac{p_1}{T_1} = \frac{p_2}{T_2}$$

or

$$T_2 = \frac{0.42 \text{ MPa } (288 \text{ R})}{0.1 \text{ MPa}} = 1210 \text{ K}$$

By definition $\Delta s = \int \delta q / T$ if we postulate a reversible process. So, for the constant-volume process 1–2,

$$s_2 - s_1 = \int \frac{\delta q}{T} = \int \frac{du + p \, dv}{T} \qquad \text{first law}$$

$$= \int \frac{du}{T} \qquad \text{constant-volume process}$$

$$= \int \frac{c_v \, dT}{T} \qquad \text{ideal gas}$$

$$= c_v \ln\left(\frac{T_2}{T_1}\right) \qquad c_v \text{ constant}$$

For the constant-pressure process 2–3,

$$s_3 - s_2 = \int \frac{\delta q}{T} = \int \frac{du + p \, dv}{T} \qquad \text{first law}$$

$$= \int \frac{d(u + pv)}{T} \qquad p \text{ constant}$$

$$= \int \frac{dh}{T} \qquad \text{definition of } h$$

$$= \int c_p \frac{dT}{T} \qquad \text{ideal gas}$$

$$= c_p \ln\left(\frac{T_2}{T_1}\right) \qquad c_p \text{ constant}$$

Finally, for the constant-temperature process 3–1,

$$s_1 - s_3 = \int \frac{\delta q}{T} = \int \frac{du + p \, dv}{T} \qquad \text{first law}$$

$$= \int \frac{p \, dv}{T} \qquad \begin{array}{l} \text{constant-temperature} \\ \text{ideal-gas process} \end{array}$$

$$= R \int \frac{dv}{v} \qquad \text{equation of state}$$

$$= R \ln\left(\frac{v_1}{v_3}\right)$$

Note that

$$\frac{V_1}{V_3} = \frac{T_1}{T_3}\frac{p_3}{p_1}$$

But $T_1 = T_3$ and $p_3 = p_2$ and, finally, $p_1/p_2 = T_1/T_2$. So

$$s_3 - s_1 = +R \ln\left(\frac{T_2}{T_1}\right)$$

Therefore, using the constants for Appendix Table A-3-3, we get

$$S_2 - S_1 = m(s_2 - s_1) = 0.0254 \text{ kg } (0.716 \text{ kJ/kg} \cdot \text{K}) \ln\left(\frac{1210}{288}\right)$$

$$= 0.0261 \text{ kJ/K}$$

$$S_3 - S_2 = m(s_3 - s_2) = 0.0254 \text{ kg } \left(\frac{1 \text{ kg}}{\text{kg} \cdot \text{K}}\right) \ln\left(\frac{288}{1210}\right)$$

$$= -0.0365 \text{ kg/K}$$

$$S_1 - S_3 = m(s_1 - s_3) = 0.01 \text{ kg/K}$$

EXAMPLE 6–14

A turbine receives air at 0.68 MPa and 430 C. The air expands irreversibly but adiabatically, leaving the turbine at 0.1 MPa and 150 C. What is Δs?

SOLUTION: To obtain a solution we must postulate a reversible process between the inlet and outlet conditions. Having done that, we may use Equation (6-21). Thus

$$\Delta s = c_p \ln\left(\frac{T_e}{T_i}\right) - R \ln\left(\frac{P_e}{P_i}\right)$$

$$= 1 \text{ kJ/(kg} \cdot \text{K)} \ln\left(\frac{423}{703}\right) - 0.284 \text{ kJ/(kg} \cdot \text{K)} \ln\left(\frac{0.1}{0.68}\right)$$

$$= (-0.50798 + 0.54441) \text{ kJ/(kg} \cdot \text{K)}$$

$$= 0.03642 \text{ kJ/(kg} \cdot \text{K)}$$

If this had been a reversible adiabatic process, the exit temperature would have to be (assuming $k = c_p/c_v = 1.4$)

$$T_e = 703\left(\frac{0.10}{0.68}\right)^{0.286} = 406 \text{ K} = 133 \text{ C}$$

PROBLEMS

6-1 Complete the following equations by inserting the proper equality sign or inequality sign.

a. For a closed system (any process):

$$W \qquad \int p \, dV$$

$$\Delta S \qquad \int \frac{\delta Q}{T}$$

$$T \, dS \qquad dU + p \, dV$$

$$T \, dS \qquad dU + \delta W$$

b. For any cycle:

$$W \qquad \oint p \, dV$$

$$\oint dS \qquad 0$$

$$\oint dS \qquad \oint \frac{\delta Q}{T}$$

$$\oint \frac{\delta Q}{T} \qquad 0$$

$$\eta \qquad 1 - \frac{T_L}{T_H}$$

$$\oint dh \qquad 0$$

6-2 A mad scientist has proposed a reversible nonflow cycle using air. The cycle consists of three processes:

1–2: constant-volume compression from 14.0 psia and 60 F to 100 psia

2–3: constant-pressure heat addition during which the volume is tripled

3–1: a process that appears as a straight line on the (p, v) diagram

Draw (p, V) and (T, S) diagrams of the cycle. Compute the net work of the cycle in Btu/lbm.

6-3 Answer the following as completely as possible. If it is impossible to obtain numerical answers, indicate whether in or out, positive or negative, or indeterminate.

a. A closed system of air is cooled reversibly at constant pressure: $\Delta S = $ _NEGATIVE_ .

b. A closed system of air is cooled irreversibly at constant pressure: $\Delta S = $ _NEGATIVE_ .

STILL HAS
AS SAME
NUMERIC VALUE

    c. A closed system of air is heated reversibly at constant pressure:
ΔS = _POSITIVE_ .

    d. A closed system of air is heated irreversibly at constant pressure: ΔS = _POSITIVE_ .

6-4 Consider the following state diagrams. Under what conditions do the cross-hatched areas in each case represent the work of the system?

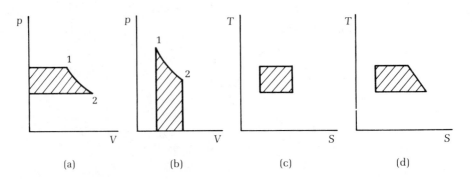

    (a)   (b)   (c)   (d)

6-5 An inventor claims he has developed a steady-flow *isothermal* turbine capable of producing 100 kW when operating at a steam flow rate of 10,600 lbm/hr between inlet conditions of 500 psia and 1000 F and an exit pressure of 14.7 psia. Heating takes place as the steam flows through the turbine. Determine: (a) the heat required and (b) the numerical value for each term in the entropy equation for the second law. Then evaluate his claim.

6-6 A 3000-ohm resistor carrying a steady current of 0.3 amp is cooled by placing it in a water bath at atmospheric pressure. The heat-transfer coefficient between the resistor and the evaporating water is 1000 Btu/hr ft^2 F. The surface area of the resistor is 1.5 in.^{2} Determine: (a) the steady-state temperature of the resistor, (b) the rate of evaporation of the water (lbm/hr), and (c) the entropy created.

6-7 Air is compressed in a reversible steady-state/steady-flow process from 15 psia and 80 F to 120 psia. The process is polytropic with n = 1.22. Calculate the work of compression per pound, the change of entropy, and the heat transfer per pound of air compressed. $\Delta h = c_p \Delta t$ $q = \Delta h + w$ $w = \frac{n}{1-n}(p_2 v_2 - p_1 v_1)$

6-8 In a compression ignition engine, air originally at 120 F is to be compressed to a temperature of 980 F. Compression obeys the law $pV^{1.34}$ = constant. Determine: (a) the compression ratio required (that is, the ratio of the volume before to the volume after compression), (b) the work of compression per pound of air, and (c) the heat transfer per pound of air.

6-9 The compression stroke for a four-stroke-cycle spark ignition engine (as in cars) is approximated as a reversible adiabatic process. Assume that the cylinder volume at bottom dead center is 400 in.^{3}, the compression ratio (V_1/V_2) is 9:1, and the cylinder is initially charged with air at 15 psia and 90 F. Determine: (a) the temperature and pressure of the air after compression and (b) the horsepower required for this compression process if engine speed is 2000 rpm. (*Note:* There is one compression stroke for every two revolutions.)

6-10 Air undergoes a steady-flow reversible adiabatic process. The initial state is 200 psia and 1500 F and the final pressure is 20 psia. Changes in kinetic and potential energy are negligible. Determine:
a. The final temperature
b. The final specific volume
c. The change in internal energy per lbm
d. The change in enthalpy per lbm
e. The work per lbm

6-11 Air undergoes a steady-flow reversible adiabatic process. The initial state is 1400 kPa and 815 C and the final pressure is 140 kPa. Changes in kinetic and potential energy are negligible. Determine:
a. The final temperature
b. The final specific volume ✓
c. The change in specific internal energy
d. The change in specific enthalpy
e. The specific work

6-12 Air at 50 psia and 90 F flows through a restriction in a pipe (ID = 2 in.). The velocity of the air upstream from the restriction is 450 ft/min. If 58 F air is desired, what must be the velocity downstream of the restriction? Comment on this method of cooling.

6-13 Air at 50 psia and 90 F flows at the rate of 1.6 lbm/sec through an insulated turbine. If the air delivers 11.5 hp to the turbine blades, at what temperature does the air leave the turbine?

6-14 Air at 50 psia and 90 F flows at the rate of 1.6 lbm/sec through an insulated turbine to an exit pressure of 14.7 psia. What is the minimum temperature attainable at exit?

6-15 Steam flows through a nozzle from inlet conditions at 200 psia and 800 F to an exit pressure of 30 psia. Flow is reversible and adiabatic. For a flow rate of 10 lbm/sec, determine the exit area if the inlet velocity is negligible.

6-16 For a new 1200-MW nuclear power plant under construction in Arkansas, the steam flow rate is 10,000,000 lbm/hr. If saturated steam enters the condenser at 1 psia and there is no subcooling of the condensate, determine the heat that will be rejected to the river water used in the condenser (in Btu/hr) *without* using the first law of thermodynamics.

6-17 A new design for a gas turbine requires the addition of heat at constant temperature as the air flows from inlet to outlet. For a flow rate of 5000 lbm/hr, inlet conditions of 500 psia and 340 F, and exit conditions of 25 psia and 340 F, determine the maximum horsepower output of the turbine. Changes in kinetic and potential energies are negligible.

6-18 A contact feedwater heater operates on the principle of mixing steam and water. Steam enters the heater at 100 psia and 98% quality. Water enters the heater at 100 psia and 80 F. As a result, 25,000 lbm/hr of water at 95 psia and 290 F leave the heater. There is no heat transfer between the heater and the surroundings. Evaluate each term in the general entropy equation for the second law.

6-19 Suppose that 3 lbm of methane (CH_4) is compressed at a constant temperature of 140 F from 20 to 100 psia in a piston/cylinder device. If compression is ideal (frictionless), determine: (a) the work required (in Btu) and (b) the heat transfer during the process (in Btu), and state whether in or out.

6-20 Air is compressed in a steady-flow, reversible process from 15 psia and 80 F to 120 psia. Determine the work and the heat transfer per pound of air compressed for each of the following processes: (a) adiabatic, (b) isothermal, (c) polytropic $(n = 1.25)$.

6-21 A Carnot cycle heat engine operating between reservoirs at 1000 and 80 K receives 500 kW-hr. Determine:
a. The thermal efficiency
b. The work done
c. The entropy change of the high and low-temperature reservoirs;
d. The entropy change of the high and low-temperature reservoirs; if the high-temperature reservoir is changed to 1500 K (heat still enters the heat engine at 1000 K)
e. The entropy change of the universe for both circumstances

6-22 A mass m_1 of a liquid at temperature T_1 is mixed with a mass m_2 of the same liquid at temperature T_2. The system is thermally insulated. If $m_1 = m_2$ and $c_{p_1} = c_{p_2}$, show that the net entropy change is

$$s_2 - s_1 = 2mc_p \ln\left\{\frac{(T_1 + T_2)/2}{\sqrt{T_1 T_2}}\right\}$$

6-23 Is the adiabatic expansion of superheated steam from 1600 F and 300 psia to 1200 F and 180 psia possible? Why?

6-24 A turbine receives steam at a pressure of 1000 psia and 1000 F and exhausts it at 3 psia. The velocity of the steam at the inlet is 50 ft/sec; at the outlet, which is 10 ft higher, the velocity is 1000 ft/sec. Assuming that the operation is reversible and adiabatic, determine the work per unit mass.

6-25 Is the adiabatic expansion of air from 25 psia and 140 F to 15 psia and 40 F possible if the specific heats are assumed constant?

6-26 Determine the efficiency of the reversible cycle in the sketch.

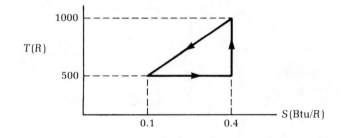

6-27 Suppose that 0.5 lbm of air in a closed system is compressed irreversibly from 15 psia and 40 F to 30 psia. During the process 8.5 Btu of heat is removed from the air and 13 Btu of work is done on the air. Determine the change in entropy of the air. (*Hint:* Assume that air is an ideal gas with c_p = 0.24 Btu/lbm-R and c_v = 0.17 Btu/lbm-R. Use $T\,ds$ relations.)

6-28 The flow rate of Freon-12 in a refrigeration cycle is 150 lbm/hr. The compressor inlet conditions are 30 psia and 20 F and the exit pressure is 175 psia. Assuming the compression process to be reversible and adiabatic, what horsepower motor is required to drive the compressor?

6-29 Determine the efficiency of the reversible cycle shown in the sketch.

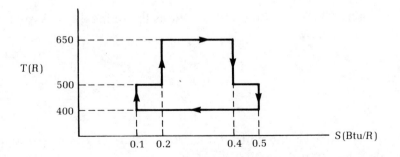

6-30 An inventor reports that he has a steady-flow refrigeration compressor that receives saturated Freon-12 vapor at 0 F and delivers it at 150 psia and 120 F. The compression is adiabatic. Would you invest in this invention? Why?

6-31 The following sketch illustrates three processes: ab, bc, and ac. Assuming constant specific heats, sketch these three processes on a (T, S) diagram. Assume that the working substance is an ideal gas.

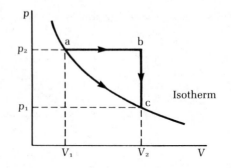

6-32 Steam at 400 psia and 600 F expands through a nozzle to 300 psia at a rate of 20,000 lbm/hr. If the process occurs isentropically ($\Delta S = 0$) and the initial velocity is very low, calculate the exit velocity.

6-33 A mass m_1 of a liquid at temperature T_1 is mixed with a mass m_2 of the same liquid but at temperature T_2. The system is thermally insulated. Show that the net entropy change is

$$S_2 - S_1 = 3mc_p \ln\left[\frac{T_f}{(T_1 T_2^2)^{1/3}}\right]$$

if $m_2 = 2m_1$ and c_p is constant. Also show that

$$T_f = \frac{T_1 + 2T_2}{3}$$

6-34 There are three $T\,ds$ relations for an ideal gas. Assuming that you know

$$T\,ds = c_v\,dT + p\,dv$$

and

$$T\,ds = c_p\,dT - v\,dp$$

derive the third form from either of these. (Hint: $pv = RT \Rightarrow dp/p + dv/v = dT/T$.)

6-35 Consider a cylinder fitted with a piston that contains saturated Freon-12 vapor at 20 F. Let this vapor be compressed in a reversible adiabatic process until the pressure is 150 lbf/in². Determine the work per pound mass for this process.

6-36 A rigid cylinder contains steam at 8 MPa and 350 C. The steam is then cooled to a pressure of 5 MPa. If the volume of the steam is 0.5 m³, calculate the heat rejected and sketch the results on a (T, S) diagram.

6-37 Steam going through an expansion valve experiences the following condition change: initially, p_1 = 0.7 MPa and x_1 = 0.96; finally, p_2 = 0.35 MPa. Calculate the entropy change per kilogram of steam.

6-38 Air in a cylinder (V_1 = 0.03 m³; p_1 = 0.1 MPa; T_1 = 10 C) is compressed reversibly at constant temperature to a pressure of 0.42 MPa. Determine the entropy change, the heat transferred, and the work done. Also sketch this process on (T, S) and (p, V) diagrams.

6-39 Calculate the work done as steam is expanded isentropically from 100 MPa and 375 C to 1 MPa.

6-40 Consider a cylinder/piston arrangement trapping air at 0.63 MPa and 550 C. Assume that it expands in a polytropic process ($pV^{1.3}$ = constant) to 0.1 MPa. What is the entropy change?

6-41 Consider a two-chambered container that is well insulated. The left chamber contains air; the right chamber is evacuated. What will be the entropy change if the volumes of the chamber are equal and the membrane separating the chambers is broken? (*Hint:* Imagine the process to take place in a polytropic fashion.) Is work done in this expansion? Why?

6-42 In the cylinders of an internal combustion engine, air is compressed reversibly from 0.1035 MPa and 23.9 C to 0.793 MPa. Calculate the work per lbm if the process is (a) adiabatic and (b) polytropic with n = 1.25.

6-43 Using Equation (6-22), deduce the process equation for a reversible adiabatic process for an ideal gas.

6-44 Compare the efficiencies of the processes indicated in the sketches.

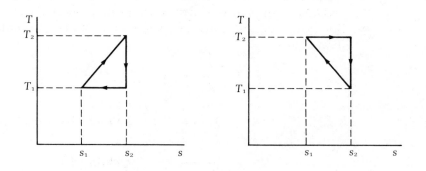

6-45 For the process indicated in the sketch, calculate the work done, the heat transferred, and the change of entropy for each process. Sketch the process on a (T, s) diagram. (*Note:* $c_p = 0.24$ Btu/lbm-R and the working substance is an ideal gas.)

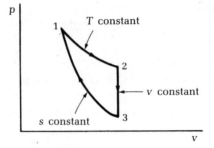

6-46 Four possible power cycles are illustrated in the following (T, s) diagrams. If each cycle operates with air in a closed system between maximums of $p = 300$ psia and $T = 1540$ F and minimums of $p = 14.7$ psia and $T = 40$ F, determine: (a) the maximum thermal efficiency of each and (b) the maximum work per unit change of entropy.

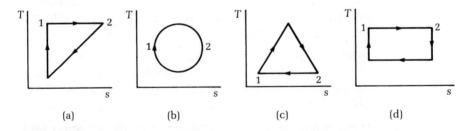

(a)	(b)	(c)	(d)

6-47 Rework Problem 6-46 if maximum conditions are 50 psia and 300 F (all else the same).

6-48 An *isothermal* steam turbine produces 450 kW when steam enters the turbine at 7 MPa and 320 C and exits at 0.7 MPa. Assume that 750,000 W of heat is added during this process. (This is a rate.) Determine the steam mass flow rate in kg/hr and the value of each term below:

$$\oint \frac{\delta Q}{AT} \, dA = \underline{\hspace{1cm}} \text{ kJ/(hr} \cdot \text{K)}$$

$$\sum_{\text{in}} (\dot{m}s) = \underline{\hspace{1cm}} \text{ kJ/(hr} \cdot \text{K)}$$

$$\sum_{\text{out}} (\dot{m}s) = \underline{\hspace{1cm}} \text{ kJ/(hr} \cdot \text{K)}$$

$$\Delta S_{irr} = \underline{\hspace{1cm}} \text{ kJ/(hr} \cdot \text{K)}$$

6-49 Compare the two cycles indicated in the sketch: 1–2–3–4 and 1–2'–3–4'. Test the validity of the Clausius inequality for these two cycles. (*Note:* 1–2' is supposed to represent an adiabatic but not isentropic expansion.)

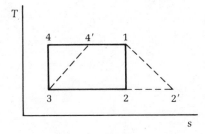

6-50 An ideal gas for which $C_v = \frac{5}{2}\bar{R}$ (the universal gas constant) is carried reversibly around the cycle in the accompanying sketch. Fill in the tables and determine the value of $\bar{R}$. Let $m = 1$ lbm and note that the numerical value of $\bar{R}$ is not required in the second table. (*Hint:* Note the geometry of the (p, V) diagram.)

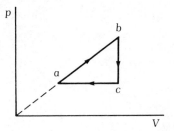

Point	p, lb/ft²	V, ft³	T, R
a	2000	300	
b	4000		
c	2000		

Path	W, ft-lbf	Q	Δu	Δs
ab		$1044\bar{R}$		$6\bar{R}\ln 2$
bc				
ca				
Sum				

2

Applications of Basic Thermodynamic Principles

Part

2

Applications of
Basic
Thermodynamic
Principles

Chapter

7

Basic Systems and Cycles

Engineering is an art as well as a science, and neither aspect can last long without the other. The art is best understood through written descriptions, pictures of equipment, and above all by visits to operating installations. The science is often expressed in terms of mathematical equations and physical concepts. The practicing engineer must understand the science within a context, however. You must know the meaning of engineering terms and be able to visualize the appearance and function of the equipment.

In the preceding chapters, we have discussed devices called turbines, nozzles, compressors, and the like. Since these devices may well have seemed like mysterious "black boxes" to you, the first part of this chapter discusses some of this elementary equipment. The remainder of the chapter presents various large-scale applications of the basic thermodynamic principles. Our discussion will be limited in all instances to basic situations; little will be said at this point concerning the many variations.

7–1 Elements of Thermal Systems

Before we discuss these devices, a word needs to be said about the efficiency of a process. In general, determining the efficiency of a process involves a comparison between the actual performance under a given circumstance and the performance in an ideal process. A vapor turbine, for example, is supposed to operate adiabatically and without friction, but friction is always present and an unavoidable heat transfer takes place between the turbine and the surroundings. Therefore, the efficiency of a turbine is defined as

$$\eta_t = \frac{W_a}{W_s} \qquad \text{(7-1)}$$

where W_a represents the actual work done by the mass of a fluid as it flows through a turbine and W_s represents the work done by the mass of a fluid as it flows through the turbine in a reversible adiabatic fashion. Figure 7-1 is an (h, s) diagram that depicts this situation. Thus

$$\eta_t = \frac{h_1 - h_{2a}}{h_1 - h_2} \qquad \text{(7-2)}$$

In compressors, assuming no effort is made to cool the gas during compression, the ideal process is isentropic, as it was in the case of the turbine. Thus if W_s represents the work done on the mass of a fluid as it flows through the compressor in a reversible adiabatic fashion and W_a represents the actual work done on the mass of a fluid as it flows through a turbine, then

$$\eta_c = \frac{W_s}{W_a} \qquad \text{(7-3)}$$

Be careful to note that both η_t and η_c range between zero and one. Figure 7-2 illustrates this situation. Thus

$$\eta_c = \frac{h_2 - h_1}{h_{2a} - h_1} \qquad \text{(7-4)}$$

As a final example, consider the nozzle. Like the preceding example, the ideal nozzle is isentropic. The difference is that the purpose of

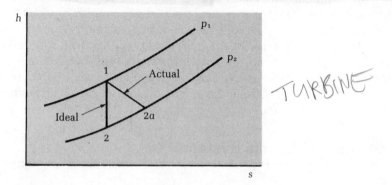

Figure 7-1. (h, s) diagram illustrating ideal and actual adiabatic expansions.

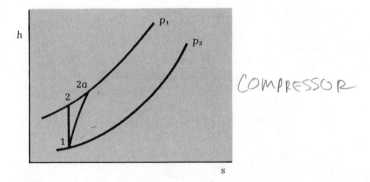

Figure 7-2. (*h, s*) diagram illustrating ideal and actual adiabatic compressions.

the nozzle is to change the kinetic energy not to do work. Thus

$$\eta_n = \frac{V_a^2}{V_s^2}$$

where V_a represents the actual exit velocity of the fluid as it leaves the nozzle and V_s represents the ideal velocity of the fluid as it leaves the nozzle if this fluid had traversed the nozzle in a reversible adiabatic fashion.

This is also an adiabatic expansion (Figure 7-1); thus

$$\eta_n = \frac{h_1 - h_{2a}}{h_1 - h_2} \tag{7-6}$$

Two important points must be noted when computing the efficiency of a device (as in these three examples, or any other device). First, to compute the isentropic cases (the ideal), one must use the same inlet conditions and exhaust pressure as the actual case. (The second exhaust parameter is the entropy.) Second, the efficiency of a device, or process, is very different from that of a cycle (recall Equation 5-2).

Expansion or Compression Work in a Cylinder

Probably the most common closed system in engineering is that involving the expansion or compression of a fluid trapped by a piston in a cylinder. Figure 7-3 shows such a system and the associated nomenclature.

First consider the expansion of a fluid behind a piston in a cylinder (Figure 7-4). The pressure in the cylinder is p and the volume of the fluid is denoted by V. The force exerted on the piston is

$$F = pA \tag{7-7}$$

where A is the surface area of the piston exposed to the fluid. The motion of the piston is in the direction of the applied force, and a

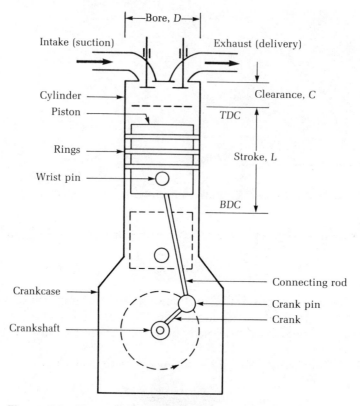

Figure 7-3. Single-acting vertical engine or compressor.

differential displacement dl may be expressed in terms of the change in volume of fluid dV as

$$dV = A\ dl \qquad\qquad (7\text{-}8)$$

Ideally, the work produced (or required, in the case of compression) is

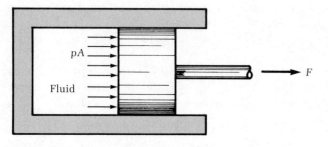

Figure 7-4. Expansion work in a cylinder.

$$W = \int F \, dx = \int pA \, dx = \int p \, dV \qquad (7\text{-}9)$$

For a real system, some (called the **lost work**, LW) of this mechanical work will be used to overcome friction. The actual useful work will be

$$W_a = W - \text{LW} \qquad (7\text{-}10)$$

The first law for a closed system also applies to the expansion (or compression) process:

$$\delta Q - \delta W = dU \qquad (7\text{-}11)$$

or

$$Q - W = (U_f - U_i) \qquad (7\text{-}12)$$

The second law for the closed system can be written

$$dS_{\text{system}} = \frac{\delta Q}{T} + dS_{\text{irr.}} = \frac{\delta Q}{T} + \frac{\delta \text{LW}}{T} \qquad (7\text{-}13)$$

or

$$S_f - S_i = \int \frac{\delta Q + \delta \text{LW}}{T} \qquad (7\text{-}14)$$

If the fluid is an ideal gas, then $\Delta U = c_v \, \Delta T$. If the expansion (or compression) process is adiabatic (as is commonly assumed), then $Q = 0$. And, consequently, if the process is reversible and adiabatic, $S_f = S_i$; if it is reversible but not adiabatic, $Q = \int_i^f T \, dS$.

Turbines, Pumps, Compressors, and Fans

A **turbine**, *whether the working fluid is a gas or a liquid, is a device in which the fluid does work against some type of blade attached to a rotating shaft.* As a result, the device produces work that may be used for some purpose. In pumps, compressors, and fans, work done on the fluid increases its pressure. Pumps are usually associated with liquids; compressors and fans are used for gases. The ratio of outlet to inlet pressure across a fan will probably be just slightly above 1, whereas for a compressor the ratio will probably be between 3 and 10. For steady flow through any of these devices the energy equation reduces to

$$q - w = h_2 - h_1 + \frac{V_2^2 - V_1^2}{2g_c} + \frac{g}{g_c}(Z_2 - Z_1) \qquad (7\text{-}15)$$

The potential-energy change across the device itself is normally negligible. For hydraulic turbines and pumps, however, it may be convenient to include part of the piping in the system under consideration and thus there may be a considerable change in elevation between inlet and outlet for the complete system.

Two other terms in Expression (7-15) require comment. First, the inclusion of heat depends on the mode of operation. If the device is not insulated, the heat gained or lost by the fluid depends on such factors as whether or not (1) a large temperature difference exists between the fluid and the surroundings, (2) a small flow velocity exists, and (3) a large surface area is present. In rotating turbomachinery (axial or centrifugal) velocities can be high, and the heat transfer normally is small compared to the shaft work. In reciprocating devices the heat transfer effects may be large. Experience enables the engineer to estimate the relative importance of heat transfer. As a second point, it is found that the change in kinetic energy is usually quite small in these devices, since the velocities at the inlet and outlet are frequently less than 200 ft/sec. In a steam turbine, the exhaust velocity is usually quite high because of the large volume of fluid at the low exhaust pressure. On the basis of the continuity equation, velocities may be kept low by selecting large flow areas (though this may not be a practical choice).

In many cases, the steady-flow first-law statement for these devices becomes

$$-w = h_2 - h_1 \qquad (7\text{-}16)$$

In this approximate solution, it is seen that the enthalpy decreases for a turbine and increases in the direction of flow for compressors and pumps.

Turbines (Steam, Gas, Hydraulic) The system illustrated in Figures 7-5, 7-6, and 7-7 is an open system. Internally a turbine converts the kinetic energy of a fluid into rotational energy (kinetic) of a shaft or wheel by means of a pressure drop. Since the fluid strikes a blade or cup on the shaft, the turbine is a type of rotary machine. Applying con-

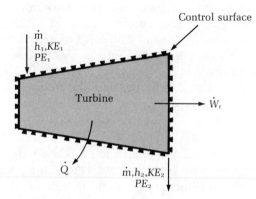

Figure 7-5. Schematic diagram of a turbine.

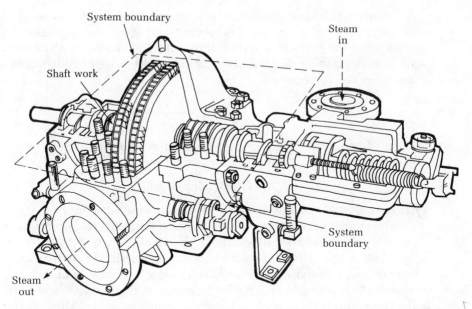

Figure 7-6. A single-stage turbine. (Reprinted by permission from Worthington Corporation)

tinuity and the first law of thermodynamics for this one-inlet/one-outlet system yields

$$h_1 + \frac{V_1^2}{2g_c} + \frac{gZ_1}{g_c} = h_2 + \frac{V_2^2}{2g_c} + \frac{gZ_2}{g_c} - q + w_t \qquad \textbf{(7-17)}$$

where w_t is the turbine work and q is the heat rejected by the turbine. Since the heat loss (radiative and convective losses to the surroundings)

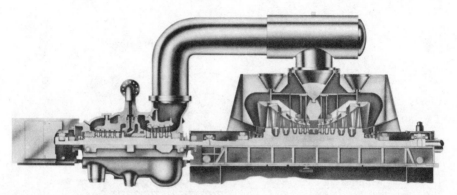

Figure 7-7. A straight noncondensing steam turbine. (Reprinted by permission from General Electric Company)

is usually quite small, the turbine is usually considered adiabatic and no significant error results. The potential-energy terms are also negligible since the vertical distance across the turbine is usually less than 10 ft. If the turbine is adiabatic and the potential energy is neglected, Equation (7-17) becomes

$$h_1 + \frac{V_1^2}{2g_c} = h_2 + \frac{V_2^2}{2g_c} + w_t \qquad (7\text{-}18)$$

If, in addition, the kinetic energies are not included, Equation (7-18) becomes (when multiplied by mass rate, $\dot{m}$)

$$\dot{W}_t = \dot{m}\,(h_1 - h_2)_s \qquad (7\text{-}19)$$

where s denotes an isentropic process (constant entropy).

The advance of waterwheel or hydraulic turbine technology is indicated by the modern Pelton wheel shown in Figure 7-8. Water taken from below the surface of a reservoir, or from a high-pressure source, is fed through a firehoselike nozzle, from which it emerges at a very high velocity (say 300 ft/sec). This jet impacts on the buckets, which turns the shaft, and is then discharged with a much lower velocity relative to the nozzle. There is a reduction in the kinetic energy of the stream as a result of this lowering of speed. This energy is transferred into work on the buckets. A modern Pelton wheel has a maximum efficiency of conversion of potential energy to useful work of about 80%.

In a Pelton wheel, which is a special case of an **impulse turbine,** the fluid stream is accelerated in a nozzle, but the flow velocity relative to the bucket does not vary much as the flow passes through the bucket. Another turbine device, the **reaction turbine,** operates completely filled with fluid and with blades on the shaft rather than buckets. The cross-sectional area for flow decreases as the flow moves between the blades (assuming the fluid is to be accelerated). This area reduction alters the pressure exerted by the fluid on the blades, thus pushing the blades and rotating the shaft.

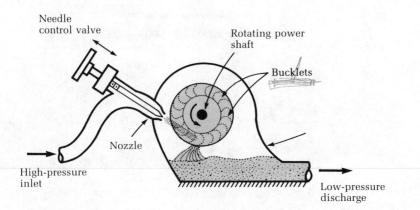

Figure 7-8. Schematic of a Pelton wheel.

EXAMPLE 7-1

A turbine receives steam at a pressure of 1000 psia and a temperature of 1000 F and exhausts it at 3 psia. The turbine inlet is 10 ft higher than the exit, the inlet steam velocity is 50 ft/sec, and the exit velocity is 1000 ft/sec. Calculate the turbine work per unit mass.

SOLUTION: Continuity:

$$\dot{m}_1 = \dot{m}_2$$

First law (assume adiabatic operation):

$$w_t = (h_1 - h_2) + \frac{V_1^2 - V_2^2}{2g_c} + \frac{g(Z_1 - Z_2)}{g_c}$$

Second law (assume reversible adiabatic operation):

$$s_1 = s_2 \qquad (\eta_t = 100\%)$$

From Appendix A-1-3:

$$h_1 = 1505.4 \text{ Btu/lbm} \qquad s_1 = 1.6530 \text{ Btu/lbm-R}$$

$$s_1 = s_2 = s_{g2} - (1 - x)s_{fg_2}$$

$$1.6530 = 1.8863 - (1 - x)\,1.6855$$

Thus

$$1 - x = 0.1380$$

$$h_2 = h_{g2} - (1 - x)h_{fg_2} = 1122.0 - (0.1380)(1013.2)$$
$$= 982.4 \text{ Btu/lbm}$$

$$w_t = (1505.4 - 982.4) + \frac{50^2 - 1000^2}{2g_c(778)} + \frac{10}{778}\left(\frac{32.17}{g_c}\right)$$

$$= (523.0 - 19.95 + 0.0128) \text{ Btu/lbm}$$

$$= 503.1 \text{ Btu/lbm}$$

Note that the total energy available for work ($|\Delta h| + |\Delta KE| + |\Delta PE|$) is 542.9 Btu/lbm. Thus the fractional contribution of each energy form is

$$\Delta h \Rightarrow \frac{523}{542.9} = 0.963$$

$$\Delta(KE) \Rightarrow \frac{19.95}{542.9} = 0.036$$

$$\Delta(PE) \Rightarrow \frac{0.0128}{542.7} = 0.001 \quad \text{NEGLECT}$$

The contribution of the potential energy is so small that it should not be included in the calculation of turbine work. The kinetic-energy contribution also has a small effect and is included only if high accuracy is desired.

EXAMPLE 7-2

Air is expanded reversibly and adiabatically in a turbine from 50 psia and 500 F to 15 psia. The turbine is insulated and the inlet velocity is small. The exit velocity is 500 ft/sec. Calculate the work output of the turbine per unit mass of air flow.

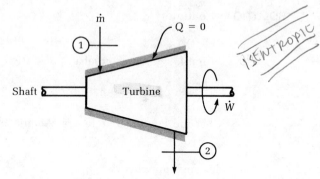

$\dot{m}$

$Q = 0$

ISENTROPIC

Shaft

Turbine

$\dot{W}$

1

2

SOLUTION: First law:

$$w_t = (h_1 - h_2) + \frac{V_1^2 - V_2^2}{2g_c}$$

$$= c_p(T_1 - T_2) + \frac{V_1^2 - V_2^2}{2g_c} \qquad \text{(ideal gas)}$$

Second law:

$$s_2 - s_1 = 0 = c_p \ln \frac{T_2}{T_1} - R \ln \frac{p_2}{p_1}$$

or

$$T_2 = T_1 \left(\frac{p_2}{p_1}\right)^{(k-1)/k} = 960 \left(\frac{15}{50}\right)^{(1.4-1)/1.4} = 680 \text{ R}$$

$$w_t = 0.24(960 - 680) + \frac{0 - 500^2}{(2)(32.2)(778)}$$

$$= 67.2 - 5.0 = 62.2 \text{ Btu/lbm}$$

EXAMPLE 7-3

A turbine receives steam at a pressure of 1 MPa and temperature of 300 C and exhausts it at 0.015 MPa. Determine the efficiency of the turbine if the actual work output is measured to be 550 kJ/kg of steam.

SOLUTION: Continuity:

$$\dot{m}_1 = \dot{m}_2$$

First law (assume adiabatic operation):

$$h_1 = h_2 + w_s$$

Second law (assume reversible adiabatic operation):

$$s_1 = s_2$$

From Appendix A-1-7:

$$h_1 = 3038.9 \text{ kJ/kg} \qquad s_1 = 6.9207 \text{ kJ/(kg} \cdot \text{K)}$$

$$s_1 = s_2 = s_{g_2} - (1 - x)s_{fg_2}$$

$$6.9207 = 8.0093 - (1 - x)\, 7.2544$$

Thus

$$1 - x = 0.1501$$

$$h_2 = h_{g_2} - (1 - x)h_{fg_2} = 2599.2 - 0.1501\,(2373.2)$$

$$= 2243.1 \text{ kJ/kg}$$

$$w_s = h_1 - h_2 = 795.8 \text{ kJ/kg}$$

But $w_a = 550$ kJ/kg, so

$$\eta_t = \frac{w_a}{w_s} = \frac{550}{795.8} = 0.691 \text{ or } 69.1\%$$

Pumps The purpose of a pump is to elevate the pressure of a liquid. Figures 7-9, 7-10, and 7-11 illustrate this open system. Again the potential-energy contribution is not considered and the pump is assumed to be adiabatic. An energy balance across the pump in the case of one inlet and one outlet yields

$$-\dot{W}_p = \dot{m}(h_2 - h_1) + \frac{\dot{m}}{2g_c}(V_2^2 - V_1^2) \qquad \text{(7-20)}$$

The kinetic energies of the fluid entering and leaving the pump are often essentially equal to each other. Thus the work is

$$-\dot{W}_p = \dot{m}(h_2 - h_1) \qquad \text{(7-21)}$$

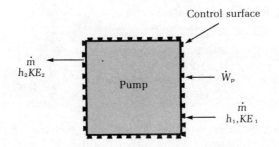

Figure 7-9. Schematic diagram of a pump (assumed adiabatic).

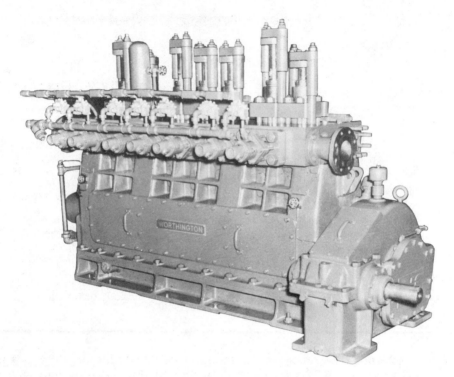

Figure 7-10. A 300-hp vertical multiple-plunger pump. (Reprinted by permission from Worthington Corporation)

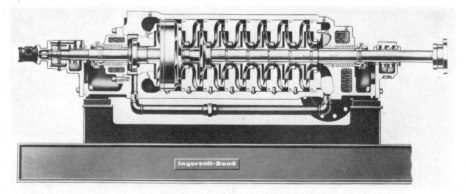

Figure 7-11. Sectional view of seven-stage, turbine-type, centrifugal pump for boiler feedwater. (Reprinted by permission from Ingersoll-Rand.)

Since, as is often the case, the exit temperature is unknown, determining h_2 may be a problem. Thus another approach is necessary. From Equation (7-21) we may deduce that

$$-\dot{W}_p = \dot{m} \int_1^2 dh$$

but for the reversible adiabatic case (remember the $T\,ds$ relations!)

$$dh = v\,dp$$

Thus

$$-\dot{W}_p = \dot{m} \int_1^2 v\,dp \qquad \text{(if reversible)} \qquad \textbf{(7-22)}$$

At low pressures and temperatures, most liquids are essentially incompressible and the specific volume is very nearly constant. As the pressure difference increases, the compressibility should be taken into account. For the incompressible case, the work becomes

$$-\dot{W}_p = \dot{m}_1 (p_2 - p_1)v \qquad \textbf{(7-23)}$$

Although it is desirable to be as accurate as possible, the magnitude and relative effect of the pump work may not demand great accuracy.

EXAMPLE 7-4

ONLY THEORETICAL !!

In a steam power plant, water enters a pump at 2 psia and 100 F and leaves the pump at 500 psia. Determine the work per lbm of water if it is assumed to be a reversible adiabatic process.

SOLUTION:

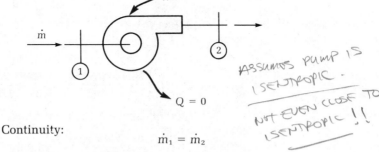

ASSUMES PUMP IS ISENTROPIC.

NOT EVEN CLOSE TO ISENTROPIC !!

Continuity:

$$\dot{m}_1 = \dot{m}_2$$

First law:

$$-\dot{W}_p = \dot{m}(h_2 - h_1) \qquad \Delta(\text{KE}) = 0$$

so

$$-w_p = \frac{\dot{W}_p}{\dot{m}} = h_2 - h_1 = v(p_2 - p_1) \qquad (v \simeq \text{constant})$$

From Appendix A-1-1:

$$v_f(100 \text{ F}) = 0.01613 \text{ ft}^3/\text{lbm} \approx v$$

$$-w_p = 0.0163 \text{ ft}^3/\text{lbm} (2 - 500) \frac{\text{lbf}}{\text{in.}^2} \frac{144 \text{ in.}^2}{\text{ft}^2} \frac{\text{Btu}}{778 \text{ ft-lbf}}$$

$$= 1.50 \text{ Btu/lbm}$$

The approximate temperature of the water leaving the pump is determined as follows:

$$h_2 = -w_p + h_1 = (1.50 + 69.36) \text{ Btu/lbm}$$

From the compressed liquid tables we get $T_e \approx 103$ F.

EXAMPLE 7-5

The discharge of a pumping system is 250 ft above the inlet. Water enters at a pressure of 20 psia and leaves at a pressure of 40 psia. The specific volume of the water is 0.016 ft³/lbm. Determine the minimum size of motor (hp) required to drive a pump handling 620 lbm/min.

SOLUTION: Continuity:

$$\dot{m}_1 = \dot{m}_2$$

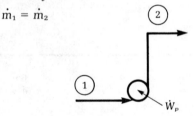

First law:

$$\dot{m}\left[\left(u_1 + p_1 v_1 + \frac{gZ_1}{g_c}\right) - \left(u_2 + p_2 v_2 + \frac{gZ_2}{g_c}\right)\right] - \dot{W} = 0$$

or

$$\dot{W}_p = (620)(60)\left[\frac{20(144)(0.016)}{778} + \frac{32.2(-250)}{32.2(778)} - \frac{40(144)(0.016)}{778}\right]$$

$$= 37,200 \,(0.0592 - 0.3213 - 0.1185)$$

$$= -14,160 \,\frac{\text{Btu}}{\text{hr}}\left(\frac{2545 \text{ hp}}{\text{Btu/hr}}\right) = -5.6 \text{ hp}$$

Alternative approach:

$$\dot{W}_{\min} = -\dot{m}[\textstyle\int v \, dp + \Delta PE]$$

$$= \frac{-37,200}{2545}\left[\frac{0.016 \,(40 - 20) \,(144)}{778} + \frac{32.2 \,(250)}{32.2 \,(778)}\right] = -5.6 \text{ hp}$$

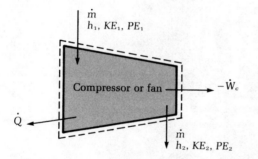

Figure 7-12. Schematic diagram of a compressor.

Compressors/Fans *Compressors are devices in which work is done on a gas to raise its pressure.* Thus the compressor does for a gas what a pump does for a liquid. A fan or blower is essentially a low-pressure compressor in which the pressure is produced for the purpose of moving the fluid. Figures 7-12 to 7-17 illustrate the physical situation. Similar to the conservation of mass and first-law analysis for a pump, the

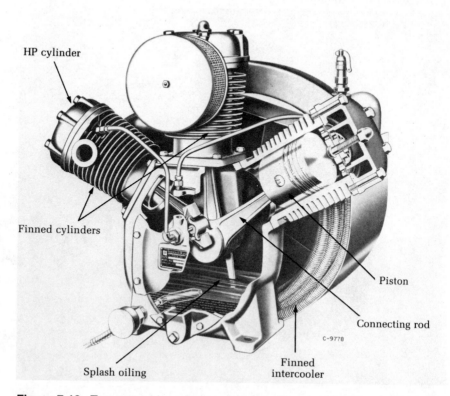

Figure 7-13. Two-stage air-cooled compressor. (Reprinted by permission from Ingersoll-Rand.)

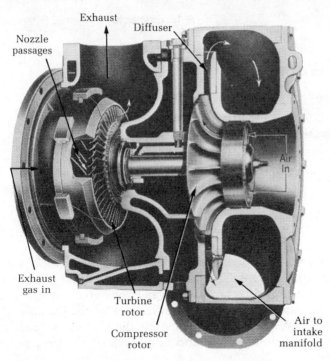

Figure 7-14. Centrifugal compressor for supercharger. (Reprinted by permission from Elliott Company.)

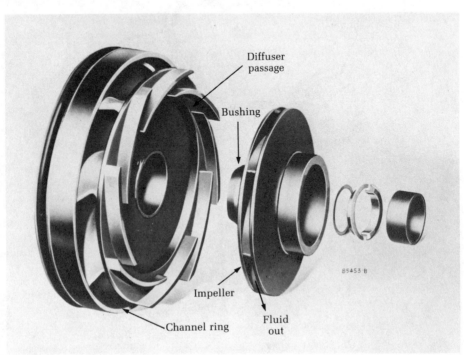

Figure 7-15. Diffuser and impeller. (Reprinted by permission from Ingersoll-Rand.)

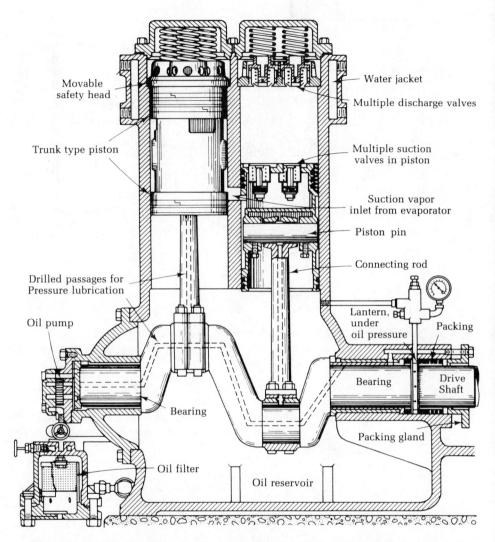

Figure 7-16. Cross-sectional view of a typical vertical, reciprocating, refrigerant compressor. (Fig. 12-14 (p. 417) in *Thermal Engineering* by C. C. Dillio and E. P. Nye (Intext). Copyright 1959 by Harper & Row, Publishers, Inc. Reprinted by permission of Harper & Row, Publishers, Inc.)

potential-energy contribution is neglected and the device is assumed to be adiabatic. Thus

$$-\dot{W}_c = \dot{m}\,(h_2 - h_1) + \frac{\dot{m}\,(V_2^2 - V_1^2)}{2g_c} \qquad (7\text{-}24)$$

EXAMPLE 7-6

Air is compressed from 14.7 psia and 79 F to 47.9 psia and 318.6 F. Determine the efficiency of the compressor and the work per lbm of air.

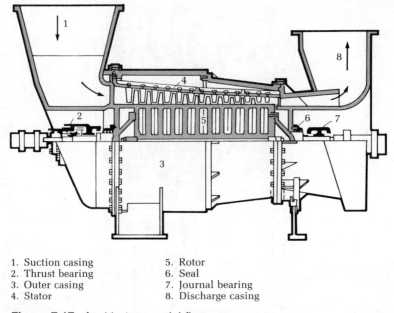

1. Suction casing
2. Thrust bearing
3. Outer casing
4. Stator
5. Rotor
6. Seal
7. Journal bearing
8. Discharge casing

Figure 7-17. An 11-stage, axial-flow, gas compressor.

SOLUTION: Assume that air is an ideal gas and assume an adiabatic process. If the information given describes a reversible adiabatic process, $\Delta s = 0$. Thus

$$\Delta s = c_p \ln\left(\frac{T_2}{T_1}\right) - R \ln\left(\frac{p_2}{p_1}\right) \quad \text{(constant } c_p \text{ and } c_v\text{)}$$

$$= 0.24 \text{ Btu/lbm-R } \ln\left(\frac{778.6}{539}\right) - 0.0684 \text{ Btu/lbm-R } \ln\left(\frac{47.9}{14.7}\right)$$

$$= 0.00747 \text{ Btu/lbm-R}$$

For a reversible adiabatic process,

$$T_2 = T_1 \left(\frac{p_2}{p_1}\right)^{(k-1)/k} = 539\left(\frac{47.9}{14.7}\right)^{0.4/1.4} = 755.4 \text{ R} = 295.4 \text{ F}$$

$$\eta_c = \frac{h_{2s} - h_1}{h_2 - h_1} = \frac{c_p(T_{2s} - T_1)}{c_p(T_2 - T_1)} = \frac{295.4 - 79}{318.6 - 79} = 0.90$$

$$-\dot{w}_c = \dot{m}(h_2 - h_1) \quad \text{if } \Delta(\text{KE}) = 0$$

$$-w_c = h_2 - h_1 = c_p(T_2 - T_1) = 0.24 \text{ Btu/lbm-R}(318.6 - 79)$$

$$w_c = -57.55 \text{ Btu/lbm}$$

HIGHER ACTUAL TEMP. THAN FOR REV. ADI, PROC

EXAMPLE 7-7

Air is compressed from 0.10 MPa and 25 C to 0.800 MPa. Calculate the work and the change in entropy if the compression takes place such that $pv^{1.25} = $ constant.

SOLUTION: For the polytropic process, assuming that air is an ideal gas,

$$T_2 = T_1\left(\frac{p_2}{p_1}\right)^{(1.25-1)/1.25}$$

$$= 298 \text{ K}\left(\frac{0.8}{0.1}\right)^{0.2} = 452 \text{ K}$$

$$w_s = -\int_1^2 v\,dp = \frac{-n}{n-1} R\,(T_2 - T_1)$$

WHY NOT
$w_s = c_p \Delta T$??
↓
BECAUSE T_2
IS THEORETICAL,
NOT ACTUAL.

$$= \frac{-1.25}{0.25}\left(0.287 \frac{\text{kJ}}{\text{kg K}}\right)(452 - 298)\text{ K}$$

$$= -221 \text{ kJ/kg}$$

THEORETICAL TEMP.!
NOT ACTUAL.

$$\Delta s = c_p \ln\left(\frac{T_2}{T_1}\right) - R \ln\left(\frac{p_2}{p_1}\right)$$

$$= 1 \ln\left(\frac{452}{298}\right) - 0.287 \ln\left(\frac{0.8}{0.1}\right)$$

$$= -0.1794 \text{ kJ/kg}$$

If the actual work had been measured to be

$$w_a = -297.2 \text{ kJ/kg}$$

the efficiency of this compressor would be

$$\eta_c = \frac{w_s}{w_a} = \frac{221}{297.2} = 0.743 \text{ or } 74.3\%$$

Heat Transfer Equipment (Heat Exchangers)

The most important steady-flow device of engineering interest is the heat exchanger. This device serves two useful purposes: to remove (or add) energy from a region of space and to change the thermodynamic state of a fluid. The automobile radiator is an example of heat removal by a heat exchanger. Modern gas turbines and electrical generators are frequently cooled internally because performance is greatly affected by the heat transfer process. In steam power plants, heat exchangers are used to remove heat from hot combustion gases and to increase the temperature and enthalpy of the steam in the power cycle. In the chemical industry, heat exchangers are used to attain certain thermodynamic states for chemical processes to be carried out.

One of the primary applications of heat exchangers is the exchange of energy between two moving fluids not in contact. The changes of kinetic and potential energy are usually negligible and no work is done. ✓ The pressure drop through a heat exchanger is small; that is, the constant-pressure assumption is valid here. A heat exchanger composed of two concentric pipes is illustrated in Figure 7-18a. Fluid A flows in the inner pipe and a second fluid, B, flows in the annular space between the pipes. Now consider a control surface placed around the entire piece of equipment (see the dashed line in Figure 7-18) and apply the first law for open systems. There is no heat transfer external to the device and no shaft work exists. Moreover, the kinetic and potential-energy changes of the fluid streams are negligible. In terms of the notation shown in Figure 7-18, the first law reduces to

$$\dot{m}_A(h_{A1} - h_{A2}) = \dot{m}_B(h_{B2} - h_{B1}) \tag{7-25}$$

Now place the boundaries around either of the two fluids. In this case not only is there a change in the enthalpy of the fluid but a heat transfer term also appears. Thus, if the heat transfer is from fluid A to fluid B, then

$$\dot{Q} = \dot{m}_B(h_{B2} - h_{B1}) \tag{7-26}$$

$$-\dot{Q} = \dot{m}_A(h_{A2} - h_{A1}) \tag{7-27}$$

The heat transfer rates are identical. Figure 7-18b shows the system boundaries in this latter case for fluid A. With care, these equations may be applied to boilers, evaporators, and condensers.

Boilers/Vapor Generators Figures 7-19, 7-20, and 7-21 illustrate a **vapor generator.** Vapor generators, often referred to as boilers, transform liquids to vapors. In a boiler operating under steady conditions, liquid is pumped into the boiler at the same mass rate as vapor leaves. Heat is supplied at a steady rate. Since the boiler does no work, and since ΔPE

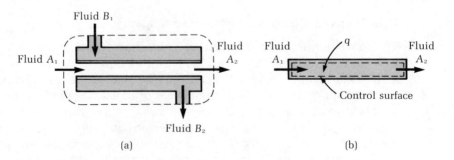

(a)   (b)

Figure 7-18. Two different control surfaces for a heat exchanger.

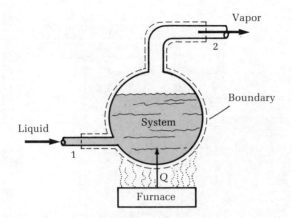

Figure 7-19. Boiler schematic.

and ΔKE from feedwater inlet 1 to steam outlet 2 are small compared to $h_1 - h_2$, the first-law equation becomes

$$(u_2 + p_2 v_2) - (u_1 + p_1 v_1) - q = 0 \qquad \textbf{(7-28)}$$

or

$$q = h_2 - h_1 \qquad \textbf{(7-29)}$$

Condensers Notice again that conservation of mass is used. In principle, a **condenser** is a boiler in reverse. In a boiler, heat is supplied to convert the liquid into vapor; in a condenser, heat is removed in order to condense the vapor into a liquid. This system, illustrated in Figure 7-22, is an open system. As with generators, if the condenser is in steady state then the amount of liquid, called a **condensate,** leaving the condenser must equal the amount of vapor entering the condenser. Thus heat is rejected as a vapor condenses. For purposes of discussion, let us assume that the working fluid is water (steam). The steam passes over one side of a heat exchanger while cooling water passes through the other. As is usually the case, the potential energy is neglected since the effect of elevation is small. The condenser is overall adiabatic, so there is no heat transfer to the surroundings. (Ideally, all heat transfer between fluids is internal.) Moreover, the difference in the velocities of the cooling water in and out is very small, so the change in kinetic energy of the cooling water is essentially zero. The first law now yields (in a two-inlet/two-outlet system)

$$\dot{m}_s (h_{s1} - h_{s2}) + \dot{m}_s \left(\frac{V_{s1}^2}{2g_c} - \frac{V_{s2}^2}{2g_c} \right) = \dot{m}_w (h_{w2} - h_{w1}) \qquad \textbf{(7-30)}$$

where the mass rates of water and steam are equal (in and out, respectively) but not to each other. Since the inlet steam velocity to the condenser is high, the entering kinetic energy must be included in the

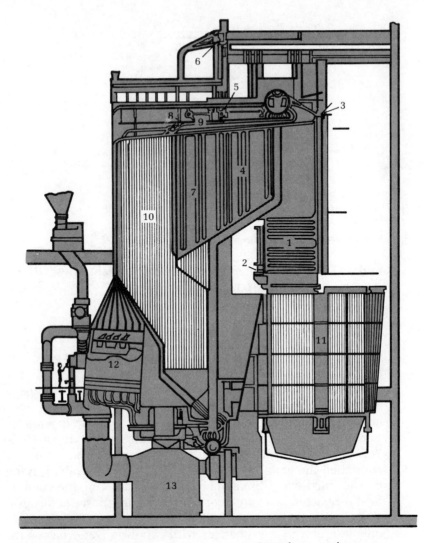

1. Economizer
2. Economizer inlet header
3. Economizer outlet header
4. Primary superheater
5. Primary superheater outlet
6. Attemperator
7. Secondary superheater
8. Secondary superheater inlet
9. Secondary superheater outlet
10. Double row screen tubes
11. Air heater
12. Cyclone furnace
13. Slag disintegrating tank

Figure 7-20. Large utility steam generator with cyclone-furnace firing. (Courtesy of Babcock & Wilcox, a McDermott Company)

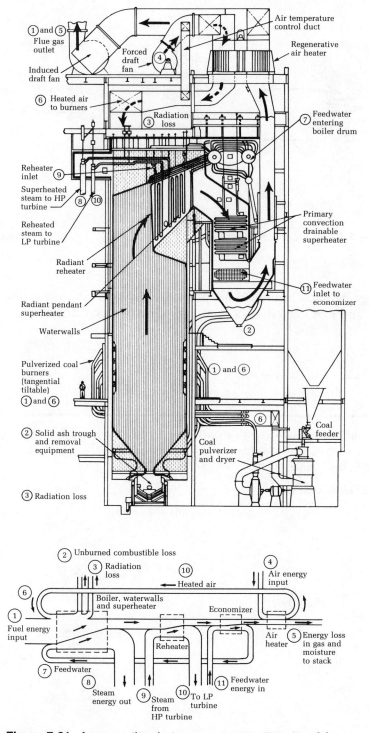

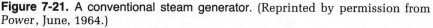

Figure 7-21. A conventional steam generator. (Reprinted by permission from *Power*, June, 1964.)

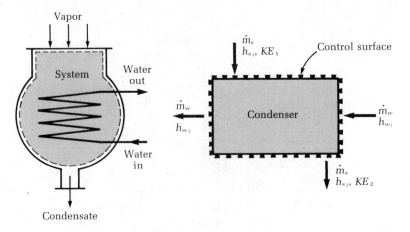

Figure 7-22. Condenser schematics.

energy balance. Applications of continuity will give estimates of the inlet and exit velocities of both the steam and condensed water. Thus

$$V_2 = V_1 \left(\frac{A_1}{A_2}\right)\left(\frac{v_2}{v_1}\right)$$ (7-31)

EXAMPLE 7-8

A condenser receives steam with the following characteristics: 10 psia, quality = 95%, $V_1 = 400$ ft/sec. The condensate exists as a saturated liquid at 10 psia. Determine the heat lost per lbm of steam.

SOLUTION: Continuity:

$$\dot{m}_s(\text{in}) = \dot{m}_s(\text{out})$$

or

$$V_2 = V_1 \left(\frac{A_1}{A_2}\right)\left(\frac{v_2}{v_1}\right)$$

First law:

$$\dot{m}_s\left(h_1 - h_2 + \frac{V_1^2 - V_2^2}{2g_c}\right) = \dot{m}_w(h_{w2} - h_{w1}) = \dot{Q}_{\text{lost}}$$

or

$$q = \frac{\dot{Q}}{\dot{m}_w} = h_1 - h_2 + \frac{V_1^2 - V_2^2}{2g_c}$$

From Appendix A-1-2:

$$h_1 = xh_g + (1 - x)h_f = [0.95\,(1143.3) + 0.05\,(161.26)] \text{ Btu/lbm}$$

$$= 1094.2 \text{ Btu/lbm}$$

$$h_2 = 161.26 \text{ Btu/lbm}$$

$$v_1 = [0.95\,(38.42) + 0.05\,(0.0166)] \text{ ft}^3\text{/lbm} = 36.5 \text{ ft}^3\text{/lbm}$$

$$v_2 = 0.01659 \text{ ft}^3\text{/lbm}$$

$$V_2 = 400 \text{ ft/sec} \left(\frac{A_1}{A_2}\right)\left(\frac{0.01659}{36.5}\right) = 0.182 \text{ ft/sec} \qquad (A_1 = A_2)$$

$$q = (1094.2 - 161.26) \text{ Btu/lbm}$$

$$+ \frac{(400 \text{ ft/sec})^2}{64.34 \text{ lbm-ft/lbf-sec}^2} \frac{\text{Btu}}{778 \text{ ft-lbf}}$$

$$= 936 \text{ Btu/lbm}$$

Combustors A **combustor** is a chamber held at a constant but high pressure that allows the burning of fuel in air. Figure 7-23 illustrates the situation. Applying continuity yields

$$\dot{m}_a + \dot{m}_f = \dot{m}_2 \tag{7-32}$$

The air/fuel ratio (m_a/m_f) is generally about $30:1$, but it can vary greatly because of the fuel, air conditions, or operating requirements.

The first law written for the combustor is

$$\dot{m}_a h_a + \dot{m}_f h_f - \dot{m}_2 h_2 = -\dot{Q} \tag{7-33}$$

where we assume that no kinetic or potential-energy changes occur. The work or power is obviously zero. If we further consider that m_f is small compared to m_a, Equation (7-33) can be written

$$\dot{m}_a (h_a - h_2) = -\dot{Q} \tag{7-34}$$

Since $\dot{Q}$ is supplied from the burning of fuel, the following equation is also true for complete combustion (see Chapter 11 for details):

$$\dot{Q} = \dot{m}_f \times HV_f \tag{7-35}$$

where HV_f is the heating value of the fuel per unit mass of fluid.

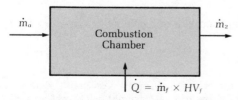

Figure 7-23. Schematic of a combustor.

EXAMPLE 7-9

Combustion of gasoline in air takes place at a constant pressure of 60 psia in an automotive gas turbine combustion chamber. The initial temperature is 340 F and the final temperature is 1500 F. The heating value of the fuel is 19,000 Btu/lbm. What is the fuel/air ratio?

SOLUTION: Neglect mass of fuel and consider only the heating effect from burning the fuel; thus

$$m_a (h_1 - h_2)_a + Q_{(\text{from fuel})} = 0$$

$$m_a c_p (T_1 - T_2)_a + m_f \times HV_f = 0$$

$$\frac{m_f}{m_a} = \frac{c_p (T_1 - T_2)}{HV_f} = \frac{0.24 (1500 - 340)}{19,000} = 0.0146$$

Nozzles and Diffusers

A single stream of fluid flowing in an enclosed duct—internal flow—can be accelerated or decelerated by an appropriate variation of the flow area. *A device that increases the velocity (and hence the kinetic energy) of a fluid at the expense of the internal energy or enthalpy and with a pressure drop in the direction of flow is called a* **nozzle.** *A* **diffuser** *is a device for increasing the pressure of a flow stream at the expense of a decrease in velocity.* These definitions apply for both subsonic and supersonic flow.* Figure 7-24 shows the general shapes of a nozzle and a diffuser under the conditions of subsonic and supersonic flow. Note that a nozzle is a converging passage for subsonic flow and

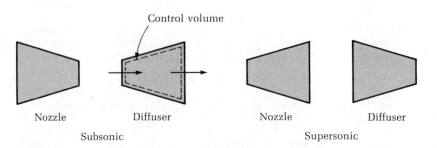

Control volume

| Nozzle | Diffuser | Nozzle | Diffuser |

Subsonic   Supersonic

Figure 7-24. General shapes of nozzles and diffuser for subsonic and supersonic flow.

---

* Subsonic flow means that the velocity of flow is less than the local velocity of sound; supersonic flow means that the velocity of flow is greater than the local velocity of sound.

a diverging passage for supersonic flow; the opposite conditions hold for a diffuser. Consequently, a converging-diverging nozzle must be used to accelerate a fluid from subsonic to supersonic velocities (Figure 7-25). Figure 7-26 illustrates an actual system that incorporates converging and diverging nozzles.

Fluids traveling through nozzles at subsonic velocities behave as our intuition would dictate. Gases traveling at sonic or supersonic speeds present complications, however. Shock waves and accelerations in diffuser sections are possible. Since both nozzle and diffusers are merely ducts, it is apparent that no shaft work is involved and the change in potential energy, if any, is negligible. The average velocity of flow through a nozzle is high; hence the fluid spends only a short time in the nozzle. For this reason, it may be assumed that there is insufficient time for heat to flow into or out of the fluid during its passage through the nozzle. Applying continuity to this one-inlet/one-outlet configuration yields in either case.

$$\dot{m}_1 = \dot{m}_2 \qquad (7\text{-}36)$$

The first law for this steady-flow situation is

$$\frac{1}{2g_c}(V_2^2 - V_1^2) + h_2 - h_1 = 0 \qquad (7\text{-}37)$$

For a liquid flowing through a nozzle, the change in specific volume Δv is negligible. If the change in internal energy Δu is also negligible, then

$$p_2v_1 - p_1v_1 + \frac{V_2^2}{2g_c} - \frac{V_1^2}{2g_c} = 0$$

or

$$\frac{V_2^2}{2g_c} - \frac{V_1^2}{2g_c} = (p_1 - p_2)v \qquad (7\text{-}38)$$

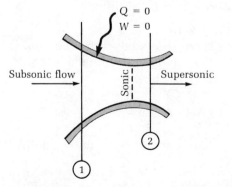

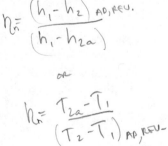

Figure 7-25. Expansion in a nozzle to high velocity.

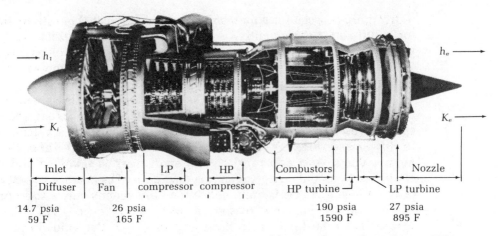

$\longrightarrow h_1$ .... $h_e \longrightarrow$

$\longrightarrow K_i$ .... $K_e \longrightarrow$

| Inlet | | LP | HP | Combustors | | Nozzle |
| Diffuser | Fan | compressor | compressor | HP turbine | LP turbine |

14.7 psia .... 26 psia .... 190 psia .... 27 psia
59 F .... 165 F .... 1590 F ...... 895 F

Figure 7-26. Turbofan jet engine. (Reprinted by permission from Pratt & Whitney Aircraft.)

If the velocity at the inlet (V_1) is essentially zero (that is, stagnation), then

$$\frac{V_2^2}{2g_c} = (p_1 - p_2)v \qquad \text{(ft-lb/lb)}$$

or

$$V_2 = \sqrt{2g_c(p_1 - p_2)v} \qquad (7\text{-}39)$$

Notice that when steam, air, or any other compressible fluid flows through a nozzle, the changes in the specific volume Δv and internal energy Δu are not negligible. The jet velocity of a compressible fluid such as steam or air is a function of the enthalpies (h_1 and h_2) of the fluid entering and leaving the nozzle. In this case Equation (7-37) may be reduced to

$$V_2 = 223.9\sqrt{h_1 - h_2} \qquad \text{(ft/sec)} \qquad (7\text{-}40)$$

The velocity of the inlet (V_1) is zero.

Frequently a stagnation state is defined as the inlet nozzle condition. **Stagnation** is an equivalent total condition such that V_1^* is zero. The resulting stagnation enthalpy h^* is

$$h_1^* = h_1 + \frac{1}{2g_c}V_1^2 \qquad (7\text{-}41)$$

and the stagnation temperature T_1^* is, for ideal gases having constant specific heats,

$$T_1^* = T_1 + \frac{1}{2c_p g_c}V_1^2$$

Thus Equation (7-37) may be rewritten as

$$V_2 = \sqrt{2g_c c_p (T_1^* - T_2)} \tag{7-42}$$

EXAMPLE 7-10

Suppose that 4 lbm/sec of air enters a diffuser at 1400 ft/sec, 600 R, and 20 psia. Assuming stagnation conditions at the exit and a nozzle efficiency of 88%, determine the exit pressure, area of the entrance, and Mach number at the entrance.

SOLUTION: From continuity:

$$\dot{m}_1 = \rho_1 V_1 A_1$$

or

$$A_1 = \frac{\dot{m} v_1}{V_1} = \frac{\dot{m} R T_1}{V_1 p_1} \qquad \text{(air is an ideal gas)}$$

$$= \frac{4 \text{ lbm/sec } 53.3 \text{ ft-lbf/lbm-R } 600 \text{ R}}{1400 \text{ ft/sec } 20 \text{ lbf/in.}^2} = 4.569 \text{ in.}^2$$

Since we have an irreversible adiabatic expansion (88% efficient),

$$\eta_D = \frac{c_p (T_2 - T_1)}{c_p (T_{2_a} - T_1)}$$

But T_{2_a} (actual) must be computed by

$$T_{2_a} = T_1 + \frac{1}{2 c_p g_c} V_1^2$$

$$= 600 \text{ R}$$

$$+ \frac{(1400 \text{ ft/sec})^2}{2 (0.24 \text{ Btu/lbm-R}) (32.2 \text{ ft-lbm/lbf-sec}^2) 778 \text{ ft-lbf/Btu}}$$

$$= 762.9 \text{ R}$$

$$0.88 = \frac{T_2 - 600}{762.98 - 600}$$

$$T_2 = 743.4 \text{ R}$$

The exit pressure is computed from the reversible adiabatic temperature:

$$p_2 = p_1 \left(\frac{T_2}{T_1}\right)^{k/(k-1)} = 20 \text{ psia} \left(\frac{743.4}{600}\right)^{1.4/0.4}$$

$$= 42.3 \text{ psia}$$

The Mach number is defined as

$$M = \frac{\text{velocity of interest}}{\text{local velocity of sound}}$$

$$\doteq \frac{V_1^{\cdot}}{\sqrt{g_c kRT}}$$

$$= \frac{1400 \text{ ft/sec}}{\sqrt{32.2 \text{ ft-lbm/lbf-sec}^2 \; 1.4(53.3 \text{ ft-lbf/lbm-R} \; 600 \text{ R}}}$$

$$= 1.16 \quad \text{(supersonic)}$$

Many flow meters use the nozzle principle as the basis for their operation and function by means of a direct or indirect measurement of velocity. In the operation of venturi and orifice-type flow meters, the ΔKE is a principal effect even if its magnitude is small.

Throttling Devices (Valves, Orifices, Capillary Tubes)

*A **throttling** process is one in which the fluid is made to flow through a restriction—for example, a partially opened valve or orifice—causing a considerable drop in the pressure of the fluid.* The main effect is a significant pressure drop without any work interactions or changes in kinetic or potential energy. Flow through a restriction such as a valve (a porous plug) fulfills the necessary conditions. (See Figure 7-27.)

Although the velocity is quite high in the region of the restriction, measurements indicate that the changes in kinetic energy across the restriction are very small. Since the control volume is rigid and no rotating shafts are present, no work is done. In most steady-flow applications, the throttling device is insulated or the heat transfer is insignificant. Thus the enthalpy change is zero, or

$$h_1 = h_2 \tag{7-43}$$

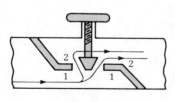

(a) A throttling valve

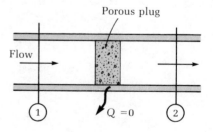

(b) A porous plug

Figure 7-27. Throttling process.

The valves in water faucets in your home are examples of throttling devices. These devices are also common in most home refrigeration units.

EXAMPLE 7-11

Refrigerant-12 is throttled in an expansion valve from saturated-liquid conditions at 150 F to a temperature of 40 F. Calculate the quality of the R-12 vapor after the throttling process.

SOLUTION: $h_1 = h_2$

$h_1 = 43.850$ Btu/lbm   (saturated liquid at 150 F)

$h_2 = h_1 = h_{f_2} + x_2 h_{fg_2}$

or

$43.850 = 17.273 + x_2(64.163)$

and

$x_2 = 0.414$

VALUE IS CONSTANT h.

$h_1 = h_2$

The process is widely used in air conditioning and refrigeration systems.

EXAMPLE 7-12

Steam enters a nozzle at 0.8 MPa and 200 C ($V \approx 0$) and exits at a pressure of 0.2 MPa. What is the exit velocity of the steam if the nozzle efficiency is 95%?

SOLUTION: From the definition, the nozzle efficiency is

$$\eta_n = \frac{V_a^2/2g_c}{V_s^2/2g_c}$$

If we assume reversible adiabatic operation, then

$$\frac{V_s^2}{2g_c} = h_1 - h_2$$

and

$$s_1 = s_2$$

From Appendix A-1-7:

$s_1 = 6.8148$ kJ/(kg·K) $h_1 = 2838.6$ kJ/kg

$s_1 = s_2 = sg_2 + x_2 s_{fg_2}$

$6.8148 = 1.5301 + x\,5.5970$

$x_2 = 0.9444$

and

$$h_{2_s} = h_{f_2} + xh_{fg_2}$$
$$= 504.7 + 0.9444\,(2201.9) = 2584.1 \text{ kJ/kg}$$

Thus

$$\frac{V_s^2}{2g_c} = 2838.6 - 2584.1 = 254.5 \text{ kJ/kg}$$

For the real process:

$$\frac{V_a^2}{2g_c} = \eta_n \left(\frac{V_s^2}{2g_c}\right) = 0.95\,(254.5) = 241.8 \text{ kJ/kg}$$

$$V_a = 22 \text{ m/s}$$

Summary of Component Operation

Table 7-1 summarizes the ideal performance of some common components of engineering systems discussed in the previous sections.

Table 7-1 Steady-Flow Energy Analyses for Ideal Performance of Common Components

System	Energy Balance	Model Process
Heater	$\dot{Q} = \dot{m}\,(h_2 - h_1)$	
Valve	$h_2 = h_1$	
Nozzle	$-\dfrac{V_2^2}{2g_c} = h_2 - h_1$	

$$\frac{V_2^2 - V_1^2}{2g_c} = h_1 - h_2$$

Table 7-1 (*continued*)

System	Energy Balance	Model Process

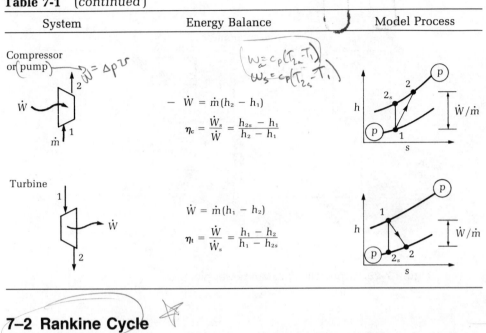

Handwritten annotations near compressor/pump:
$\dot{W} = \Delta p \bar{v}$

$$\left(\begin{array}{l} W = c_p(T_2 - T_1) \\ W_s = c_p(T_{2s} - T_1) \end{array}\right)$$

Compressor or pump:

$$-\dot{W} = \dot{m}(h_2 - h_1)$$

$$\eta_c = \frac{\dot{W}_s}{\dot{W}} = \frac{h_{2s} - h_1}{h_2 - h_1}$$

Turbine:

$$\dot{W} = \dot{m}(h_1 - h_2)$$

$$\eta_t = \frac{\dot{W}}{\dot{W}_s} = \frac{h_1 - h_2}{h_1 - h_{2s}}$$

7–2 Rankine Cycle

Although the basic components of power plants are by now familiar to you, the seeming simplicity of these elements can mask the complexity of the actual systems power generation. Every power plant is composed of many interacting systems. In a steam power plant, there are systems for handling fuel and waste, for transporting air, steam, and water, and for distributing electrical power.

Knowledge of power plant components is important to the engineer, but this is not the whole story. The engineer must be able to see beyond the construction drawing to the equipment itself, to understand the relationship of the individual parts to the power plant as a whole, and to relate engineering knowledge to the operating equipment.

The steam power plant operates on the **Rankine cycle** (Figure 7-28). Steam is produced in the steam generator (boiler) at high pressure and temperature. Figures 7-29 and 7-30 depict the boiler or steam-generating unit (SGU). In the turbine the steam does work as it expands to a very low pressure. The steam leaving the turbine enters the condenser where it is condensed. The pump then draws the condensate from the condenser and builds up sufficient pressure to force it into the steam generator. Because the pump feeds water into the steam generator, this water is known as **feedwater.** In the steam generator the water is turned into steam and then superheated.

The steam turbine is the heart of the steam power plant. For typical fossil fuel power plants, the turbine receives steam at pressures

Handwritten margin note (left):
ISENTROPIC
$s_1 = s_2$

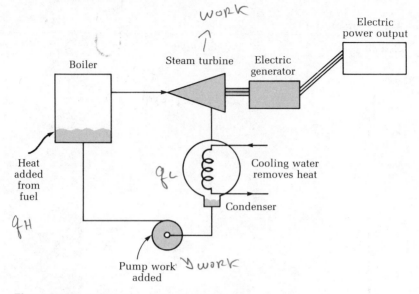

Figure 7-28. Schematic of steam power plant.

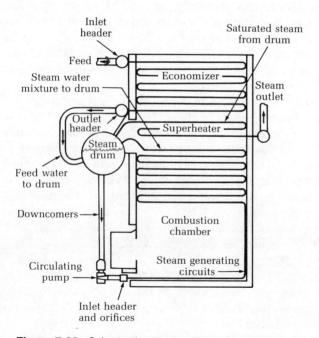

Figure 7-29. Schematic arrangement of controlled circulation boiler. (Courtesy of Babcock & Wilcox, a McDermott Company.)

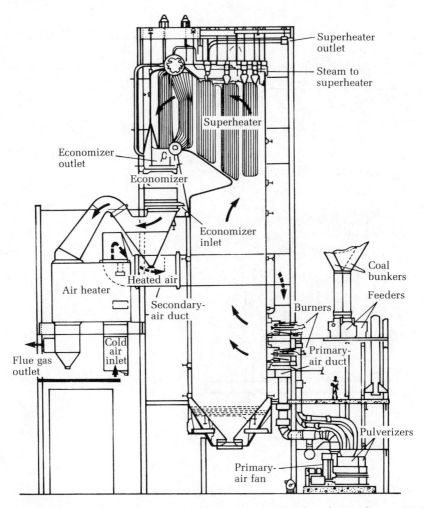

Figure 7-30. A large steam generator. (Courtesy of Babcock & Wilcox, a McDermott Company.)

normally ranging from 2400 to 3500 psia and at temperatures around 1000 F. Upon entering the turbine the steam, with velocities of perhaps 1500 to 2500 ft/sec, expands through 20 to 30 rows of nozzles, blades, and wheels, depending on the steam conditions (see Figures 7-31 to 7-35).

Figure 7-36 illustrates the elements of a simple condenser. A large condenser may contain as many as 100,000 tubes that are 1 in. in diameter and perhaps 40 ft long. Steam from the turbine surrounds the tubes and it condenses. The condensate is relatively free of minerals and dissolved gases and hence is pumped back into the steam generator where it is turned into steam.

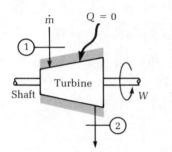

Figure 7-31. Simple turbine schematic.

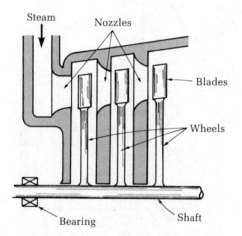

Figure 7-32. Elements of a steam turbine.

The Cycle

The Rankine cycle or vapor turbine cycle overcomes many of the operational difficulties encountered with the Carnot cycle when the working fluid is a vapor. In this cycle, the heating and cooling processes occur at constant pressure. Figure 7-37a and b illustrates the Rankine cycle on a (p, v) and a (T, s) diagram; Figure 7-37c shows an energy flow diagram for the cycle.

Starting the cycle from state 1, the fluid enters the vapor generator as a subcooled liquid at pressure p_2. The energy supplied in the vapor generator raises the state of the liquid from that of a subcooled liquid to that of a saturated liquid and, further, to that of a saturated vapor at state 2. The vapor leaves the vapor generator at state 2 and enters a turbine, where it expands isentropically to state 3. It enters the condenser at this point and is condensed at constant pressure from state 3 to state 4. At state 4, the liquid is a saturated liquid at the pressure of the condenser. The liquid cannot enter the vapor generator, which is at a higher pressure, until its pressure is raised by a pump to that of the

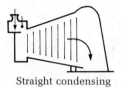
Straight condensing

Condensing bleeder

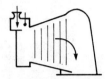
Low-pressure condensing

Single-extraction condensing

Double-extraction condensing

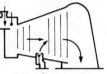
Mixed-pressure

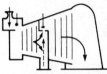
Extraction-induction

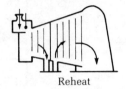
Reheat

Noncondensing,
or superposed

Noncondensing bleeder,
or superposed bleeder

Single-extraction
noncondensing

Double-extraction
noncondensing

Figure 7-33. Basic turbine types for a variety of power and process steam demands.

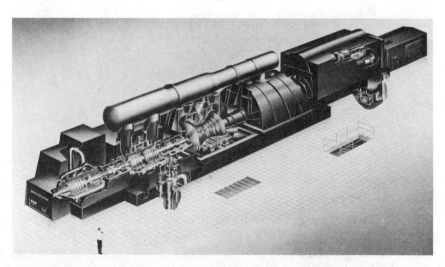

Figure 7-34. A large steam turbine. (Reprinted by permission from General Electric Company.)

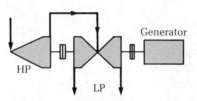

Tandem compound, double flow

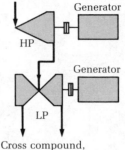

Cross compound,
double flow

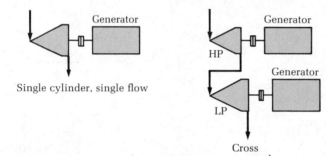

Single cylinder, single flow

Cross
compound

Figure 7-35. Plan-view arrangements of turbine units.

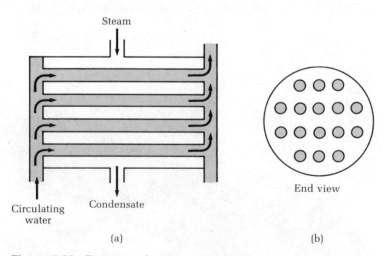

Steam

Circulating
water

Condensate

(a)

End view

(b)

Figure 7-36. Elements of a simple condenser.

vapor generator. The liquid is now subcooled at state 1 and the cycle is complete.

If steam entered the turbine as a saturated vapor, its moisture content would be too high as it passed through the turbine, resulting in

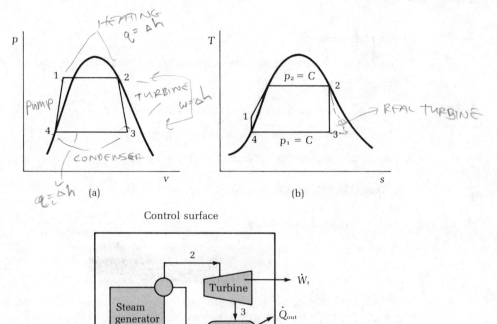

(a)

(b)

Control surface

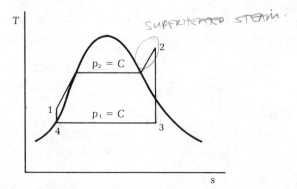

(c)

Figure 7-37. The Rankine cycle: (*a*) (*p, v*) diagram; (*b*) (*T, s*) diagram; (*c*) schematic diagram.

erosion of the blades. The Rankine cycle is characterized by constant-pressure heating, however, so the vapor may be easily superheated to a much higher temperature. Figure 7-38 illustrates the superheating shift that prevents a high moisture content when steam enters the turbine.

Figure 7-38. Rankine cycle with superheating.

Thermal Efficiency

In the steam power plant, the characteristics of a heat engine are evident:

1. Heat addition at high temperature (boiler)
2. Heat rejection at low temperature (condenser)
3. Net work output (work out of turbine minus work into pump)

The **thermal efficiency** of a heat engine is defined as

$$\text{Thermal efficiency} = \frac{\text{net work output}}{\text{heat input at high temperature}}$$

$$\eta = \frac{\dot{W}_{net}}{\dot{Q}_{in}} = \frac{\dot{W}_t - \dot{W}_p}{\dot{m}(h_2 - h_1)} \, {}^* \tag{7-44}$$

This efficiency has definite economic significance because the heat input at the high temperature represents the energy that must be purchased (fuel, uranium, and so forth) and the net work output represents what we get for the purchase. Large steam power plants can achieve efficiencies on the order of 40%. Keep in mind that this efficiency is for the cycle (a series of processes). Moreover, all the components have efficiencies as well.

Table 7-2 shows why it is desirable to have as low a turbine exhaust pressure as possible. The theoretical work per pound mass of steam for an exhaust pressure of 0.25 psia is about 40% greater than would be obtained if the steam were exhausted to the atmosphere. In terms of dollars, this means that for a unit consuming 250 tons of coal per hour, there could be a savings of approximately 100 tons of coal per hour. If the cost of coal is $30/ton, this would amount to a savings of more than $25 million per year.

Table 7-2 Theoretical Turbine Work

Exhaust Pressure, psia	Steam Temperature, F	Work, Btu/lbm of Steam
0.25	59.3	664.7
0.50	79.6	635.6
1.00	101.7	604.7
2.00	126.0	571.8
14.696	212.0	463.2

* Absolute values are to be used when W is letter subscripted

Improvements to the Cycle

Many efforts have been made to increase the efficiency of the Rankine cycle. From our previous study of the Carnot cycle, it is obvious that by increasing pressure p_1 (that is, T_1 and T_2) or lowering pressure p_4 (that is, T_3 and T_4), the efficiency could be increased. The processes used to increase the efficiency of the cycle are operations called **reheating** and **regenerating** (Chapter 8). Today's central station power plant may make use of several of these stages, as illustrated in Figure 7-39.

EXAMPLE 7-13

Determine the efficiency of a Rankine cycle utilizing steam as the working fluid. The condenser pressure is 3 lbf/in.^{2}. The steam generator conditions are 1000 psia and 1000 F.

SOLUTION: To determine the efficiency, we must use

$$\eta = \frac{\dot{W}_t - \dot{W}_p}{\dot{m}\,(h_2 - h_1)}$$

→ NET WORK
→ HEAT INPUT

From the problem setup, it must be assumed that there are no kinetic-energy contributions. Find W_p first:

$$-w_p = v\,\Delta p$$

SAT.
⇒ LIQ. WATER AT INLET TEMP.

$$= 0.01630 \text{ ft}^3/\text{lbm } (1000 - 3)\,\frac{\text{lbf}}{\text{in.}^2}\,\frac{144 \text{ in.}^2}{\text{ft}^2}\,\frac{\text{Btu}}{778 \text{ ft-lbf}}$$

$$= 3.00 \text{ Btu/lbm}$$

Applying the first law to the turbine yields

$$w_t = h_2 - h_3 + (V_2^2 - V_3^2)/2g_c$$

and

$$h_2 = 1505.4 \text{ Btu/lbm} \qquad \text{(from the superheated tables)}$$

To obtain a value for h_3, we must also assume reversible adiabatic expansion in the turbine ($s_2 = s_3$). Thus

$$s_2 = 1.6530 \text{ Btu/lbm-R} = s_3$$

$$1.6530 \text{ Btu/lbm-R} = x_3 s_{g_3} + (1 - x_3)s_{f_3} \quad \} \; 3 \text{ PSIA}$$

$$= 1.8862\,x_3 + (1 - x)\,0.2008$$

$$x_3 = 0.8613$$

$$h_3 = x_3 h_{g_3} + (1 - x_3)h_{f_3}$$

$$= [0.8613\,(1122.6) + (0.1387)\,109.37] \text{ Btu/lbm}$$

$$= 982.06 \text{ Btu/lbm}$$

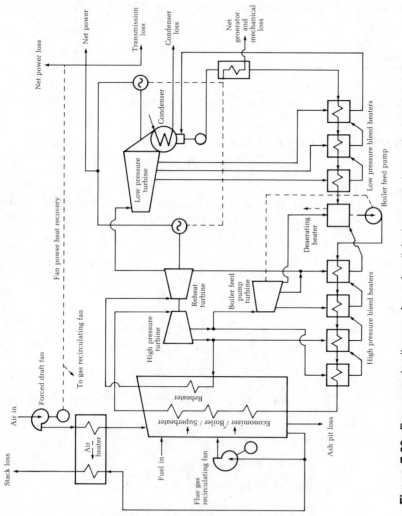

Figure 7-39. Power cycle diagram for a fossil fuel power plant—single reheat, eight-stage regenerative feed heating (3515 psia, 1000 F/1000 F steam). (Courtesy of Babcock & Wilcox, a McDermott Company.)

$$w_t = (1505.4 - 982.1) \text{ Btu/lbm} = 523.3 \text{ Btu/lbm}$$

$$q_{in} = h_2 - h_1 = h_2 - (h_4 + w_p) = (1505.4 - 109.37 + 3) \text{ Btu/lbm}$$

$$= 1399 \text{ Btu/lbm}$$

Thus

$$\eta = \frac{523.3 - 3}{1399} = 0.3719$$

EXAMPLE 7-14

Compare the operation of two steam power plants. In both plants the boiler pressure is 4.2 MPa and the condenser pressure is 0.0035 MPa. Plant A operates on a Carnot cycle using wet steam; plant B operates on a Rankine cycle with saturated steam at the turbine inlet.

SOLUTION: For plant A, the given pressures dictate the temperatures. From Appendix A-1-6 the temperatures are easily determined (see the diagram):

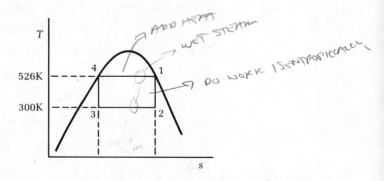

The η_{Carnot} is

$$1 - \frac{T_2}{T_1} = \frac{526 - 300}{526} = 0.430 \text{ or } 43\%$$

The heat q_H is

$$h_1 - h_4 = h_{fg} \quad (4.2 \text{ MPa})$$

$$= 1698 \text{ kJ/kg}$$

Thus the work done is

$$w = \eta \, q_H = 734 \text{ kJ/kg}$$

For plant B, the cycle must be revised:

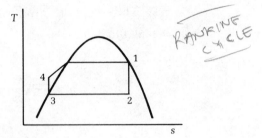

From Appendix A-1-6:

$h_1 = 2800$ kJ/kg $h_3 = 112$ kJ/kg $s_1 = 6.049$ kJ/(kg $\cdot$ K)

We can determine h_2 by noting that process 1–2 is reversible adiabatic. Thus

$$s_1 = s_2 = s_{f_2} + x s_{fg_2} = 0.391 + x_2 8.13$$

or

$$x_2 = 0.696$$

And

$$h_2 = h_{f_2} + x_2 h_{fg_2}$$
$$= 112 + 0.696\,(2438) = 1808 \text{ kJ/kg}$$

The pump work follows directly from the preceding discussion:

$$-w_p = v\,(p_4 - p_3) = 0.001\,(4.2 - 0.0035)$$
$$= 4.2 \text{ kJ/kg}$$

For the turbine work,

$$w_t = h_1 - h_2 = 2800 - 1808 = 992 \text{ kJ/kg}$$

Thus

$$\eta = \frac{w_t - w_p}{h_1 - h_4} = \frac{w_t - w_p}{(h_1 - h_3) - (h_4 - h_3)}$$

$$= \frac{w_t - w_p}{(h_1 - h_3) - w_p} = \frac{992 - 4.2}{(2800 - 112) - 4.2}$$

$$= 0.368 \text{ or } 36.8\%$$

The net work is therefore

$$(992 - 4.2) \text{ kJ/kg} = 987.8 \text{ kJ/kg}$$

As a matter of interest, if the turbine efficiency were 80%, the work of the turbine would be reduced to 794 kJ/kg and the plant efficiency would be reduced to 29%.

7–3 Brayton Cycle

The Cycle

The simple gas turbine is modeled by the Brayton (Joule) cycle. The open-cycle version of this turbine (new air continuously employed) uses a combustion process to add heat to the air as it passes through the system. The closed-cycle version (same air recycled) resorts to a simple heat transfer process to accomplish the same end. In the open cycle, the more common mode, atmospheric air is drawn into the compressor, heat is added, and the fluid expands through the turbine and exhausts

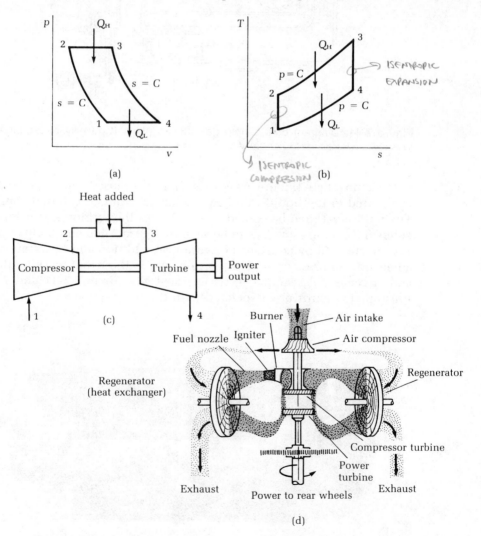

Figure 7-40. The Brayton cycle; (*a*) (*p, v*) diagram; (*b*) (*T, s*) diagram; (*c*) schematic diagram of an open system; (*d*) an automotive gas turbine schematic.

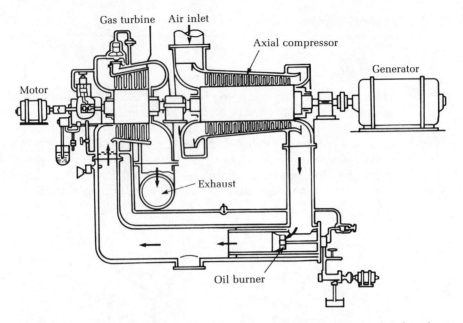

Figure 7-41. Diagram of·a simple gas turbine plant. (Courtesy of Babcock & Wilcox, a McDermott Company.)

to the atmosphere (Figure 7-40c and d). In the closed cycle, the heat is transferred to the liquid in a heat exchanger from an external source. The fluid must then be cooled after it leaves the turbine and before it enters the compressor. As can be seen in Figure 7-40a and b, this cycle is characterized by constant pressure heat addition and rejection, and adiabatic compression and expansion. Air is the usual working fluid and is generally assumed to be an ideal gas. Figures 7-41 and 7-42 illustrate the hardware associated with the Brayton cycle.

Figure 7-42. A gas turbine used for aircraft propulsion. (Reprinted by permission from Pratt & Whitney Aircraft.)

Thermal Efficiency

The thermal efficiency follows directly from the definition:

$$\eta = \frac{\text{output}}{\text{input}} = \frac{W_{net}}{Q_{in}}$$

$$= 1 - \frac{Q_L}{Q_H}$$

$$= 1 - \frac{c_p(T_4 - T_1)}{c_p(T_3 - T_2)}$$

$$= 1 - \frac{T_1(T_4/T_1 - 1)}{T_2(T_3/T_2 - 1)} \tag{7-45}$$

To simplify this expression, recall that

$$\frac{p_3}{p_4} = \frac{p_2}{p_1}$$

and that for isentropic processes

$$\frac{p_2}{p_1} = \left(\frac{T_2}{T_1}\right)^{k/(k-1)} = \frac{p_3}{p_4} = \left(\frac{T_3}{T_4}\right)^{k/(k-1)}$$

or

$$\frac{T_4}{T_1} = \frac{T_3}{T_2}$$

Using this in Equation (7-45) yields

$$\eta = 1 - \frac{T_1}{T_2}$$

→ *ISENTROPIC PRESSURE RATIO*

$$= 1 - \left(\frac{p_2}{p_1}\right)^{(1-k)/k} \tag{7-46}$$

where p_2/p_1 is the isentropic pressure ratio. Thus the Brayton cycle efficiency increases as this pressure ratio increases. Again recall that the components of the system also have efficiencies.

At this point one might ask: If the temperature extremes are given (T_3 and T_1 are limited by materials), is there a temperature T_2 (or T_4) such that the performance is optimum (maximum)? To answer this question, note that the net work is

$$W_{net} = mc_p(T_3 - T_4) - mc_p(T_2 - T_1) \tag{7-47}$$

W *TURBINE* W *COMPRESSOR*

but

$$T_4 = \frac{T_3 T_1}{T_2}$$

$$W_{net} = mc_p \left(T_3 - \frac{T_3 T_1}{T_2} - T_2 + T_1 \right)$$

For the maximum work differentiate with respect to T_2, and set the result equal to zero.

$$\frac{dW_{net}}{dT_2} = 0$$

This yields

$$T_2 = \sqrt{T_1 T_3}$$

Then

$$W_{net} = mc_p \left(T_3 - 2\sqrt{T_1 T_3} + T_1 \right) \tag{7-48}$$

EXAMPLE 7-15

Air enters a compressor at 14.7 psia and 70 F and leaves with a pressure of 73.5 psia. Assuming a Brayton cycle, no kinetic or potential-energy change, and a maximum operating temperature of 2000 R, determine:

1. p and T at each point of the cycle

2. w_c, w_t, and η_B

SOLUTION:

From the problem statement,

$$p_1 = p_4 = 14.7 \text{ psia} \qquad T_1 = 530 \text{ R} \qquad T_4 = ?$$
$$p_2 = p_3 = 73.5 \text{ psia} \qquad T_3 = 2000 \text{ R} \qquad T_2 = ?$$

Thus we need T_2 and T_4, so

$$\frac{T_2}{T_1} = \left(\frac{p_2}{p_1} \right)^{(k-1)/k} \qquad \text{and} \qquad \frac{T_4}{T_3} = \left(\frac{p_4}{p_3} \right)^{(k-1)/k}$$

$$T_2 = 530 \text{ R} \left(\frac{73.5}{14.7} \right)^{0.4/1.4} = 839.4 \text{ R}$$

$$T_4 = 2000 \left(\frac{14.7}{73.5} \right)^{0.4/1.4} = 1262.8 \text{ R}$$

Consider the compressor to determine w_c:

$$w_c = h_2 - h_1 = c_p(T_2 - T_1)$$

$$= 0.24 \text{ Btu/lbm-R } (839.4 - 530) \text{ R}$$

$$= 74.26 \text{ Btu/lbm}$$

And w_t is next:

$$w_t = h_3 - h_4 = c_p(T_3 - T_4)$$

$$= 0.24 \text{ Btu/lbm-R } (2000 - 1262.8) \text{ R}$$

$$= 176.93 \text{ Btu/lbm}$$

And, finally, the cycle efficiency:

$$\eta_B = 1 - \frac{1}{(p_2/p_1)^{(k-1)/k}} = 1 - \frac{1}{(5)^{0.4/1.4}} = 0.369$$

EXAMPLE 7-16

A gas power plant operates on a Brayton cycle. The maximum and minimum temperatures and pressures are 1200 K, 0.38 MPa, and 290 K, 0.095 MPa. Determine the power output of the turbine and the fraction of the power from the turbine used to operate the compressor of a plant whose net output is 40 000 kW.

SOLUTION: The accompanying sketch illustrates the cycle:

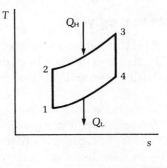

$$p_1 = 0.095 \text{ MPa}$$

$$p_3 = 0.38 \text{ MPa}$$

$$T_1 = 290 \text{ K}$$

$$T_3 = 1200 \text{ K}$$

We must assume that the gas is ideal and, for convenience, that the specific heats are constant over this temperature range. Now, since $s_1 = s_2$,

$$T_2 = T_1 \left(\frac{p_2}{p_1}\right)^{(k-1)/k} = 290 \left(\frac{38}{9.5}\right)^{0.286} = 431.1 \text{ K}$$

$$w_c = h_2 - h_1 = c_p(T_2 - T_1) = 141.6 \text{ kJ/kg}$$

Similarly, $s_3 = s_4$, so

$$T_4 = T_3\left(\frac{p_4}{p_3}\right)^{(k-1)/k} = 807.2 \text{ K}$$

and

$$w_t = h_3 - h_4 = c_p(T_3 - T_4) = 394.2 \text{ kJ/kg}$$

Recall that $\dot{W} = \dot{m}w$, where w is the net work. Thus

$$\dot{m} = \frac{40\,000 \text{ kW}}{(394.2 - 141.6) \text{ kJ/kg}}$$

$$= \frac{40{,}000 \text{ kJ/s}}{252.6 \text{ kJ/kg}} = 158.4 \text{ kg/s}$$

Thus

$$\dot{W}_t = \dot{m}w_t = 158.4 \text{ kg/s} (394.2 \text{ kJ/kg}) = 62.42 \text{ MW}$$

The ratio of the compressor power to the turbine power is

$$\frac{w_c}{w_t} = \frac{141.6}{394.2} = 0.359 \text{ or } 35.9\%$$

Improvements to the Cycle

As with the Rankine cycle, there are many methods to improve the efficiency of the Brayton cycle. These procedures include regeneration, multistaging with intercooling, and multistage expansion with reheating.

Effort is being expended to develop the gas turbine/steam turbine power plant. Several power companies have used this system in conjunction with older, low-pressure plants. It was found that after many years of operation, the steam generator (boiler) had so deteriorated that it was unusable. Since the steam turbine and condenser were still serviceable, however, a power-producing gas turbine unit was installed instead of replacing the steam generator. The installation was such that the hot exhaust gases from the gas turbine were used to generate steam for the old steam turbine (Figure 7-43). Not only was the total power output increased with this setup but the thermal efficiency of the two units was much higher than that of the old steam power plant (but not better than that of modern steam power plants).

Until the 1970s, the maximum gas temperature for stationary gas turbine/steam turbine power plants was around 1600 F. The overall efficiency of the unit is of course limited by the allowable operating

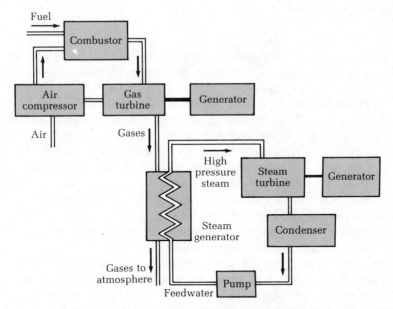

Figure 7-43. Gas turbine/steam turbine power plant (STAG).

temperatures for the gas turbine. Much effort is being devoted to finding a way to use gas with temperatures of at least 2200 to 2400 F by using materials such as ceramics.

7–4 Air-Standard Cycles

There are obvious difficulties in modeling a real process too closely. In the case of an internal combustion engine, for example, air and fuel are changed to combustion products. Moreover, air is taken in at low temperature and the combustion products are rejected at high temperature. (That is, it is an open system and chemical changes occur in the working fluid.) These and many other difficulties are evaded by modeling the actual system with an ideal system. In this system, it is assumed that:

1. We are dealing with a closed system.

2. Air (an ideal gas) is the working substance.

3. The combustion process is replaced by a heat transfer operation.

4. There are no combustion products.

5. Specific heats are constant.

6. All processes are reversible.

This procedure is called the **air-standard analysis.**

Otto Cycle

*The idealized approximation of the spark ignition (SI) internal com-
bustion engine is the* **air-standard Otto cycle.** Figure 7-44 illustrates the
operating characteristics of the cycle; Figure 7-45 presents the idealized
(p, v) and (T, s) diagrams.

The operation of the four-stroke cycle is outlined in Figure 7-44. In
(1), the piston is in the top dead center (also called head-end dead-
center) position. Intake valve *I* is open while exhaust valve *E* is closed.
Pressure is ideally atmospheric. The suction stroke is represented by
line 0–1 in Figure 7-45. In (2), the cylinder is full of working substance
and both valves are closed. Isentropic compression is 1–2; the piston
returns to the head-end position. In (3), instantaneous combustion (2–3)
occurs; heat is supplied at constant volume. Isentropic expansion is
represented by 3–4. In (4), the piston is in the bottom dead center (also
called crank-end dead center) position; the exhaust valve *E* opens, gases
flow out of the cylinder, and the pressure drops instantly to atmospheric. The piston then makes the exhaust stroke (1–0), pushing more
gases from the cylinder. The exhaust valve then closes, the intake valve
opens, and the cycle repeats itself. Figure 7-46 illustrates the hardware
for this cycle.

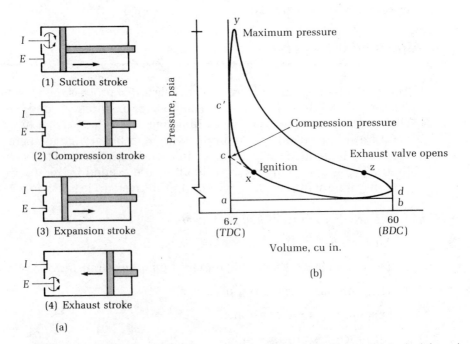

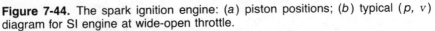

Figure 7-44. The spark ignition engine: (*a*) piston positions; (*b*) typical (*p, v*)
diagram for SI engine at wide-open throttle.

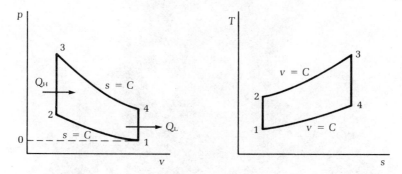

Figure 7-45. (p, v) and (T, s) diagrams of the air-standard Otto cycle.

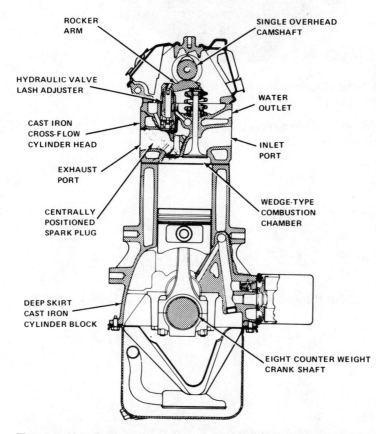

Figure 7-46a. Schematic and hardware for air-standard Otto cycle: cross section of overhead-valve, in-line, automotive SI engine.

Figure 7-46b. Cross section of a V-8, overhead-valve, automotive engine. (Courtesy of Chevrolet Motor Division.)

To determine the efficiency of the air-standard Otto cycle, we apply the definition:

$$\eta = \frac{W}{Q_H} = 1 - \frac{Q_L}{Q_H}$$

$$= 1 - \frac{mc_v(T_4 - T_1)}{mc_v(T_3 - T_2)}$$

$$= 1 - \frac{T_1(T_4/T_1 - 1)}{T_2(T_3/T_2 - 1)} \tag{7-49}$$

Since two of the processes are isentropic and $V_1 = V_4$ and $V_2 = V_3$,

$$\frac{T_2}{T_1} = \left(\frac{V_1}{V_2}\right)^{k-1} = \frac{T_3}{T_4} = \left(\frac{V_1}{V_3}\right)^{k-1}$$

Thus

$$\frac{T_3}{T_2} = \frac{T_4}{T_1}$$

and the efficiency becomes

$$\eta = 1 - \frac{T_1}{T_2} = 1 - \left(\frac{V_1}{V_2}\right)^{1-k} \tag{7-50}$$

Thus the efficiency is a function of the isentropic compression ratio r_v $(= V_1/V_2 = V_4/V_3)$.

EXAMPLE 7-17

An air-standard Otto cycle (compression ratio of 9) absorbs 1000 Btu/lbm of air. If at the beginning of the compression stroke the pressure is 14.7 psia and the temperature is 80 F, determine:

1. p and T at each point of the cycle

2. η

SOLUTION:

POINT 1: From the problem statement, p_1 = 14.7 psia and T_1 = 540 R.

POINT 2: Since (1–2) and (3–4) are reversible adiabatic processes,

$$T_2 = T_1 \left(\frac{V_1}{V_2}\right)^{k-1} = 540\,(9)^{0.4}\ R = 1300.4\ R$$

and

$$p_2 = p_1 \left(\frac{V_1}{V_2}\right)^{k} = 14.7\ \text{psia}\ (9)^{1.4} = 318.6\ \text{psia}$$

POINT 3: Since 1000 Btu/lbm = $c_v\,(T_3 - T_2)$, we get

$$T_3 = 7148\ R$$

Process (2–3) is a constant-volume process, so

$$\frac{p_3}{p_2} = \frac{T_3}{T_2}$$

Hence

$$p_3 = p_2 \left(\frac{7148}{1300}\right) = 1751\ \text{psia}$$

POINT 4:
$$T_4 = T_3 \left(\frac{V_2}{V_1}\right)^{k-1} = 2968\ R$$

$$p_4 = p_3 \left(\frac{V_2}{V_1}\right)^{k} = 80.8\ \text{psia}$$

and finally

$$\eta = 1 - \frac{1}{(r_v)^{k-1}} = 1 - \frac{1}{(9)^{0.4}}$$

$$= 0.585$$

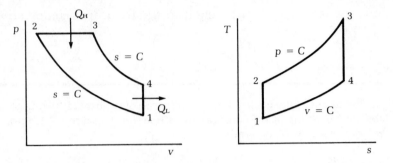

Figure 7-47. (p, v) and (T, s) diagrams of the air-standard diesel cycle.

Diesel Cycle

The **air-standard diesel cycle** is the ideal approximation of the diesel (compression-ignition) engine. Figure 7-47 illustrates both the (p, v) and the (T, s) diagrams of this cycle. In this cycle, heat is added at constant pressure but rejected at constant volume. Figure 7-48 illustrates the cylinder arrangement in an internal combustion engine.

Again, we apply the basic definition to determine the efficiency:

$$\eta = \frac{W}{Q_H} = 1 - \frac{Q_L}{Q_H} = 1 - \frac{mc_v\,(T_4 - T_1)}{mc_p\,(T_3 - T_2)}$$

$$= 1 - \frac{T_1\,(T_4/T_1 - 1)}{kT_2\,(T_3/T_2 - 1)} \qquad k = c_p/c_v \qquad \textbf{(7-51)}$$

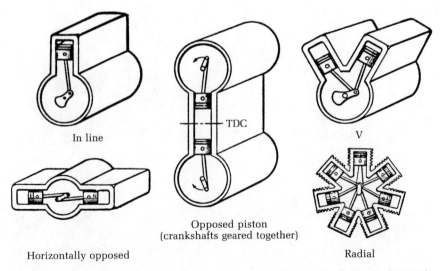

In line

Horizontally opposed

TDC

Opposed piston
(crankshafts geared together)

V

Radial

Figure 7-48. Common cylinder arrangements used in multicylinder reciprocating IC engines.

Again, $V_1/V_2 = r_v$ is the isentropic compression ratio, so

$$\eta = 1 - \frac{T_4/T_1 - 1}{k\,(T_3/T_2 - 1)\,r_v^{\,k-1}} \tag{7-52}$$

EXAMPLE 7-18

The compression ratio of an air-standard diesel cycle is 15. At the beginning of the compression stroke, the pressure is 14.7 psia and the temperature is 80 F. The maximum workable temperature is 4500 F. What is the thermal efficiency?

SOLUTION: To determine η, we must obtain the temperature at each point of the system as well as the compression ratio. Take

$$T_3 = 4500 \text{ R}$$

$$T_1 = 540 \text{ R}$$

$$T_2 = T_1 \left(\frac{V_1}{V_2}\right)^{k-1} = 540\,(15)^{0.4} = 1595.3 \text{ R}$$

$$T_4 = ?$$

Using process (3–4)

$$T_4 = T_3 \left(\frac{V_3}{V_4}\right)^{k-1} = T_3 \left(\frac{V_3}{V_1}\right)^{k-1}$$

$$= T_3 \left[\frac{V_2}{V_1}\left(\frac{T_3}{T_1}\right)\right]^{k-1}$$

$$= 4500 \text{ R} \left[\frac{1}{15}\left(\frac{4500}{1595.3}\right)\right]^{0.4} = 2306 \text{ R}$$

Thus

$$\eta = 1 - \frac{T_4/T_1 - 1}{k\,(T_3/T_2 - 1)\,r_v^{\,k-1}}$$

$$= 1 - \frac{2306/540 - 1}{[\,(4500/1595) - 1\,]\,(15)^{0.4}\,1.4}$$

$$= 0.434$$

Other Cycles

Other air-standard cycles are becoming more important—especially the air-standard Stirling and Ericsson cycles (both of which are regenerative). These two cycles are represented in Figures 7-49 and 7-50. Table 7-3 presents the salient points of all the cycles we have been discussing.

Table 7-3 Cycle Information

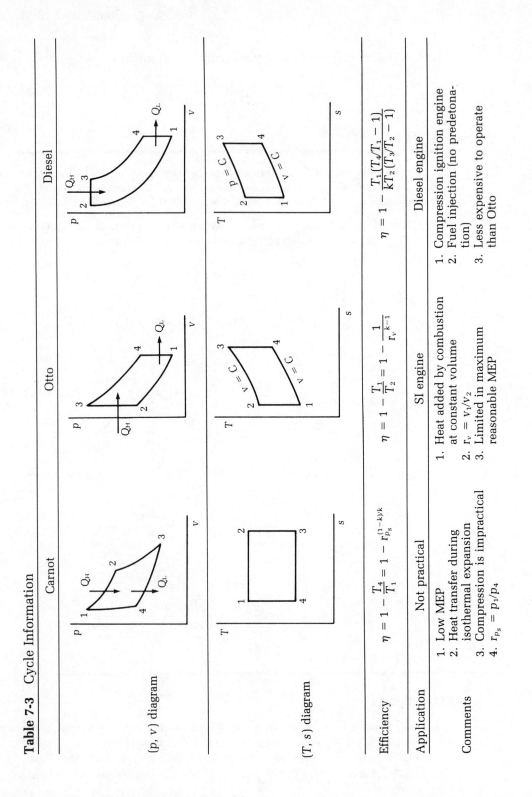

	Carnot	Otto	Diesel
(p, v) diagram			
(T, s) diagram			
Efficiency	$\eta = 1 - \dfrac{T_4}{T_1} = 1 - r_{P_S}^{(1-k)/k}$	$\eta = 1 - \dfrac{T_1}{T_2} = 1 - \dfrac{1}{r_v^{k-1}}$	$\eta = 1 - \dfrac{T_1(T_4/T_1 - 1)}{kT_2(T_3/T_2 - 1)}$
Application	Not practical	SI engine	Diesel engine
Comments	1. Low MEP 2. Heat transfer during isothermal expansion 3. Compression is impractical 4. $r_{P_S} = p_1/p_4$	1. Heat added by combustion at constant volume 2. $r_v = v_1/v_2$ 3. Limited in maximum reasonable MEP	1. Compression ignition engine 2. Fuel injection (no predetonation) 3. Less expensive to operate than Otto

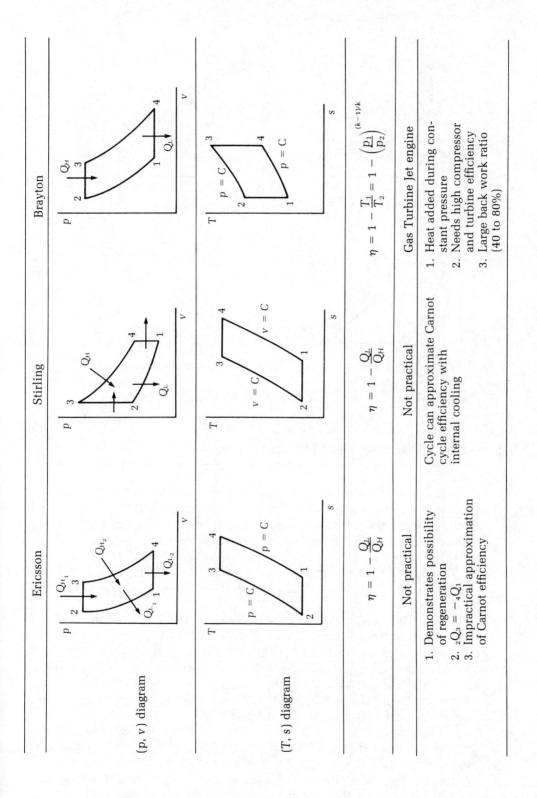

	Ericsson	Stirling	Brayton
(p, v) diagram	Q_{H_1}, Q_{H_2}, Q_{L_1}, Q_{L_2} (states 2, 3, 4, 1)	Q_H, Q_L (states 2, 3, 4, 1)	Q_H, Q_L (states 2, 3, 4, 1)
(T, s) diagram	$p = C$, $p = C$ (states 2, 3, 4, 1)	$v = C$, $v = C$ (states 2, 3, 4, 1)	$p = C$, $p = C$ (states 2, 3, 4, 1)
	$\eta = 1 - \dfrac{Q_L}{Q_H}$	$\eta = 1 - \dfrac{Q_L}{Q_H}$	$\eta = 1 - \dfrac{T_1}{T_2} = 1 - \left(\dfrac{p_1}{p_2}\right)^{(k-1)/k}$
	Not practical	Not practical	Gas Turbine Jet engine
	1. Demonstrates possibility of regeneration 2. $_2Q_3 = {}_4Q_1$ 3. Impractical approximation of Carnot efficiency	Cycle can approximate Carnot cycle efficiency with internal cooling	1. Heat added during constant pressure 2. Needs high compressor and turbine efficiency 3. Large back work ratio (40 to 80%)

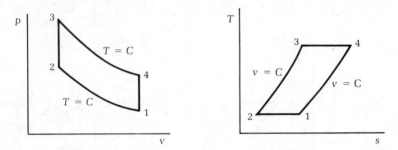

Figure 7-49. (p, v) and (T, s) diagrams of the Stirling cycle.

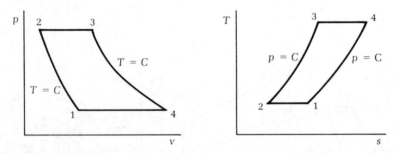

Figure 7-50. (p, v) and (T, s) diagrams of the Ericsson cycle.

7–5 Refrigerator and Heat Pump Cycles

The purpose of a **refrigeration system** is to remove heat from a space that is at a lower temperature than the surroundings. But refrigeration systems have also been developed whose purpose is to put heat into a volume that is at a higher temperature than the surroundings. In this case the device is called a **heat pump,** but its thermodynamic execution is identical to the refrigerator; the major difference is the coefficient of performance. Figure 7-51 illustrates these two applications schematically.

In the beginning, refrigeration was accomplished by the melting of ice (a noncyclic process). Today the refrigerator operates as a cyclic process in which work must be supplied to transfer the heat. That is, the work must be done on the system. In this section we discuss a few refrigeration (heat pump) systems to give you at least a nodding acquaintance with them. A schematic of a refrigeration system, air conditioner, or heat pump is shown in Figure 7-52.

The energy-conversion objective of the air conditioner or refrigeration cycle is entirely different from that of the steam power plant. In the power cycle, the objective is to obtain work as output from a heat input. In the refrigeration cycle, the objective is to obtain a cooling effect. The energy that we must purchase is the work input to the

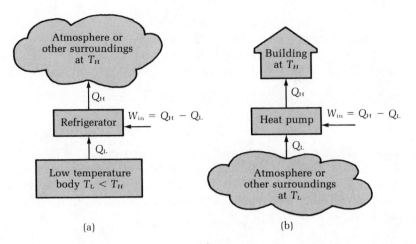

Figure 7-51. Refrigeration cycle application; (a) as a refrigerator and (b) as a heat pump.

compressor. Thus we measure our objective in this case by a coefficient of performance (η_R):*

$$\text{COP}_c = \frac{\text{cooling effect}}{\text{work input}} \qquad (7\text{-}53)$$

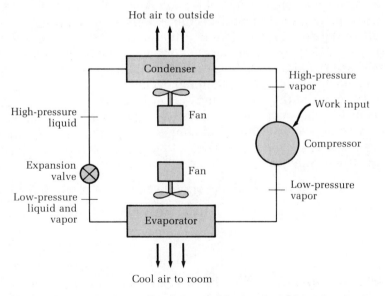

Figure 7-52. Schematic of an air conditioning or refrigeration system.

* In industry the common expressions for these coefficients are

$$\text{COP}_c = \eta_R \qquad \text{for cooling (refrigerator)}$$

and

$$\text{COP}_h = \eta_{HP} \qquad \text{for heating (heat pump)}$$

Actual refrigeration devices can have a COP on the order of 3.

The refrigeration cycle is called a heat pump when it takes the heat at the low outside temperature and pumps it up to the high temperature room. The coefficient of performance of a heat pump is defined as the ratio

$$\text{COP}_h = \frac{\text{heating effect}}{\text{work input}} \tag{7-54}$$

and under the proper conditions can be as much as 4 or 5. A typical house, for example, might require 100,000 Btu/hr of heat input. With a heat pump operating at a COP of 4, the electric power input would only be 25,000 Btu/hr, or about 7.3 kW. Direct electric heating would require 29.3 kW.

Vapor-Compression Cycle

Figure 7-53 illustrates schematically the basic vapor-compression cycle as well as the (T, s) and (p, h) diagrams of this refrigeration cycle. Process (1–2) takes a low-pressure saturated vapor and compresses it in a reversible adiabatic process. In process (2–3) heat is rejected from the vapor until it exists as a high-pressure saturated liquid. The action of the expansion value is to reduce the pressure (yielding wet steam). Finally, process (4–1) completes the vaporization, returning the system to its initial conditions. Note that the process (3–4) is an irreversible process ($\Delta s \neq 0$). If this process had been reversible, the cycle would be a reversed Rankine cycle (i.e., an irreversible expansion value replaced the reversible pump of the Rankine cycle).

Historically the phrase "tons of refrigeration" (i.e., the number of tons of ice that melts in one day) was used to describe the size of a refrigeration system. This is, of course, an insufficient definition from an engineering standpoint. Today this phrase is still used, but it is explicitly defined as

$$1 \text{ ton of refrigeration} = 12,000 \text{ Btu of refrigeration/hr}$$

$$= 200 \text{ Btu/min}$$

$$= 3.514 \text{ kw}$$

$$= 4.712 \text{ hp}$$

It may be instructive to list the thermal analysis associated with each component of this cycle. That is, as per our usual procedure, one must isolate each component and analyze it. Thus, for the compressor

$$\dot{m}\left[\left(h_1 + \frac{V_1^2}{2g_c}\right) - \left(h_2 + \frac{V_2^2}{2g_c}\right)\right] + {}_1Q_2 - {}_1W_2 = 0, \quad W_C = {}_1W_2, \quad Q_C = 0$$

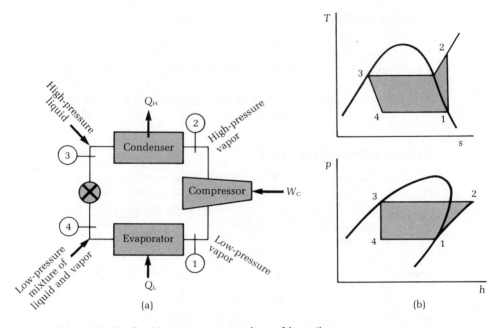

Figure 7-53. Basic vapor-compression refrigeration.

For the condenser:

$$\dot{m}\left[\left(h_2 + \frac{V_2^2}{2g_c}\right) - \left(h_3 + \frac{V_3^2}{2g_c}\right)\right] + {_2}Q_3 = 0, \; Q_H = {_2}Q_3$$

For the expansion value:

$$h_3 - h_4 = 0$$

For the evaporator:

$$\dot{m}\left[\left(h_4 + \frac{V_4^2}{2g_c}\right) - \left(h_1 + \frac{V_1^2}{2g_c}\right)\right] + {_4}Q_1 = 0, \; Q_L = {_4}Q_1$$

For the overall cycle:

Overall: $\quad Q_L + W_C = Q_C + Q_H \quad$ or $\quad ({_1}Q_2 + {_2}Q_3 + {_4}Q_1) - ({_1}W_2) = 0$

Thus the overall or cycle coefficient of performance (efficiency) depends upon the purpose:

$$\text{Coefficient of performance for cooling:} \quad \frac{Q_L}{W_C}$$

$$\text{Coefficient of performance for heating:} \quad \frac{Q_H}{W_C}$$

Applying the first law to this cycle yields

$$_1Q_2 + {}_2Q_3 + {}_4Q_1 = {}_1W_2$$

Finally, many types of refrigerants are available as the working fluid of this device. The brand name most familiar to us is Freon, but others do exist. Care must be taken when selecting a refrigerant, primarily in obtaining the desired evaporation temperature (and pressure).

Ammonia-Absorption Cycle

An ammonia-absorption refrigeration cycle is shown schematically in Figure 7-54. Note from this diagram that the compression process (from the vaporization pressure to the condensing pressure) is accomplished by energy supplied by heat not work. This point, then, is the primary difference between this system and the vapor compression cycle. That is, you will note that the refrigerant is ammonia and that condenser, the expansion valve and evaporator, of this system is set up just like the vapor compression cycle—the compression is more complicated.

The ammonia vapor leaves the evaporator and enters the absorber. In this device the vapor is mixed with water. The concentrated aqua-ammonia is pumped to the generator-heat exchanger (A–B). At the same time, the low concentration aqua-ammonia is returned to the absorber (C) where it is reconcentrated and recycled. Heat supplied to the generator-heat exchanger boils the mixture of ammonia and water to

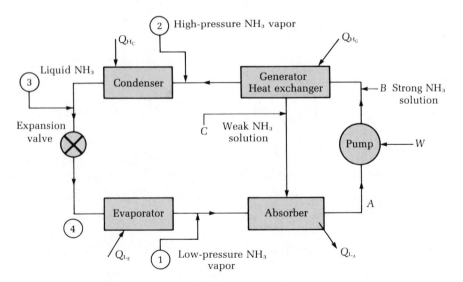

Figure 7-54. The ammonia-absorption refrigeration cycle.

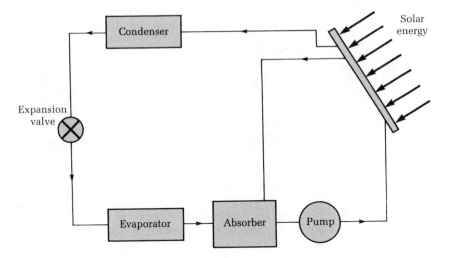

Figure 7-55. Elements of an absorption system for solar air conditioning.

make the weak ammonia solution. Almost pure ammonia vapor proceeds to the condenser. Since more equipment is involved in an absorption system than in the vapor-compression cycle, it is economically feasible only when a source of heat is available that would otherwise be wasted. Thus the ammonia absorption cycle offers significant potential for use with solar energy—as illustrated in Figure 7-55.

Air-Standard Cycle

The air-standard refrigeration cycle is most often modeled by the reversed Brayton cycle. The main uses of this system are aircraft cooling and the liquefaction of air and other gases. Figure 7-56 illustrates this system schematically and shows a (T, s) diagram of the reversed Brayton cycle. Note from this diagram that it represents the closed-cycle version.

In process (1–2) air is compressed. Process (2–3) represents a heat transfer (to the ambient air outside the refrigerated space), thus cooling the air. The purpose of the expander is to reduce the pressure of the air to that required to enter the compressor. This process also results in a decrease in temperature. In process (4–1) heat is absorbed by the air from the refrigerated space (cooling this space). Since this is a refrigerator, the coefficient of performance involves the net work (W) and the heat absorbed (Q_L). Thus

$$\text{COP}_c \; (\eta_R) = \frac{Q_L}{W} = \frac{\text{area } (1\text{–}4\text{–}5\text{–}6\text{–}1)}{\text{area } (1\text{–}2\text{–}3\text{–}4\text{–}1)}$$

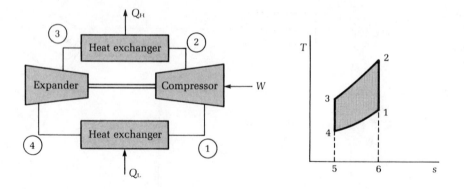

Figure 7-56. The air-standard refrigeration cycle.

Heat Pumps

Heat pumps for air conditioning service may be classified according to:

1. Type of heat source and sink

2. Heating and cooling distribution fluid

3. Type of thermodynamic cycle

4. Type of building structure

5. Size and configuration

The more common types are listed in Table 7-4.

The **air-to-air heat pump** is the most common type. It is particularly suitable for factory-built unitary heat pumps and is widely used for residential and commercial applications. The first diagram in Table 7-4 is typical of the refrigeration circuit employed. In air-to-air heat pump systems, as shown in the second diagram of Table 7-4, the air circuits may be interchanged by means of dampers (motor-driven or manually operated) to obtain either heated or cooled air for the conditioned space. With this system, one heat-exchanger coil is always the evaporator and the other is always the condenser. The conditioned air passes over the evaporator during the cooling cycle and the outdoor air passes over the condenser. The change from cooling to heating is accomplished by positioning the dampers.

A **water-to-air heat pump** uses water as a heat source and sink; it uses air to transmit heat to or from the conditioned space. **Air-to-water heat pumps** are commonly used in large buildings where zone control is necessary; they are also employed for the production of hot or cold water in industrial applications.

Earth-to-air **heat pumps** may employ direct expansion of the refrigerant in an embedded coil, or they may be of the indirect type (described under the water-to-air type). An *earth-to-water* **heat pump** (not shown in Table 7-4) may be like the earth-to-air type shown, except for the substitution of a refrigerant-water heat exchanger for the finned coil shown on the indoor side. It may also take a form similar to the water-to-water system shown, in which case a secondary-fluid ground coil is used. Some heat pumps that use earth as the heat source and sink are essentially of the water-to-air type. An antifreeze solution is pumped through a circuit consisting of the chiller-condenser and a pipe coil embedded in the earth. Earth source/sink systems are seldom used today.

A *water-to-water* **heat pump** uses water as the heat source and sink for both cooling and heating. Heating/cooling changeover may be accomplished in the refrigerant circuit, but in many cases it is more convenient to perform the switching in the water circuits.

There are other types of heat pumps in addition to those listed in Table 7-4. One uses solar energy as a source of heat; its refrigerant circuit may resemble the water-to-air, air-to-air, or other types, depending on the form of solar collector and the means of heating and cooling distribution. Another variation is the use of more than one heat source. Some heat pumps use air as the primary heat source but can be used to extract heat from water (from a well or storage tank) during periods of insufficient solar radiation. Any thermodynamic cycle that is capable of producing a cooling effect may theoretically be used as a heat pump.

As Table 7-4 shows, various media can be described as heat sources and sinks. The subject of media is worth pursuing further here, since the most practical choice for an application will be influenced primarily by geographic location, climatic conditions, initial cost, availability, and type of structure.

Air Outdoor air offers a universal heat-source/heat-sink medium for the heat pump. Extended-surface, forced-convection, heat-transfer coils are normally used to transfer the heat between air and refrigerant. Typically, these surfaces are 50 to 100% larger than the corresponding surface on the indoor side of heat pumps using air as the distributive medium. The volume of outdoor air handled is also usually greater in about the same proportions. The temperature difference during heating operation and the outdoor air and the evaporating refrigerant is generally in the range of from 10 to 25 degrees.

When selecting or designing an air-source heat pump, two factors in particular must be taken into consideration: local variations in tem-

Table 7-4 Common Heat Pump Types

Heat Source and Sink	Distribution Fluid	Thermal Cycle*	Diagram → Heating → Cooling → Heating and Cooling
Air	Air	Refrigerant changeover	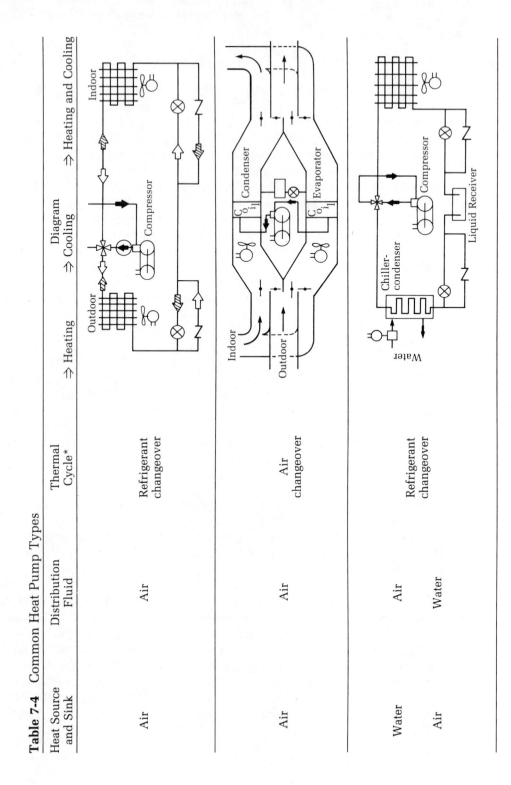
Air	Air	Air changeover	
Water Air	Air Water	Refrigerant changeover	

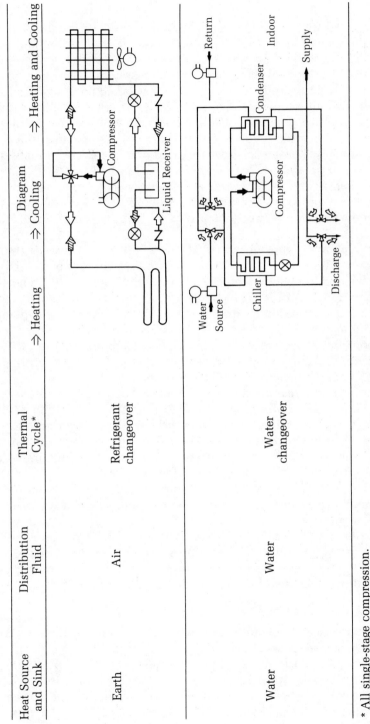

Heat Source and Sink	Distribution Fluid	Thermal Cycle*	Diagram
Earth	Air	Refrigerant changeover	
Water	Water	Water changeover	

* All single-stage compression.
Source: ASHRAE 1976 Systems Handbook. Reprinted by permission from ASHRAE.

perature and the formation of frost. As the outdoor temperature goes down, the heating capacity of an air-source heat pump decreases. Selecting the equipment for an outdoor heating design temperature is therefore more critical than for a fuel-fired system. Care must be exercised in sizing the equipment for as low a balance point as possible for heating without having excessive cooling capacity during the summer.

Early applications of air-source heat pumps followed commercial refrigeration practice involving relatively wide fin spacing (four to five fins per inch). Theory suggested that this practice would minimize the frequency of defrosting, but experience has proved that, with effective hot gas defrosting, much closer fin spacing can be tolerated and will obtain the advantage of a reduction in the size and bulk of the system. In current practice, a fin spacing of 8 to 13 per inch is widely used.

Water   Water may represent a satisfactory, and in many cases an ideal, heat source. Well water is particularly attractive because of its relatively high and nearly constant temperature—generally about 50 F in northern areas and 60 F and higher in the south. Abundant sources of suitable water are becoming increasingly scarce, however, and the application of this type of system is rather limited. Even if sufficient water is available from wells, the water may corrode heat exchangers or induce scale formation. Other considerations are the costs of drilling, piping, and pumping and the means for disposal of used water.

Surface or stream water may be used, but under reduced winter temperatures the cooling spread between inlet and outlet must be limited to prevent freeze-up in the water chiller that is absorbing the heat. In Europe, large heat pump systems have been in operation in areas where winter stream temperatures of about 35 F permit inlet-outlet temperature spreads of only 1 or 2 degrees. Under certain industrial circumstances, waste process water, such as spent warm water in laundries and warm condenser water, may be a source for specialized heat pump operations.

Water-refrigerant heat exchangers generally take the form of direct-expansion water coolers—either the shell-and-coil type or the shell-and-tube type. They are circuited to permit use as a refrigerant condenser during the heating cycle and as a refrigerant evaporator during the cooling cycle.

Earth   Earth as a heat source and sink, by heat transfer through buried coils, has not been used extensively. This neglect may be attributed to installation expense, ground area requirements, and the difficulty of predicting performance. Soil composition varies widely from wet clay to sandy soil and has a predominant effect on thermal properties and overall performance. The heat transfer process in the soil is primarily one of bulk temperature change with time (transient heat flow). **Thermal diffusivity** is a dominant factor that is hard to determine at different building sites. The moisture content likewise has an influence, since energy in the ground may also be transported via moisture travel in the soil. Earth coils, usually spaced horizontally at 3 to 6 ft intervals, are submerged 3 to 6 ft below the surface. Although a lower depth might be preferred, excavation costs demand a compromise. Wherever the site may be, mean undisturbed ground temperatures generally follow the mean annual climatic temperature.

Solar Although it is still largely in the research stage, the use of solar energy as a heat source, either on a primary basis or in combination with other sources, is attracting increasing interest. The principal advantage of using solar radiation as a heat source for heat pumps is that, when available, it provides heat at higher temperatures than other sources, thus resulting in an increase in coefficient of performance. Compared to a solar heating system without a heat pump, collector efficiency and capacity are materially increased due to the lower collector temperature required.

Research and development in solar-source heat pumps has been concerned with two basic types of systems: direct and indirect. In the direct system, refrigerant evaporator tubes are embedded in a solar collector, usually of the flat-plate type. Research has shown that when the collector has no glass cover plates, the same collector surface can also function to extract heat from the outdoor air. The same surface may then be employed as a condenser using outdoor air as a heat sink for cooling. The refrigeration circuit used may resemble that shown in Table 7-4 for an earth-to-air heat pump.

An indirect system employs another fluid, either water or air, which is circulated through the solar collector. When air is used, the first system shown in Table 7-4 for an air-to-air heat pump may be utilized. The collector is added in such a way that (1) the collector can serve as an outdoor-air preheater, (2) the outdoor-air loop can be closed so that all source heat is derived from the sun, or (3) the collector can be disconnected and the outdoor air used as the source or sink. When water is circulated through the collector, the heat pump circuit may be of either the water-to-air or water-to-water type illustrated in Table 7-4. On heat pump systems employing solar energy as the only heat source, either an alternative heating system or a means of storing heat is required during periods of insufficient solar radiation.

EXAMPLE 7-19

A vapor-compression refrigeration system uses Freon-12 as the refrigerant. The temperature of Freon-12 in the evaporation is -10 F and in the condenser it is 96 F. The circulation rate of Freon-12 is 300 lbm/hr. Determine the coefficient of performance.

SOLUTION: Consider the compressor:

$$w_c = h_2 - h_1 \quad \text{(first law)}$$

$$s_2 = s_1 \quad \text{(second law)}$$

Consulting Appendix Table A-2-1 for Freon, we get

$$h_1 = 76.196 \text{ Btu/lbm}$$
$$s_1 = 0.16989 \text{ Btu/lbm-R} \quad \text{VAPOR}$$
$$p_1 = 19.189 \text{ psia}$$
$$p_2 = 124.70 \text{ psia}$$
$$s_2 = s_1$$

Thus

$$h_2 = 90.449 \text{ Btu/lbm}$$
$$T_2 = 101.5 \text{ F}$$

and

$$w_c = 14.253 \text{ Btu/lbm}$$

For the expansion value,

$$h_3 = h_4 = 30.14 \text{ Btu/lbm}$$

Finally, for the evaporation we obtain

$$q_L = h_1 - h_3 = 46.056 \text{ Btu/lbm} \quad \text{(first law)}$$

Hence

$$\text{COP}_c = \frac{q_L}{w_c} = 3.23$$

Note that cooling capacity is

$$\frac{q_L \dot{m}}{12,000 \text{ Btu/hr}} = 1.15 \text{ tons of refrigeration}$$

7–6 Other Applications

There are an unlimited number of other systems to which the rules of thermodynamics may be applied. All you need is a volume of space designated as the one of interest and you can apply the simple rules. All fields are touched to a greater or lesser extent by engineers attempting to understand, build, and improve with the aid of thermodynamic principles. The following is just a partial list of the devices and processes that can be analyzed:

Electric motors

Electric generators

Batteries

Fuel cells

Thermoelectric devices

Energy-collecting devices (solar collectors) and processes

Energy-producing devices and processes

Human skin, teeth, organs

Hemodialysis

Magnetohydrodynamics

Electrodynamics

Airplanes, automobiles, submarines

The field is open; the tools are available; the promise is great.

PROBLEMS

7-1 A nonflow process occurs for which the pressure changes according to the equation $p = 288v + 900$. In this relation, the pressure is expressed in psia and the specific volume in ft³/lbm. If the initial specific volume is 10 ft³/lbm and the final volume is 20 ft³/lbm, compute the work done per pound (Btu/lbm).

7-2 A steady-flow process occurs for which the pressure changes according to the equation $p = 288v + 900$, with p in psia and v in ft³/lbm. The specific volume changes from 10 ft³/lbm at inlet to 20 ft³/lbm at outlet. There are no changes in kinetic or potential energy. Determine: (a) the mechanical work done between inlet and outlet (Btu/lbm) and (b) the flow work done at inlet and outlet (Btu/lbm).

7-3 Water is pumped through pipes embedded in the concrete of a large dam. Water pressure is 100 psia at inlet and 20 psia at outlet. In picking up the heat of hydration of the curing concrete, the water increases in temperature from 50 to 100 F. During curing, the heat of hydration for a section of the dam is 140,000 Btu/hr.

Determine: (a) the required water flow rate for the section (lbm/hr) and (b) the minimum size of motor needed to drive the pump (hp). Neglect changes in kinetic and potential energy.

7-4 A water pump is to deliver 160 lbm/hr at a pressure of 1000 psia when inlet conditions are 15 psia and 100 F. Neglecting changes in kinetic and potential energy, find the minimum size of motor (hp) required to drive the pump. Work this problem in two different ways.

7-5 The water level in College Hills is 400 ft below the surface. You have to install a well pump that will deliver 15 gal/min of water (8.33 lbm/gal and 0.016 ft³/lbm) at a pressure of 30 psig. What horsepower motor should you use?

7-6 A booster pump is used to move water from the basement equipment room to the 13th floor of an apartment building at the rate of 800 lbm/min. The elevation change is 130 ft between the basement and the 13th floor. Determine the minimum size of pump (hp) required.

7-7 The discharge of a pump is 3 m above the inlet. Water enters at a pressure of 138 kPa and leaves at a pressure of 1380 kPa. The specific volume of the water is 0.001 m³/kg. If there is no heat transfer and no change in kinetic or internal energy, what is the specific work?

7-8 Tests performed on a residential air-conditioning system yielded the following data:

> Refrigerant: R-12
>
> Evaporating pressure: 50 psia
>
> Condensing pressure: 200 psia
>
> Actual air cooling effect: 32,450 Btu/hr
>
> Power meter reading: 5.76 kW

Determine both actual and ideal performance: (a) COP; (b) EER; (c) hp/ton.

7-9 A gas turbine unit for power production has compressor inlet conditions of 15 psia and 60 F with a flow rate of 12,500 ft³/min. The pressure ratio across the compressor is 6. Compressor efficiency is 80%. The maximum allowable temperature in the system is 2250 F. The unit burns fuel oil with a heating value of 139,000 Btu/gal at a cost of $1.10/gal. Determine the hourly fuel cost ($/hr).

7-10 Suppose that 2000 ft³/min of air at 14.7 psia and 60 F enters a fan with negligible inlet velocity. The discharge duct from the fan has a cross-sectional area of 3 ft². The process across the fan is isentropic (reversible and adiabatic) with a fan discharge pressure of 14.8 psia. Determine: (a) the velocity in the discharge duct (fpm) and (b) the size of motor required to drive the fan (hp).

7-11 A fan is used to provide fresh air to the welding area in an industrial plant. The fan takes in outside air at 80 F and 14.7 psia at the rate of 1200 ft³/min with negligible inlet velocity. In the 10 ft² duct leaving the fan, air pressure is 1 psig. If the process is assumed to be reversible and adiabatic (isentropic), determine the size of motor (hp) needed to drive the fan.

7-12 Steam at 400 psia and 600 F expands through a nozzle to 300 psia at the rate of 20,000 lbm/hr. If the process occurs reversibly and adiabatically and the initial velocity is low, calculate: (a) the velocity leaving the nozzle and (b) the exit area of the nozzle.

7-13 Methane enters a compressor at 15 psia and 40 F with a velocity of 200 ft/sec through a cross-sectional area of 0.60 ft². The methane is compressed frictionlessly, steadily, and adiabatically to 30 psia. The discharge velocity is very low. Determine the size of motor (hp) needed to operate the compressor. (CH₄, C_p = 0.531 Btu/lbmR)

7-14 A centrifugal pump delivers liquid nitrogen at −240 F at the rate of 100 lbm/sec. The nitrogen enters the pump as liquid at 15 psia and the discharge pressure is 500 psia. Determine the minimum size of motor (hp) needed to drive this pump.

7-15 An engine operating on the Otto cycle has an air/fuel ratio of 15/1 by weight. The fuel has a heating value of 18,000 Btu/lbm. At the start of compression, the air is at 70 F and 14.7 psia. The compression ratio (v_1/v_2) is 6/1. Determine the temperature after combustion (T_3).

7-16 In an Otto cycle (spark ignition) engine, air is compressed adiabatically from 14.7 psia and 80 F. The compression ratio (v_1/v_2) is 8/1. Determine: (a) the temperature at the end of compression and (b) the work required for compression per pound of air (Btu).

7-17 In a diesel engine, air is compressed until it reaches the self-ignition temperature of the fuel (1075 F). The air is initially at 60 F and 14.7 psia. Determine: (a) the compression ratio (v_1/v_2) and final pressure and (b) the work of compression per pound of air. Assume reversible and adiabatic compression.

7-18 After being heated in the combustion chamber of a jet engine, air at low velocity, 100 psia, and 1600 F enters the jet nozzle. Determine the maximum velocity (ft/sec) that can be obtained from the nozzle if the air discharges from it at 11 psia.

7-19 An ideal gas turbine unit for power production has compressor inlet conditions of 15 psia and 60 F with a flow rate of 12,500 ft³/min. The pressure ratio across the compressor is 6/1. The maximum allowable temperature in the system is 2250 F. The unit burns fuel oil with a heating value of 139,000 Btu/gal at a cost of $1.10/gal. Determine:
a. Compressor power requirements (hp)
b. Rating of power plant (kW)
c. Thermal efficiency (%)
d. Hourly fuel cost ($/hr)

7-20 Steam flows at the rate of 12,000 lbm/min through a turbine from 500 psia and 700 F to an exhaust pressure of 1 psia. Determine the ideal output of the turbine (hp). If the specific entropy increases between inlet and outlet of the turbine by 0.1 Btu/lbm-R, determine the turbine efficiency.

7-21 In a conventional power plant, 1,500,000 lbm/hr of steam enters a turbine at 1000 F and 500 psia. The steam expands adiabatically to 1 psia with 98% quality. Determine: (a) the turbine rating (kW) and (b) the turbine efficiency (%).

7-22 A simple steam power plant burns coal that has a heating value of 11,480 Btu/lbm. Steam leaves the SGU at 500 psia and 900 F. Saturated steam leaves the turbine and enters the condenser at 2 psia. Determine: (a) the thermal efficiency of the cycle (%) and (b) the turbine efficiency (%).

7-23 During the operation of a simple steam power plant, the steam flow rate is 650,000 lbm/hr with turbine inlet conditions of 500 psia and 1000 F and turbine exhaust pressure of 1.0 psia. Turbine efficiency is 91%. Determine: (a) turbine output (kW), (b) each term in the second-law entropy equation for the turbine process, and (c) the thermal efficiency of the plant (neglecting pump work).

7-24 During the operation of a simple steam power plant, the steam flow rate is 500,000 lbm/hr with turbine inlet conditions of 500 psia and 1000 F and turbine exhaust (condenser inlet) conditions of 1.0 psia (90% quality). Determine:
a. Turbine output (kW)
b. Condenser heat rejection rate (Btu/hr)
c. Turbine efficiency (%)
d. Each term in the second-law entropy equation for the condensing process
e. Approximate pump work (kW)
f. Thermal efficiency of the plant (%)
g. Ideal thermal efficiency of the plant (%)

7-25 A simple nuclear steam power plant has a rated capacity of 500 000 kW when operating between limiting conditions of 500

psia and 600 F at the SGU outlet (also the turbine inlet) and 1 psia at the turbine outlet. Steam leaves the turbine with a quality of 90% and enters the condenser. Neglecting pump work, determine:
a. Heat added to the nuclear steam generating unit (Btu/hr)
b. Thermal pollution (Btu/hr)
c. Thermal efficiency of the power plant (%)
d. Maximum thermal efficiency possible for the plant (%)
e. Turbine efficiency (%)

7-26 For the steam power plant shown in the diagram below, determine the following list of quantities, assuming both the turbine and the pump are adiabatic and there are no kinetic or potential energy changes. Note $h[=]$ Btu/lbm and $s[=]$ Btu/lbm R
a. Each term in the second-law entropy equation applied to the turbine
b. Each term in the second-law entropy equation applied to the condenser
c. Turbine efficiency
d. Thermal efficiency of the cycle

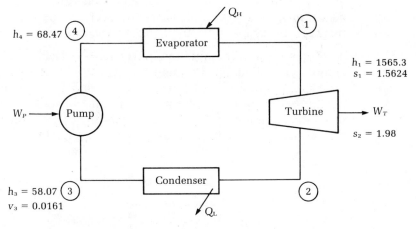

$p_1 = p_4 = 3500 \text{ lbf/in.}^2, \ p_2 = p_3 = 1 \text{ lbf/in.}^2, \ T_1 = 1200 \text{ F}, \ T_3 = 90 \text{ F}$

7-27 Steam is supplied to a turbine at 100 lbf/in.^{2}, 600 F. After produc-

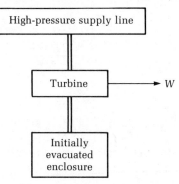

ing some work (W of the turbine) the steam is exhausted into an initially evacuated enclosure whose volume is 1000 ft³. The turbine stops producing work when the exit conditions are 100 lbf/in.², 550 F. If this process is adiabatic, calculate:

(a) The turbine work
(b) The entropy created (Btu/R)

7-28 A refrigerator utilizes Freon-12 as the refrigerant and handles 200 lbm/hr. The condensing temperature is 110 F and the evaporating temperature is 5 F. For a cooling effect of 11,000 Btu/hr, determine the minimum size of motor (hp) required to drive the compressor.

7-29 Freon-12 enters the condenser of a vapor-compression refrigeration system at 175 psia and 140 F and leaves as saturated liquid at 120 F. The mass flow rate of the refrigerant is 4.8 lbm/min. The heat is rejected to the surrounding air, which is at 90 F. Determine: (a) the heat rejection rate (Btu/hr) and (b) the separate and overall entropy changes per hour (Freon and surroundings).

7-30 Freon-12 enters the evaporator of a freezer at −20 F with a quality of 85%. The refrigerant leaves as saturated vapor. Determine the heat transfer per pound of refrigerant (a) by using the first law of thermodynamics and (b) by using the second law.

7-31 A refrigeration unit employing refrigerant-12 is shown in the accompanying diagram. Condensing pressure is 216 psia; evaporator temperature is −10 F. The unit is rated at 66,000 Btu/hr for cooling. Determine: (a) the minimum size of motor required to drive the compressor, (b) the corresponding COP$_c$ of the unit, and (c) the output of the system as a heat pump (Btu/hr).

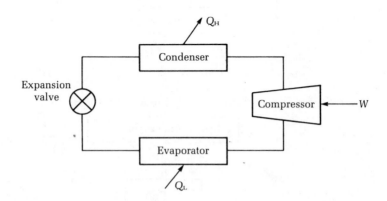

7-32 A heat pump using Freon-12 (shown in the sketch) is to be used for winter space heating of a residence. The building heat loss is 65,000 Btu/hr. The compressor process ideally will be reversible and adiabatic.

a. Determine the work required by the compressor (hp).
b. Determine the COP for heating.
c. If the same work is supplied to the compressor in the summer and the operating conditions of the refrigeration system remain unchanged, determine the rating of the unit as an air conditioner for cooling (Btu/hr).
d. Determine the COP for cooling.
e. Determine the EER for cooling (Btu/hr/W).

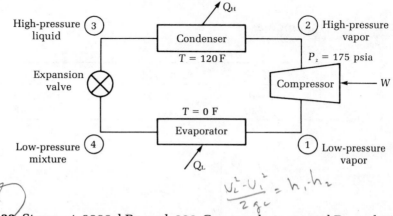

$$\frac{V_2^2 - V_1^2}{2 g_c} = h_1 \cdot h_2$$

7-33 Steam at 2000 kPa and 290 C expands to 1400 kPa and 245 C through a nozzle. If the entering velocity is 100 m/s and the ratio of specific heats (k) is 1.3, determine: (a) the exit velocity and (b) the nozzle efficiency. $_{m\,1.0}$ $419 \frac{m}{s}$

7-34 For the basic refrigeration cycle indicated in the sketch, determine η_R if the working fluid is assumed to be Freon-12.

$$T_2 = T_3 = 50 \text{ C}$$

$$T_1 = T_4 = -30 \text{ C}$$

$$h_4 = h_{f_3}$$

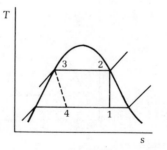

7-35 Consider a vapor-compression refrigeration cycle using ammonia as a working fluid. For the conditions stated: (a) make a component sketch and (b) determine η_R.

$$T_1 = T_4 = -10 \text{ C}$$
$$x_1 = 0.9$$
$$T_2 = T_3 = 20 \text{ C}$$
$$h_2 = 1732.3 \text{ kJ/kg}$$
$$h_3 = h_4$$

For ammonia:

T, C	h_f, kJ/kg	h_g, kJ/kg
−10	372.751	1669.152
20	512.381	1699.548

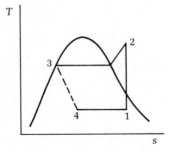

7-36 Determine the refrigeration capacity in tons and η_R for a refrigeration cycle using Freon-12 as the working fluid. The relevant data are listed below with the figure. Begin by making a component sketch.

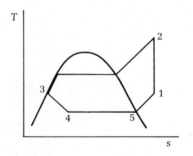

$$\dot{m} = 300 \text{ lbm/hr}$$

$p_1 = 25$ psia	$T_1 = 20$ F
$p_2 = 200$ psia	$T_2 = 170$ F
$p_5 = 29.3$ psia	$T_5 = 10$ F
	$T_3 = 100$ F

7-37 For the vapor-compression cycle indicated in the accompanying diagram, determine the cycle efficiency. The working fluid is Freon-12.

$$T_1 = 10 \text{ F}$$
$$T_3 = 110 \text{ F}$$
$$p_2 = 150 \text{ psia}$$

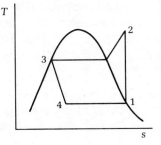

7-38 For the air-standard refrigeration cycle of the sketch (a) sketch a component schematic, (b) determine η_R, and (c) determine $\dot{m}$ for 12,000 Btu/hr of refrigeration.

$p_1 = p_4 = 14.7$ psia	$T_1 = 460$ R
$p_2 = p_3 = 80$ psia	$T_3 = 520$ R

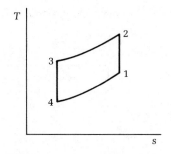

7-39 Consider a Carnot cycle refrigerator using Freon-12 as the working

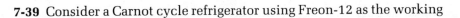

fluid. In the heat rejection portion of the cycle (at 100 F), the Freon changes from a saturated vapor to a saturated liquid. Heat input to the Freon occurs at 0 F.

a. Indicate this cycle on (p, v) and (T, S) diagrams.

b. Determine the qualities of the heat input process.

c. What is the coefficient of performance?

7-40 For the physical conditions of Problem 7-39 assume a heat pump and then rework the problem.

7-41 From a reservoir at 1200 F, a Carnot heat engine receives 600 Btu while rejecting heat at 100 F. Determine: (a) the net work and the efficiency and (b) the entropy change of the high- and low-temperature reservoirs.

7-42 Steam expands in a turbine operating on a Rankine cycle from 5000 kPa and 400 C to 40 kPa. Determine the power output of the steam if it is supplied at a rate of 136 kg/s.

7-43 The inlet conditions of 100 kPa and 20 C exist in an air-standard Brayton cycle. Determine (per lbm): (a) the compressor work, (b) the heat added, (c) the turbine work, (d) the cycle efficiency for a pressure ratio of 7/1 and the entering gas turbine temperature of 800 C.

7-44 An air-standard Otto cycle with a compression ratio of 7/1 has a heat input of 2100 kJ/kg. If the initial conditions of 100 kPa and 15 C are considered, determine the net work.

7-45 The following data are for a simple steam power plant (see sketch):

$$p_1 = 10.0 \text{ psia} \qquad T_1 = 160 \text{ F}$$

$$p_2 = 500 \text{ psia}$$

$$p_3 = 480 \text{ psia} \qquad T_3 = 800 \text{ F}$$

$$p_4 = 10 \text{ psia}$$

Steam flow rate = 225,000 lbm/hr

Turbine efficiency = 87%

ASSUME.

$T_1 = T_{SAT'D}$ (INSTEAD OF 160°)

IGNORE ΔKE TERMS

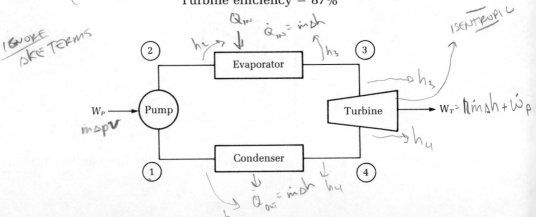

Determine:
a. Pipe size between condenser and pump if the velocity is not to exceed 25 ft/sec
b. Minimum pump work (hp)
c. Output of plant (kW)
d. Thermal pollution (condenser heat rejection) (Btu/hr)
e. Heat input at boiler (Btu/hr)
f. Thermal efficiency (%)
g. Maximum possible thermal efficiency (%)
h. Fuel cost ($/hr) using coal with a heating value of 10,400 Btu/lbm and a cost of $33/ton

7-46 The cycle shown in the sketch is used for air conditioning aircraft and has air as the working fluid. Considering the compression process as ideal, determine:
a. Net work required (hp) per ton of refrigeration (12,000 Btu/hr)
b. Heat rejected at the heat exchanger (Btu/hr)
c. Turbine efficiency (%)
d. Coefficient of performance

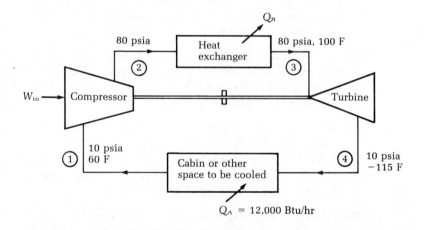

7-47 A four-stroke compression-ignition (diesel) engine, for which fuel is injected during the compression stroke, takes in ambient air at 60 F and 14.7 psia. The engine runs at 1900 rpm. Its fuel/air ratio is 0.058 with a fuel having a heating value of 19,000 Btu/lbm. The six cylinders have a 4-in. bore and a 4.5-in. stroke. For an assumed overall thermal efficiency of 28%, determine the output (hp). Let $r_c = V_1/V_2 = 12$ and $V_{CD} = V_4/V_3 = 3$. How much fuel is used per hour?

7-48 The power output of a steam turbine is 30 000 kW. Determine the rate of steam flow in the turbine if the inlet conditions are 0.8 MPa and 250 C, the outlet conditions are 0.1 MPa, and the expansion is reversible adiabatic.

7-49 Steam enters a diffuser at 700 m/s, 0.2 MPa, and 200 C. It leaves the diffuser at 70 m/s. Assuming reversible adiabatic operation, what would be the final pressure and temperature? (*Hint:* The (h, s) diagram of Chapter 2 may be helpful.)

7-50 Gas enters a turbine at 550 C and 0.5 MPa and leaves at 0.1 MPa. The entropy change is 0.174 kJ/(kg · K) (only approximately adiabatic). What is the temperature of the gas leaving the turbine if you assume that the gas is ideal with c_p = 1.11 kJ/(kg · K) and c_v = 0.835 kJ/(kg · K)?

7-51 Air is compressed through a pressure ratio of 4 : 1. Assume that the process is steady flow and adiabatic and that the temperature increases by a factor of 1.65. Calculate the entropy change.

7-52 Water enters a pump at 0.01 MPa and 35 C and leaves at 5 MPa. For reversible adiabatic operation, calculate the work done and the exit temperature.

7-53 Air is compressed through a pressure ratio of 8 : 1 in a steady-flow process. If the inlet conditions are 0.1 MPa and 25 C, calculate the work, the heat transfer, and the entropy change per kg if the process is polytropic $(n$ = 1.25). Show on (T, S) and (p, v) diagrams how this process changes if it is adiabatic.

7-54 Calculate the exit velocity and temperature (if superheated) or quality (if saturated) of a nozzle whose efficiency is 95%. Steam enters at 0.8 MPa and 200 C and leaves at 0.2 MPa. Assume that the entrance velocity is zero.

7-55 In a power plant operating on a Rankine cycle, steam at 0.4 MPa and quality 100% enters the turbine while its pressure is reduced to 0.0035 MPa in the condenser. Determine the cycle efficiency. How is the efficiency changed if the turbine inlet conditions are changed to 4.0 MPa and 350 C?

7-56 Air is compressed from 0.1013 MPa and 15 C to 0.7 MPa. Determine the power required to process 0.3 m³/min at the outlet if the operation is (a) polytropic $(n$ = 1.25) and (b) isentropic.

7-57 Consider the converging-diverging nozzle shown in the sketch. For the conditions stated, calculate the throat and exit cross-sectional areas if air is the fluid. (*Hint:* Recall continuity and let k = 1.4.)

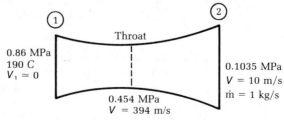

7-58 Consider a Rankine cycle using Freon-12 as the working fluid. Saturated vapor leaves the boiler at 85 C while the condenser temperature is 40 C. What is the cycle efficiency?

7-59 Consider a Brayton cycle using air as the working fluid. At the compressor inlet the air has the conditions of 0.102 MPa and 15 C while the pressure is increased to 0.612 MPa at the compressor outlet. If the maximum cycle temperature is 800 C, what is the cycle efficiency?

7-60 Assume that your car engine has a compression ratio of 8/1. If the ambient conditions are 0.1 MPa and 15 C, determine the cycle efficiency if 1800 kJ/kg of energy is transferred to the air every time your engine turns over. What is the heat rejected to the atmosphere?

7-61 Rework Problem 7-60 assuming a diesel engine with a compression ratio of 16/1.

7-62 Work may be computed from the relation

$$w = -\int_1^2 v \, dp$$

for a reversible adiabatic steady-flow process (a pump). For the usual assumption of pump operation, prove that if the working fluid is an ideal gas, the relation can be manipulated to be

$$w = \frac{kRT_1}{k-1}\left[1 - \left(\frac{p_2}{p_1}\right)^{(k-1)/k}\right]$$

7-63 Consider the air-standard refrigeration cycle of Figure 7-56. If air enters the compressor at 0.1 MPa (−20 C) and experiences a compression to 0.5 MPa, determine the cycle efficiency if the air that enters the expander has a temperature of 15 C.

7-64 The air for burning coal in a magnetohydrodynamic (MHD) power

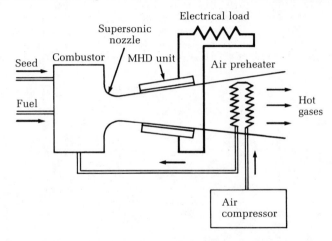

plant is to be heated from 500 to 3100 F before entering the MHD combustion chamber (see the sketch). Determine the amount of heat to be added to the air per hour if 400 tons/hr of coal is to be burned with the air ($12\frac{1}{2}$ lbm of air per pound of coal).

7-65 Geothermal water is available at 450 F. (See the system shown in the diagram.) It is then cooled to 350 F and used to boil a secondary fluid at a temperature of 340 F. The secondary fluid vapor passes through a turbine and is condensed at 100 F. The efficiency of the turbine is 60% that of a Carnot cycle engine operating under the same temperature conditions. If the specific heat of the hot water is 1.2 Btu/lb-F, determine the amount of water required per minute to heat the secondary fluid and produce 10^5 kW of electrical energy.

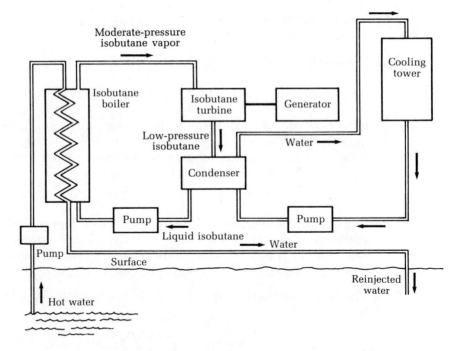

7-66 The accompanying sketch illustrates a hydroelectric generating

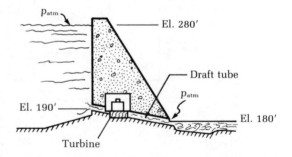

station. The diameter of the draft tube is 12 ft. When operating at full capacity the system accommodates a volume flow rate of 25×10^3 ft³/min. Determine the pressures at the inlet and exit to the turbine and the maximum power generation rate that could be developed by the turbine if it operates reversibly.

7-67 A pumped-storage power plant is shown in the following sketch. It is designed to produce an average of 350 000 kW of electrical power in an 8-hr period with an average elevation difference of 85 ft. The overall efficiency is 70%. What volume of water must be stored?

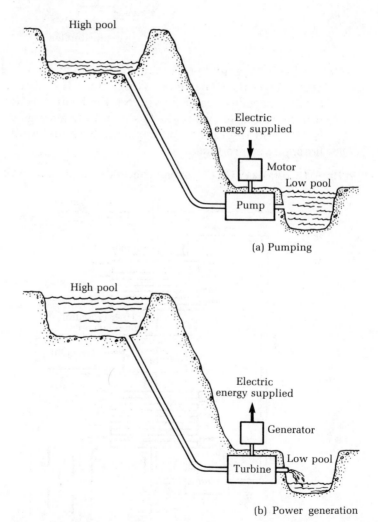

(a) Pumping

(b) Power generation

7-68 Assume that the efficiency of the generator shown in the sketch is 90%. If the flow of the river is 1000 ft³/sec, determine the average power output.

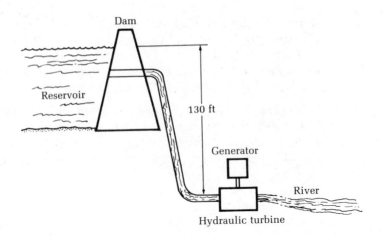

7-69 The basin for a tidal power turbine is 1 mi wide, 4 mi long, and 4 ft deep. When the tide comes in (high tide), the basin is filled. When the tide recedes, water flows from the basin through the turbine. If it takes 6 hr to empty the basin and the average working height is $3\frac{1}{2}$ ft and the turbine efficiency is 90%, determine the average horsepower produced.

7-70 A 125 000-kW OTEC (ocean thermal energy conversion) plant is to

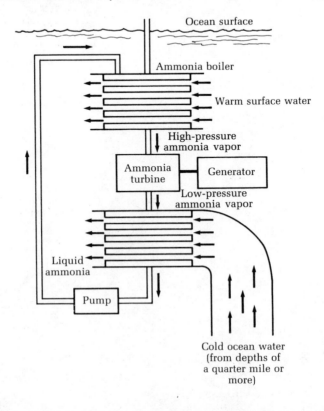

be installed in the Gulf Stream off the coast of Florida (see the sketch). Assume that the surface water temperature is 78 F and the temperature at a depth of 2500 ft is 43 F. If the efficiency of the plant is 45% that of the Carnot cycle plant for the given temperatures, how much heat is given up per second by the warm surface water?

7-71 A solar water heater is designed to heat water from 30 to 85 C. (See the sketch.) Even though the bottom portion of the device is insulated, some power is radiated back out of the glass top (about 250 W/m²). On a clear day the irradiation from the sun is about 1100 W/m². How many square feet would be required to produce a hot-water flow of 10 gal/min?

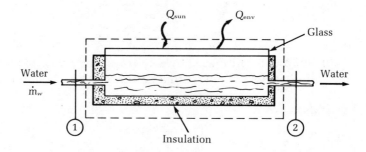

7-72 A fan is tested in the constant-area duct shown in the diagram. The duct is well insulated. If the duct area is 0.60 m², determine the actual power requirement and the fan efficiency (the ratio of the reversible to the actual power requirement) for the conditions shown.

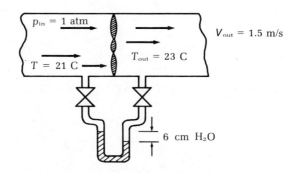

7-73 A normal human heart pumps 5000 cm³/min of blood to the body. Blood returns to the heart through the vena cava (superior and inferior) at a gauge pressure of 10 mm Hg. Blood is pumped to the lungs through the pulmonary artery at pressures varying between 24 and 29 mm Hg. Blood is pumped to the body (via the aorta) at pressures varying between 80 and 125 mm Hg. If the diameters of the aorta, vena cava, and pulmonary arteries are about 1 cm, determine the minimum rate of work done by the heart. Assume the average blood density is 1 g/cm³.

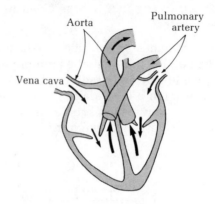

7-74 For a windmill (as shown in the accompanying figure) prove that the maximum power that could be taken from the air by the mill is

$$W_{max} = \frac{\pi D^2 \rho V^2}{8g_c}$$

Notice that this expression indicates that the power varies as the cube of the wind velocity and as the square of the windmill diameter.

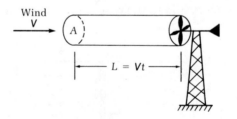

7-75 Consider the closed feedwater heater of a steam power plant

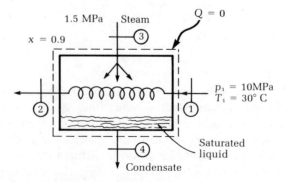

shown in the sketch. The high-pressure liquid water enters heating tubes and is heated up to the saturation temperature of the steam, which is condensing. Calculate the mass of steam required per unit mass of incoming liquid.

7-76 For the cycle discussed in Problem 6-45, determine the thermal efficiency.

7-77 For the Stirling cycle in the sketch, determine whether the efficiency is

$$\eta = \frac{R(T_2 - T_1) \ln(v_3/v_2)}{c_v(T_2 - T_1) + RT_2 \ln(v_3/v_2)}$$

The working fluid may be assumed to be an ideal gas.

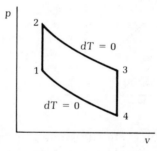

7-78 For the Ericsson cycle in the sketch, determine whether the efficiency is

$$\eta = \frac{R(T_2 - T_1) \ln(v_3/v_2)}{c_v(T_2 - T_1) + RT_2 \ln(v_3/v_2)}$$

The working fluid may be assumed to be an ideal gas.

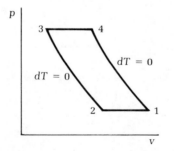

7-79 The following (p, v) and (T, s) diagrams represent the dual cycle—the result of a combination of the Otto and Diesel cycle processes. Note that processes 2–3 and 5–1 are constant volume whereas 3–4 is constant pressure and 1–2 and 4–5 are isentropic. Assuming that air is the working fluid, determine the efficiency η

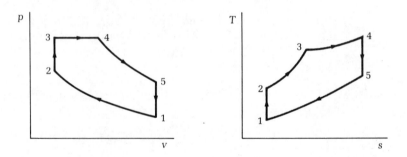

7-80 The following (p, v) diagrams represent the Brown (a) and Lenoir (b) cycles. Compare the efficiencies of these cycles.

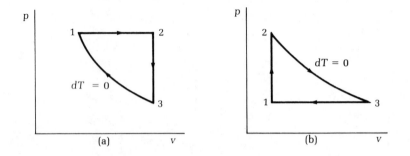

7-81 For the two cycles illustrated in Problem 7-80, make sketches of (T, s) diagrams. Assume 1 kg of working fluid (air) in executing these cycles. Calculate all of the heat transfer values if the maximum pressure change is 0.01 MPa (that is, 0.001 to 0.011 MPa) while the minimum volume in both cases is 0.006 m³.

8

Some Cycle Improvements

In Chapter 7 we discussed the Rankine and Brayton cycles, but only the basic situations. In this chapter we look at some improvements. These improvements include reheating and regeneration of the Rankine cycle and regeneration of the Brayton cycle.

8–1 Review of Basic Information

Figures 8-1 to 8-4 summarize the basic information from Chapters 5 and 7 that deals with the major cycles. Figure 8-1 presents heat engine and refrigerator–heat pump information; included are general definitions of thermal efficiency (coefficient of performance) and the expression that results if the governing cycle is Carnot. Figure 8-2 presents the relevant information for a basic steam power plant operating on a Rankine cycle. Figure 8-3 presents the same information for the basic gas turbine system. The air-standard Otto cycle is presented in Figure 8-4.

Figure 8-1. Schematic of heat engine and refrigerator–heat pump information.

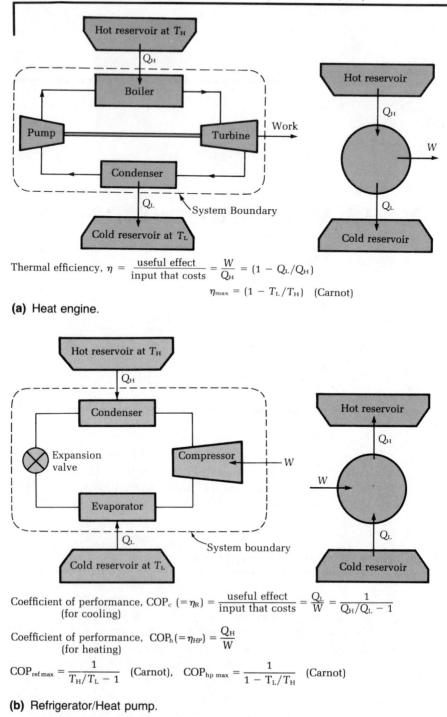

Thermal efficiency, $\eta = \dfrac{\text{useful effect}}{\text{input that costs}} = \dfrac{W}{Q_H} = (1 - Q_L/Q_H)$

$$\eta_{max} = (1 - T_L/T_H) \quad \text{(Carnot)}$$

(a) Heat engine.

Coefficient of performance, $\text{COP}_c \ (= \eta_R) = \dfrac{\text{useful effect}}{\text{input that costs}} = \dfrac{Q_L}{W} = \dfrac{1}{Q_H/Q_L - 1}$
(for cooling)

Coefficient of performance, $\text{COP}_h (= \eta_{HP}) = \dfrac{Q_H}{W}$
(for heating)

$\text{COP}_{\text{ref max}} = \dfrac{1}{T_H/T_L - 1}$ (Carnot), $\quad \text{COP}_{\text{hp max}} = \dfrac{1}{1 - T_L/T_H}$ (Carnot)

(b) Refrigerator/Heat pump.

Figure 8-2. Basic steam power plant (Rankine).

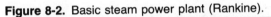

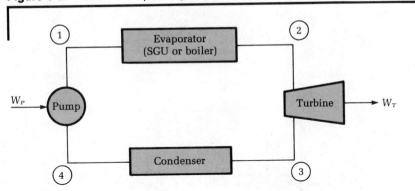

Ideal Rankine cycle:
4–1: reversible adiabatic pumping process in the pump
1–2: constant-pressure transfer of heat in the boiler
2–3: reversible adiabatic expansion in the turbine (or other prime mover such as a steam engine)
1–4: constant-pressure transfer of heat in the condenser

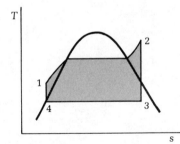

SGU:

$$\dot{m}\left[\left(h_1 + \frac{V_1^2}{2g_c}\right) - \left(h_2 + \frac{V_2^2}{2g_c}\right)\right] + {}_1\dot{Q}_2 = 0 \quad \text{where } Q_A = {}_1Q_2$$

Turbine:

$$\dot{m}\left[\left(h_2 + \frac{V_2^2}{2g_c}\right) - \left(h_3 + \frac{V_3^2}{2g_c}\right)\right] + {}_2\dot{Q}_3 - {}_2\dot{W}_3 = 0 \quad \text{where } W_t = {}_2W_3$$

Condenser:

$$\dot{m}\left[\left(h_3 + \frac{V_3^2}{2g_c}\right) - \left(h_4 + \frac{V_4^2}{2g_c}\right)\right] + {}_3\dot{Q}_4 = 0 \quad \text{where } Q_R = -{}_3Q_4$$

Pump:

$$\dot{m}\left[\left(h_4 + \frac{V_4^2}{2g_c}\right) - \left(h_1 + \frac{V_1^2}{2g_c}\right)\right] - {}_4\dot{W}_1 = 0 \quad \text{where } W_P = -{}_4W_1$$

$$W_p \approx v_4(p_1 - p_4)\frac{144}{778}$$

(Continued on next page.)

Figure 8-2. (*continued*)

Overall:
$$Q_A + W_P = Q_t + W_t + Q_R$$

or
$$(_1Q_2 + {_2}Q_3 + {_3}Q_4) - (_2W_3 + {_4}W_1) = 0$$

Thermal efficiency:
$$\eta = \frac{W_t - W_p}{Q_A} \times 100$$

Steam-generating unit: economizer, boiler, superheater

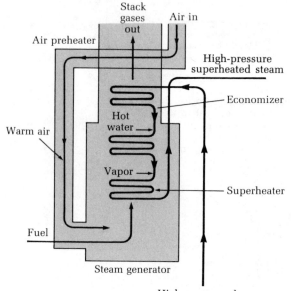

High-pressure, low-
temperature water to boiler

Figure 8-3. Basic gas turbine system (Brayton or Joule).

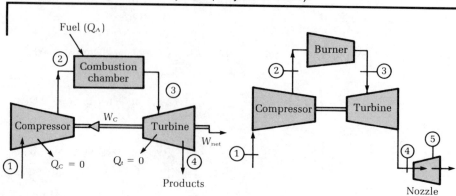

A gas turbine operating on the Brayton cycle

Gas-turbine cycle for a jet engine

Figure 8-3. (*continued*)

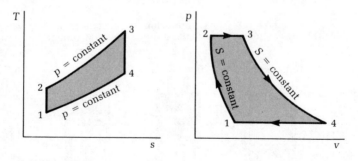

The air-standard Brayton cycle

Power plant:
1. No nozzle
2. $W_{net} = W_t - W_c$

3. Thermal efficiency $= \dfrac{W_{net}}{Q_A} \times 100$

Turbojet:
1. No W_{net}; $W_t = W_c$.
2. High V_5 is desired quantity for jet propulsion.

Ideal Brayton cycle:
Compressor: reversible and adiabatic; $S = c$; $T_2 = T_1(p_2/p_1)^{(k-1)/k}$
Turbine: reversible and adiabatic; $S = c$; $T_4 = T_3(p_4/p_3)^{(k-1)/k}$
Nozzle: reversible and adiabatic; $S = c$; $T_5 = T_4(p_5/p_4)^{(k-1)/k}$
Combustion chamber: $p = c$, reversible

$$p_a v_a = RT_a \qquad h_a - h_b = c_p(T_a - T_b)$$

Air: $c_p = 0.24$; $R = 53.3$; $k = 1.4$

Compressor: $\qquad \dot{m}\left[\left(h_1 + \dfrac{V_1^2}{2g_c}\right) - \left(h_2 + \dfrac{V_2^2}{2g_c}\right)\right] + {}_1\dot{Q}_2 - {}_1\dot{W}_2 = 0 \qquad$ where $W_c = -{}_1W_2$

Combustion chamber: $\qquad \dot{m}\left[\left(h_2 + \dfrac{V_2^2}{2g_c}\right) - \left(h_3 + \dfrac{V_3^2}{2g_c}\right)\right] + {}_2\dot{Q}_3 = 0 \qquad$ where $Q_a = {}_2\dot{Q}_3$

$$Q_a = \dot{m}_{fuel}\, HV_{fuel}$$

Turbine: $\qquad \dot{m}\left[\left(h_3 + \dfrac{V_3^2}{2g_c}\right) - \left(h_4 + \dfrac{V_4^2}{2g_c}\right)\right] + {}_3\dot{Q}_4 - {}_3\dot{W}_4 = 0 \qquad$ where $W_t = {}_3W_4$

Nozzle: $\qquad \dot{m}\left[\left(h_4 + \dfrac{V_4^2}{2g_c}\right) - \left(h_5 + \dfrac{V_5^2}{2g_c}\right)\right] = 0$

Figure 8-4. Air-standard spark ignition cycle (Otto).

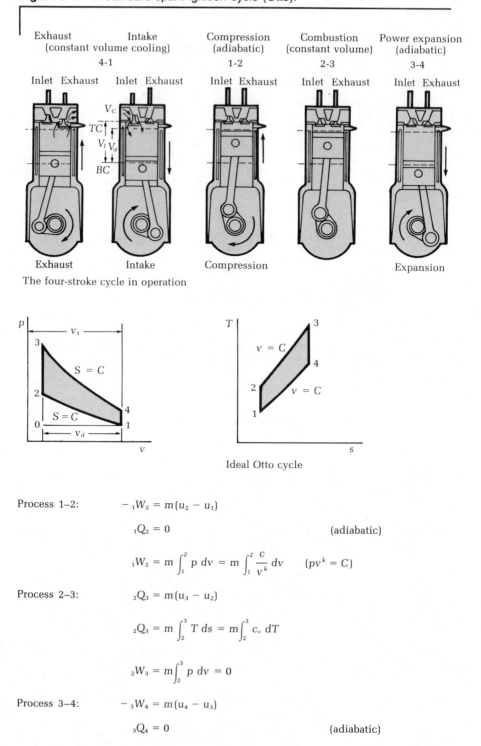

The four-stroke cycle in operation

Ideal Otto cycle

Process 1–2: $- {_1}W_2 = m(u_2 - u_1)$

$${_1}Q_2 = 0 \qquad\qquad\qquad \text{(adiabatic)}$$

$${_1}W_2 = m\int_1^2 p\,dv = m\int_1^2 \frac{C}{v^k}\,dv \qquad (pv^k = C)$$

Process 2–3: $_2Q_3 = m(u_3 - u_2)$

$${_2}Q_3 = m\int_2^3 T\,ds = m\int_2^3 c_v\,dT$$

$${_2}W_3 = m\int_2^3 p\,dv = 0$$

Process 3–4: $- {_3}W_4 = m(u_4 - u_3)$

$${_3}Q_4 = 0 \qquad\qquad\qquad \text{(adiabatic)}$$

Figure 8-4. (*continued*)

$$_3W_4 = m \int_3^4 p\, dv = m \int_3^4 \frac{C}{v^k}\, dv \qquad (pv^k = C)$$

Process 4–1:

$$_4Q_1 = m(u_1 - u_4)$$

$$_4Q_1 = m \int_4^1 T\, ds = m \int_4^1 c_v\, dT$$

$$_4W_1 = m \int_4^1 p\, dv = 0$$

Overall:

$$W_{\text{net}} = {}_1W_4 + {}_2W_3 + {}_3W_4 + {}_4W_1 = {}_1W_2 + {}_3W_4$$

$$Q_{\text{net}} = {}_1Q_2 + {}_2Q_3 + {}_3Q_4 + {}_4Q_1 = {}_2Q_3 + {}_4Q_1$$

$$W_{\text{net}} = Q_{\text{net}}$$

Thermal efficiency

$$= \frac{W_{\text{net}}}{_2Q_3} \times 100$$

Compression ratio

$$= \frac{v_1}{v_2} = \frac{v_4}{v_3}$$

8–2 Improving the Rankine Cycle

The two primary improvement procedures for the Rankine cycle are reheating and regenerating.

Reheating

The reheat of the Rankine cycle was developed to take advantage of higher efficiency at higher pressures while avoiding excessive moisture in the low-pressure portion of the turbine. Figure 8-5 illustrates this

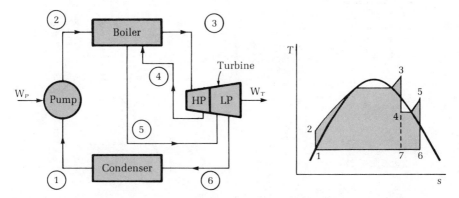

Figure 8-5. The ideal reheat cycle.

situation schematically and on a (T, s) diagram. The steam is expanded to some intermediate pressure in the turbine (4—the high-pressure side of the turbine to the exhaust pressure (6). The real advantage of this procedure is not a great increase in the efficiency but, rather, a lowering side of the turbine to the exhause pressure (6). The real advantage of this procedure is not a great increase in the efficiency but, rather, a lowering of the moisture content of the exhaust vapor. (Compare the quality of point 6 with point 7; $x_6 > x_7$.) From the (T, s) diagram you can see that to increase the efficiency you must design to high reheat pressure and temperature. This is dangerous and costly.

Regenerating

The regenerative procedure for the Rankine cycle involves the use of feedwater heaters. Figure 8-6 illustrates an ideal situation. The procedure may best be understood by looking at the (T, s) diagram in Figure 8-6. During process 1–2–3, the working fluid (water) is heated. If the energy needed to heat this liquid to the saturated state could be supplied by waste heat from the system, energy costs could be saved. Thus if the liquid circulates around the turbine casing after leaving the pump (2), it is possible to transfer heat from the vapor as it flows through the turbine (4–5) to the liquid flowing around the turbine (2–3). Since this is an ideal system, the reversible heat transfer (4–5) just compensates (1–2–3). That is, areas 2–3-b-a- 2 and 5–4-d-c -5 are equal and congruent and represent the heat transferred to the liquid and from the vapor. With a little effort, it may be noted that the heat rejected from the ideal cycle 5–c-b-a -1-1'-5 is exactly equal to that rejected by 5'–d-c-b -1'-5-5'. Similarly, the input for the ideal cycle is 4–5'-d-c-b -1'-3-4.

It would be unwise to try to implement this idealized regenerative

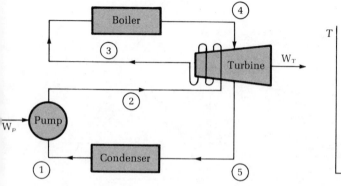
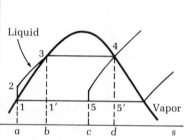

Figure 8-6. The ideal regeneration cycle.

Rankine cycle, even though the efficiency is exactly equal to the Carnot cycle. There are two major drawbacks to the implementation:

1. The heat transfer from the turbine to the liquid feedwater is not 100% efficient.

2. Because of the increased heat transfer from the turbine, the quality of the steam leaving the turbine is low. (It is wetter.)

As a result, a usable Rankine regenerative cycle (Figure 8-7) may be built which uses only a portion of the vapor from the high-pressure side of the turbine (3). It enters a feedwater heater. The remainder of the vapor continues to condense as it proceeds to the low-pressure side of the turbine (4). This portion then enters the condenser. The saturated liquid (5) is pumped into the feedwater heater (6). The heat transfer which occurs in this feedwater heater between the higher temperature steam (from 3) and this saturated liquid (from 6) should result in a saturated liquid (only) at a higher temperature (7). From this position (7) the cycle is like the original Rankine cycle in that a second pump is used to raise the pressure for the boiler, and so on.

One should notice immediately that the (T, s) diagram of Figure 8-7 is not entirely accurate. This is because there is no convenient way to point out on the diagram that the mass flow is not the same for the various paths. That is, the heat transfer to the working fluid is the area under the process 1–2. The heat transfer from the working fluid occurs to only a portion of it (that went through the condenser) and is the area under the process 4–3. The other portion leaving the turbine has heat transfer to it.

Modern steam power plants have one or more stages of reheating and a number of regenerative heaters. Figure 8-8 illustrates such a plant.

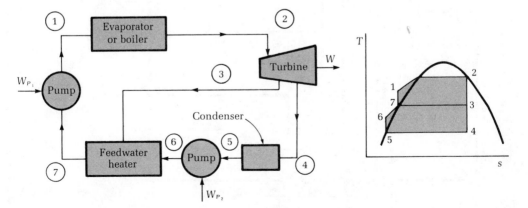

Figure 8-7. Regenerative cycle with open feedwater heater.

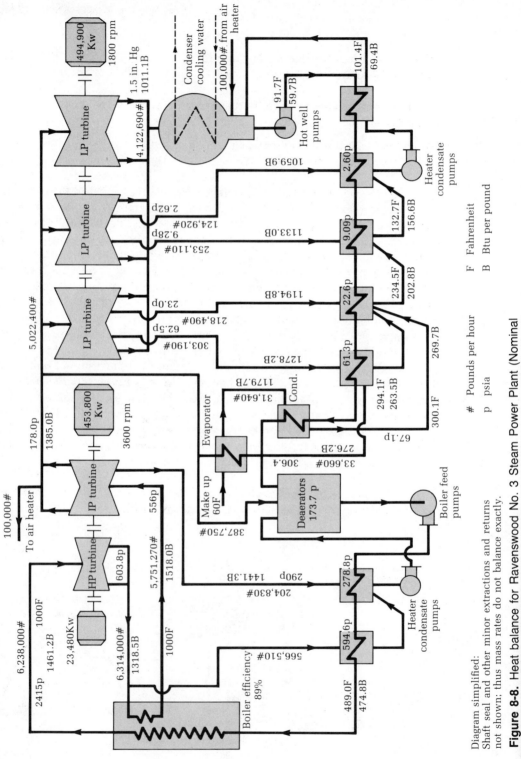

Figure 8-8. Heat balance for Ravenswood No. 3 Steam Power Plant (Nominal ... Consolidated Edison Co. of New York

Diagram simplified:
Shaft seal and other minor extractions and returns
not shown; thus mass rates do not balance exactly.

\# Pounds per hour
p psia
F Fahrenheit
B Btu per pound

EXAMPLE 8-1

Determine the cycle efficiency of a steam power plant that operates on a Rankine reheat cycle. Steam enters the high-pressure turbine at 700 psia, 800 F. The exhaust condition of this turbine is 60 psia. Upon reheating to 800 F, the steam enters the low-pressure turbine which has an exhaust pressure of 1 psia.

SOLUTION:

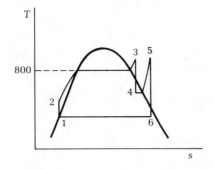

Consider a control surface around the turbine:

First law: $w_t = (h_3 - h_4) + (h_5 - h_6)$

Second law: $s_4 = s_3$

$$s_6 = s_5.$$

First we must find w_t. From Appendix A-1-1, we get $h_3 = 1403.7$ Btu/lbm and $s_3 = 1.6154$ Btu/lbm-F. Thus

$$s_4 = s_3 = 1.6154 = 1.6444 - (1 - x_4)1.2170$$

$$= 1.6154 = 1.6440 - (1 - x_4)1.2167$$

Hence

$$x_4 = 0.9764$$

$$h_4 = 1177.6 - 0.02351(915.4) = 1156.1$$

$$h_5 = 1431.3 \text{ Btu/lbm} \quad \text{and} \quad s_5 = 1.9024 \text{ Btu/lbm-F}$$

$$s_5 = s_6 = 1.9024 = 1.9781 - (1 - x_6)1.8455$$

Hence

$$x_6 = 0.9589$$

$$h_6 = 1105.8 - 0.0411(1036.1) = 1063.3$$

$$w_t = (1403.7 - 1156.1) + (1431.3 - 1063.3) = 615.6 \text{ Btu/lbm}$$

Next w_p must be obtained:

First law: $-w_p = h_2 - h_1$

Second law: $s_2 = s_1$

Since $s_2 = s_1$,

$$h_1 - h_2 = \int_1^2 v \, dp = v(p_2 - p_1)$$

Therefore,

$$-w_p = v(p_2 - p_1) = 0.01614(700 - 1)\frac{144}{778} = 2.088 \text{ Btu/lbm}$$

$$h_2 = 69.7 + 2.088 = 71.79$$

Thus the net work $(w_t - w_p) = 613.5$ Btu/lbm. To determine the heat input, we must consider the boiler. Thus

$$q_H = (h_3 - h_2) + (h_5 - h_4)$$

$$= (1403.7 - 71.8) + (1431.3 - 1156.1) = 1607 \text{ Btu/lbm}$$

Hence

$$\eta_{\text{thermal}} = \frac{w_{\text{net}}}{q_H} = 0.382$$

If this problem were reworked without reheating, the efficiency would be 0.376 and the moisture content at 1 psia would be 0.1969—not much change in efficiency, but a large decrease in the moisture content with reheating.

EXAMPLE 8-2

Compare the effect of reheating on a steam power plant's efficiency (Rankine cycle). In both cases, assume that the boiler pressure is 4.2 MPa while the condenser pressure is 0.0035 MPa. Also, superheating of the steam occurs to 500 C. For case A, consider simple superheating; for case B, include reheating from a saturated vapor to the initial turbine temperature. All processes are reversible in both cases.

SOLUTION:

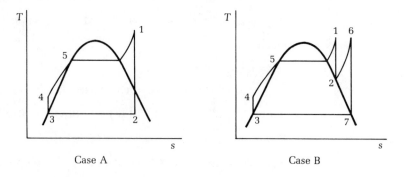

Case A   Case B

Consider the control surface around the turbine for case A.

First law: $w_t = h_1 - h_2$

Second law: $s_1 = s_2$

First we must find w_t. From Appendix A-1-5, we get

$$h_1 = 3442.6 \text{ kJ/kg} \qquad s_1 = 7.066 \text{ kJ/kg}$$

Thus

$$s_1 = s_2 = s_{f_2} + x_2 s_{fg_2} = 7.066 = 0.391 + x_2 8.13$$

Hence

$$x_2 = 0.821$$

$$h_2 = h_{f_2} + x_2 h_{fg_2} = 4771.6 \text{ kJ/kg}$$

and

$$w_t = h_1 - h_2 = 1329 \text{ kJ/kg}$$

Next we determine the heat input q_H:

First law: $q_H = h_1 - h_3 = 3330.6 \text{ kJ/kg}$

For convenience, we neglect the pump term. Thus

$$\eta = \frac{w_t}{q_H} = \frac{h_1 - h_2}{h_1 - h_3}$$

$$= \frac{1329.6}{3330.6} = 0.399 \qquad \text{or } 39.9\%$$

For case B, h_1, h_3, and s_1 have the same values they did for case A. The corresponding point 2 in this case is much different. That is, h_2 must be determined from the tables as the point where $s_1 = s_2 = s_g$. From Appendix A-1-5, we see that this occurs at a pressure of approximately 0.23 MPa. The resulting $h_g = h_2 = 2713 \text{ kJ/kg}$. The work of the turbine is again determined by the first law:

$$w_t = (h_1 - h_2) + (h_6 - h_7)$$

Again from Appendix A-1-7:

$$h_6(0.23 \text{ MPa}; 500 \text{ C}) = 3487 \text{ kJ/kg}$$

and

$$h_7(0.0035 \text{ MPa}; s = 8.513 \text{ kJ/kg} \cdot \text{K}) = 2550 \text{ kJ/kg}$$

So

$$w_t = 1667 \text{ kJ/kg}$$

Again let us determine the heat in by use of the first law:

$$q_H = (h_1 - h_3) + (h_6 - h_2)$$

$$= 4105 \text{ kJ/kg}$$

Again neglecting the pump term, we get

$$\eta = \frac{w_t}{q_H} = \frac{1667}{4105} = 0.406 \text{ or } 40.6\%$$

The reheating effect is very small.

EXAMPLE 8-3

Determine the cycle efficiency of a steam power plant that operates on a regenerative Rankine cycle. Steam enters the turbine at 700 psia, 800 F. Some of the steam is extracted when the pressure is 60 psia and put into the feedwater heater (also at 60 psia). The remainder of the steam in the turbine is exhausted at 1 psia. A saturated liquid leaves the feedwater heater.

SOLUTION:

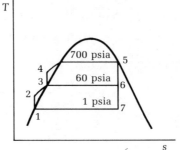

For pump w_{p1}:

First law: $-w_{p1} = h_2 - h_1$

Second law: $s_2 = s_1$

Therefore

$$h_2 - h_1 = v(p_2 - p_1)$$

$$-w_{p1} = v(p_2 - p_1) = 0.01614(60 - 1)\frac{144}{778}$$

$$= 0.2 \text{ Btu/lbm}$$

$$h_2 = h_1 - w_{p1} = (69.7 + 0.2) \text{ Btu/lbm} = 69.9 \text{ Btu/lbm}$$

$$h_3 = 262.2 \text{ Btu/lbm}$$

For the turbine:

First law: $w_t = (h_5 - h_6) + (1 - m_1)(h_6 - h_7)$

Second law: $s_5 = s_6 = s_7$

As in Example 8-1 (h_6 here $= h_4$ in Example 8-1):

$$h_6 = 1156.1 \text{ Btu/lbm}$$

and with a little effort you may determine that $h_7 = 901.8$ Btu/lbm and $(1 - x_7) = 0.1969$. For the feedwater heater:

First law: $m_1 h_6 + (1 - m_1)h_2 = h_3$

$$m_1(1156.1) + (1 - m_1)69.9 = 262.2$$

Hence

$$m_1 = 0.1772$$

Thus

$$w_t = (h_5 - h_6) + (1 - m_1)(h_6 - h_7)$$

$$= (1403.7 - 1156.1) + (1.0 - 0.1772)(1156.1 - 901.8)$$

$$= 456.1 \text{ Btu/lbm}$$

For the high-pressure pump w_{p2}:

First law: $-w_{p2} = h_4 - h_3$

Second law: $s_2 = s_3$

Hence

$$-w_{p2} = v(p_4 - p_3) = 0.01738(700 - 60)\frac{144}{778} = 2.059 \text{ Btu/lbm}$$

$$w_{net} = w_t + (1 - m_1)w_{p1} + w_{p2} = 456.1 - 0.08229(0.2) - 2.06$$

$$= 453.9 \text{ Btu/lbm}$$

The heat input to the boiler is

$$q_H = h_5 - h_4 = 1402.9 - 264.3 = 1139.4 \text{ Btu/lbm}$$

$$\eta_{thermal} = \frac{w_{net}}{q_H} = \frac{453.9}{1139} = 0.398$$

EXAMPLE 8-4

Compare the effect of regeneration on a steam power plant's efficiency (Rankine cycle). In both cases assume that the boiler pressure is 4.2 MPa while the condenser pressure is 0.0035 MPa. For case A, consider a simple Rankine cycle; for case B, include one feedwater heater (see the sketch).

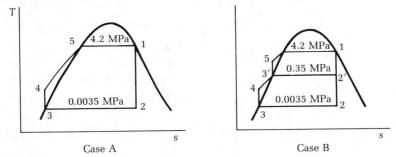

Case A   Case B

SOLUTION: Case A has already been considered as Example 7-14. The resulting efficiency was 36.8%. Thus let us proceed to case B. In both cases, $T_1 = 253$ C and $T_2 = 26.7$ C. And from Appendix A-1-6, we get

$$T_2' = 139 \text{ C}$$

Appealing again to the first law, we may determine the fraction of the mass going through regeneration. For the heater:

$$m_1 h_2 + (1 - m_1)h_4 = h_3'$$

$$m_1 = \frac{h_3' - h_4}{h_2' - h_4}$$

From the appendix,

$$h_3' = 584 \text{ kJ/kg} \qquad h_3 = 112 \text{ kJ/kg}$$

and

$$s_2' = s_1 = s_2 = 6.049 \text{ kJ/(kg} \cdot \text{K)}$$

With a little effort you may deduce that

$$s_2' = 0.829 \text{ kJ/(kg} \cdot \text{K)}$$

$$s_2 = 0.696 \text{ kJ/(kg} \cdot \text{K)}$$

Hence

$$h_2' = h_{f_2'} + x_2' h_{fg_{2'}} = 2364 \text{ kJ/kg}$$

$$h_2 = h_{f_2} + x_2 h_{fg_2} = 1808 \text{ kJ/kg}$$

Therefore

$$m_1 = \frac{584 - 112}{2364 - 112} = 0.21$$

Thus the turbine work, using the first law, is

$$w_t = (h_1 - h_2') + (1 - m_1)(h_2' - h_2)$$

$$= (2800 - 2364) + (1 - 0.21)(2364 - 1808)$$

$$= 876 \text{ kJ/kg}$$

The heat into the boiler is

$$q_H = h_1 - h_3' \qquad \text{(the first law again)}$$

$$= 2800 - 584 = 2216 \text{ kJ/kg}$$

Neglecting the work of the two pumps, we obtain the efficiency:

$$\eta = \frac{w_t}{q_H} = \frac{876}{2216} = 0.396 \text{ or } 39.6\%$$

The regeneration effect is small.

8–3 Improving the Brayton Cycle

Like the Rankine cycle, the Brayton cycle may be improved by various means including regeneration, multistaging with intercooling, and multistage expansion with reheating. In the following paragraphs we discuss these improvements.

Regeneration

Figure 8-9 shows a simple gas turbine cycle with regeneration (that is, an ideal air-standard cycle with regeneration) by means of a schematic and a (T, s) diagram. From the (T, s) diagram, one can see that after the compression process, (1–2), the working fluid (gas) temperature is increased in the regenerator to a temperature (3) which is equal to the temperature of the turbine exhaust gas (5). Ideally the required energy for this temperature increase comes from heat transfer from this exhaust gas. The temperature of the gas is further increased to (4) by an external source. After expansion through the turbine, the gas is partially cooled (6) in the regenerator and finally reduced to the compressor inlet temperature in a cooler. Ideally the heat transfer areas 1–2–3–9–10–1 and 6–5–7–8 should be equal.

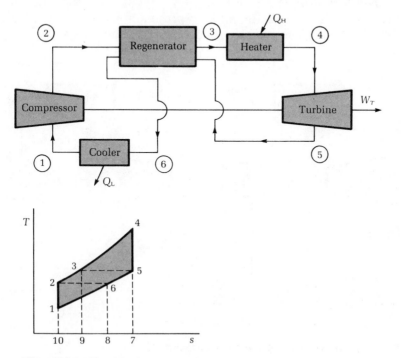

Figure 8-9. The ideal regenerative cycle.

Other Improvements

Figure 8-10 illustrates the other improvements used in gas turbine systems—especially multistaging with intercooling and multistage expansion with reheating. Since detailed discussion of these subjects is beyond the scope of this book, only a few comments will be made here.

Figures 8-10 and 8-11 illustrate an improved gas turbine cycle with two stages of compression, two stages of expansion, and regeneration. For the following list of conditions the maximum efficiency would result (ideally):

1. $T_1 = T_3$

2. $T_6 = T_8$

3. $T_5 = T_9$

4. $p_1/p_2 = p_3/p_4$

5. $p_7/p_6 = p_9/p_8$

It may be noted in passing that as the number of expansions and compressions increase, the representative (T, s) diagram will look like that of an Ericsson cycle (Table 7-3). That is, the number of points in the upper and lower portions of Figure 8-11 would increase, whereas the temperature differences (e.g., $T_6 - T_7$) would decrease. In the limit these little "saw teeth" might be represented by constant temperature lines (isotherm heat transfer processes—like the Carnot cycle).

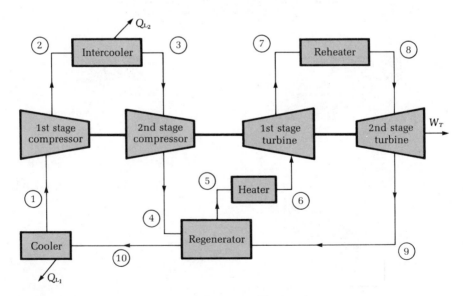

Figure 8-10. Multi-improvements of a gas turbine cycle.

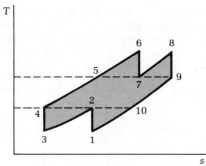

Figure 8-11. Two-stage ideal gas turbine with intercooling, reheating, and regeneration.

EXAMPLE 8-5

For the ideally regenerated Brayton cycle indicated in the sketch, determine the efficiency of this ideal gas turbine. Notice that the energy to raise the temperature of the gas from (2) to (e) is provided by the regenerator.

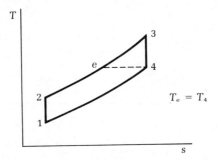

SOLUTION: By definition:

$$\eta_{\text{thermal}} = \frac{W_{\text{net}}}{q_H} = \frac{W_t - W_c}{q_H}$$

$$q_H = c_p(T_3 - T_e)$$

$$W_t = c_p(T_3 - T_4)$$

But $T_4 = T_e$ and therefore the externally provided heat, $q_H = W_t$.

$$\eta_{\text{thermal}} = 1 - \frac{W_c}{W_t} = 1 - \frac{c_p(T_2 - T_1)}{c_p(T_3 - T_4)}$$

$$= 1 - \frac{T_1(T_2/T_1 - 1)}{T_3(1 - T_4/T_3)} = \frac{T_1}{T_3} \frac{(p_2/p_1)^{(k-1)/k} - 1}{1 - (p_2/p_2)^{(k-1)/k}}$$

$$= 1 - \frac{T_1}{T_3}\left(\frac{p_2}{p_1}\right)^{(k-1)/k}$$

Thus the thermal efficiency of the ideal cycle with regeneration depends not only on the pressure ratio but also on the ratio of the minimum to maximum temperature. Notice that, in contrast to the basic Brayton cycle, the efficiency decreases with an increase in pressure ratio. See the sketch.

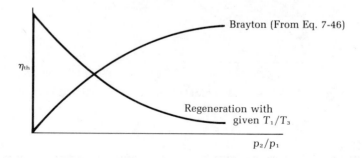

Thus if η_b is the efficiency of the basic Brayton cycle and $\eta_{thermal}$ is the efficiency of the regenerated Brayton cycle, then

$$\eta_{thermal} = 1 - \frac{T_1}{T_3}\left(\frac{1}{1 - \eta_b}\right)$$

$$= 1 - \frac{T_1}{T_3}(1 + \eta_b + \eta_b^2 + \cdots)$$

EXAMPLE 8-6

Compare the effect of an ideal regenerator on a gas turbine (Brayton) cycle's efficiency. In both cases assume that the gas enters the compressor at 0.1 MPa and 15 C and leaves at 0.5 MPa. The maximum cycle temperature is 900 C. For case A, consider a simple gas turbine; for case B, include the ideal regeneration. Assume that $k = 1.4$.

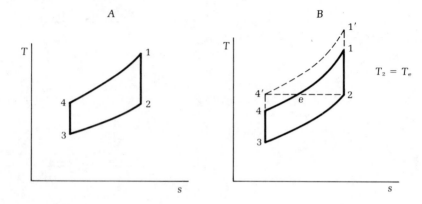

SOLUTION: For case A, note that

$$p_3 = p_2 = 0.1 \text{ MPa} \quad \text{and} \quad p_4 = p_1 = 0.5 \text{ MPa}$$

$$T_3 = 15 \text{ C} = 288 \text{ K}$$

$$T_1 = 900 \text{ C} = 1173 \text{ K}$$

Thus

$$\eta = 1 - \left(\frac{p_3}{p_4}\right)^{(k-1)/k} = 1 - \left(\frac{1}{5}\right)^{0.286} = 0.369 \text{ or } 36.9\%$$

For case B, we must determine $T_2 = T_e$, T_4, and the net work of the cycle. To accomplish that task, recall that process 1–2 (the turbine) is reversible adiabatic. Thus

$$\left(\frac{p_1}{p_2}\right)^{(k-1)/k} = \frac{T_1}{T_2}$$

$$T_2 = 740.4 \text{ K}$$

The same type of relation is true for process 3–4 (the compressor). Thus

$$\left(\frac{p_4}{p_3}\right)^{(k-1)/k} = \frac{T_4}{T_3}$$

$$T_4 = 456.6 \text{ K}$$

Applying the first law to the turbine and the compressor; we get

$$w_t = h_1 - h_2 = c_p(T_2)$$

$$= 434 \text{ kJ/kg}$$

and

$$-w_c = h_4 - h_3 = c_p(T_4 - T_3)$$

$$= 169 \text{ kJ/kg}$$

and

$$w_{net} = w_t + w_c = 265 \text{ kJ/kg}$$

Applying the first law to the heat exchanger yields

$$q_H = h_1 - h_e = c_p(T_1 - T_e)$$

$$= 434 \text{ kJ/kg}$$

The cycle efficiency again follows from the definition:

$$\eta = \frac{w_{net}}{q_H} = \frac{265}{434}$$

$$= 0.611 \text{ or } 61.1\%$$

Notice that the ideal regeneration has a strong effect on the efficiency. Unfortunately, as you might expect, this perfect heat exchange cannot be implemented.

PROBLEMS

8-1 For the conditions indicated on the accompanying sketch (assuming steam as the working fluid): (a) make a component schematic, (b) determine the cycle efficiency, and (c) determine the moisture content leaving the low-pressure turbine.

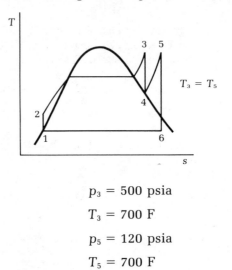

$$p_3 = 500 \text{ psia}$$

$$T_3 = 700 \text{ F}$$

$$p_5 = 120 \text{ psia}$$

$$T_5 = 700 \text{ F}$$

$$p_6 = 2 \text{ psia}$$

8-2 For the conditions shown in the sketch, determine the cycle efficiency and make a component schematic. Assume that steam is the working fluid.

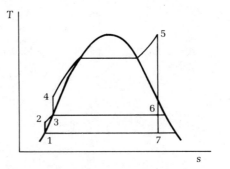

$$p_1 = 1 \text{ psia}$$

$$T_1 = T_7 = 101.7 \text{ F}$$

$$p_2 = p_3 = p_6 = 140 \text{ psia}$$

$$T_3 = T_6 = 353.1 \text{ F}$$

$$p_4 = p_5 = 2500 \text{ psia}$$

$$T_5 = 1000 \text{ F}$$

8-3 Consider a combined reheated and regenerated Rankine cycle in which the net power output of the turbine is 10^5 kW. See the sketch for other pertinent parameters.

a. Make a component schematic.

b. Determine the size of motor (hp) required to drive each pump.

c. Determine what diameter of pipe is needed if the flow velocity from the turbine to the condenser is 400 ft/sec.

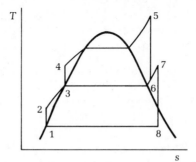

$$p_1 = p_8 = 1 \text{ psia}$$

$$T_1 = T_{sat}$$

$$p_2 = p_3 = p_6 = p_7 = 90 \text{ psia}$$

$$p_4 = p_5 = 1200 \text{ psia}$$

$$T_5 = 1000 \text{ F}$$

$$T_7 = 700 \text{ F}$$

8-4 Consider an ideal regenerator in an ideal air-standard Brayton cycle with the characteristics illustrated below. Determine the cycle efficiency and make a component schematic.

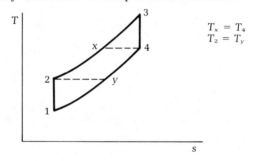

$$p_1 = 14.7 \text{ psia}$$
$$T_1 = 520 \text{ R}$$
$$T_3 = 1960 \text{ R}$$
$$T_4 = 1319 \text{ R}$$
$$p_2 = 58.8 \text{ psia}$$
$$w_t = 162 \text{ Btu/lbm}$$
$$-w_c = 84 \text{ Btu/lbm}$$

8-5 Using the figure for Problem 8-2, determine the cycle efficiency for the following conditions:
a. Input turbine conditions; 1000 psia, 800 F
b. Open feedwater heater at 90 psia
c. Condenser pressure of 1 psia

8-6 For the conditions stated on the accompanying diagram and table, assume that the working fluid is air. Determine: (a) the net work and (b) the thermal efficiency.

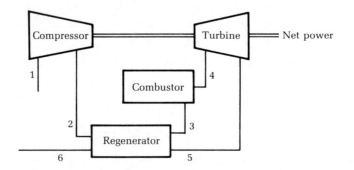

State	p, kPa	T, C
1	100	20
2	800	259
3	800	360
4	800	874
5	100	360
6	100	259

8-7 Using the figure for Problem 8-1, determine (a) the net work output and (b) the thermal efficiency for the following conditions:

1. $p_3 = 3000 \text{ kPa}$; $T_3 = T_5 = 400 \text{ C}$.

2. $p_4 = 500$ kPa; $T_4 = 180$ C.

3. $p_1 = p_6 = 2$ kPa; $x_6 = 0.99$.

4. The feed pump is reversible and adiabatic.

8-8 For the actual reheat–regenerative cycle shown below, fill in the blanks and compute: (a) the efficiency of the high-pressure turbine and (b) the cycle thermal efficiency.

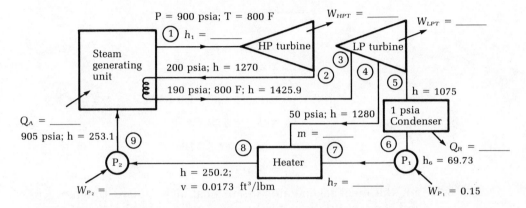

8-9 For the conditions indicated with the sketch, assume that steam is the working fluid. Determine: (a) the cycle efficiency and (b) the moisture content of the steam leaving the low-pressure turbine.

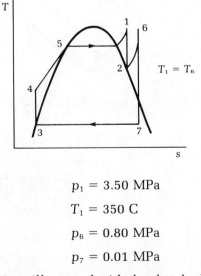

$$p_1 = 3.50 \text{ MPa}$$

$$T_1 = 350 \text{ C}$$

$$p_6 = 0.80 \text{ MPa}$$

$$p_7 = 0.01 \text{ MPa}$$

8-10 For the conditions illustrated with the sketch, determine the cycle efficiency assuming steam as the working fluid.

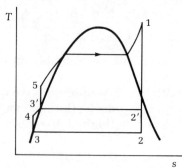

$$p_3 = 0.01 \text{ MPa}$$

$$T_3 = T_2 = 45.8 \text{ C}$$

$$p_4 = p_{3'} = p_{2'} = 1.0 \text{ MPa}$$

$$T_{3'} = T_{2'} = 180 \text{ C}$$

$$p_5 = p_1 = 17.5 \text{ MPa}$$

$$T_1 = 550 \text{ C}$$

8-11 For the ideal regenerator in an ideal air-standard Brayton cycle and the conditions listed below, determine the cycle efficiency.

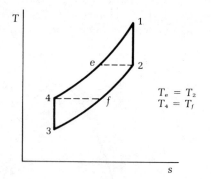

$$p_3 = 0.1 \text{ MPa}$$

$$T_3 = 271.1 \text{ C}$$

$$T_1 = 1071 \text{ C}$$

$$T_2 = 715 \text{ C}$$

$$p_4 = 0.40 \text{ MPa}$$

$$w_t = 376.8 \text{ kJ/kg}$$

$$w_c = 195.4 \text{ kJ/kg}$$

8-12 Consider a combined reheated and regenerated Rankine cycle in which the net power output of the high-pressure turbine is 10^5 kW. See the sketch for other pertinent parameters. Determine the diameter of pipe needed if the flow velocity from the low-pressure turbine to the condenser is 122 m/s.

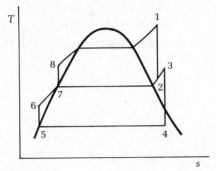

$$p_1 = p_8 = 8.5 \text{ MPa}$$

$$p_2 = p_3 = p_7 = p_6 = 0.6 \text{ MPa}$$

$$p_4 = p_5 = 0.01 \text{ MPa}$$

$$T_1 = 550 \text{ C}$$

$$T_3 = 370 \text{ C}$$

$$T_5 = T_{sat}$$

8-13 The efficiency of the turbine in the power plant cycle indicated in the sketch is 85%. For the conditions stated, determine the cycle efficiency. (Notice that the pump efficiency is 100%.)

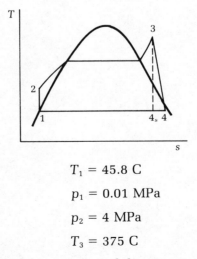

$$T_1 = 45.8 \text{ C}$$

$$p_1 = 0.01 \text{ MPa}$$

$$p_2 = 4 \text{ MPa}$$

$$T_3 = 375 \text{ C}$$

8-14 The pressure and temperature of the air entering the compressor (pressure ratio 4:1) of the Brayton cycle are 0.1 MPa and 20 C. For a flow rate of 545 kg/min, the maximum cycle temperature is 900 C. Determine: (a) the compressor work, (b) the turbine work, and (c) the thermal efficiency of the cycle. Assume constant specific heat.

8-15 Suppose you included an ideal regenerator in the cycle described in Problem 8-14. What would be the change in the cycle's thermal efficiency?

Chapter

9

Availability and Irreversibility

Whereas the first law of thermodynamics concerns the quantity of energy, the second law deals with the quality. Since work (organized energy) is the highest-quality (or lowest-entropy) form of interaction among systems, it is also the most valuable form of energy and should therefore be the index used to rank energy-conversion processes. The second law permits the definition of a property called the available energy—the maximum amount of theoretical work that can be produced from the energy of a system.* Unlike energy, available energy can be consumed in a process. The goal of energy conservation, therefore, should be to minimize the consumption of available energy. In this chapter we examine this property called availability.

9–1 General Concepts

In thermodymanics, availability of energy is taken to mean availability of energy for performing *work* (for example, mechanical or electrical). The limitation placed by the second law on the transformation of heat

---

* Scientists have presented the concept of available energy under a variety of names: availability, available work, energy utilizable, essergy, exergy, potential energy, and useful energy. The concept of availability, introduced by Gibbs, was later popularized in this country by Keenan, Obert, Gaggioli, Evans, Tribus, Coad, and Sussman (see the Bibliography).

into work has the following significance: Although a given quantity of heat may be available for heating purposes, only a certain portion of this heat is capable of performing work. Hence the availability of energy, thermodynamically speaking, is less than the quantity of heat under consideration.

Energy in the form of shaft work, or in a form completely convertible into such work by ideal processes, is called **available energy.** Energy that is in part convertible and in part nonconvertible into shaft work is said to be made up of an available part and an unavailable part. The available part is sometimes called the **availability** and the unavailable part the **unavailability** of the energy.

The common method of energy accounting completely ignores the quality of energy. It restricts itself to keeping track of the quantity, which according to the first law of thermodynamics will never change. A better system of energy accounting would be based on second-law principles, which assert that work is the highest-quality form of energy.

The figure of merit most widely used to evaluate alternative uses of energy is the **first-law efficiency,** commonly called thermal efficiency or coefficient of performance. (Recall that COP is usually used when discussing the thermal efficiency of a reversed cycle.) It is conveniently defined as *the useful energy effect divided by the energy input required to achieve the effect.* This efficiency is far from adequate and is also confusing. To illustrate the problem of this first-law concept of efficiency, consider the common situation for space heating, where there exists a combustion flame at an average temperature of 2000 F; an outdoor ambient temperature of 20 F; and the desired indoor space temperature of 70 F. This indoor temperature is obtained by burning the fuel with a given heating characteristic: the heating value (HV) of the fuel at the flame temperature. Using a Carnot engine to operate a Carnot heat pump would yield the result that the heat to the indoor space is

$$\eta_{HP}\,\eta_{HE}\text{HV} = \frac{530}{530 - 480} \times \frac{1980}{2460} \times \text{HV}$$

$$\doteq 8.5 \quad (\text{HV})$$

(Heating value is discussed in detail in Chapter 11.)

Thus if the fuel has a heating value of 140,000 Btu/gal, the theoretical maximum amount of heat to the space from the fuel would not be 140,000 Btu/gal but 1,194,600 Btu/gal $\approx$ 8.5 times 140,000 Btu/lbm. The higher value could be referred to as the thermal availability. Claiming that an oil furnace has an efficiency of 80% (100% efficiency is the best use of energy) implies that 80% efficiency yields 112,000 Btu (0.80 $\times$ 140,000). This represents only 9.3% of the energy theoretically available for heating! Example 9-1 demonstrates the concept for a solar heating system.

EXAMPLE 9-1

A solar heat pump is designed to operate as shown in the accompanying sketch. Solar energy is used as the heat source for the boiler in a Rankine power system operating with R-22 as the working fluid. The turbine output drives the compressor in a regular R-12 heat pump system. For the conditions shown, determine the minimum square footage of solar collectors if the heat loss from the house is 45,000 Btu/hr.

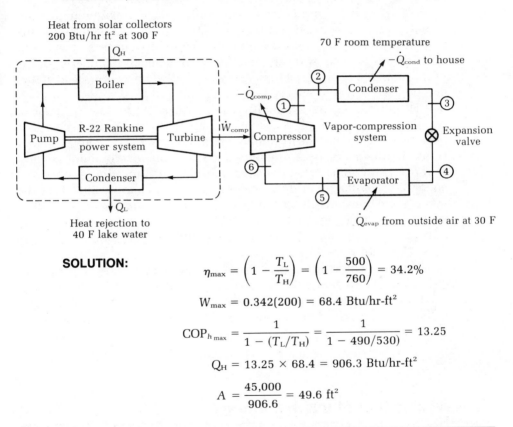

Heat from solar collectors
200 Btu/hr ft² at 300 F

70 F room temperature

Heat rejection to
40 F lake water

$\dot{Q}_{evap}$ from outside air at 30 F

SOLUTION:

$$\eta_{max} = \left(1 - \frac{T_L}{T_H}\right) = \left(1 - \frac{500}{760}\right) = 34.2\%$$

$$W_{max} = 0.342(200) = 68.4 \ \text{Btu/hr-ft}^2$$

$$COP_{h\,max} = \frac{1}{1 - (T_L/T_H)} = \frac{1}{1 - 490/530} = 13.25$$

$$Q_H = 13.25 \times 68.4 = 906.3 \ \text{Btu/hr-ft}^2$$

$$A = \frac{45,000}{906.6} = 49.6 \ \text{ft}^2$$

The fundamental difficulty with the first-law efficiency is its reliance on energy as a basic unit of measure. Since energy is a property that cannot be consumed, the first-law efficiency is an ambiguous and inadequate measure of energy-use effectiveness. It is the available-energy content of a substance, not its energy content, that truly represents the potential of the substance to cause change. Available energy is the only rational basis for evaluating (1) fuels and resources, (2) process, device, and system efficiencies, (3) dissipations and their costs, and (4) the value and cost of system outputs.

It is *available* energy that drives processes—and, in so doing, it is literally used up. "Energy converters" such as engines, power plants, and HVAC systems take available energy in one form and convert it, in part, to another form; the part that is not converted is used up to accomplish the conversion.

A **second-law efficiency** *can be defined as the ratio of the minimum amount of available energy required to perform a task to the available energy actually consumed.* Whatever the conversion process, the theoretical upper limit of this second-law efficiency is 100%, which corresponds to the ideal case with no dissipation or losses. Maximizing the second-law efficiency necessarily minimizes consumption of a fuel or other energy source, since availability is an extensive property and is proportional to the mass of fuel. A waste of availability is a waste of fuel and hence a waste of energy resources.

To determine the availability, it is necessary to specify a set of conditions under which the energy content is regarded as zero. It is usually assumed that the available-energy content of any body, whether solid, liquid, or gas, is zero when the body is chemically inert, when it is at rest (zero velocity) at the surface of the earth (minimum potential energy), and when it has the same pressure p_0 and temperature T_0 as the atmosphere that constitutes the receiver. Similar requirements could be set forth regarding magnetic, electrical, and surface effects if these are relevant to the problem. Unless specified otherwise, common values are **$T_0 = 77$ F and $p_0 = 14.7$ psia.**

The available-energy input to a region or medium is commonly divided into two general types: (1) the available energy of the system itself at a given state (availability of system) and (2) the available energy in the heat transferred to the region during a change of state (availability of heat).

9–2 Available Part of Internal Energy

To determine the amount of work (availability) that can be performed solely because of the internal energy possessed by the system at a given state, consider a nonflow system for which $\Delta KE = \Delta PE = 0$. Allow the system to come to equilibrium with the surroundings reversibly, neither absorbing nor rejecting available energy other than the work. Thus we are restricting heat transfer to that with the surroundings, since this heat is in no sense available. The first law can be written

$$\delta Q - \delta W = dU \tag{9-1}$$

or on a unit mass basis

$$\delta q - \delta w = du \tag{9-2}$$

For a reversible process,

$$\delta q = T \, ds \tag{9-3}$$

which upon substituting in (9-2) yields

$$T \, ds - \delta w = du \tag{9-4}$$

Integration of (9-4) between the existing state (no subscripts) and the reference (or dead) state zero, with reversible isothermal heat transfer with the surroundings at T_0,

$$\int_s^{s_0} T_0 \, ds - \int_0^w \delta w = \int_u^{u_0} du \tag{9-5}$$

yields

$$T_0(s_0 - s) - w = u_0 - u \tag{9-6}$$

or

$$w_{\text{gross}} = (u - u_0) - T_0(s - s_0) \tag{9-7}$$

The work done on the atmosphere is unavailable for use in driving a rotating element (mechanical work) and is therefore to be deducted from the gross availability. The work on the atmosphere is evaluated as

$$w_{\text{atm}} = \int p \, dv = p_0 \int_v^{v_0} dv = p_0(v_0 - v) \tag{9-8}$$

Subtracting (9-8) from (9-7) yields the availability of the internal energy:

$$u_{\text{av}} = (u - u_0) - T_0(s - s_0) - p_0(v_0 - v) \tag{9-9}$$

9–3 Available Part of Kinetic and Potential Energy

Since both kinetic energy and potential energy are "mechanical" forms of energy, they are wholly available for conversion to rotary shaft work. Thus

$$KE_{\text{av}} = KE = \frac{V^2}{2g_c} \tag{9-10}$$

if V is taken relative to V_0. Moreover,

$$PE_{\text{av}} = PE = \frac{g}{g_c} Z \tag{9-11}$$

if Z is taken relative to Z_0.

9–4 Available Part of Flow Work

When the fluid is flowing, the availability of each unit of mass in the stream is augmented by the amount of work that could be delivered by virtue of the flow—that is, the displacement or flow work, pv, less the work that must be expended on the atmosphere, $\int_0^v p_0\, dv = p_0 v$. The amount $(p - p_0)v$ can be delivered to things other than the medium and is thus referred to as the available part of the flow work.

9–5 Availability of Closed Systems

Since matter can possess internal, kinetic, and potential energy, the availability of a system of fixed mass (neglecting chemical reactions) is the sum of the available parts of these three forms of energy. The availability of a state for a closed system is commonly designated ϕ and can be written as

$$\phi = u_{av} + KE_{av} + PE_{av} \tag{9-12}$$

or

$$\phi = (u - u_0) - T_0(s - s_0) - p_0(v_0 - v) + \frac{V^2}{2g_c} + \frac{g}{g_c}Z \tag{9-13}$$

9–6 Availability in Steady Flow

The availability of a system in steady flow, ψ, consists of the available parts of internal energy, kinetic energy, and potential energy augmented by the available part of the flow work at that location:

$$\psi = (u - u_0) - T_0(s - s_0) - p_0(v_0 - v) + \frac{V^2}{2g_c} + \frac{g}{g_c}Z + (p - p_0)v$$

which reduces to

$$\psi = (h - h_0) - T_0(s - s_0) + \frac{V^2}{2g_c} + \frac{g}{g_c}Z \tag{9-14}$$

9–7 Availability of Heat

The Kelvin–Planck statement of the second law says that heat cannot be completely converted into work; a fraction of the heat must be rejected to a second reservoir at a lower temperature than the heat source. When the lower-temperature reservoir is at the dead-state temperature, T_0, the amount of heat that can be converted into work is the availability of the heat.

The Carnot engine and cycle afford a convenient means for determining the availability of energy in the form of heat. Consider the Carnot cycle shown in Figure 9-1. From the second law and its corollaries (Chapter 6), the maximum thermal efficiency of the cycle is

$$\eta_{max} = \left(1 - \frac{T_L}{T_H}\right) \tag{9-15}$$

where $\eta \equiv W/Q_H$.

The maximum work obtainable from an engine operating between T_H and T_L is thus

$$W_{max} = Q_H\left(1 - \frac{T_L}{T_H}\right)$$

$$= Q_H - T_L\frac{Q_H}{T_H} \tag{9-16}$$

If $T_L = T_0$, the dead-state temperature, then the available part of the heat, Q, is determined as

$$W_{max} = Q_{av} = Q_H - T_0\frac{Q_H}{T_H} \tag{9-17}$$

In many applications, however, the heat transfer does not occur at a constant temperature T_H but rather at varying temperatures as the energy transfer occurs. Thus for the general case Equation (9-17) should be replaced with

$$Q_{av} = Q - T_0\int_{rev}\frac{\delta Q}{T} \tag{9-18}$$

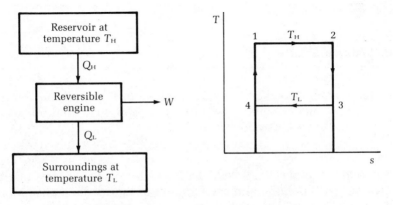

Figure 9-1. Carnot engine operating between T_H and T_L.

where Q = actual amount of heat
Q_{av} = availability of heat
T_0 = dead-state temperature
T = temperature of system

The portion of the heat that was rejected ($Q_L = Q - Q_{av}$) is referred to as the unavailable energy or the unavailable part of the heat. Table 9-1 summarizes the expressions for the available part of each form of energy (neglecting chemical and nuclear).

Table 9-1 Available Energy

Availability of System

Available part of internal energy:

$$u_{av} = (u - u_0) - T_0(s - s_0) - p_0(v_0 - v)$$

Available part of kinetic energy:

$$KE_{av} = KE = \frac{V^2}{2g_c} \quad \text{if } V \text{ is taken relative to } V_0$$

Available part of potential energy:

$$PE_{av} = PE = \frac{g}{g_c} Z \quad \text{if } Z \text{ is taken relative to } Z_0$$

Available part of flow work:

$$W_{f_{av}} = (p - p_0)v$$

Availability of Heat

Available part of heat:

$$Q_{av} = Q_{in} - T_0 \int_{1_{rev}}^{2} \frac{\delta Q}{T} = Q_{in} - T_0 (s_2 - s_1)_{rev}$$

9–8 Reversible Work

Reversible work *is the maximum useful work that may be obtained for a given change in state—including heat supplied from other systems but excluding the work done on the surroundings. That is, if the initial and final states of a system are specified, the reversible work refers to the maximum work that can be done by the system as it goes from the initial to the final state. It is evident that this concept of reversible work involves both the first and second laws of thermodynamics. Moreover, it is clear that the work will be maximum only if the process is entirely* reversible. Note that the reversible work is not only a function of the initial and final states of the system; it also depends on the temperature of the surroundings.

Two cases are of particular interest: the steady-flow process and the closed system. For the steady-flow process, the reversible work (per pound of flowing fluid) is

$$W_{rev} = \psi_1 - \psi_2 + Q_{av} \qquad \text{(9-19)}$$

For the closed system, the reversible work (per pound mass) is

$$W_{rev} = \phi_1 - \phi_2 + {}_1Q_2 \qquad \text{(9-20)}$$

9–9 Irreversibility and Lost Work

Many processes lead to a loss of available energy—for example, heat transfer through a finite temperature difference, mixing of two substances, all kinds of friction, and electric current flow through a resistance. In fact, *all* actual processes lead to a loss of available energy. These processes are called irreversible because they result in a permanent and irretrievable loss of available energy. The amount of available energy lost is called the **irreversibility** of the process, and good engineers are usually on guard against unnecessary irreversibility. Some irreversibility is always necessary in real systems. For example, to produce steam with hot gases through a heat transfer area of reasonable size, a large temperature difference is required; the dropping of energy from flame temperature to steam temperature is an irreversibility that must be accepted if one wants to have a boiler of finite size and cost. The engineer is always faced with a compromise and must try to balance the disadvantages of irreversibility with other factors—usually cost, time, and size.

Since every actual process has irreversibilities associated with it, the actual work W for a given change of state is always less than the corresponding reversible work, W_{rev}:

$$W \leq W_{rev}$$

This leads us to a definition of the irreversibility of a process. The irreversibility I for a given process is defined by the relation

$$\delta I = \delta W_{rev} - \delta W$$

$$I = W_{rev} - W \qquad \text{(9-21)}$$

In words this equation states that the actual work is less than the reversible work by the amount of the irreversibility. Irreversibility or **available energy degraded** is the decrease in available energy due to irreversibilities; it is equal to the reversible work minus the actual work for a process. **Entropy production** or **entropy growth,** ΔS_{in}, is the increase in entropy resulting from irreversibility.

The general expression for the second law of thermodynamics in terms of entropy is useful in determining most of the quantities associated with irreversibility:

$$dS_{system} = \frac{\delta Q}{T} + \delta m_{in} s_{in} - \delta m_{out} s_{out} + dS_{in} \qquad (9\text{-}22)$$

where dS_{in} is the entropy increase due to irreversibility.

Because every real process has some irreversibilities, it may be helpful to present an analytical explanation of the effect on the change of entropy. The idea of lost work (LW) may be used as a conceptual illustration. Recall that the irreversibility is a permanent loss, whereas the lost work is work lost during a particular process. A portion of this work may still be available for conversion into work.

To clarify this idea, recall for a reversible process in a closed system

$$\delta W_{rev} = p \, dV \qquad (1\text{-}11)$$

Note that this is the maximum possible work; the actual work would be less. Thus

$$p \, dV = \delta W + \delta(LW) \qquad (9\text{-}23)$$

where δW is the actual work and $\delta(LW)$ is the work that is lost. Using the first $T\,ds$ relation from Equation (6-14) in Equation (9-23) yields

$$T \, dS = dU + \delta W + \delta(LW) \qquad (9\text{-}24)$$

The first law for a closed system (with no kinetic or potential energy changes) is

$$dU = \delta Q - \delta W \qquad (9\text{-}25)$$

where the Q and W terms are actual heat and actual work quantities. Using Equation (9-25) in Equation (9-24) yeilds

$$T \, dS = \delta Q + \delta(LW) \qquad (9\text{-}26)$$

or
$$dS = \frac{\delta Q}{T} + \frac{\delta(LW)}{T} \qquad (9\text{-}27)$$

Two points should be noted:

1. $\delta Q_{rev} - \delta W_{rev} = T \, dS - p \, dV = \delta Q - \delta W \qquad (9\text{-}28)$

This is not surprising since $(\delta Q - \delta W)$ is a point function.

2. For an adiabatic process $\quad dS = \dfrac{\delta(LW)}{T} \qquad (9\text{-}29)$

Again, this is not surprising in that irreversibilities are responsible for any increase in entropy.

Using Equation (9-29), transforms Equation (9-22) into

$$dS_{system} = \frac{\delta Q + \delta LW}{T} + \delta m_{in} s_{in} - \delta m_{out} s_{out} \qquad (9\text{-}30)$$

The lost work can be expressed as $\qquad LW = \int T\, dS_{in} \qquad (9\text{-}31)$

whereas the irreversibility is $\qquad I = \int T_0\, dS_{in} = T_0\, \Delta S_{in}$

$$(9\text{-}32)$$

9–10 Closure

As a measure of performance when dealing with a second-law analysis, the *effectiveness* of a work-producing device is often determined, rather than (or in addition to) its efficiency. **Effectiveness** *is defined as the ratio of the actual work to the reversible work and compares the actual to the ideal performance for operation between the same two states, without the constraint of following a particular process.* Efficiency, on the other hand, compares the actual performance of a machine to the performance that would have been achieved had the specified process been reversible. (Note the constraint on the process.)

EXAMPLE 9-2

A heat engine operates on the Carnot cycle between temperatures of 1000 R and the dead-state value of 500 R, with an entropy change of 1.0 Btu/R as shown in the sketch. Complete the following table:

Quantity	Value	Area on (T, S) Diagram
Q_{in}	_____ Btu	_____
Q_{out}	_____ Btu	_____
W_{out}	_____ Btu	_____
$\eta_{thermal}$	_____ %	_____
$Q_{in(av)}$	_____ Btu	_____
$Q_{in(unav)}$	_____ Btu	_____
$Q_{out(av)}$	_____ Btu	_____
$Q_{out(unav)}$	_____ Btu	_____
LW	_____ Btu	_____
LW_{av}	_____ Btu	_____
LW_{unav}	_____ Btu	_____
I	_____ Btu	_____

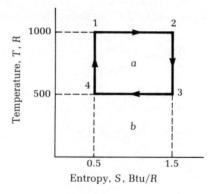

center
Entropy, S, Btu/R

SOLUTION:

Quantity	Value	Area on (T, S) Diagram
Q_{in}	1000 Btu	$a + b$
Q_{out}	500 Btu	b
W_{out}	500 Btu	a
$\eta_{thermal}$	50 %	
$Q_{in(av)}$	500 Btu	a
$Q_{in(unav)}$	500 Btu	b
$Q_{out(av)}$	0 Btu	—
$Q_{out(unav)}$	500 Btu	b
LW	0 Btu	—
LW_{av}	0 Btu	—
LW_{unav}	0 Btu	—
I	0 Btu	—

EXAMPLE 9-3

Consider the heat engine of Example 9-2 between temperatures of 1000 and 600 R, as shown in the sketch. The dead state is at 500 R. Complete the following table:

Quantity	Value	Area on (T, S) Diagram
Q_{in}	_____ Btu	_____
Q_{out}	_____ Btu	_____
W_{out}	_____ Btu	_____
$\eta_{thermal}$	_____ %	

Quantity	Value	Area on (T, S) Diagram
$Q_{in(av)}$	_____ Btu	_____
$Q_{in(unav)}$	_____ Btu	_____
$Q_{out(av)}$	_____ Btu	_____
$Q_{out(unav)}$	_____ Btu	_____
LW	_____ Btu	_____
LW_{av}	_____ Btu	_____
LW_{unav}	_____ Btu	_____
I	_____ Btu	_____

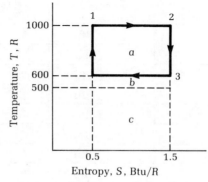

SOLUTION:

Quantity	Value	Area on (T, S) Diagram
Q_{in}	1000 Btu	$a + b + c$
Q_{out}	600 Btu	$b + c$
W_{out}	400 Btu	a
$\eta_{thermal}$	40 %	
$Q_{in(av)}$	500 Btu	$a + b$
$Q_{in(unav)}$	500 Btu	c
$Q_{out(av)}$	100 Btu	b
$Q_{out(unav)}$	500 Btu	c
LW	0 Btu	—
LW_{av}	0 Btu	—
LW_{unav}	0 Btu	—
I	0 Btu	—

EXAMPLE 9-4

A heat engine operates on a cycle similar to the Carnot cycle except that the adiabatic expansion process (2–3) is not frictionless but rather results in an entropy creation of 0.1 Btu/R. The engine operates between temperatures of 1000 R and the dead-state value 500 R with an entropy increase due to the heat addition of 1.0 Btu/R, as shown in the sketch. Complete the following table:

Quantity	Value	Area on (T, S) Diagram
Q_{in}	_____ Btu	_____
Q_{out}	_____ Btu	_____
W_{out}	_____ Btu	_____
$\eta_{thermal}$	_____ %	
$Q_{in(av)}$	_____ Btu	_____
$Q_{in(unav)}$	_____ Btu	_____
$Q_{out(av)}$	_____ Btu	_____
$Q_{out(unav)}$	_____ Btu	_____
LW	_____ Btu	_____
LW_{av}	_____ Btu	_____
LW_{unav}	_____ Btu	_____
I	_____ Btu	_____

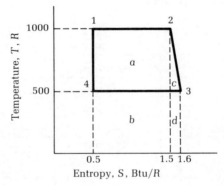

SOLUTION:

Quantity	Value	Area on (T, S) Diagram
Q_{in}	1000 Btu	$a+b$
Q_{out}	550 Btu	$b+d$
W_{out}	450 Btu	$a-d$
$\eta_{thermal}$	45 %	

Quantity	Value	Area on (T, S) Diagram
$Q_{in(av)}$	500 Btu	a
$Q_{in(unav)}$	500 Btu	b
$Q_{out(av)}$	0 Btu	—
$Q_{out(unav)}$	550 Btu	$b+d$
LW	50 Btu	d
LW_{av}	0 Btu	—
LW_{unav}	50 Btu	d
I	50 Btu	d

EXAMPLE 9-5

The heat engine of Example 9-4 operates between temperatures of 1000 and 600 R, as shown in the sketch. The dead state is at 500 R. Complete the following table:

Quantity	Value	Area on (T, S) Diagram
Q_{in}	_____ Btu	_____
Q_{out}	_____ Btu	_____
W_{out}	_____ Btu	_____
$\eta_{thermal}$	_____ %	
$Q_{in(av)}$	_____ Btu	_____
$Q_{in(unav)}$	_____ Btu	_____
$Q_{out(av)}$	_____ Btu	_____
$Q_{out(unav)}$	_____ Btu	_____
LW	_____ Btu	_____
LW_{av}	_____ Btu	_____
LW_{unav}	_____ Btu	_____
I	_____ Btu	_____

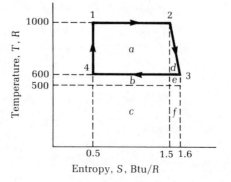

SOLUTION:

Quantity	Value	Area on (T, S) Diagram
Q_{in}	1000 Btu	$a + b$
Q_{out}	660 Btu	$b + c + e + f$
W_{out}	340 Btu	$a - e - f$
$\eta_{thermal}$	34 %	
$Q_{in(av)}$	500 Btu	$a + b$
$Q_{in(unav)}$	500 Btu	c
$Q_{out(av)}$	110 Btu	$b + e$
$Q_{out(unav)}$	550 Btu	$c + f$
LW	60 Btu	$e + f$
LW_{av}	10 Btu	e
LW_{unav}	50 Btu	f
I	50 Btu	f

EXAMPLE 9-6

An isothermal steam turbine produces 600 hp with the steam entering at 1000 psia and 600 F and exiting at 100 psia. Heat is added during the process at the rate of 2,500,000 Btu/hr. Determine:

1. Steam flow rate (lbm/hr)

2. Availability at inlet (Btu/lbm)

3. Availability at outlet (Btu/lbm)

4. Availability of the heat added (Btu/lbm)

5. Reversible work (Btu/lbm)

6. Turbine effectiveness (%)

7. Irreversibility of the process (Btu/lbm)

SOLUTION: $p_1 = 1000$ psia $\quad T_1 = 600$ F $\quad h_1 = 1248.8 \quad s_1 = 1.4450$

$p_2 = 100$ psia $\quad T_2 = 600$ F $\quad h_2 = 1329.3 \quad s_2 = 1.7582$

$p_0 = 14.7$ psia $\quad T_0 = 77$ F $\quad h_0 = 45.1 \quad s_0 = 0.0877$

$$m(h_1 - h_2) + Q - W = 0$$

$$\dot{m}(1248.8 - 1329.3) + 2,500,000 - 600(2545) = 0$$

1. $\dot{m} = 12,087$ lbm/hr

$$\frac{\dot{Q}}{\dot{m}} = \frac{2,500,000}{12,087} = 206.8 \text{ Btu/lbm}$$

$$\frac{\dot{W}}{\dot{m}} = \frac{600(2545)}{12,087} = 126.3 \text{ Btu/lbm}$$

2. $\psi_1 = (h_1 - h_0) - T_0(s_1 - s_0)$
$\quad = (1329.3 - 45.1) - 537(1.7582 - 0.0877)$
$\quad = 387.1 \text{ Btu/lbm}$

3. $\psi_2 = (h_2 - h_0) - T_0(s_2 - s_0)$
$\quad = (1329.3 - 45.1) - 537(1.7582 - 0.0877)$
$\quad = 387.1 \text{ Btu/lbm}$

4. $Q_{av} = Q - T_0\left(\dfrac{Q}{T}\right) = 206.8 - 537\left(\dfrac{206.8}{1060}\right) = 102 \text{ Btu/lbm}$

5. $W_{rev} = \psi_1 - \psi_2 + Q_{av} = 474.8 - 387.1 + 102 = 189.7 \text{ Btu/lbm}$

6. Effectiveness $= \dfrac{W_{act}}{W_{rev}} \times 100 = \dfrac{126.3}{189.7} \times 100 = 66.6\%$

7. $I = W_{rev} - W_{act} = 189.7 - 126.3 = 63.4 \text{ Btu/lbm}$

Check:

$$\underbrace{m_f s_f - m_i s_i}_{0} = \underbrace{\int \frac{\delta Q}{T}}_{0.1951} + \underbrace{\sum (ms)_{in}}_{1.4450} - \underbrace{\sum (ms)_{out}}_{1.7582} + \Delta S_{irr}$$

$$\Delta S_{irr} = +0.1181$$

$$I = T_0 \, \Delta S_{irr} = 537(0.1181) = 63.4 \text{ Btu/lbm}$$

PROBLEMS

Unless specified otherwise, $T_0 = 77$ F and $p_0 = 14.7$ psia in the following problems.

9-1 Dead state is 14.7 psia and 60 F. Air expands adiabatically across a valve with negligible changes in kinetic and potential energy for the steady-flow process. The air upstream of the valve has a temperature of 300 F and a pressure of 200 psia. After the valve the pressure is 15 psia. Determine:
a. Availability per pound of air before valve
b. Availability per pound of air after valve
c. Reversible work across valve (Btu/lbm)
d. Lost work (Btu/lbm)
e. Irreversibility (Btu/lbm)
f. Maximum useful work
g. Actual mechanical work

9-2 Air flows adiabatically through a nozzle from inlet conditions of 65 psia and 1400 F (negligible velocity) to an exhaust velocity of 2600 ft/sec at the discharge pressure of 14.0 psia. Determine:
a. Nozzle efficiency (%)
b. Availability at exhaust (Btu/lbm)
c. Irreversibility of process (Btu/lbm)

9-3 An isothermal air turbine is designed to operate on 37,500 ft³/min of air entering at 150 psia and 350 F while exhausting at 14.7 psia. The turbine is rated at 1450 hp. Determine:
a. Mass flow rate (lbm/hr)
b. Heat added (Btu/lbm)
c. Availability at inlet (Btu/lbm)
d. Availability at outlet (Btu/lbm)
e. Availability of heat added (Btu/lbm)
f. Reversible work (Btu/lbm)
g. Ideal work (Btu/lbm)
h. Turbine efficiency (%)
i. Turbine effectiveness (%)
j. Irreversibility (Btu/lbm)

9-4 Water (initial quality of zero) at 200 psia receives heat at the rate of 500 Btu/lbm while the pressure remains constant. No useful mechanical work is obtained during this boiling process. Determine (on a per pound basis assuming the dead state is 60 F, 14.7 psia):
a. T, h, s, v, and u at initial and final states
b. Entropy change due to heat transfer
c. Entropy change due to irreversibility
d. Change in availability if steady-flow process
e. Change in availability if closed system
f. Lost work
g. Irreversibility
h. Maximum useful work
i. Complete (T, s) diagram with significance of areas labeled

9-5 Find the change in availability of the system corresponding to the following processes. The system is 1 lbm of H₂O initially at 200 psia and 500 F. (Take the dead state as 14.7 psia, 60 F.)
a. The system is confined at constant pressure by a piston and is heated until its volume is doubled.
b. The system expands reversibly and adiabatically behind a piston until its volume is doubled.
c. It expands reversibly and isothermally behind a piston until its volume is doubled.
d. It expands adiabatically into an adjacent chamber that is initially evacuated. The final pressure is 100 psia. No work is done.

9-6 A design for a turbine has been proposed involving the reversible, isothermal, steady flow of 20,000 lbm/hr of steam through the turbine. Saturated vapor at 250 psia enters the turbine and the steam leaves at 10 psia. These are the *proposed* conditions.

During the first qualification test on the turbine, however, it was determined that to maintain the proposed outlet conditions, the actual amount of heat required was 250 Btu/lbm of steam with a reduction in the power output of the isothermal turbine.

For the second test, conducted at the same inlet conditions as before, the heat supplied was 100 Btu/lbm. During this test, in which the exit pressure was not controlled, the isothermal turbine operated at an efficiency of 95%.

Determine for each of the three cases:
a. Power output of the turbine (kW)
b. Available energy content at inlet per lbm of steam
c. Unavailability at inlet per lbm of steam
d. Available energy content at exit per lbm of steam
e. Unavailable energy content at exit per lbm of steam
f. Turbine efficiency
g. Available energy in the heat added per lbm of steam
h. Unavailable part of the heat added per lbm of steam
i. Reversible work per lbm of steam
j. Maximum useful work per lbm of steam
k. Entropy production per lbm of steam
l. Available energy degraded per lbm of steam
m. Irreversibility per lbm of steam
n. Lost work per lbm of steam
o. Effectiveness of the turbine

Sketch the (T,S) diagrams and give the meaning of the various areas.

9-7 In the solar space and hot water heating system of sketch (a), auxiliary energy is added to the main storage tank whenever its

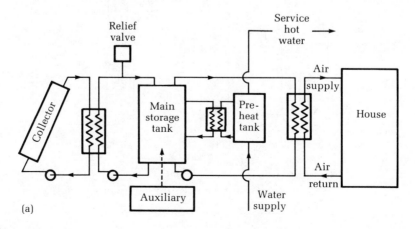

(a)

temperature falls below 65 F. Based on second-law concepts, would you expect this system to perform better, worse, or the same as the system shown in sketch (b), in which a separate auxiliary energy source for service hot water is required? Why?

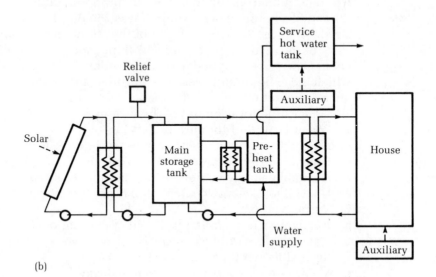

(b)

9-8 Two common nuclear power plant systems are shown in the following sketches. Discuss the relative merits of the two systems from the viewpoint of effective use of available energy.

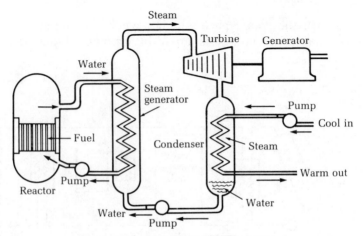

Pressurized water reactor (PWR) system

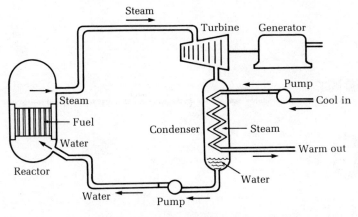

Boiling water reactor (BWR) system

9-9 Assume that in a steam power plant (Rankine cycle), there is a pressure and temperature drop between the boiler and the turbine. For example, at the boiler exit we get 3.5 MPa and 370 C while at the turbine entrance we measure 3.25 MPa and 340 C. What is the irreversibility of this process if the ambient temperature is taken to be 25 C?

9-10 In Problem 9-9, it is found that 0.22 kJ/kg of work is done by the turbine if the exhaust pressure is 0.01 MPa. What is the reversible work and the irreversibility for this actual process? (*Hint:* A (T, s) diagram of the ideal and actual processes may be helpful.)

Chapter

10

Mixtures and Psychrometrics

To this point our discussion has been limited to working fluids that are pure substances (homogeneous and unchanging in chemical composition). In fact, we have loosely included air as a pure substance although it is not. This limitation, of course, precludes any sort of reaction (oxidation). As an engineer, however, you may have to make thermodynamic calculations involving a mixture of different homogeneous gases. Since an infinite number of mixtures are possible today, tabulations like the steam tables are virtually impossible. The usual procedure is to learn to determine the thermodynamic properties of a mixture of ideal gases from the properties of the individual components. It is then a natural step to psychrometrics and later the combustion process. This chapter introduces psychrometrics.

10–1 Mixtures

Ideal Gases

Mixtures of different gases are encountered in many engineering applications. Air is a good example of such a mixture. Since the individual gases can often be approximated as ideal gases, the study of mixtures of ideal gases and their properties is of considerable importance. The first properties we will examine are pressure, volume, and temperature.

Each constituent gas in a mixture has its own pressure, called the

partial pressure of the particular gas. The **Gibbs—Dalton law** (also called the Dalton law or law of additive pressures) states that in a mixture of ideal gases, the pressure of the mixture is equal to the sum of the partial pressures of the individual constituent gases. In equation form:

$$p_m = p_1 + p_2 + p_3 + \cdots \tag{10-1}$$

where p_m is the total pressure of the mixture of gases 1, 2, 3, and so on, and p_1, p_2, p_3, and so on, are the partial pressures. In a mixture of ideal gases the partial pressure of each constituent equals the pressure which that constituent would exert if it existed alone at the temperature and volume of the mixture.

With regard to volume, on the other hand, we know from experience and experiment that generally, in mixtures of gases, each constituent gas behaves as though the other gases were not present—each gas occupies the total volume of the mixture at the mixture's temperature at its partial pressure. If V_m is the volume of the mixture, then

$$V_m = V_1 + V_2 + V_3 + \cdots \tag{10-2}$$

for the volumes of the constituents. *However, the volume of a mixture of ideal gases equals the sum of the volumes of its constituents if each existed alone at the temperature and pressure of the mixture.* This statement is known as Amagat's law, Leduc's law, or the law of additive volumes.

This law may be applied to real as well as ideal gas mixture; it is exactly true for the ideal mixture and approximately true for real gas mixtures. This law, like the Gibbs–Dalton law has been experimentally verified and may be used whenever the mixture temperature is greater than any constituent critical temperature.

If T_m is the temperature of the mixture, then

$$T_m = T_1 = T_2 = T_3 = \cdots \tag{10-3}$$

for the temperature relationship.

A volume fraction definition is needed when a volumetric analysis is to be used. Thus

$$\frac{V_1(p_m, T_m)}{V_m} = \frac{\text{volume of gas 1 existing alone at } p_m, T_m}{\text{volume of mixture at } p_m, T_m} \tag{10-4}$$

Later, when mole fraction is defined, you will be able to prove that the mole fraction and the volume fraction of an ideal gas mixture are identical. Again, this rule is exactly true for ideal gas and only approximate for real gases.

It can be easily seen that the mass of a mixture is just the sum of the masses of the constituents. Thus

$$m_m = m_1 + m_2 + m_3 + \cdots \tag{10-5}$$

Again the subscripts m, 1, 2, and so on represent mixture, constituent 1, and the like, respectively, An analysis based on mass is called a gravimetric analysis. Note that by rearranging Equation (10-5), we get

$$1 = \frac{m_1 + m_2 + m_3 + \cdots}{m_m}$$

The ratio m_1/m_m is called the **mass fraction of constituent 1** (g_1). Thus

$$g_1 = \frac{m_1}{m_1 + m_2 + m_3 + \cdots} = \frac{m_1}{m_m} \tag{10-6}$$

Recall that 1 mole of a substance is a mass of that substance numerically equal to its molecular weight. Thus the mole is a unit of mass and, regardless of phase, 1 mole of a substance always has the same mass. Moreover, according to Avogadro's law: *Equal volumes of perfect gases held under exactly the same temperature and pressure have equal numbers of molecules.* This law can be misleading, however, since 1 mole of hydrogen (H_2) has a mass of say 2 kg while 1 mole of oxygen (O_2) has a mass of 32 kg and occupies the same volume if both gases are at the same temperature—that is, the mole is not a volume measurement. Further, the total number of moles in a mixture is defined as the sum of the number of moles of its constituents:

$$n_m = n_1 + n_2 + n_3 + \cdots \tag{10-7}$$

The mole fraction $\bar{X}$ is defined as n/n_m, and

$$M_m = \bar{X}_1 M_1 + \bar{X}_2 M_2 + \bar{X}_3 M_3 + \cdots \tag{10-8}$$

where M_m is called the **apparent (or average) molecular weight** of the mixture and M_1, M_2, and so forth are the molecular weights of each component.

Notice that for $p_1 = p_2 = p_3 = \cdots$ and $T_1 = T_2 = T_3 = \cdots$, the ideal-gas equation indicates that V_1 is directly proportional to n_1 and so on. Thus, from Equation (10-6),

$$m_1 = g_1(m_1 + m_2 + m_3 + \cdots)$$

And since $m_1 = n_1 M_1$ and so on,

$$n_1 M_1 = g_1(n_1 M_1 + n_2 M_2 + n_3 M_3 + \cdots)$$

and

$$V_1 M_1 = g_1(V_1 M_1 + V_2 M_2 + V_3 M_3 + \cdots)$$

Dividing by V_m and rearranging yields

$$g_1 = \frac{\bar{X}_1 M_1}{\bar{X}_1 M_1 + \bar{X}_2 M_2 + \bar{X}_3 M_3 + \cdots} \tag{10-9}$$

In a similar fashion

$$\bar{X}_1 = \frac{g_1/M_1}{g_1/M_1 + g_2/M_2 + g_3/M_3 + \cdots} \tag{10-10}$$

Thus, once one analysis is complete the other may also be carried out.

The second part of the Gibbs–Dalton law can be taken as a basic definition: *Some properties of an ideal-gas mixture are equal to the sums of those properties for each component when it occupies the total volume by itself; these properties are internal energy, enthalpy, and entropy.* Hence

$$U_m = U_1 + U_2 + U_3 + \cdots \tag{10-11}$$

$$H_m = H_1 + H_2 + H_3 + \cdots \tag{10-12}$$

$$S_m = S_1 + S_2 + S_3 + \cdots \tag{10-13}$$

And further:

$$u_m = \frac{U_m}{m_m} = \frac{m_1 u_1 + m_2 u_2 + m_3 u_3 + \cdots}{m_m}$$

$$= g_1 u_1 + g_2 u_2 + g_3 u_3 + \cdots \tag{10-14}$$

$$h_m = g_1 h_1 + g_2 h_2 + g_3 h_3 + \cdots \tag{10-15}$$

$$s_m = g_1 s_1 + g_2 s_2 + g_3 s_3 + \cdots \tag{10-16}$$

Recall that the specific heat at constant volume is defined as

$$c_v = \frac{\partial u}{\partial T}\bigg|_v$$

Thus, from Equation (10-14),

$$c_{v_m} = \frac{\partial u_m}{\partial T}\bigg|_{v_m} = g_1 c_{v_1} + g_2 c_{v_2} + g_3 c_{v_3} + \cdots \tag{10-17}$$

or

$$m c_{v_m} = m_1 c_{v_1} + m_2 c_{v_2} + m_3 c_{v_3} + \cdots \tag{10-18}$$

Similarly,

$$c_{p_m} = g_1 c_{p_1} + g_2 c_{p_2} + g_3 c_{p_3} + \cdots \tag{10-19}$$

and

$$R_m = g_1 R_1 + g_2 R_2 + g_3 R_3 + \cdots \tag{10-20}$$

From Equations (10-14) to (10-20) the specific values of internal energy, enthalpy, entropy, specific heats (both constant volume and constant pressure) and the gas constant of a mixture are the mass weighted averages of the corresponding constituent values (the sum of the constituent property multiplied by the mass fraction). The individual constituent properties are to be evaluated at the mixture temperature and volume—or at the mixture temperature and the constituent partial pressure.

EXAMPLE 10-1

Consider a gas mixture of 20% hydrogen, 50% nitrogen, and 30% carbon dioxide (percentages are by mass). The mixture is at 10 psia and 70 F. Determine for this mixture:

1. The partial pressure of each component

2. The gas constant

3. The molecular weight

4. The enthalpy (per lbm)

5. The internal energy (per lbm)

6. The entropy (per lbm)

SOLUTION: Let subscript 1 represent hydrogen, 2 nitrogen, and 3 carbon dioxide. Then:

	M_i	c_p, kJ/(kg $\cdot$ K)	c_v, kJ/(kg $\cdot$ K)	$g_i = m_i/M_i$	$n_i = m_i/M_i$	$\bar{X}_i = n_i/n$
Hydrogen (H_2)	2	14.28	10.13	0.2	0.100	0.8019
Nitrogen (N_2)	28	1.04	0.741	0.5	0.0179	0.1434
Carbon dioxide (CO_2)	44	0.85	0.661	0.3	0.00682	0.0547
				1 kg	0.1247 mole	1.00

1. $p_1 = \bar{X}_1 p = 8.019$ psia; $p_2 = \bar{X}_2 p = 1.434$ psia; and $p_3 = \bar{X}_2 p = 0.547$ psia

2. $R_m = g_1 R_1 + g_2 R_2 + g_3 R_3$ (Remember: $c_p - c_v = R$.)
   $= [0.2(4.15) + 0.5(0.299) + 0.3(0.189)]$ kJ/(kg $\cdot$ K)
   $= 1.0357$ kJ/(kg $\cdot$ K) $= 0.247$ Btu/lbm-R

3. $M_m = \dfrac{M}{n} = \dfrac{1 \text{ lbm}}{0.1247} = 8.02$ kg/(kg-mole)

4. $h = g_1 h_1 + g_2 h_2 + g_3 h_3$. Recall that $h - h_o = c_p(T - T_o)$. So, for convenience, let $h_o = 0$ at $T_o = 0$ R. Thus,

$$h = (g_1 c_{p_1} + g_2 c_{p_2} + g_3 c_{p_3})T$$

$$= [0.2(14.28) + 0.5(1.04) + 0.3(0.85)] \, 530 \text{ kJ/kg}$$

$$= 1924.42 \text{ kJ/kg} = 255.34 \text{ Btu/lbm}$$

5. Let us also assume that $u_o = 0$ at $T_o = 0$ R so that

$$u = g_1 u_1 + g_2 u_2 + g_3 u_3 = (g_1 c_{v_1} + g_2 c_{v_2} + g_3 c_{v_3})T$$

$$= [0.2(10.13) + 0.5(0.741) + 0.3(0.661)] \, 530$$

$$= 1375.24 \text{ kJ/kg} = 182.46 \text{ Btu/lbm}$$

6. $s = g_1 s_1 + g_2 s_2 + g_3 s_3$. If $s = 0$ at 0 F and 1 atm, then

$$s = c_p \ln\frac{T}{T_o} - R \ln(p/p_o)$$

$$= c_p(0.14165) - R(-0.3853)$$

Thus

$$s = 0.14165(g_1 c_{p_1} + g_2 c_{p_2} + g_3 c_{p_3}) + 0.3853(g_1 R_1 + g_2 R_2 + g_3 R_3)$$

$$= 0.14165[(2.856) + (0.52) + (0.255)]$$

$$+ 0.3853[(0.830) + (0.149) + (0.0567)]$$

$$= 0.91339 \text{ kJ/kg} \cdot \text{K} = 0.21814 \text{ Btu/lbm} - \text{R}.$$

Real Gases

Many times real gases are not treated as ideal. Thus the ideal-gas equation of state does not adequately represent the gas and we must resort to approximations. In applying these approximations, the Gibbs–Dalton law is valid:

$$p = p_1 + p_2 + \cdots \tag{10-1}$$

The problem is that p_1/p is not equal to $\bar{X}_1$ (as it is for the ideal-gas case).

The usual approach is to use a compressibility factor. For a mixture at pressure p, volume V, and temperature T,

$$pV = nZ\bar{R}T \tag{10-21}$$

where n is the number of moles of mixture and Z is the compressibility factor of the mixture ($Z = Z(p, T, \bar{X})$). Each component of the mixture must obey a similar expression:

$$p_i V = n_i Z_i \bar{R}T \tag{10-22}$$

Adding all component equations of state yields

$$(p_1 + p_2 + \cdots)V = (n_1 Z_1 + n_2 Z_2 + \cdots)\bar{R}T \tag{10-23}$$

But from the Gibbs–Dalton law

$$pV = (n_1 Z_1 + n_2 Z_2 + \cdots)\bar{R}T$$

$$= n_m Z \bar{R}T$$

Hence

$$Z = \frac{n_1 Z_1 + n_2 Z_2 + \cdots}{n_1 + n_2 + \cdots}$$

$$= \bar{X}_1 Z_1 + \bar{X}_2 Z_2 + \cdots \tag{10-24}$$

EXAMPLE 10-2

Consider a number of Clausius gases. The equation of state of each gas is of the form $p(V - nb) = n\bar{R}T$. Determine an expression for the overall mixture.

SOLUTION: To attack this problem, we allow $\bar{X}_1$ moles of gas 1. Thus

$$p_1(V - n_1 b) = n_1 \bar{R}T$$

and, for $\bar{X}_2$ moles of gas 2,

$$p_2(V - n_2 b) = n_2 \bar{R} T$$

and so forth.

Solving each component equation of state for its partial pressure p_1, then adding and using the Gibbs–Dalton law, yields

$$p = p_1 + p_2 + \cdots$$

$$= \bar{R}T\left(\frac{n_1}{V - n_1 b} + \frac{n_2}{V - n_2 b} + \cdots\right)$$

Hence

$$p = \bar{R}T \sum\left(\frac{n_i}{V - n_i b}\right) = \bar{R}T \sum \frac{\bar{X}_i}{V - \bar{X}_i b}$$

Note that if $b = 0$ (ideal gases), then

$$p = n\left(\frac{\bar{R}T}{V}\right)$$

That is, the pressure is directly proportional to the number of moles.

Closure

Before completing this section, we need to comment on air. It is generally assumed that dry air has a definite composition. The proportions by volume are 21% oxygen, 78% nitrogen, and not quite 1% argon and then traces of carbon dioxide, hydrogen, helium, krypton, neon, ozone, and xenon. This is the volumetric analysis. The corresponding approximate gravimetric figures are 23% oxygen, 76% nitrogen, and slightly more than 1% of other gases such as argon. For convenience, we will use volumetrically 21% oxygen and 79% nitrogen ($N_2/O_2 = 3.76$) and gravimetrically 23.2% oxygen and 76.8% nitrogen ($N_2/O_2 = 3.31$). Unfortunately for engineers, however, dry air is rarely found; another component, water vapor, is usually present. Since the water vapor in air is usually superheated, we may approximate this water vapor as an ideal gas. There are cases in which great care must be taken—for this approximation to be appropriate, none of the vapor may condense or solidify.

10–2 Psychrometrics

Basic Definitions

Psychrometrics *is the science involving thermodynamic properties of moist air and the effect of atmospheric moisture on materials and human comfort.* As it applies in this text, the definition must be broadened to include the method of controlling the thermal properties of moist air. Many new terms are used in the science of psychrometrics; the main ones are defined in the following paragraphs.

Dry-bulb temperature (T) is the temperature of air as registered by an ordinary thermometer. **Thermodynamic wet-bulb temperature** $(T^\ast)$ is the temperature to which water (liquid or solid) can bring air as the air is brought to saturation adiabatically (at the same temperature $T^\ast$) while the pressure p is kept constant. This is accomplished by evaporating water into moist air at some given dry-bulb temperature T and humidity ratio W until saturated. Figure 10-1 illustrates the apparatus for measuring $T^\ast$.

The **humidity ratio** (or, alternatively, the mixing ratio) W of a moist air sample is defined as the ratio of the mass of water vapor to the mass of dry air contained in the sample:

$$W = \frac{m_w}{m_a}$$

$$= \frac{R_a p_w}{R_w p_a} = \frac{53.3\, p_w}{85.7\, p_a}$$

$$= 0.6219 \frac{p_w}{p - p_w} = 0.6219 \frac{\overline{X}_w}{\overline{X}_a} \qquad (10\text{-}25a)$$

The humidity ratio has also been related empirically to the wet bulb condition by

$$W = \frac{(1093 - 0.556 T^\ast) W_s^\ast - 0.240(T - T^\ast)}{1093 + 0.444 T - T^\ast} \qquad (10\text{-}25b)$$

where $W_s^\ast$ is evaluated at $T^\ast$ and p_w is the water vapor partial pressure.

The **degree of saturation** (μ) is the ratio of the humidity ratio W to the humidity ratio W_s of saturated air at the same temperature and pressure:

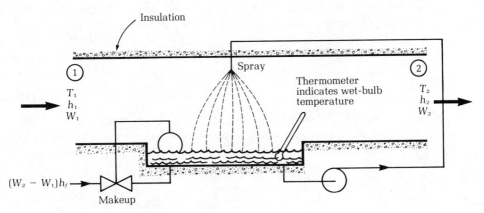

Figure 10-1. An adiabatic saturator.

$$\mu = \frac{W}{W_s}\bigg|_{T,\,p} \qquad (10\text{-}26)$$

Relative humidity (ϕ) is the ratio of the mole fraction of water vapor $\bar{X}_w$ in a moist air sample to the mole fraction $\bar{X}_{ws}$ in an air sample that is saturated at the same temperature and pressure:

$$\phi = \frac{\bar{X}_w}{\bar{X}_{ws}}\bigg|_{T,\,p} \qquad (10\text{-}27)$$

$$\phi = \frac{p_w}{p_{ws}}\bigg|_{T,\,p} \qquad (10\text{-}28)$$

The term p_{ws} represents the saturation pressure of water vapor at the given temperature T.

Dew-point temperature (T_d) is the temperature of moist air that is saturated at the same pressure p and has the same humidity ratio W as the sample of moist air. For the temperature range of 32 to 150 F,

$$T_d = 79.047 + 30.5790\alpha + 1.8893\alpha^2 \qquad (10\text{-}29)$$

For temperatures below 32 F,

$$T_d = 71.98 + 24.873\alpha + 0.8927\alpha^2 \qquad (10\text{-}30)$$

where $\alpha = \log_e(p_w)$ and p_w is in in. Hg.

The **volume** (v) of a moist air mixture is expressed in terms of a unit mass of dry air, with the relation $p = p_a + p_w$,

$$v = \frac{R_a T}{p - p_w} \qquad (10\text{-}31)$$

where $R_a = \bar{R}/28.9645 = 53.352$ ft-lbf/lbm $-$ R. To estimate v at temperatures below about 150 F, the volume v of moist air per pound of dry air may be computed by

$$v = v_a + \mu v_{as} \qquad (10\text{-}32)$$

where v_a is the specific volume of dry air and v_{as} is the difference in specific volumes of dry air and water-saturated air.

The **enthalpy** of a mixture of perfect gases is equal to the sum of the individual partial enthalpies of the components. The enthalpy of moist air is thus

$$h = h_a + W h_g \qquad (10\text{-}33)$$

where h_a is the specific enthalpy for dry air and h_g is the specific enthalpy for saturated water vapor at the temperature of the mixture. Approximately,

$$h_a = 0.240T \qquad (\text{Btu/lbm}) \qquad (10\text{-}34)$$

and

$$h_g = 1061 + 0.444T \qquad (\text{Btu/lbm}) \qquad (10\text{-}35)$$

where T is the dry-bulb temperature (F). The moist air enthalpy then becomes

$$h = 0.240T + W(1061 + 0.444T) \tag{10-36}$$

where h has units of Btu/lbm dry air. At temperatures below 150 F, the enthalpy and entropy h of moist air per pound of dry air may be computed by

$$h = h_a + \mu h_{as} \tag{10-37}$$

$$s = s_a + \mu s_{as} \tag{10-38}$$

The Psychrometric Chart

The ASHRAE psychrometric chart is convenient for solving numerous process problems involving moist air. Processes performed with air can be plotted on the chart for quick visualization as well as for determining changes in significant properties such as temperature, humidity ratio, and enthalpy for the process. Figure 10-2 is an abridgement of ASHRAE Psychrometric Chart 1.

Figure 10-3 shows some of the basic air-conditioning processes. *Sensible heating only* (C) or *sensible cooling only* (G) shows a change in dry-bulb temperature with no change in humidity ratio. For either sensible-heat-change process, the temperature changes but not the

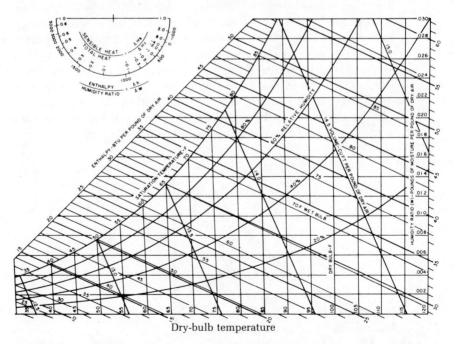

Figure 10-2. ASHRAE normal psychrometric chart. (Reprinted with permission from the 1977 Fundamentals Volume, ASHRAE HANDBOOK & Product Directory.)

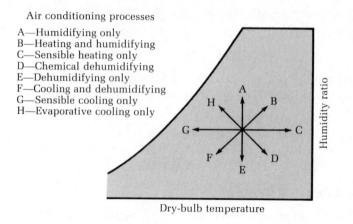

Air conditioning processes

A—Humidifying only
B—Heating and humidifying
C—Sensible heating only
D—Chemical dehumidifying
E—Dehumidifying only
F—Cooling and dehumidifying
G—Sensible cooling only
H—Evaporative cooling only

Figure 10-3. Psychrometric representations of basic air-conditioning processes.

moisture content of the air. *Humidifying only* (A) or *dehumidifying only* (E) shows a change in humidity ratio with no change in dry-bulb temperature. For these latent heat processes, the moisture content of the air is changed but not the temperature. *Cooling and dehumidifying* (F) result in a reduction of both the dry-bulb temperature and the humidity ratio. Cooling coils generally perform this type of process. *Heating and humidifying* (B) result in an increase of both the dry-bulb temperature and the humidity ratio. *Chemical dehumidifying* (D) is a process in which moisture from the air is adsorbed or absorbed by a hygroscopic material. Generally the process occurs at constant enthalpy. *Evaporative cooling only* (H) is an adiabatic heat transfer process in which the wet-bulb temperature of the air remains constant but the dry-bulb temperature drops as the humidity rises. *Adiabatic mixing* of air at one condition with air at another condition is represented on the psychrometric chart by a straight line drawn between the points representing the two air conditions as shown on Figure 10-4.

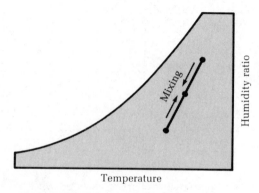

Figure 10-4. Adiabatic mixing.

EXAMPLE 10-3

A room contains an air/water vapor mixture at standard atmospheric pressure, 60% relative humidity, and 85 F temperature. The room has dimensions of 10 ft $\times$ 20 ft $\times$ 8 ft. Calculate:

1. The humidity ratio

2. The dew-point temperature

3. The mass of dry air

4. The mass of water vapor

5. The enthalpy of the mixture

6. The mass of water vapor condensed from the mixture if the air/water vapor mixture is cooled at constant pressure to 50 F

SOLUTION: From Appendix A-1-1 (at 85 F):

$$p_{ws} = 0.59604 \text{ psia}$$

1. Recall that

$$W = 0.62198 \frac{p_w}{p - p_w}$$

and

$$\phi = \frac{p_w}{p_{ws}} \bigg|_{T, p}$$

Solving for p_w from the last equation, we get

$$p_w = \phi p_{ws} = 0.60(0.59604) = 0.358 \text{ psia}$$

Using the first equation yields

$$W = 0.62198 \frac{0.358}{14.696 - 0.358} = 0.0152 \text{ lbm vapor/lbm air}$$

2. The dew-point temperature is defined by

$$W_s(p, T_d) = w$$

The temperature of moist air that is saturated at the same pressure and has the same W (or p_w) is the corresponding saturation temperature for water vapor at a saturation pressure of 0.358 psia. From Appendix A-1-1, we see that this temperature is

$$T_d = 69.6 \text{ F}$$

3. Using the equation of state for air, we get

$$p_a V = n_a \bar{R} T$$

Recall that

$$n_a = \frac{m_a}{M_a}$$

Then

$$p_a V = m_a \left(\frac{\bar{R}}{M_a} \right) T$$

Therefore

$$m_a = \frac{p_a V}{(\bar{R}/M_a)T}$$

$$= \frac{(14.696 - 0.358)(10 \times 20 \times 8)(144)}{(1545/28.96)(85 + 460)}$$

$$= 116 \text{ lbm}$$

4. Likewise

$$m_w = \frac{p_w V}{R/M_w T}$$

$$= \frac{0.358(10 \times 20 \times 8)(144)}{(1545/18)(85 + 460)}$$

$$= 1.762 \text{ lbm}$$

or

$$W = \frac{m_w}{m_a}$$

$$m_w = W m_a = 0.0152(116) = 1.76 \text{ lbm}$$

5. Now

$$h = 0.24T + W(1061 + 0.444T)$$

$$= 0.24(85) + 0.0152(1061 + 0.444 \times 85)$$

$$= 37.1 \text{ Btu/lbm air}$$

6. At 50 F the mixture is saturated, since this temperature is less than the dew-point temperature for the mixture. From Equation (10-28) and Appendix A-1-1 we obtain

$$\phi_w = \frac{p_{w2}}{P_{ws_2}} = 1.0$$

and

$$p_{w2} = p_{ws_2} = 0.17799 \text{ psia}$$

At the cooled condition,

$$W_2 = 0.62198 \frac{p_{w2}}{p - p_{w2}} = 0.62198 \frac{0.178}{14.696 - 0.178}$$

$$= 0.00762 \text{ lbm vapor/lbm air}$$

The water vapor condensed from the air/water vapor mixture is

$$(W_1 - W_2)m_a = (0.0152 - 0.00762)116 = 0.88 \text{ lbm}$$

EXAMPLE 10-4

A cooling and dehumidifying coil receives in a steady-flow process an air/water vapor mixture at 16 psia, 95 F, and 83% relative humidity and discharges it at 14.7 psia, 50 F, and 96% relative humidity. The condensate leaves the unit at 50 F. Calculate the heat transfer per pound of dry air flowing through the unit.

SOLUTION: Recall the first law of thermodynamics for a steady-state/steady-flow system:

$$_1\dot{Q}_2 + \left(\frac{V_1^2}{2g_c} + \frac{gZ_1}{g_c} + h_1 \right) \dot{m}_{a_1} = {_1}\dot{W}_2 + \left(\frac{V_2^2}{2g_c} + \frac{gZ_2}{g_c} + h_2 \right) \dot{m}_{a_2}$$

The continuity equation is

$$m = A\rho V$$

For the air flowing through the apparatus this becomes

$$m_{a_1} = m_{a_2} = m_a$$

For the water vapor this becomes

$$m_{w_1} = m_{w_2} + m_{cond}$$

where the subscript "cond" stands for condensate. Neglecting any kinetic or potential-energy changes of the flowing fluid and noting that there is no mechanical work being done on or by the system, the first-law equation reduces to

$$_1\dot{Q}_2 = m_a(h_2 - h_1) + (m_{cond})(h_{cond})$$

Using Equation (10-33), we get the enthalpy terms

$$\frac{_1Q_2}{m_a} = h_{a_2} - h_{a_1} + W_2 h_{w_2} - W_1 h_{w_1} + \frac{m_{cond}}{m_a} h_{cond}$$

From the continuity equation this becomes

$$\frac{_1Q_2}{m_a} = h_{a_2} - h_{a_1} + W_2 h_{w_2} - W_1 h_{w_1} + (W_1 - W_2)h_{cond}$$

From Appendix A-1-1, we get

$$h_{w2} = 1083.06 \text{ Btu/lbm}$$

$$h_{w1} = 1102.59 \text{ Btu/lbm}$$

$$h_{cond} = 18.07 \text{ Btu/lbm}$$

Using Equations (10-28) and (10-25a), we can calculate W_1 and

W_2:

$$p_{w_1} = \phi_1 p_{ws_1} = 0.83(0.8156) = 0.678 \text{ psia}$$

$$W_1 = 0.622 \frac{p_{w_1}}{p_1 - p_{w_1}} = 0.622 \frac{0.678}{16 - 0.678} = 0.0275 \text{ lbm/lbm}$$

$$p_{w_2} = 0.86(0.178) = 0.171 \text{ psia}$$

$$W_2 = 0.622 \frac{p_{w_2}}{p_2 - p_{w_2}} = 0.622 \frac{0.0171}{14.7 - 0.0171} = 0.0073 \text{ lbm/lbm}$$

Substituting into the energy equation yields

$$\frac{_1Q_2}{m_a} = 0.24(50 - 95) + 0.0073(1083.1) - 0.0275(1102.6)$$

$$+ (0.0275 - 0.0073)18.07$$

$$= -10.8 + 7.91 - 30.3 + 0.36 = 32.8 \text{ Btu/lbm air}$$

EXAMPLE 10-5

Determine by thermodynamic analysis the humidity ratio, relative humidity, and enthalpy for an air/water vapor mixture at 14.7 psia with a dry-bulb temperature of 90 F and a thermodynamic wet-bulb temperature of 76 F. Also evaluate these properties of the mixture by using the ASHRAE psychrometric chart (Figure 10-2).

SOLUTION: The humidity ratio can be found by applying Equation (10-25b):

$$W = \frac{(1093 - 0.556T^*)W_s^* - 0.24(T - T^*)}{1093 - 0.44T - T^*}$$

where $T = 90$ F and $T^* = 76$ F. From Appendix A-1-1 we get

$$W_s^* = 0.01948$$

$$W = \frac{[1093 - 0.556(76)]0.01948 - 0.24(90 - 76)}{1093 + 0.444(90) - 76}$$

$$= 0.01625$$

And from Equation (10-25b) we get

$$p_w = \frac{pW}{0.62198 + W} = \frac{14.7(0.01625)}{0.622 + 0.01625}$$

$$= 0.375 \text{ psia}$$

$$p_{w_s} = 0.6983 \text{ psia} \quad \text{(from Appendix A-1-1)}$$

$$\phi = \frac{p_w}{p_{w_s}} \times 100 = 53.7\%$$

Using $p_w = 0.375$ psia and finding the corresponding value of saturation temperature from the Appendix gives the dew-point temperature:

$$T_d = 70.9 \text{ F}$$

Using Equation (10-36), we can find the enthalpy:

$$h = 0.24T + W(1061 + 0.444T)$$

$$= 0.24(90) + 0.01625[1061 + 0.444(90)]$$

$$= 39.5 \text{ Btu/lbm}$$

From the psychrometric chart (Figure 10-2) the following values can be found:

$$W = 0.0163 \text{ lbm water/lbm air}$$

$$\phi = 53.5\%$$

$$h = 39.6 \text{ Btu/lbm}$$

EXAMPLE 10-6

An air/water vapor mixture at standard barometric pressure is at 100 F and 50% relative humidity. Calculate: (1) specific volume, (2) enthalpy, and (3) entropy.

SOLUTION: By combining Equations (10-25a), (10-26), and (10-27) we get

$$\phi = \frac{\mu}{1 - (1 - \mu)\bar{X}_{ws}}$$

And, recalling that $\bar{X}_w = p_w/p$,

$$\phi = \frac{\mu}{1 - (1 - \mu)p_{ws}/p}$$

Solving for μ yields

$$\mu = \phi\frac{1 - (p_{ws}/p)}{1 - \phi(p_{ws}/p)}$$

Using Appendix A-1-1 for $T = 100$ F, we get $p_{ws} = 0.949$ psia. Therefore

$$\mu = 0.50\frac{1 - (0.949/14.69)}{1 - 0.50(0.949/14.69)} = 0.483$$

1. From Equation (10-33) we get

$$v = v_a + \mu v_{as}$$

From the Appendix,

$$v_a = 14.106 \text{ ft}^3/\text{lbm air}$$

$$v_{as} = 0.975 \text{ ft}^3/\text{lbm air}$$

$$v = 14.106 + 0.483(0.975) = 14,578 \text{ ft}^3/\text{lbm air}$$

2. From Equation (10-37) we get

$$h = h_a + \mu h_{as}$$

From the Appendix,

$$h_a = 24.029 \text{ Btu/lbm air}$$

$$h_{as} = 47.70 \text{ Btu/lbm air}$$

$$h = 24.029 + 0.483(47.70) = 47.08 \text{ ft}^3/\text{lbm air}$$

3. From Equation (10-38) we get

$$s = s_a + \mu s_{as}$$

From the Appendix,

$$s_a = 0.04529 \text{ Btu/F-lbm air}$$

$$s_{as} = 0.09016 \text{ Btu/F-lbm air}$$

$$s = 0.04729 + 0.483(0.09016) = 0.0908 \text{ Btu/F-lbm air}$$

PROBLEMS

10-1 A 0.5-m³ rigid vessel contains 1 kg of carbon monoxide and 1.5 kg of air at 15 C. The gravimetric analysis of air is the standard values (23.3% O_2 and 76.7% N_2). What are the partial pressures of each component?

10-2 A volumetric analysis of a mixture of ideal gases is as follows:

N_2:	80%
CO_2:	10%
O_2:	6%
CO:	4%

Recalculate these quantities on a mass basis. What is the gas constant and specific heat at constant pressure?

10-3 Assume a gas mixture of O_2, N_2, and CO_2 to the following mole relations: 5.5, 3, and 1.5 moles respectively. Make a volumetric analysis. Determine the mass and molecular weight of the mixture.

10-4 For air containing 75.53% N_2, 23.14% O_2, 1.28% Ar, and 0.05% CO_2, by mass, determine the gas constant and molecular weight of this air. How do these values compare if the mass analysis is 76.7% N_2 and 23.3% O_2?

10-5 A mixture of 15% CO_2, 12% O_2, and 73% N_2 (by volume) is to be expanded. The volume ratio is 6:1 and the corresponding temperature change is 1800 to 1400 F. What is the entropy change?

10-6 In a 3-ft^3 rigid vessel is a 50–50 mixture of N_2 and CO (by volume). Determine the mass of each component if $T = 65$ F and $p = 30$ psia.

10-7 A room is 20 ft × 12 ft × 8 ft and contains an air/water vapor mixture at 80 F. The barometric pressure is standard and the partial pressure of the water vapor is measured to be 0.2 psia. Calculate:
a. Relative humidity
b. Humidity ratio
c. Dew-point temperature
d. Pounds of water vapor contained in the room

10-8 Given room conditions of 75 F (dry bulb) and 60% relative humidity, determine the following for the air/vapor mixture without using the ASHRAE psychrometric chart:
a. Humidity ratio
b. Enthalpy
c. Dew-point temperature
d. Specific volume
e. Degree of saturation

10-9 For the conditions of Problem 10-8, use the ASHRAE psychrometric chart (Figure 10-2) to find:
a. Wet-bulb temperature
b. Enthalpy
c. Humidity ratio

10-10 Using the ASHRAE psychrometric chart (Figure 10-2), complete the following table:

Dry-Bulb Temperature, F	Wet-Bulb Temperature, F	Dew-Point Temperature, F	Humidity Ratio, lbm/lbm	Enthalpy, Btu/lbm	Relative Humidity, %	Specific Volume, ft^3/lbm
85	60					
75		50				
				30	60	
	70		0.01143			
		82		50		

10-11 Using the ASHRAE psychrometric chart (Figure 10-2), complete the following table:

Dry-Bulb Temperature, F	Wet-Bulb Temperature, F	Dew-Point Temperature, F	Humidity Ratio, lbm/lbm	Relative Humidity, %	Enthalpy, Btu/lbm	Specific Volume, ft³/lbm
80						13.8
70	55					
100		70				
				40	40	
			0.01			13.8
	60	40				
40				20		
		60			30	
85			0.012			
80	80					

10-12 Complete the following table by using the psychrometric chart (Figure 10-2):

Dry-Bulb, F	Wet-Bulb, F	Dew-Point, F	Humidity Ratio, lbm/lbm	Relative Humidity, %	Enthalpy, Btu/lbm air	Specific Volume, ft³/lbm air
90	75					
105					35	
		65		30		
			0.022			14.5
45	45					

10-13 Complete the following table:

Dry-Bulb Temperature, C	Wet-Bulb Temperature, C	Dew-Point Temperature, C	Humidity Ratio, kg/kg	Relative Humidity, %	Enthalpy, kJ/kg air	Specific Volume, m³/kg air
26.5						0.86
21	13					
38		21				
				40	95	
			0.01			0.85
	16	4				
4				20		
		16			70	
30			0.012			
27	27					

10-14 Determine without using the psychrometric chart the humidity ratio and relative humidity of an air/water vapor mixture with a dry-bulb temperature of 90 F and thermodynamic wet-bulb temperature of 78 F. The barometric pressure is 14.7 psia. Check your result by using the psychrometric chart (Figure 10-2).

10-15 One of the many methods used for drying air is to cool it below the dew-point temperature so that condensation or freezing of the moisture takes place. To what temperature must atmospheric air be cooled in order to have a humidity ratio of 0.000017 lbm/lbm? To what temperature must this air be cooled if its pressure is 10 atm?

10-16 One method of removing moisture from atmospheric air is to cool the air so that the moisture condenses or freezes out. Suppose an experiment requires a humidity ratio of 0.0001. To what temperature must the air be cooled at a pressure of 0.1 MPa in order to achieve this humidity?

10-17 A room of dimensions 4 m × 6 m × 2.4 m contains an air/water vapor mixture at a total pressure of 100 kPa and a temperature of 25 C. The partial pressure of the water vapor is 1.4 kPa. Calculate:

a. Humidity ratio
b. Dew point
c. Total mass of water vapor in the room

10-18 The air conditions at the intake of an air compressor are 70 F temperature, 50% relative humidity, and 14.7 psia pressure. The air is compressed to 50 psia and sent to an intercooler. If condensation of water vapor from the air is to be prevented, what is the lowest temperature to which the air can be cooled in the intercooler?

10-19 Humid air enters a dehumidifier with an enthalpy of 21.6 Btu/lbm of dry air and 1100 Btu/lbm of water vapor. There is 0.02 lbm of vapor per pound of dry air at entrance and 0.009 lbm of vapor per pound of dry air at exit. The dry air at exit has an enthalpy of 13.2 Btu/lbm; the vapor at exit has an enthalpy of 1085 Btu/lbm. Condensate with an enthalpy of 22 Btu/lbm leaves. The rate of flow of dry air is 287 lbm/min. Determine: (a) the amount of moisture removed from the air (lbm/min) and (b) the rate of heat removal required.

10-20 Air is supplied to a room from the outside, where the temperature is 20 F and the relative humidity is 60%. The room is to be maintained at 70 F and 50% relative humidity. How many pounds of water must be supplied per pound of air supplied to the room?

10-21 Air is heated to 80 F, without the addition of water, from 60 F dry-bulb and 50 F wet-bulb temperature. Use the psychrometric chart to find:
a. Relative humidity of the original mixture
b. Original dew-point temperature
c. Original specific humidity
d. Initial enthalpy
e. Final enthalpy
f. Heat added
g. Final relative humidity

10-22 Saturated air at 40 F is first preheated and then saturated adiabatically. This saturated air is then heated to a final condition of 105 F and 28% relative humidity. To what temperature must the air initially be heated in the preheat coil?

10-23 Atmospheric air at 100 F dry-bulb and 65 F wet-bulb temperature is humidified adiabatically with steam. The supply steam contains 10% moisture and is at 16 psia. What is the dry-bulb temperature of the humidified air if enough steam is added to bring the air to 70% relative humidity?

10-24 The summer design conditions in New Orleans are 95 F dry-bulb and 80 F wet-bulb temperature. In Tucson they are 105 F dry-bulb and 72 F wet-bulb temperature. What is the lowest air temperature that could theoretically be attained in an evaporative cooler at the summer design conditions in these two cities?

10-25 Air at 29.92 in. Hg enters an adiabatic saturator at 80 F dry-bulb and 66 F wet-bulb temperature. Water is supplied at 66 F. Find without using the psychrometric chart the humidity ratio, degree of saturation, enthalpy, and specific volume of entering air.

10-26 An air/water vapor mixture enters an air conditioning unit at a pressure of 150 kPa, a temperature of 30 C, and a relative humidity of 80%. The mass of dry air entering is 1 kg/s. The air/vapor mixture leaves the air conditioning unit at 125 kPa, 10 C, 100% relative humidity. The moisture condensed leaves at 10 C. Determine the heat transfer rate for the process.

10-27 Air at 40 C and 300 kPa, with a relative humidity of 35%, is to be expanded in a reversible adiabatic nozzle. To how low a pressure can the gas be expanded if no condensation is to take place? What is the exit velocity at this condition?

10-28 Using basic definitions and Dalton's law of partial pressure, show that Equation (10-32) reduces to

$$v = \frac{R_a T}{p - p_w}$$

10-29 In an air conditioning unit, 71,000 ft³/min of air enters at 80 F (dry bulb), 60% relative humidity, and standard atmospheric pressure. The leaving condition of the air is 57 F (dry bulb) and 90% humidity. Calculate:
a. Cooling capacity of the air conditioning unit (Btu/hr)
b. Rate of water removal from the unit
c. Sensible heat load on the conditioner (Btu/hr)
d. Latent heat load on the conditioner (Btu/hr)
e. Dew point of the air leaving the conditioner

10-30 Suppose that 4 lbm of air at 80 F (dry bulb) and 50% relative humidity is mixed with 1 lbm of air at 60 F and 50% relative humidity. Determine: (a) the relative humidity of the mixture and (b) the dew-point temperature of the mixture.

10-31 Air is compressed in a compressor from 85 F, 60% relative humidity, and 14.7 psia to 60 psia and then cooled in an intercooler before entering a second stage of compression. What is the minimum temperature to which the air can be cooled so that condensation does not take place?

10-32 Suppose that 4000 ft³/min of an air/water vapor mixture at 84 F dry-bulb and 70 F wet-bulb temperature enters a perfect refrigeration coil. The air leaves the coil at 53 F. How many Btu/hr of refrigeration are required?

10-33 Air at 40 F dry-bulb and 35 F wet-bulb temperature is mixed with air at 100 F dry-bulb and 77 F wet-bulb temperature in the ratio of 2 lbm of cool air to 1 lbm of warm air. Compute the humidity ratio and enthalpy of the mixed air.

10-34 Outdoor air at 90 F dry-bulb and 78 F wet-bulb temperature is mixed with return air at 75 F and 52% relative humidity. There are 1000 lbm of outdoor air for every 5000 lbm of return air. What are the dry-bulb and wet-bulb temperatures for the mixed air stream?

10-35 In a mixing process of two streams of air, 10,000 ft³/min of air at 75 F and 50% relative humidity mixes with 4000 ft³/min of air at 98 F dry-bulb and 78 F wet-bulb temperature. Calculate the following conditions after mixing at atmospheric pressure:
a. Dry-bulb temperature
b. Humidity ratio
c. Relative humidity
d. Enthalpy
e. Dew-point temperature

10-36 Determine the humidity ratio and relative humidity of an air/water vapor mixture that has a dry-bulb temperature of 30 C, an adiabatic saturation temperature of 25 C, and a pressure of 100 kPa.

10-37 An air/water vapor mixture at 100 kPa, 35 C, and 70% relative humidity is contained in a 0.5-m³ closed tank. The tank is cooled until the water begins to condense. Determine the temperature at which condensation begins and the heat transfer for the process.

10-38 A room is to be maintained at 76 F and 40% relative humidity. Air is to be supplied at 39 F to absorb 100,000 Btu/hr sensible heat and 35 lbm of moisture per hour. How many pounds of dry air per hour are required? What should be the dew-point temperature and relative humidity of the supply air?

10-39 Moist air enters a chamber at 40 F dry-bulb and 36 F wet-bulb temperature at a rate of 3000 ft³/min. In passing through the chamber, the air absorbs sensible heat at a rate of 116,000 Btu/hr and picks up 83 lbm/hr of saturated steam at 230 F. Determine the dry-bulb and wet-bulb temperatures of the leaving air.

10-40 In an auditorium maintained at a temperature not to exceed 77 F, and at a relative humidity not to exceed 55%, a sensible-heat load of 350,000 Btu and 1,000,000 grains of moisture per hour must be removed. Air is supplied to the auditorium at 67 F.

a. How many pounds of air per hour must be supplied?
b. What is the dew-point temperature of the entering air, and what is its relative humidity?
c. How much latent heat load is picked up in the auditorium?
d. What is the sensible-heat ratio?

10-41 A meeting hall is maintained at 75 F dry-bulb and 65 F wet-bulb temperature. The barometric pressure is 29.92 in. Hg. The space has a load of 200,000 Btu/hr (sensible) and 200,000 Btu/hr (latent). The temperature of the supply air to the space cannot be lower than 65 F (dry bulb).
a. How many pounds of air per hour must be supplied?
b. What is the required wet-bulb temperature of the supply air?
c. What is the sensible-heat factor?

10-42 A structure to be air conditioned has a sensible-heat load of 20,000 Btu/hr at a time when the total load is 100,000 Btu/hr. If the inside state is to be at 80 F and 50% relative humidity, is it possible to meet the load conditions by supplying air to the room at 100 F and 60% relative humidity? If not, discuss the direction in which the inside state would be expected to move if such air were supplied.

10-43 A flow rate of 30,000 lbm/hr of conditioned air at 60 F and 85% relative humidity enters a space that has a sensible load of 120,000 Btu/hr and a latent load of 30,000 Btu/hr.
a. What dry-bulb and wet-bulb temperatures are in the space?
b. If a mixture of 50% return air and 50% outdoor air at 98 F dry-bulb and 77 F wet-bulb temperature enters the air conditioner, what is the refrigeration load?

10-44 An air/water vapor mixture enters a heater-humidifier unit at 5 C, 100 kPa, and 50% relative humidity. The flow rate of dry air is 0.1 kg/s. Liquid water at 10 C is sprayed into the mixture at the rate of 0.0022 kg/s. The mixture leaves the unit at 30 C and 100 kPa. Calculate: (a) the relative humidity at the outlet and (b) the rate of heat transfer to the unit.

10-45 A room is being maintained at 75 F and 50% relative humidity. The outside air conditions are at this time 40 F and 50% relative humidity. Return air from the room is cooled and dehumidified by mixing it with fresh ventilation air from the outside. The total air flow to the room is 60% outdoor and 40% return air—by mass. Determine the temperature, relative humidity, and humidity content of the mixed air going to the room. For the cooling–dehumidifying process, calculate total heat removal, latent heat removal, and sensible-heat removal.

10-46 A room with a sensible load of 20,000 Btu/hr is maintained at

75 F and 50% relative humidity. Outdoor air at 95 F and 80 F wet-bulb temperature is mixed with the room return air. The outdoor air that is mixed is 25% by mass of the total flow going to the conditioner. This air is then cooled and dehumidified by a coil and leaves the coil saturated at 50 F, at which condition it is on the condition line for the room. The air is then mixed with some room return air so that the temperature of the air entering the room is at 60 F. Find:

a. The air conditioning processes on the psychrometric chart (Figure 10-2)

b. Ratio of latent to sensible load

c. Air flow rate

d. Percent by mass of room return air mixed with air leaving the cooling coil

10-47 An air/water vapor mixture at 14.7 psia, 85 F, and 50% relative humidity is contained in a 15-ft³ tank. At what temperature will condensation begin? If the tank and mixture are cooled an additional 15 F, how much water will condense from the mixture?

10-48 Suppose that 1000 ft³/min of air at 14.7 psia, 90 F, and 60% relative humidity is passed over a coil with a mean surface temperature of 40 F. A spray on the coil assures that the leaving air is saturated at the coil temperature. What is the required cooling capacity of the coil?

10-49 An air/vapor mixture at 100 F dry-bulb temperature contains 0.02 lbm water vapor per pound of dry air. The barometric pressure is 28.561 in. Hg. Calculate the relative humidity, dew-point temperature, and degree of saturation.

10-50 Air enters a space at 20 F and 80% relative humidity. Within the space, sensible heat is added at the rate of 45,000 Btu/hr and latent heat is added at the rate of 20,000 Btu/hr. The conditions to be maintained inside the space are 50 F and 75% relative humidity. What must be the air exhaust rate (lbm/hr) from the space to maintain a 50 F temperature? What must be the air exhaust rate (lbm/hr) from the space to maintain a 75% relative humidity? Discuss the difference.

10-51 Moist air at a low pressure of 11 psia is flowing through a duct at a low velocity of 200 ft/min. The duct is 1 ft in diameter and has negligible heat transfer to the surroundings. The dry-bulb temperature is 85 F and the wet-bulb temperature is 70 F. Calculate: (a) the humidity ratio (lbm vapor/lbm), (b) the dew-point temperature, and (c) the relative humidity (%).

10-52 If an air compressor takes in moist air (at about 90% relative humidity) at room temperature and pressure and compresses it to 120 psig (and slightly higher temperature), would you expect some condensation to occur? Why? If yes, where would the condensation form? How would you remove it?

10-53 Does a sling psychrometer give an accurate reading of the adiabatic saturation temperature? Explain.

10-54 At an altitude of 5000 ft, a sling psychrometer reads 80 F dry-bulb and 67 F wet-bulb temperature. Determine correct values of relative humidity and enthalpy from the psychrometric chart. Compare these values to the corresponding values for the same readings at sea level. Include a schematic psychrometric chart in your solution.

10-55 The average person gives off sensible heat at the rate of 250 Btu/hr and perspires and respirates about 0.27 lbm/hr of moisture. Estimate the sensible and latent load for a room with 25 people. (The lights give off 9000 Btu/hr.) If the room conditions are to be 78 F and 50% relative humidity, what flow rate of air would be required if the supply air comes in at 63 F? What would be the supply air's relative humidity?

10-56 A space in an industrial building has a sensible heat loss in winter of 200,000 Btu/hr and a negligible latent-heat load. (Latent losses to outside are made up by latent gains in the space.) The space is to be maintained precisely at 75 F and 50% relative humidity. Due to the nature of the process, 100% outside air is required for ventilation. The outdoor air conditions can be taken as saturated air at 20 F. The amount of ventilation air is 7000 SCFM and the air is to be preheated, humidified with an adiabatic saturator to the desired humidity, and then reheated. (SCFM represents ft³/min at standard density of 0.075 lbm/ft³.) The temperature out of the adiabatic saturator is to be maintained at 60 F dry-bulb temperature. Determine:
a. Temperature of the air entering the space to be heated (F)
b. Heat supplied to the preheat coil (Btu/hr)
c. Heat supplied to the reheat coil (Btu/hr)
d. Amount of humidification (gal/min)

10-57 An air-conditioned room with an occupancy of 20 people has a sensible heat load of 200,000 Btu/hr and a latent load of 50,000 Btu/hr. It is maintained at 76 F dry-bulb and 64 F wet-bulb temperature. On a mass basis, 25% outside air is mixed with return air. Outside air is at 95 F dry-bulb and 76 F wet-bulb temperature. Conditioned air leaves the apparatus and enters the room at 60 F dry-bulb temperature. Neglect any temperature change due to the fan.

a. Draw and label the schematic flow diagram for the complete system.
b. Complete the following table:

Point	Dry-Bulb Temperature, F	ϕ,%	h, Btu/lbm	W, lbm/lbm	m_a, lbm/hr	SCFM	ft³/min
OA							
r							
m							
s							

c. Plot and draw all processes on a psychrometric chart.

d. Specify the fan size (SCFM).

e. Determine the size of refrigeration unit needed (Btu/hr and tons).

f. What percentage of the required refrigeration is for (1) sensible cooling and (2) for dehumidification?

g. What percentage of the required refrigeration is due to outside air load?

Chapter

11

Combustion

To this point, our study of thermodynamics has been restricted to nonreacting systems. Since chemical compositions did not change at any point in the process, the analysis did not involve any chemical reactions. This restriction is now lifted and a particular type of chemical reaction will be considered: combustion. Although this is just an introduction to the subject, it is included because of its importance to engineers.

11–1 Fundamentals

Combustion *is defined as a chemical process in which an oxidant is rapidly reacted with a fuel to liberate thermal energy—generally in the form of high-temperature gases.* For our applications, the oxidant for combustion is oxygen in the air. The fuels to be used are conventional hydrocarbons in either elemental form or compounds. Their complete combustion produces carbon dioxide and water; small quantities of carbon monoxide, partially reacted flue gases and aerosols, sulfur (SO_2 or SO_3), ash, and inert gases are also released.

The combustion process is a complex chemical reaction with many intermediate products. Our study will ignore these intermediate products; only the initial reactants and final products will be considered. It must be remembered, however, that these intermediate reactions would be extremely important in a detailed study of combustion.

In **complete combustion,** all hydrogen and carbon in the fuel is oxidized to H_2O and CO_2. Generally, it is necessary to supply an amount of oxygen (air) beyond that theoretically required to oxidize the fuel. Excess oxygen or air is usually expressed as a percentage of the air theoretically required to oxidize the fuel completely (**theoretical air**).

In **stoichiometric** or **theoretical combustion,** fuel is reacted with the exact amount of oxygen required to oxidize all the fuel to CO_2, H_2O, and SO_2. Exhaust gas from this type of combustion contains no incompletely oxidized fuel or oxygen. The percentage of CO_2 contained in products of stoichiometric combustion is the maximum attainable; it is referred to as **stoichiometric CO_2, ultimate CO_2,** or **maximum theoretical percentage of carbon dioxide.** Since theoretical combustion is seldom realized in practice, economy (and safety) require that combustion equipment operate with excess air so that fuel is not wasted. In fact, combustion equipment is designed and operated to attain complete (no wasted fuel) but not stoichiometric combustion.

Incomplete combustion occurs when a fuel is not completely oxidized in the process. For example, a hydrocarbon may not completely oxidize to carbon dioxide and water. It may form instead a partially oxidized compound such as carbon monoxide. From an engineering standpoint, this form of combustion represents inefficient fuel use— and from an environmental standpoint it is hazardous to your health because of the CO and other pollutants.

Oxygen for combustion is obtained from air. For calculation purposes, the nitrogen is assumed to pass through the process unchanged chemically. Table 11-1 gives the oxygen and air requirements for theoretical combustion of common fuels.

Combustion results in the release of thermal energy (heat). The quantity of heat generated by complete combustion of a unit of a specific fuel is constant if the reactants and the products are maintained at a constant temperature and pressure. This constant is known as the **heating value, heat of combustion, calorific value,** or **enthalpy of combustion** of that fuel. This quantity results from the fact that the enthalpy of the reactants and that of the products (at the same temperature) are not the same. This is, the heating value is

$$HV = H(\text{products})_{T_0, p_0} - H(\text{reactants})_{T_0, p_0} \qquad (11\text{-}1)$$

Therefore, the heating value of a fuel may be determined directly by measuring the heat released during combustion of a known quantity of the fuel.

Higher heating value is the term used when water vapor in the products of fuel combustion is condensed and the latent heat of vaporization of water is included in the fuel's heating value. Conversely, **lower heating value** is obtained when latent heat of vaporization is not included. When the heating value of a fuel is specified without designating higher or lower, it generally means the higher heating value.

Table 11-1 Combustion Reactions of Common Fuel Constituents

Constituent	Molecular Symbol	Combustion Reactions	Theoretical Oxygen and Air Requirements			
			lbm/lbm Fuel*		ft³/ft³ Fuel	
			O_2	Air	O_2	Air
Carbon (to CO)	C	$C + 0.5O_2 \rightarrow CO$	1.33	5.75	—	—
Carbon (to CO_2)	C	$C + O_2 \rightarrow CO_2$	2.66	11.51	—	—
Carbon monoxide	CO	$CO + 0.5O_2 \rightarrow CO_2$	0.57	2.47	0.50	2.39
Hydrogen	H_2	$H_2 + 0.5O_2 \rightarrow H_2O$	7.94	34.28	0.50	2.39
Methane	CH_4	$CH_4 + 2O_2 \rightarrow CO_2 + 2H_2O$	3.99	17.24	2.00	9.57
Ethane	C_2H_6	$C_2H_6 + 3.5O_2 \rightarrow 2CO_2 + 3H_2O$	3.72	16.09	3.50	16.75
Propane	C_3H_8	$C_3H_8 + 5O_2 \rightarrow 3CO_2 + 4H_2O$	3.63	15.68	5.00	23.95
Butane	C_4H_{10}	$C_4H_{10} + 6.5O_2 \rightarrow 4CO_2 + 5H_2O$	3.58	15.47	6.50	31.14
—	C_nH_{2n+2}	$C_nH_{2n+2} + (1.5n + 0.5)O_2 \rightarrow nCO_2 + (n + 1)H_2O$	—	—	$1.5n + 0.5$	$7.18n + 2.39$
Ethylene	C_2H_4	$C_2H_4 + 3O_2 \rightarrow 2CO_2 + 2H_2O$	3.42	14.78	3.00	14.38
Acetylene	C_2H_2	$C_2H_2 + 2.5\alpha_2 \rightarrow 2CO_2 + H_2O$	3.07	13.27	2.50	11.96
—	C_nH_{2m}	$C_nH_{2m} + (n + 0.5m)O_2 \rightarrow nCO_2 + mH_2O$	—	—	$n + 0.5m$	$4.78n + 2.39m$
Sulfur (to SO_2)	S	$S + O_2 \rightarrow SO_2$	1.00	4.31	—	—
Sulfur (to SO_3)	S	$S + 1.5O_2 \rightarrow SO_3$	1.50	6.47	—	—

* Atomic weights: H = 1.008; C = 12.01; O = 16.00; S = 32.06.

Table 11-2 Heating Values of Components of Common Fuels

Substance	Molecular Symbol	Higher Heating Values*		Lower Heating Values*	
		Btu/lbm	Btu/lb-mole	Btu/lbm	Btu/lb-mole
Carbon (to CO)	C	3,950	47,400	3,950	47,400
Carbon (to CO₂)	C	14,093	169,116	14,093	169,116
Carbon monoxide	CO	4,347	121,716	4,347	121,716
Hydrogen	H₂	61,095	122,190	51,023	102,046
Methane	CH₄	23,875	382,000	21,495	343,920
Ethane	C₂H₆	22,323	669,690	20,418	612,540
Propane	C₃H₈	21,669	953,436	19,937	877,228
Butane	C₄H₁₀	21,321	1,236,618	19,678	1,141,324
Ethylene	C₂H₄	21,636	605,808	20,275	567,700
Propylene	C₃H₆	21,048	884,016	19,687	826,854
Acetylene	C₂H₂	21,502	559,052	20,769	539,994
Sulfur (to SO₂)	S	3,980	127,360	3,980	127,360
Sulfur (to SO₃)	S	5,940	—†	—†	190,080
Hydrogen sulfide	H₂S	7,097	241,298	6,537	222,258

* All values corrected to 60 F, 30 in. Hg dry. For gases saturated with water vapor at 60 F, deduct 1.74% of the Btu value.
† Values not available.

Heating values are usually expressed in Btu/ft³ for gaseous fuels, Btu/gal for liquid fuels, and Btu/lbm for solid fuels. Heating values are generally given in relation to a certain reference temperature, usually 60, 68, or 77 F, depending on industry practice. Heating values of several substances in common fuels are given in Table 11-2; Table 11-3 lists typical heating values for various energy sources.

With incomplete combustion, the fuel is not completely oxidized and the heat released is less than the heating value of the fuel, implying lower combustion efficiency. Thus not all heat available for release during combustion is used effectively. The largest heat loss is in the form of increased temperature of exhaust gases above the temperature of incoming air and fuel. There are generally other heat losses, includ-

Table 11-3 Typical Heating Values of Energy Sources

Material	Heating Value as Fired
Solids	(Btu/lbm)
Anthracite coal	13,000
Bituminous coal	12,000
Subbituminous coal	9,000
Lignite coal	6,900
Coke	11,000
Newspapers	8,000
Brown paper	7,300
Corrugated board	7,000
Magazines	5,300
Waxed milk cartons	11,400
Asphalt or tar	17,000
Typical urban refuse	5,000
Corn cobs	8,000
Rags	7,500
Wood	9,000
Liquids	(Btu/gal)
Fuel oil	
Grade 1	135,000
Grade 2	140,000
Grade 6	154,000
Kerosene	133,000
Gasoline	111,000
Methy alcohol	68,000
Ethyl alcohol	88,000
LPG	91,000
Gases	(Btu/ft³)
Natural gas	1,000
Commercial propane	2,500
Commercial butane	3,200
Acetylene	1,500
Methane	950
Biogas	500

Table 11-4 Ignition Temperature in Air (At Pressure of 1 Atm)*

Combustible	Formula	Temperature, F
Sulfur	S	470
Charcoal	C	650
Fixed carbon (bituminous coal)	C	765
Fixed carbon (semibituminous coal)	C	870
Fixed carbon (anthracite)	C	840–1115
Acetylene	C_2H_2	580–825
Ethane	C_2H_6	880–1165
Ethylene	C_2H_4	900–1020
Hydrogen	H_2	1065–1095
Methane	CH_4	1170–1380
Carbon monoxide	CO	1130–1215
Kerosene	—	490–560
Gasoline	—	500–800

* Rounded-off values and ranges.

ing radiation and convection heat transfer from outer walls of combustion equipment to the environment.

Though not directly a part of our study, **ignition temperature** is, of course, a variable important to a study of combustion. As a matter of interest, Table 11-4 is presented to illustrate this temperature (in air) for various substances.

11–2 Chemical Reactions

During a combustion process the mass of each constituent remains constant. Hence solving the chemical reaction equation involves the conservation of mass. The three most prevalent components in hydrocarbon fuels are completely combusted by the following reactions:

Reactants   Products

$$C + O_2 \rightarrow CO_2 \qquad \text{(11-2)}$$

$$H_2 + 0.5O_2 \rightarrow H_2O \qquad \text{(11-3)}$$

$$S + O_2 \rightarrow SO_2 \qquad \text{(11-4)}$$

Recalling the principle presented in Chapter 10 on mixtures, it is possible to write these reaction equations in several ways. Thus Equation (11-2) may be interpreted as

$$1 \text{ mol C} + 1 \text{ mol } O_2 = 1 \text{ mol } CO_2 \qquad \text{(11-2}a\text{)}$$

$$12 \text{ lbm C} + 32 \text{ lbm } O_2 = 44 \text{ lbm } CO_2 \qquad \text{(11-2}b\text{)}$$

$$1 \text{ lbm C} + (32 \div 12) \text{ lbm } O_2 = (44 \div 12) \text{ lbm } CO_2 \qquad \text{(11-2}c\text{)}$$

$$359 \text{ ft}^3 \text{ C} + 359 \text{ ft}^3 \text{ } O_2 = 359 \text{ ft}^3 \text{ } CO_2 \qquad \text{(11-2}d\text{)}$$

$$1 \text{ vol } C + 1 \text{ vol } O_2 = 1 \text{ vol } CO_2 \qquad \textbf{(11-2e)}$$

Notice how each equation balances. There are the same number of atoms of each element and the same mass of reacting substances on each side of the equality sign but not necessarily the same number of molecules, moles, or volumes. Thus one molecule of carbon plus one molecule of oxygen gives only one molecule of carbon dioxide and two moles of hydrogen plus one mole of oxygen gives only two moles of water vapor. It will be evident from a consideration of the mole–volume relationship that percentage by volume is numerically the same as percentage by mole. Similar interpretations may be made to describe Equations (11-3) and (11-4), but with the corresponding approximate atomic weights.

When air is involved in combustion, we must include the inert nitrogen (air = 21% O_2 and 79% N_2 by volume). That is, for each mole of oxygen there are 3.76 (= 79/21) moles of nitrogen (total of 4.76 moles). Thus the chemical equation for the theoretical combustion of ethane with air can be written

$$C_2H_6 + (3.5)\,O_2 + 3.5\,(3.76)\,N_2 \rightarrow 2CO_2 + 3H_2O + 13.16N_2 \qquad \textbf{(11-5)}$$

Recall that more than the theoretical air quantity is usually supplied. Thus for 150% theoretical air, this reaction would be

$$C_2H_6 + 3.5\,(1.5)\,O_2 + 3.5\,(3.76)\,(1.5)\,N_2$$
$$\rightarrow 2CO_2 + 3H_2O + (3.5)\,O_2 + 19.74N_2 \qquad \textbf{(11-6)}$$

Note that this is complete combustion (not theoretical) and one of the products is oxygen (excess air). This excess air is defined in percentage as

$$\text{Percentage of excess air} = \frac{\text{air supplied} - \text{theoretical air}}{\text{theoretical air}} \times 100 \qquad \textbf{(11-7)}$$

In this example we have 50% excess air. The excess air level at which a combustion process operates significantly affects its overall efficiency. Too much excess air lowers its temperature; too little may cause incomplete combustion. The highest combustion efficiency is obtained when just enough excess air is supplied and properly mixed with combustible gases to ensure complete combustion. The general practice is to supply from 5 to 50% excess air; the exact amount depends on the fuel and other factors.

Another term used often in industry is the **air/fuel ratio (A/F).** This ratio may be expressed on a mass basis (the usual) or on a mole basis. The theoretical A/F is, of course, for situations in which the air supplied produces theoretical combustion. Thus the A/F for Equation (11-5) is

$$A/F = \frac{3.5 + 13.16}{1} = 16.66 \text{ moles of air/mole of fuel}$$

and

$$A/F = \frac{16.66\,(28.97)}{30} = 16.09 \text{ lbm of air/lbm of fuel}$$

where 28.97 is the molecular weight of air and 30 is the molecular weight of ethane. For the reaction of Equation (11-6), on a mole basis A/F = 24.99 moles/mole fuel and on a mass basis A/F = 24.135 lbm/lbm fuel.

How do we know whether a higher heating value or lower heating value is indicated in a reaction equation? In Equation (11-5) it is tacitly assumed that all the constituents are vapor—indicating that a lower heating value is involved. If the water vapor had condensed, the higher heating value would be involved. To be specific:

$C_2H_6\,(g) + (3.5)\,O_2\,(g) + (13.16)\,N_2\,(g)$
$\quad \rightarrow 2CO_2\,(g) + 3H_2O\,(g) + (13.16)\,N_2\,(g)$ (lower heating value)

whereas

$C_2H_6\,(g) + (3.5)\,O_2\,(g) + (3.16)\,N_2\,(g)$
$\quad \rightarrow 2CO_2\,(g) + 3H_2O\,(l) + (13.16)\,N_2\,(g)$ (higher heating value)

where (g) represents gas phase and (l) represents liquid phase.

EXAMPLE 11-1

Carbon burns with 160% theoretical air. Combustion goes to completion. Determine:

1. The air/fuel ratio by mass

2. The percentage by mole of each product (the so-called Orsat analysis) and dew point of the products.

SOLUTION: Theoretical:　$C + O_2 + 3.76\,N_2 \rightarrow CO_2 + 3.76\,N_2$

Actual: $C + 1.6\,O_2 + 1.6\,(3.76)\,N_2 \rightarrow CO_2 + 0.6\,O_2 + 1.6\,(3.76)\,N_2$

1. $A/F = \dfrac{1.6\,(32) + 1.6\,(3.76)\,28}{12} = 18.3$ lbm air/lbm fuel

2. The total number of moles of the product is 7.62.

CO_2:　1.0 mole ÷ 7.62 → 13.1%

O_2:　　0.6 mole ÷ 7.62 → 7.9%　　(Orsat)

N_2:　6.02 moles ÷ 7.62 → $\underline{\quad 79.0\%\quad}$
$\qquad\qquad\qquad\qquad\qquad\quad 100.0\%$

Note that there is no H_2O in the products. Hence no dew point is possible.

EXAMPLE 11-2

The flue gas analysis of a hydrocarbon fuel on a percent by volume dry basis shows: $CO_2 = 12.4\%$, $O_2 = 32.7\%$, $CO = 0.1\%$, $H_2 = 0.2\%$, and $N_2 = 84.1\%$. Determine the air/fuel ratio by volume.

SOLUTION: $C_xH_y + \dfrac{84.1}{3.76}O_2 + 84.1N_2 \rightarrow 12.4CO_2 + 3.2O_2 + 0.1CO + 0.2H_2$

$$+ 84.1N_2 + zH_2O$$

$$x = 12.4 + 0.1 = 12.5$$

$$22.4 = 12.4 + 3.2 + 0.05 + \frac{z}{2}$$

$$z = 13.5$$

$$y = 0.4 + 27 = 27.4$$

$$C_{12.5}H_{27.4} + 22.4O_2 + 84.1N_2 \rightarrow 12.4CO_2 + 3.2O_2 + 0.1CO$$

$$+ 0.2H_2 + 84.1N_2 + 13.5H_2O$$

$$A/F = \frac{22.4 \text{ moles } O_2 + 84.1 \text{ moles } N_2}{1 \text{ mole fuel}}$$

$$A/F = \frac{106.5}{1} \text{ by volume}$$

11–3 The First Law for Open Systems and Combustion

The first law of thermodynamics for a steady-state/steady-flow process is directly applicable to the combustion process. The reactants are the quantities in and the products are the quantities out. In most instances, changes in kinetic energy and potential energy may be neglected.

If there is no work done in the action, the form of the first law is

$$Q + \sum_{\text{react}} H = \sum_{\text{prod}} H \tag{11-8}$$

Usually the problem is presented on the basis of 1 mole of fuel. Thus it may be convenient to rearrange the summation in terms of number of moles (n) and modal enthalpy ($\bar{h}$). Hence

$$Q + \sum_{\text{react}} n\bar{h} = \sum_{\text{prod}} n\bar{h} \tag{11-9}$$

Whether the problem is to be solved by using a "per mass" or a "per mole" approach, the summation must be done carefully. If the reactants and products are at the same temperature (T_0), then

$$Q = \sum_{\text{prod}} n\bar{h} - \sum_{\text{react}} n\bar{h}$$

$$= -\Delta H_R (T_0) \tag{11-10}$$

where ΔH_R is the enthalpy of reaction (at constant pressure also) or the heating value (see Table 11-2) and T_0 is a reference temperature.

If, on the other hand, the reactants and products are all at different temperatures, the calculation procedure is much more complicated. The complication results from the necessity to account for the enthalpy difference resulting from the differences in temperature (and pressure), T_0, and those of the constituents. That is, the T_0 heating values are available for only a few temperature values but the variety of combustion constituent temperatures is infinite. Thus

$$Q = -\Delta H_R + \sum_{\substack{i \\ \text{prod}}} n_i (\bar{h}_i - \bar{h}_0) - \sum_{\substack{j \\ \text{react}}} n_j (\bar{h}_j - \bar{h}_0) \tag{11-11}$$

where $\bar{h}_0$ is the enthalpy of a constituent at the temperature at which ΔH_R is available and the $(\bar{h}_i, \bar{h}_j)$ are the enthalpies of the constituents at the given temperatures. Note that Equation (11-11) reduces to Equation (11-10) for constant-temperature combustion.

Now that we recall the first law for an open system we may easily relate higher and lower heating values—directly from the definition. For this constant-pressure process,

$$\text{HHV} = \text{LHV} + mh_{fg} \tag{11-12}$$

EXAMPLE 11-3

Consider the theoretical combustion of ethane in a steady-flow process. Determine the heat transfer per lb-mole and per lbm of fuel in a combustion chamber for the following cases:

1. The products and reactants are the same temperature and pressure: 60 F and 14.7 psia.

2. The air and ethane enter at 40 and 140 F respectively and the products all leave at 840 F.

SOLUTION: The combustion equation is

$$C_2H_6 + (3.5)O_2 + 3.5(3.76)N_2 \rightarrow 2CO_2 + 3H_2O + 13.16N_2$$

1. All constituents are at the same temperature and pressure. The first law for this situation is

$$Q = -\Delta H_R$$

This value can be looked up on Table 11-2 directly: $-20,418$ Btu/lbm or $-612,540$ Btu/lb-mole.

2. According to our first-law analysis,

$$Q = -\Delta H_R + \sum_{\text{prod}} n_i (\bar{h}_i - \bar{h}_0) - \sum_{\text{react}} n_j (\bar{h}_j - \bar{h}_0)$$

$$= \Delta H_R + 2[\bar{h}(1300) - \bar{h}(520)]_{CO_2} + 3[\bar{h}(1300) - \bar{h}(520)]_{H_2O}$$

$$+ 13.16 [\bar{h}(1300) - \bar{h}(520)]_{N_2}$$

$$- 1[\bar{h}(600) - \bar{h}(520)]_{C_2H_6} - 3.5[\bar{h}(500) - \bar{h}(520)]_{O_2}$$

$$- 13.6[\bar{h}(500) - \bar{h}(520)]_{N_2}$$

At this point, we can go no further unless we have a property table of gases. Tables 11-5 and 11-6 are short versions of what we need. Thus

$$Q = -612,540 + 2(12,137 - 3881) + 3(10,715 - 4122)$$

$$+ 13.16(9154 - 3611)$$

$$- (5982 - 4974) - 3.5(3466 - 3606)$$

$$- 13.16(3472 - 3611)$$

$$= -501,992 \text{ Btu/lb-mole of } C_2H_6$$

$$= -16,733 \text{ Btu/lbm of } C_2H_6$$

Table 11-5 Properties of Ideal Gases ($\bar{h}$ in Btu/mole)

T, R	CO	CO$_2$	H$_2$	H$_2$O	N$_2$	O$_2$
0	0	0	0	0	0	0
300	2082	2108	2064	2368	2082	2074
400	2777	2875	2710	3164	2777	2769
500	3472	3706	3386	3962	3472	3466
537	3730	4030	3640	4258	3730	3725
600	4168	4601	4076	4765	4168	4168
700	4866	5552	4770	5575	4865	4879
800	5568	6553	5467	6397	5564	5602
900	6276	7598	6165	7231	6268	6338
1000	6992	8682	6865	8079	6978	7088
1100	7717	9803	7565	8942	7695	7850
1200	8451	10955	8266	9820	8420	8626
1300	9195	12137	8969	10715	9154	9413
1400	9948	13345	9674	11625	9897	10210
1500	10711	14576	10382	12551	10649	11017
1600	11483	15829	11903	13495	11410	11833
1700	12264	17101	11807	14455	12179	12656
1800	13053	18392	12527	15433	12956	13486
1900	13850	19698	13251	16428	13742	14322
2000	14653	21019	13980	17439	14534	15164
2200	16279	23699	15454	19511	16140	16863
2400	17927	26424	16951	21646	17768	18579

(Table 11-5 continues on next page.)

Table 11-5 (Continued)

T, R	CO	CO$_2$	H$_2$	H$_2$O	N$_2$	O$_2$
2600	19594	29187	18470	23840	19416	20311
2800	21277	31983	20012	26088	21081	22058
3000	22973	34807	21577	28386	22762	23818
3200	24681	37655	23164	30730	24455	25591
3400	26399	40524	24772	33116	26160	27376
3600	28127	43411	26399	35540	27874	29174
3800	29862	46314	28043	37999	29598	30984
4000	31605	49231	29704	40489	31329	32806
4200	33354	52162	31380	43008	33068	34640
4400	35109	55105	33071	45554	34813	36485
4600	36869	58060	34776	48124	36564	38341
4800	38634	61025	36493	50716	38320	40209
5000	40403	64000	38223	53327	40080	42086

Source: Gas Tables, by Joseph H. Keenan and Joseph Kaye, Copyright © 1948 by John Wiley & Sons, Inc., as it appears in Jones/Hawkins, Copyright © 1960 by John Wiley & Sons, Inc. Reprinted by permission of John Wiley & Sons, Inc.

With the first law for open systems and combustion in hand, let us pause for a moment to explain how the data presented earlier as Table 11-2 come about. To do so we must define another enthalpy. The enthalpy of the products of combustion at 77 F and 1 atm pressure (with the enthalpy of elements assumed to be zero) is called the **enthalpy of formation** (equal to the heat of reaction at the base temperature to 77 F). Table 11-7 gives the enthalpy of formation of several substances. Thus, using the first law, we obtain

$$\Delta H_R = \left(\sum H_f^0 \right)_{prod} - \left(\sum H_f^0 \right)_{react} \qquad (11\text{-}13)$$

The physical process is illustrated in Figure 11-1.

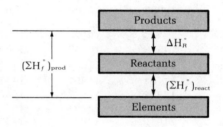

Figure 11-1. Enthalpy of combustion.

Table 11-6 Enthalpy of Eight Gases at Low Pressure (in Btu/mole)

T, R	Methane CH_4	Ethane C_2H_6	Propane C_3H_8	n-Butane C_4H_{10}	n-Octane C_8H_{18}	Methanol CH_3OH	Ammonia NH_3	Hydrazine N_2H_4
0	0	0	0	0	0	0	0	0
537	4312	5137	6318	8356	15715	4912	4325	5059
600	4875	5982	7501	9929	18803	5603	4871	5886
700	5791	7422	9556	12669	24204	6756	5766	7309
800	6791	9042	11880	15734	30196	8052	6703	8971
1000	9012	12752	17259	22805	43941	10890	8710	12227
1200	11545	17077	23493	30976	59757	14121	10888	15916
1400	14370	21924	30458	37258	77243	17695	13227	19866
1600	17461	27214	38055	49903	96145	21571	15715	24046
1800	20794	32882	46176	60404	116312	25714	18342	28416
2000	24334	38894	54944	71473	137490	30091	21098	32966

Source: Methanol data from E. V. Ivash, J. C. M. Li, and K. S. Pitzer, *Journal of Chemical Physics*, vol. 23, no. 10, October 1955, pp. 1814-1818. Ammonia and hydrazine data from K. A. Kobe and R. H. Harrison, *Petroleum Refiner*, vol. 33, no. 11, November 1954, pp. 161-164. Other data from *Selected Values of Properties of Hydrocarbons*, National Bureau of Standards Circular C461, 1947.

Table 11-7 Enthalpy of Formation

Compound	State	$\overline{h}_f^o$, Btu/lb-mole	Compound	State	$\overline{h}_f^o$, Btu/lb-mole
Carbon (C)	Graphite	0	Ethylene (C_2H_2) (ethene)	Gas	22,480
	Diamond	820			
Carbon dioxide (CO_2)	Gas	−169,180			
Carbon monoxide (CO)	Gas	− 47,520	Ethane (C_2H_6)	Gas	− 36,400
Water (H_2O)	Gas	−103,970	Propane (C_3H_8)	Gas	− 44,650
	Liquid	−122,890	Butane (C_4H_{10})	Gas	− 54,230
Sulfur dioxide (SO_2)	Gas	−127,640	Benzene (C_6H_6)	Gas	35,650
Methane (CH_4)	Gas	− 32,180	Octane (C_8H_{18})	Gas	− 89,620
Acetylene (C_2H_3)	Gas	97,490		Liquid	−107,460

EXAMPLE 11-4

Determine the magnitude of the enthalpy of reaction (higher value) for CH_4 (methane) gas.

SOLUTION:
$$CH_4(g) + 2O_2(g) \rightarrow CO_2(g) + 2H_2O(l)$$
$$\underset{-32,179}{} \quad + \quad \underset{0}{} \quad \quad \underset{-169,182}{} \quad \underset{-122,891}{}$$

$$\Delta H_R = [(1)(-169,182) + (2)(-122,891)]$$
$$- [(1)(-32,179)] = 382,800 \text{ Btu/lb-mole}$$

(*Note:* Table 11-2 gives the heating value as 382,000 Btu/lb-mole.)

11–4 The First Law for Closed Systems and Combustion

As we have seen, the terms of the first law must be interpreted differently for a closed system and an open system. For a closed system in which there is no change in kinetic and potential energy

$$q - w = u_2 - u_1 \tag{11-14}$$

And if no work is done, then

$$q = u_2 - u_1 \tag{11-15}$$

From your experience, you know that if we require the temperature at the end of combustion to be the same as that at the beginning, heat must be removed (that is, $q < 0$ and $u_2 < u_1$). As with the open system, the constant-temperature condition produces a special case (an internal energy of reaction). For $T_1 = T_2 = T_0$,

$$Q = -\Delta U_R \tag{11-16}$$

This quantity, like ΔH_R, must be measured.

If the reactants are at different temperatures, we again must account for the energy differences due to the temperature difference. The similarity between the open analysis and this case is direct. Thus

$$Q = -\Delta U_R + \sum_{\text{prod}} n_i [\bar{u}_i(T_i) - \bar{u}(T_0)] \tag{11-17}$$

$$- \sum_{\text{react}} n_i [\bar{u}(T_1) - \bar{u}(T_0)]$$

and since $U = H - pV$,

$$Q = -[\Delta H_R - \Delta(p\bar{V})_R] + \sum_{\text{prod}} n_i [\bar{h}_i(T_i) - \bar{h}(T_0) - (p\bar{v})_i + (p\bar{v})_0]$$

$$- \sum_{\text{react}} n_i [\bar{h}_i(T_i) - \bar{h}(T_0) - (p\bar{v})_i + (p\bar{v})_0] \tag{11-18}$$

For closed systems the relation between the higher and lower heating value may be determined directly from definitions, recalling that this is a constant-volume process. Thus

$$HHV = LHV + mu_{fg} \tag{11-19}$$

EXAMPLE 11-5

Determine the heat transfer in the constant-volume combustion of 1 lbm of carbon. The reactants are at 100 F and the products are at 400 F. The chemical reaction equation is

$$C + (1.5)O_2 \rightarrow CO_2 + (0.5)O_2$$

SOLUTION: The reaction equation indicates that 1 mole of C + 1.5 mole of $O_2 \rightarrow$ 1 mole of CO_2 and 0.5 mole of O_2, or 12 lbm of C + 48 lbm of $O_2 \rightarrow$ 44 lbm of CO_2 and 16 lbm of O_2. In our case,

$$1 \text{ lbm C} + 4 \text{ lbm } O_2 \rightarrow 3.67 \text{ lbm } CO_2 + 1.33 \text{ lbm } O_2$$

Since this is a closed system, the first law must be in the form of Equation (11-17):

$$Q = -\Delta U_R + [u(400 \text{ F}) - u(60 \text{ F})]_{CO_2} + [u(400 \text{ F}) - u(60 \text{ F})]_{O_2}$$
$$- [u(100 \text{ F}) - u(60 \text{ F})]_C - [u(100 \text{ F}) - u(60 \text{ F})]_{O_2}$$

The 60 F of this equation accounts for ΔH_R of Table 11-2 being reference to 60 F.

$$\Delta H_R = 14,093 \text{ Btu/lbm}$$
$$\Delta U_R = \Delta H_R - \Delta(pV)_R$$
$$\doteq \Delta H_R$$

For convenience let C, O_2, and CO_2 be ideal gases (that is, $\Delta u \doteq mc_v \Delta T$). Thus, from Appendix A-3-3, $c_v(CO_2) \doteq 0.159$ Btu/lbm-R and $c_v(O_2) = 0.158$ Btu/lbm-R. Estimate $c_v(C) = 0.17$ Btu/lbm-R.

$$Q = -14,093 + 3.67(0.159)(340) + 1.33(0.158)(340)$$
$$- (0.17)(40) - 4(0.158)(40)$$
$$= -13,855 \text{ Btu/lbm}$$

PROBLEMS

Assume for these problems that air is 79% N_2 and 21% O_2 by volume.

11-1 Set up the necessary combustion equations and determine the weight of air required to burn 1 lbm of pure carbon to equal weights of CO and CO_2.

11-2 The gravimetric analysis of a gaseous mixture is $CO_2 = 32\%$, $O_2 = 54.5\%$, and $N_2 = 11.5\%$. The mixture is at a pressure of 3 psia. Determine: (a) the volumetric analysis and (b) the partial pressure of each component.

11-3 A liquid petroleum fuel, C_2H_5OH, is burned in a space heater at atmospheric pressure.

    a. For combustion with 20% excess air, determine the air/fuel ratio by weight, the weight of water formed by combustion per pound of fuel, and the dew point of the combustion products.

    b. For combustion with 80% theoretical air, determine the dry analysis of the exhaust gases in percentage by volume.

11-4 Find the air/fuel ratio by weight when benzene, C_6H_6, burns with theoretical air, and determine the dew point at atmospheric pressure of the combustion products if an air/fuel ratio of 20:1 by weight is used.

11-5 A diesel engine used 30 lbm of fuel per hour when the brake output is 75 hp. If the heating value of the fuel is 19,600 Btu/lbm, what is the brake thermal efficiency of the engine?

11-6 Methane, CH_4, is burned with air at atmospheric pressure. The Orsat analysis of the flue gas gives $CO_2 = 10.00\%$, $O_2 = 2.41\%$, $CO = 0.52\%$, and $N_2 = 87.07\%$. Balance the combustion equation and determine the air/fuel ratio, the percentage of theoretical air, and the percentage of excess air.

11-7 Fuel oil composed of $C_{16}H_{32}$ is burned with the chemically correct air/fuel ratio. Find:

    a. Pounds of moisture formed per pound of fuel

    b. Partial pressure of the water vapor (psia)

    c. Percentage of CO_2 in the stack gases on an Orsat basis

    d. Volume of exhaust gases in ft^3/lbm of oil if the gas is at 500 F and 14.8 psia

11-8 Determine the composition of a hydrocarbon fuel if the Orsat analysis gives $CO_2 = 8.0\%$, $CO = 1.0\%$, $O_2 = 8.7\%$, and $N_2 = 82.3\%$.

11-9 Determine the air/fuel ratio by mass when a liquid fuel of 16% hydrogen and 84% carbon by mass is burned with 15% excess air.

11-10 Compute the compositions of the flue gases (percentage by volume on a dry basis—same as Orsat) resulting from the combustion of C_8H_{18} with 84% theoretical air.

11-11 A liquid petroleum fuel having a hydrogen/carbon ratio by weight of 0.169 is burned in a heater with an air/fuel ratio of 17 by weight. Determine: (a) the volumetric analysis of the exhaust gases on both wet and dry bases and (b) the dew point of the exhaust gas.

11-12 Natural gas with a volumetric composition of 93.32% methane, 4.17% ethane, 0.69% propane, 0.19% butane, 0.05% pentane, 0.98% carbon dioxide, and 0.61% nitrogen burns with 30% excess air. Calculate the volume of dry air at 60 F and 30 in. Hg used to burn 1000 ft^3 of gas at 68 F and 29.92 in. Hg, and find the dew point of the combustion products.

11-13 A representative No. 4 fuel oil has a gravity of 25° API and the following composition: C, 87.4%; H, 10.7%; S, 1.2%; N, 0.2% moisture, 0%; solids, 0.5%. (a) Estimate its higher heating value. (b) Compute the weight of air required to burn, theoretically, 1 gal of the fuel.

11-14 A furnace in El Paso, Texas, uses an average of 250 ft^3/min of natural gas during 6000 hr per year of operation. Estimate the annual savings for fan power alone if the excess air is reduced from 25 to 20% when electric energy costs 6.0 mills per kW-hr, motor efficiency is 90%, and the fan requires 2.1 bhp per 1000 ft^3/min. The theoretical air/fuel ratio by volume is 10:1.

11-15 The following data were taken from a test on an oil-fired furnace:

Fuel rate: 20 gal oil/hr

Specific gravity of fuel oil: 0.89

Percentage by weight of hydrogen in fuel: 14.7%

Temperature of fuel for combustion: 80 F

Temperature of entering combustion air: 80 F

Relative humidity of entering air: 45%

Temperature of flue gases leaving furnace: 550 F

(a) Calculate the heat loss in water vapor in products formed by combustion. (b) Calculate the heat loss in water vapor in the combustion air.

11-16 An office building requires 2,750,000,000 Btu of heat for the winter season. Compute the seasonal heating costs if the following fuels are used:
a. Bituminous coal: 13,500 Btu/lb; $38/ton
b. No. 2 fuel oil: 138,000 Btu/gal; 99¢/gal

Assume that the conversion efficiency is 75% for the oil and 61% for the coal.

11-17 An auditorium requires $2.9(10^9)$ kJ of heat for the winter season. Compute the seasonal heating costs if the following fuels are used:

a. Bituminous coal: 8.726 kJ/kg; 5¢/kg

b. No. 2 fuel oil: 38 457.1 kJ/L; 26.1¢/L

Assume that the conversion efficiency is 75% for the oil and 71% for the coal.

11-18 A diesel engine uses 14 kg of fuel per hour when the output is 75 hp. If the heating value of the fuel is 45 000 kJ/kg, what is the thermal efficiency of the engine?

11-19 Fuel oil composed of $C_{16}H_{32}$ is burned with the chemically correct air/fuel ratio. Find:

a. Kilograms of moisture formed per kilogram of fuel

b. Partial pressure of the water vapor (MPa)

c. Volume of exhaust gases (m^3) per kilogram of oil if the gas is at 260 C and 0.1 MPa

12

Fundamentals of Heat Transfer

Now that your introduction to the science of thermodynamics is complete, you are ready for the art and science of heat transfer. To convince you that this subject is not magic, the first part of the chapter is designed to awaken your need to associate intuition and mathematical statements of physical processes. The purpose of the second part of the chapter is to acquaint you with conduction. The view will be one-dimensional and for the most part steady-state. Two special subjects, fins and heat exchangers, are also discussed.

Because of the great mathematical difficulty associated with a thorough study of convection, a straightforward approach is used in the third part of the chapter. The point of view will be limited to energy interchange between a fluid and a solid. In fact, our whole effort will be to determine the convection coefficient (h). Once this coefficient is determined, the convective heat transfer follows directly from Newton's law of cooling.

Probably the most important mode of heat transfer in your life is thermal radiation. This radiation has been presenting life-giving energy to earth for eons. Though this mode of heat transfer is slightly difficult conceptually (because it is nonlinear with temperature), you must understand it. The last part of the chapter presents the basic information for determining radiative heat transfer.

12–1 Introduction to Heat Transfer

Heat transfer is commonly thought of as the science that treats energy transit within or between systems where the driving potential is a temperature difference. Heat transfer is a transport phenomenon—a general classification that includes mass transfer, momentum transfer or fluid friction, and electrical conduction. All transport phenomena have similar rate equations; that is, the flux (quantity per area per time) is proportional to a **potential difference.** In the case of heat transfer by conduction and convection, the potential difference is the temperature difference. In addition to explaining how heat may be transferred, the science of heat transfer is also used to predict the rate at which this transfer takes place. Heat transfer deals with nonequilibrium processes and is not to be confused with thermodynamics, which deals with systems in equilibrium. (Thermodynamics is used to predict the amount of energy required to change a system from one equilibrium state to another and does not predict how fast a change will take place.) Thus heat transfer supplements the first and second laws of thermodynamics by providing additional information that may be used to establish energy transfer rates.

From a thermodynamic viewpoint, the amount of heat transferred during a process equals the stored energy change plus the work done. It is evident that this type of analysis considers neither the mechanism nor the time required to transfer the heat. Thermodynamics prescribes how much heat is supplied to or rejected from a system during a process between equilibrium states without thought of whether or how this may be accomplished. Although it is of great practical importance, the question of how long it would take to transfer a specified amount of heat does not usually enter into thermodynamic analysis.

From an engineering viewpoint, the rate of heat transfer at a specified temperature difference may be a key problem. To estimate the cost, the feasibility, and the size of equipment necessary to transfer a specified amount of heat in a given time, the engineer must make a detailed heat transfer analysis. The physical size of heaters, refrigerators, typical heat exchangers, and solar collectors depends not only on the amount of heat to be transmitted, but also on the rate at which the heat is to be transferred under given conditions. The successful operation of equipment may depend on the cooling of parts by removing heat continuously at a rapid rate from a surface. Furthermore, a heat transfer analysis must be made to avoid damage to the equipment.

For years, heat transfer has been recognized as a necessary field of study having special arts and techniques that are useful in other specialties. The architect or civil engineer, concerned mainly with the erection of structures, benefits from sufficient heat transfer theory to insulate and heat buildings properly, to deice bridges, and to provide for proper curing of concrete. The power engineer, in the design of ever more efficient central stations, needs to understand the action of condensers and boilers and the means of preventing heat loss from system piping. Chemical engineers use many types of fluid heat exchangers and often pioneer in this phase of heat transfer technology. Supersonic aircraft now have interrelated thermal systems that control the environments (temperatures) from engine oil to the pilot's G-suit.

Moreover, design engineers today are vitally interested in the reliability of electronic components. Extreme component temperatures, both high and low, are prime causes of equipment failure. The origin of these thermal problems is the continuing need to design components that are lighter and smaller. These "improved" components have great utility because of the savings in weight and volume. But, since lighter, smaller electronic components are required to perform the same task as their heavier, larger predecessors, it is not uncommon to subject these components to excessively high temperatures. These high temperatures, in turn, result in performance degradation and accelerated failure. For example, reliability considerations demand that semiconductor devices (such as transistors) be maintained at fairly low temperatures in order to obtain the required life expectancy. Long life is important since the transistor, beginning in 1948, has immense utility in many applications such as communications equipment, audio and high-frequency amplifiers, switching circuits, electronic computers, and control systems. The low-temperature requirement may prevent a phenomenon known as thermal runaway. This occurs when junction temperature increases beyond a certain level, which in turn causes an increase in cutoff current. This increases the collector current, which raises the transistor function power dissipation. The increase in power dissipation further increases the junction temperature. This cycle continues until the transistor essentially destroys itself.

These varied examples indicate that every branch of engineering encounters heat transfer problems that have no solution using thermodynamics alone. They require an analysis based on the science of heat transfer.

Modes of Heat Transfer

Heat transfer *can be defined as the transmission of energy from one region to another primarily as a result of a temperature difference.* Since differences in temperatures exist all over the universe, heat transfer is as universal as gravity. Unlike gravity, however, heat transfer is governed not by a single relationship but by a combination of various independent laws of physics.

Heat transfer is generally divided into three distinct modes: conduction, convection, and radiation. Strictly speaking, only conduction and radiation should be classified as heat transfer processes, since only these two mechanisms obey the definition just given. Convection does not comply with the definition of heat transfer since mass transport is also involved. Convection does transmit energy from regions of higher temperature to regions of lower temperature, however, so the term *heat transfer by convection* is generally accepted.

In the following pages each mode of heat transfer is described separately and used to analyze problems. It must be emphasized that in most situations heat is transferred by several modes acting simultaneously. It is particularly important to the practicing engineer to be aware of the interrelation of the various modes of heat transfer, since when one mode dominates, we must be aware of the approximate solutions obtainable by neglecting all other modes.

Conduction   *Conduction is a process by which heat is transferred from a region of higher temperature to a region of lower temperature within a medium (solid, liquid, or gaseous) or between different media in direct physical contact where there is no heat transfer by mass movement.* In conduction heat transfer, the energy is transmitted by direct intermolecular collision. (There is no appreciable overall displacement of the molecules.) According to the kinetic theory, the temperature of an element of matter is directly proportional to the mean kinetic energy of its constituent molecules. The energy possessed by an element of matter due to the velocity and relative position of the molecules is called **internal energy.** Thus the more rapidly the molecules are vibrating, the greater will be the temperature and the internal energy of an element. When molecules in one region have a mean kinetic energy greater than that of molecules in an adjacent region (that is, a difference in temperature), the molecules possessing the greater energy will lose part of their energy to the molecules in the lower-temperature region. This transfer of energy takes place by elastic collision (as in fluids) or by diffusion of electrons, as in metals.

The theory of heat transfer by thermal conduction was first proposed by Jean B. Fourier (1768–1830) in a noted work, published in 1822 in Paris, entitled *Theorie analytique de la chaleur.* The pioneering experimental work was done by J. B. Biot. Biot's works, entitled *Bibliotheque britannique* and *Traité de physique,* were published in 1804 and 1816 respectively. Fourier's generalization of Biot's empirical information has the form*

$$\frac{q}{A} = -k\nabla T \qquad\qquad \textbf{(12-1)}$$

---

* $q[=]$ Btu/hr or W. Although apparently inconvenient, this change in nomenclature is consistent with common usage. Later another change, $\dot{q}[=]$Btu/hr-ft³, will be introduced in the discussion of the General Coordination Equation.

where the symbol ∇ represents the vector gradient operation ($\partial/\partial x$ in the x direction). It must be noted in passing that this is not truly a law but an equation used to define the thermal conductivity k.

Convection   Convection *is a heat transfer process whereby energy is transported by the combined action of conduction, energy storage, and mixing motion.* Convection is the mechanism of heat transfer between a solid surface and a liquid or a gas. In contrast to thermal energy transport by conduction, thermal convection involves energy transfer by mixing and diffusion in addition to conduction. To appreciate the complexity of this mode, consider the case of heat transfer to a fluid flowing inside a pipe. If the fluid in the pipe is flowing very fast, three different flow regions exist. Immediately adjacent to the wall is a **laminar sublayer**; the heat transfer occurs by thermal conduction. Outside the laminar sublayer is the transition region called the **buffer layer.** In this region, both eddy mixing and conduction effects are significant. Beyond the buffer layer and extending to the center of the pipe is the **turbulent region** where the dominant mechanism of transfer is eddy mixing. In low-velocity flow in small tubes, the entire flow may be **laminar** and there is no transition or eddy region. And finally, when the flow is produced by sources external to the heat transfer region (by a pump, for example) the heat transfer is termed **forced convection.** If the fluid flow is a result of nonhomogeneous densities arising from temperature variations, the heat transfer is termed **free or natural convection.**

The complex phenomena of convection were first analyzed successfully by Sir Isaac Newton in 1701. He proposed the general Newton rate equation (called Newton's law of cooling):

$$q = hA(T_w - T_\infty) \qquad (12\text{-}2)$$

where T_w represents the surface or wall temperature and T_∞ represents the liquid or gas temperature. Newton lumped considerations of fluid motion, fluid conductivity, and the role of turbulent eddies into a single factor: h. Known as the **convective heat transfer coefficient, film coefficient,** or **convection coefficient,** h depends on a large number of properties and parameters. In fact, Equation (12-2) is not truly a law but an equation used to define h.

Radiation   Radiation *is a process by which the net heat transfer from a high-temperature body to a lower-temperature body occurs when the bodies are separated in space.* The term *radiation* is generally applied to all electromagnetic-wave phenomena, but the science of heat transfer limits the study to phenomena that are the result of temperature. In thermal conduction and convection the transfer of heat requires a transport medium and is affected primarily by temperature difference and somewhat by temperature level. For radiant heat transfer, however, a change in energy form takes place—from internal energy at the source to electromagnetic energy for transmission through the intervening space, then back to internal energy at the receiver. The heat transferred by radiation increases rapidly with temperature. All bodies emit radiant heat continuously, and the intensity depends on the temperature and the nature of the surface. Radiant heat is emitted by a body in the form of finite batches, or **quanta,** of energy. Their motion in space is similar to the propagation of light and can be described by the wave theory. When these quanta encounter an object, a fraction of their energy is absorbed at its surface. Heat transfer by radiation becomes increasingly important as the temperature of an object rises. In engineering practice, radiant heat transfer rates involving temperatures approximating those of the atmosphere may often be neglected.

The mathematical expression that describes thermal radiation heat transfer is called the **Stefan–Boltzmann law.** This law has the form

$$\frac{q}{A} = \sigma T^4 \tag{12-3}$$

This expression was deduced by Stefan in 1879 from experimental data obtained by Tindall. Boltzmann, in 1884, using classic thermodynamics, derived the expression and placed it on firm theoretical ground.

The Art of Heat Transfer

In heat transfer, as in other branches of engineering, the successful solution of a problem requires assumptions and idealizations that approximately describe physical phenomena. That is, in order to make progress it is necessary to put the problem into the form of an equation that can be solved. When it becomes necessary to make an assumption or approximation to solve a problem, the engineer must rely on ingenuity and past experience. There are no simple guidelines. An assumption valid for one problem may be misleading in another. However, the first requirement for making sound engineering assumptions is a complete physical understanding of the problem. This requires familiarity not only with the laws describing the modes of heat transfer but also with those of fluid mechanics, physics, and mathematics.

In the solution of heat transfer problems, it is necessary to recognize the modes of heat transfer that dominate and to determine whether a process is steady or unsteady. When the rate of heat flow in a system does not vary with time (the temperature at any point does not change), steady-state conditions prevail. Under these conditions, the rate of heat transfer to any point of the system must be exactly equal to the rate of heat transfer from that point. (No change in internal energy can take place.) In this chapter the majority of heat transfer problems are concerned with steady-state situations. The heat flow in a system is transient, or unsteady, when the temperatures at various points in the system change with time. Unsteady heat transfer problems are more complex than those of steady state and are often solved only by approximate methods.

12–2 Thermal Conduction

Thermal conduction (or simply **conduction**) *is the mode of heat transfer whereby energy is transported between parts of a stationary medium or between two media in direct physical contact.* In gases, conduction is due to the elastic collision of molecules; in liquids and electrically nonconducting solids, it is due primarily to longitudinal oscillations of the lattice structure. Thermal conduction in metals takes place in the same manner as electrical conduction—that is, through the motions of free electrons.

Fourier's Law

The fundamentals of pure-conduction energy transfer were firmly established by Jean Fourier. **Fourier's law of conduction** (sometimes referred to as the Fourier–Biot law) may be stated: The heat transfer rate by conduction past any plane is given by

$$q = -kA\frac{\partial T}{\partial n} \qquad (12\text{-}4)$$

where A is the area normal to the flow of heat and $\partial T/\partial n$ is the temperature gradient in the direction of the heat flow.

The thermal conductivity of the material, k, is always positive. The minus sign is inserted to satisfy the second law of thermodynamics—that is, heat must flow downhill on the temperature scale, as indicated in the coordinate system of Figure 12-1. Notice that the second law requires the heat transfer rate to be a vectorlike quantity. The thermal conductivity k is the specific property of matter that indicates a stationary material's ability to transfer heat expressed as energy transferred per unit time per unit area per unit temperature gradient. The units of this quantity become power per unit length per degree.* Order of magnitude values for the thermal conductivity of various materials are listed as Table 12-1. Although there are always exceptions, thermal conductivity generally varies in the following fashion:

$$
\begin{array}{lll}
\text{Gases at low density:} & k \uparrow & \text{as} \quad T \uparrow \\
\text{Liquids:} & k \downarrow & \text{as} \quad T \uparrow \\
\text{Pure metals:} & k \downarrow & \text{as} \quad T \uparrow \\
\text{Nonmetals:} & k \uparrow & \text{as} \quad T \uparrow
\end{array}
$$

Figure 12-2 illustrates the variation of thermal conductivity with temperature.

---

* Common units are Btu/hr-ft²-F/ft—often shortened to Btu/hr-ft-F (English) or W/(m · C) (SI).

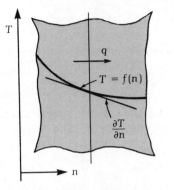

Figure 12-1. Direction of heat flow.

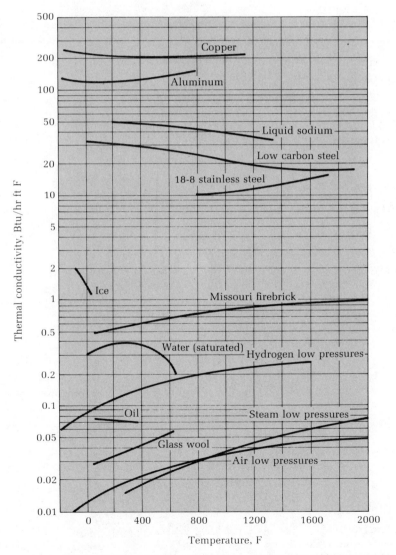

Figure 12-2. Variation in thermal conductivity with temperature.

Table 12-1 Order of Magnitude of Thermal Conductivity

Material	k, Btu/hr-ft-F	k, W/(m · C)
Gases	0.004–0.11	0.007–0.173
Insulating materials	0.02–0.12	0.004–0.208
Nonmetallic liquids	0.05–0.40	0.086–0.692
Nonmetallic solids (brick, stone, concrete)	0.02–2.0	0.004–2.595
Liquid metals	5.0–50	8.650–77.85
Alloys	8.0–70	13.84–121.1
Pure metals	30–240	51.90–415.2

EXAMPLE 12-1

The temperature distribution within a stone wall, 1 ft thick, at a certain instant of time is given by the following relation:

$$T = 80 + 5x - 25x^2$$

where T and x are in degrees F and feet, respectively. The wall area is 5 ft^2 and wall properties are: $k = 0.40$ Btu/hr-ft-F; $\rho = 136$ lbm/ft^3; $c = 0.2$ Btu/lbm-F. Determine:

1. The heat entering or leaving each face (Btu/hr)

2. Whether the wall is heating up or cooling down

3. The maximum and minimum temperatures within the wall

SOLUTION:
$$\frac{dT}{dx} = 5 - 50x$$

1. $q_{x=0} = -kA\left(\dfrac{dT}{dx}\right)_{x=0} = -0.40(5)(5)$

$$= -10 \text{ Btu/hr} \leftarrow$$

$q_{x=L} = -kA\left(\dfrac{dT}{dx}\right)_{x=1'} = -0.40(5)(5 - 50)$ } direction indicated by sign

$$= 90 \text{ Btu/hr} \rightarrow$$

2. Cooling down—heat is flowing out both faces.

3. $\dfrac{dT}{dx} = 0 = 5 - 50x$

$$x = \frac{1}{10} \text{ ft}$$

$$T_{x=1/10} = 80 + \left(\frac{5}{10}\right) - \left(\frac{25}{10^2}\right) = 80.25 \text{ F (max)}$$

$$T_{x=0} = 80 \text{ F}$$

$$T_{x=1'} = 80 + 5 - 25 = 60 \text{ F} \cdot (\text{min})$$

Applying Fourier's Law

Single-Layer Structure In many situations Fourier's law (Equation 12-4) may be used directly to obtain heat transfer information. Let us consider some special cases in which we limit our study to one-dimensional, steady-state conduction. Under these constraints Fourier's law may be integrated as follows:

$$q \int \frac{dn}{A} = -\int k \, dt \tag{12-5}$$

The variation of the thermal conductivity with temperature may be expressed as

$$k = f(t)$$

Thus

$$q \int_{n_1}^{n_2} \frac{dn}{A} = -\int_{T_1}^{T_2} f(t) \, dt \tag{12-6}$$

If the right-hand member of this equation is multiplied and divided by the temperature difference $T_2 - T_1$, then Equation (12-6) becomes

$$q \int_{n_1}^{n_2} \frac{dn}{A} = \frac{-\int_{T_1}^{T_2} f(t) \, dt}{T_2 - T_1} \; T_2 - T_1 \tag{12-7}$$

But the term

$$\frac{\int_{T_1}^{T_2} f(t) \, dt}{T_2 - T_1}$$

is the mean value of $f(t)$ between T_1 and T_2. Let us call it k_m, the mean value of k over this temperature range. Therefore

$$q \int_{n_1}^{n_2} \frac{dn}{A} = -k_m (T_2 - T_1)$$

or

$$q = \frac{k_m (T_1 - T_2)}{\int_{n_1}^{n_2} (dn/A)} \tag{12-8}$$

Hence, without error, the term $-\int_{T_1}^{T_2} k\, dt$ has been replaced by $k_m(T_1 - T_2)$ irrespective of the relation between A and n.

In the case of a single rectangular wall d ft wide whose outside surfaces are maintained at constant temperatures T_1 and T_2, the area A is constant (Figure 12-3). Here n_1 is zero and n_2 is d; hence the integration of Equation (12-8) yields the formula

$$q = \frac{A k_m(T_1 - T_2)}{d} \tag{12-9}$$

where A = area (ft^2) taken perpendicular to direction of flow
k_m = mean thermal conductivity for temperature range
$T_1 - T_2$ = temperature difference (F or R)

Equation (12-9) applies to any body having a constant cross section and heat flowing in one direction. It may be used to approximate the heat transfer in a plane wall if its length and width, perpendicular to the direction of heat flow, are large compared to the dimension in the direction of heat flow—thus making the conduction through the edges small enough to neglect.

Let us now reconsider Equation (12-8) but with a different geometric situation: the long cylinder (see Figure (12-4). Fourier's law may be rewritten as

$$q_r = -kA_r \frac{dT}{dr} \tag{12-10}$$

which evolves into the following form of Equation (12-8):

$$q_r = \frac{k_m(T_1 - T_2)}{\int_{r_1}^{r_2} dr/2\pi Lr}$$

$$= \frac{2\pi k_m L(T_1 - T_2)}{\ln(r_2/r_1)} \tag{12-11}$$

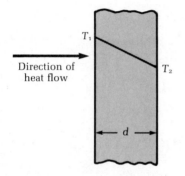

Figure 12-3. Single-plane wall.

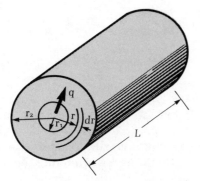

Figure 12-4. One-dimensional heat flow through a hollow cylinder.

Comparing Equations (12-8) and (12-11), one may easily see that the geometry of the situation drastically affects the form of the heat transfer expression.

Simple Composite Structures In engineering practice one rarely encounters a single-layer structure. Much more often, many layers are encountered. In the home, for example, there is a composite structure: inner wall, support structure, insulation, and outside siding. To analyze this situation, consider a simple two-layer composite wall but complicate matters by allowing both outer surfaces to be subject to convection heat transfer conditions. Figure 12-5 illustrates the situation. Let us further suppose that we know the constants, k_1, k_2, d_1, d_2, h_h, h_c, T_h, and T_c; we are asked to determine the heat transfer (or heat flux, q/A). From our previous experience we know

$$T_h - T_1 = \frac{q_1/A}{h_h} \tag{12-12}$$

$$T_1 - T_2 = \frac{q_2}{A}\left(\frac{d_1}{k_1}\right) \tag{12-13}$$

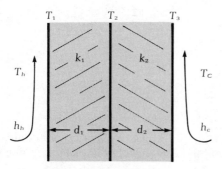

Figure 12-5. Simple composite rectangular wall.

$$T_2 - T_3 = \frac{q_3}{A}\left(\frac{d_2}{k_2}\right) \tag{12-14}$$

$$T_3 - T_c = \frac{q_4/A}{h_c} \tag{12-15}$$

Unfortunately, we do not know q_1, q_2, q_3, q_4, T_1, T_2, or T_3. That is, we have four equations and seemingly seven unknowns. By recalling the first law of thermodynamics, we note that $q_1/A = q_2/A = q_3/A = q_4/A = q/A$. Thus we now have four equations with four unknowns. By simply adding Equation (12-12) to (12-15) and rearranging, we obtain

$$\frac{q}{A} = \frac{T_h - T_c}{1/h_h + d_1/k_1 + d_2/k_2 + 1/h_c} \tag{12-16}$$

Therefore, by measuring the temperatures, physical properties, and dimensions of the materials of this composite structure, one may determine the heat transferred per unit area and time (the flux).

EXAMPLE 12-2

A thick-walled pipe with inside and outside diameters of 2 and 4 cm, respectively, is used to transport superheated steam. Assuming that the inside wall temperature is 300 C, determine the outside wall temperature if 700 W/m is lost from the pipe and:

1. $k = 19$ W/(m·C) (steel)

2. $k = 0.2$ W/(m·C) (ceramic)

SOLUTION: Using Equation (12-11), we get

$$T_2 = T_1 + q\frac{\ln{(r_2/r_1)}}{2\pi kL}$$

$$= 300\ C + \frac{-700\ \text{W/m}}{2\pi k}\ln\left(\frac{4\ \text{cm}}{2\ \text{cm}}\right) = 300\ C - \frac{77.22\ \text{W/m}}{k}$$

CASE 1: $k = 19$ W/(m·C)

$T_2 = 296\ C$

CASE 2: $k = 0.2$ W/(m·C)

$T_2 = -86.1\ C$

Does this seem right? It should, since the lower conductivity would require a larger temperature difference to dissipate the same power.

EXAMPLE 12-3

Prove that the temperature profile of Figure 12-3 is linear.

SOLUTION: Recall Fourier's law:

$$q = -kA\frac{dT}{dx}$$

Putting this in integral form yields

$$\frac{q}{A}\int dx = -\int k\,dT$$

Instead of integrating from zero to d and T_1 to T_2, let us integrate from zero to $x < d$ and T_1 to $T > T_2$, yielding

$$\frac{q}{A}\int dx = -\frac{\int_{T_1}^{T} k\,dT}{T - T_1}(T - T_1) = -k_m(T - T_1)$$

But from Equation 12-9 we may eliminate q/A. Thus

$$T = T_1 + \frac{x}{d}(T_2 - T_1)$$

EXAMPLE 12-4

The exposed basement wall of a residence is constructed of 9-in. concrete ($k = 1.04$ Btu/hr-ft-F). Outside air is at 105 F and the outside convective coefficient is 4.0 Btu/hr-ft²-F. Inside air is at 75 F and the inside convective coefficient is 1.6 Btu/hr-ft² F. Determine:

1. The heat flow through the wall (per ft²)

2. The outside surface temperature of the concrete

SOLUTION: 1. $q = \dfrac{T_i - T_o}{1/h_iA + L/kA + 1/h_oA} = \dfrac{75 - 105}{1/1.6 + (9/12)/1.04 + 1/4.0}$

$$= \frac{-30}{\underbrace{0.625 + 0.721 + 0.25}_{1.596}}$$

$$= -18.8 \text{ Btu/hr-ft}^2$$

2. $q = \dfrac{T_o - T_{wo}}{1/h_oA} = 18.8 = \dfrac{105 - T_{wo}}{1/4.0}$

$$T_{wo} = 105 - 4.7 = 100.3 \text{ F}$$

EXAMPLE 12-5

An electric wire is $\frac{1}{8}$ in. in diameter and is covered with an electrical insulation that is $\frac{3}{32}$ in. thick and has a k of 0.4 Btu/hr-ft-F. Four watts of electrical energy are dissipated into heat by the resistance of the wire per foot of length. If this heat is given up to the surrounding air at 80 F, determine the surface temperature of the wire insulation and its maximum temperature. Take $h = 3$ Btu/hr-ft²-F.

SOLUTION: $\qquad r_1 = \frac{1}{16}$ in. $= \frac{2}{32}$ in.

$$r_2 = \frac{5}{32} \text{ in.}$$

$$q = \frac{T_1 - T_{air}}{\dfrac{\ln (r_2/r_1)}{2\pi kl} + \dfrac{1}{h_o A_o}}$$

$$A_o = 2\pi r_2 l$$

$$4(3.413)l = \frac{T_1 - 80}{\dfrac{\ln (5/2)}{2\pi (0.4)(l)} + \dfrac{1}{3[\pi 5/(16 \times 12)l]}} ; \quad T_{max} = T_1; \ T_{surf} = T_2$$

$$T_1 = 141.55 \text{ F}$$

Also

$$q = \frac{T_1 - T_2}{\dfrac{\ln (r_2/r_1)}{2\pi kl}} = \frac{T_2 - T_{air}}{1/h_o A_o}$$

$$4(3.413)l = \frac{T_2 - 80}{\dfrac{1}{3\{\pi[5/(16 \times 12)](l)\}}}$$

$$T_2 = 135.62 \text{ F}$$

EXAMPLE 12-6

The exposed basement wall of a residence is constructed of 0.229-m concrete ($k = 1.8$ W/(m · C). The outside air is 40.56 C and the outside convective coefficient is 22.74 W/(m² · C). The inside air is at 23.89 C; the inside convection coefficient is 9.09 W/(m² · C). If the total heat gained by the residence (thus the wall) is 10.340 kW, what is the surface area of this wall?

SOLUTION: $\qquad q = \dfrac{T_i - T_o}{1/h_i A + L/kA + 1/h_o A}$

or

$$A = \frac{q(1/h_1 + L/K + 1/h_o)}{(T_i - T_o)}$$

$$= \frac{10.340 \text{ kW}}{16.67 \text{ C}} \left(\frac{1}{22.74 \text{ W/(m}^2 \cdot \text{C)}} + \frac{0.229 \text{ m}}{1.86/(\text{m} \cdot \text{C})} + \frac{1}{9.09 \text{ W/(m}^2 \cdot \text{C)}} \right)$$

$$= 620.276 \text{ W/C } (0.04397 + 0.12311 + 0.11001) \text{ (m}^2 \cdot \text{C)/W}$$

$$= 171.88 \text{ m}^2$$

General Conduction Equation

The second important equation from Fourier's work is the **general conduction equation.** This equation, deduced from the first law of thermodynamics, permits the investigator to predict the temperature distribution in a multidimensional body whether the body has reached a steady-state temperature or is cooling or heating (transient). Appendix B-6 presents in detail the uses of Fourier's law and the first law of thermodynamics to obtain the following general relations for the temperature distribution in stationary materials.

Rectangular Coordinates Equation (12-17) presents the general conduction equation in rectangular coordinates; this equation is consistent with Figure 12-6.

$$\frac{\partial}{\partial x}\left(k\frac{\partial T}{\partial x}\right) + \frac{\partial}{\partial y}\left(k\frac{\partial T}{\partial y}\right) + \frac{\partial}{\partial z}\left(k\frac{\partial T}{\partial z}\right) + \dot{q} = \rho c \frac{\partial T}{\partial t} \qquad \textbf{(12-17)}$$

where $\dot{q}$ = energy generated per unit volume
c = specific heat of material
ρ = density

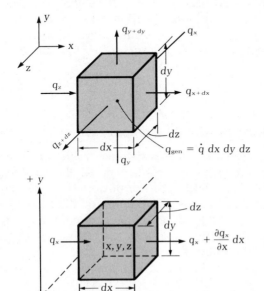

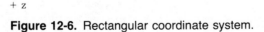

Figure 12-6. Rectangular coordinate system.

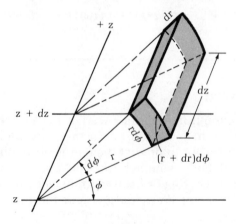

Figure 12-7. Cylindrical coordinate system.

For constant thermal conductivity, this equation becomes

$$\frac{\partial^2 T}{\partial x^2} + \frac{\partial^2 T}{\partial y^2} + \frac{\partial^2 T}{\partial z^2} + \frac{\dot{q}}{k} = \frac{1}{\alpha}\frac{\partial T}{\partial t}$$ (12-18)

where $\alpha = k/\rho c$, the thermal diffusivity. Recall that the specific heat c is the specific property of a material that gives the quantity of energy required to raise a unit mass of the material by one degree in temperature.

Cylindrical Coordinates (Uniform k) In cylindrical coordinates (Figure 12-7) the general conduction equation takes a slightly different form. Equation (12-19) presents the same information in cylindrical coordinates that Equation (12-18) presents for rectangular coordinates.

$$\frac{\partial^2 T}{\partial r^2} + \frac{1}{r}\frac{\partial T}{\partial r} + \frac{1}{r^2}\frac{\partial^2 T}{\partial \phi^2} + \frac{\partial^2 T}{\partial z^2} + \frac{\dot{q}}{k} = \frac{1}{\alpha}\frac{\partial T}{\partial t}$$ (12-19)

Analogy Between Heat Flow and Electrical Flow

The heat transfer through a thermal resistance (Equation 12-4) is analogous to the flow of electric current through an electrical resistance, (Ohm's law, $I = \overline{V}/R$) since both types of flow obey similar equations. To simulate heat flow using an electrical network, the T is equivalent to $\overline{V}$, the R_{thermal} is equivalent to R, and the q is equivalent to i. The electrical schematic for a hollow cylinder with convection on both the inner and outer surfaces follows:

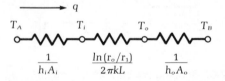

All quantities must be properly scaled. Notice that this electrical analog may not be used for systems in which there is generation or systems that are not steady state.

One-Dimensional Steady-State Conduction

Fourier's law (Equation 12-4) and the general conduction equation (Equations 12-18 and 12-19) have been solved for a large number of heat transfer situations. The resulting solutions for some common cases are presented in Tables 12-2 and 12-3. Several of the cases involve convective conditions at the surfaces.

Overall Heat Transfer

As indicated earlier, most steady-state heat transfer problems encountered in practice involve more than one heat transfer mode. It is convenient to combine these various heat transfer coefficients into an overall coefficient so that the total heat transfer may be calculated from the terminal temperatures. The solution to the problem is simpler if we employ the concept of the thermal circuit and thermal resistance.

To emphasize this fact, consider an application in which heat is transferred from one fluid to another by a steady-state process: from a warmer fluid to a solid wall, through the solid wall, and thence to a colder fluid. It is customary to employ an overall coefficient of heat transfer U based on the overall difference between the bulk temperatures of the two fluids, $T_1 - T_2$, defined as follows:

$$q = UA(T_1 - T_2) \tag{12-20}$$

where A is the surface area. Equation (12-20) is a definition of U; the surface area A upon which U is based is arbitrary and should always be specified in referring to U.

The temperature drops across each part of the heat flow path are as follows:

$$T_1 - T_{s1} = qR_1$$

$$T_{s1} - T_{s2} = qR_2$$

$$T_{s2} - T_2 = qR_3$$

where T_{s1} and T_{s2} are the surface temperatures of the wall on the warm side and cold side, respectively, and R_1, R_2, and R_3 are the thermal resistances. Since the same quantity of heat flows through each thermal resistance, these equations can be combined to give

$$\frac{T_1 - T_2}{q} = \frac{1}{UA} = R_1 + R_2 + R_3$$

As demonstrated above, the equations involved are analogous to those

Table 12-2 Steady One-Dimensional Heat Conduction

$$q = \frac{\Delta T_{overall}}{\Sigma R_{thermal}} = UA \Delta T_{total} \qquad UA = \frac{1}{R_1 + R_2 + \cdots + R_n}$$

Plane wall without internal generation:

$$q = \frac{kA(T_1 - T_2)}{L}$$

$$R = \frac{L}{kA}$$

$$T = T_1 - \frac{T_1 - T_2}{L}x$$

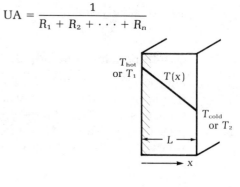

Multilayer wall without internal generation:

$$q = \frac{T_1 - T_{n+1}}{\dfrac{L_1}{k_1 A} + \dfrac{L_2}{k_2 A} + \cdots + \dfrac{L_n}{k_n A}}$$

$$R = \frac{L_1}{k_1 A} + \frac{L_2}{k_2 A} + \cdots + \frac{L_n}{k_n A}$$

where n = number of layers

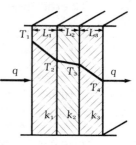

Multilayer wall without internal generation (convection at surfaces):

$$q = \frac{T_i - T_o}{1/h_i A + L_1/k_1 A + \cdots + L_n/k_n A + 1/h_o A}$$

$$R = \frac{1}{h_i A} + \frac{L_1}{k_1 A} + \cdots + \frac{L_n}{k_n A} + \frac{1}{h_o A}$$

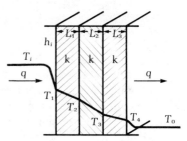

Plane wall with heat sources:

$$\frac{d^2 T}{dx^2} + \frac{\dot{q}}{k} = 0$$

With uniformly distributed heat sources:

$$T = -\frac{\dot{q}}{2k}x^2 + C_1 x + C_2$$

$$q_x = -kA \left(\frac{dT}{dx} \right)_x$$

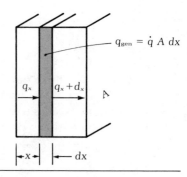

Table 12-3 Steady One-Dimensional Cylindrical Heat Conduction

Single hollow cylinder without internal generation:

$$q = \frac{T_1 - T_o}{\ln(r_o/r_i)/2\pi kL} = \frac{2\pi kL(T_i - T_o)}{\ln(r_o/r_i)}$$

$$R_{\text{thermal}} = \frac{\ln(r_o/r_i)}{2\pi kL}$$

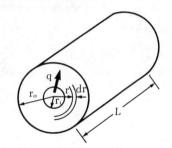

Composite cylinder without internal generation:

$$q = \frac{T_1 - T_{n+1}}{\dfrac{\ln(r_2/r_1)}{2\pi k_1 L} + \dfrac{\ln(r_3/r_2)}{2\pi k_2 L} + \cdots + \dfrac{\ln(r_{n+1}/r_n)}{2\pi k_n L}}$$

$$R = \frac{\ln(r_2/r_1)}{2\pi k_1 L} + \frac{\ln(r_3/r_2)}{2\pi k_2 L} + \cdots + \frac{\ln(r_{n+1}/r_n)}{2\pi k_n L}$$

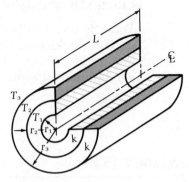

Composite cylinder without internal generation (with convection):

$$q = \frac{T_i - T_o}{\dfrac{1}{h_i A_i} + \dfrac{\ln(r_2/r_1)}{2\pi k_1 L} + \cdots + \dfrac{\ln(r_{n+1}/r_n)}{2\pi k_n L} + \dfrac{1}{h_o A_o}}$$

$$R = \frac{1}{h_i A_i} + \frac{\ln(r_2/r_1)}{2\pi k_1 L} + \cdots + \frac{\ln(r_{n+1}/r_n)}{2\pi k_n L} + \frac{1}{h_o A_o}$$

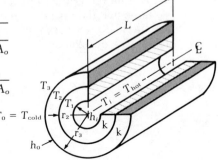

Cylinder with heat sources:

$$\frac{d^2 T}{dr^2} + \frac{1}{r}\frac{dT}{dr} + \frac{\dot{q}}{k} = 0$$

With uniformly distributed heat sources:

$$T = \frac{-\dot{q}r^2}{4k} + C_1 \ln(r) + C_2$$

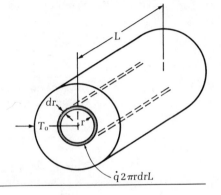

391

for electrical circuits—that is, when there is a thermal current flowing through several resistances in series, the resistances are additive:

$$R_0 = R_1 + R_2 + R_3 + \cdots + R_n$$

Similarly, conductance is the reciprocal of resistance. For heat flow through several resistances in parallel, the conductances are additive:

$$C = \frac{1}{R_0} + \frac{1}{R_1} + \frac{1}{R_2} + \frac{1}{R_3} + \cdots + \frac{1}{R_n}$$

Critical Thickness of Insulation

There is a critical thickness of insulation for a cylinder (pipe). The **critical thickness** of insulation is the thickness, usually expressed as the outer radius, of a single layer of insulation that results in the **maximum** heat transfer rate:

$$r_{o\,crit} = \frac{k}{h}$$

EXAMPLE 12-7

Determine the critical thickness of insulation on a pipe of outside radius r_o. Assume that the inside insulation surface temperature is known, as well as all the thermal physical properties and T_∞, the gas temperature outside the insulation (see the sketch).

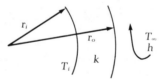

SOLUTION: From our previous discussion

$$q = \frac{2\pi L (T_i - T_\infty)}{\dfrac{\ln (r_o/r_i)}{k} + \dfrac{1}{r_o h}}$$

To determine the critical thickness, we must extremize q with respect to r_o. That is,

$$\frac{dq}{dr_o} = 0$$

The result is $r_{o\,crit} = k/h$.

The relationship between q and r_o is best represented by the following sketch:

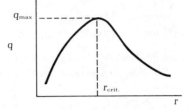

If we are considering a steel pipe of 1 cm outside diameter and estimate $h = 5.60\ \text{W/(m}^2 \cdot \text{C)}$ on the outside surface of the asbestos insulation ($k \doteq 0.208\ \text{W/(m} \cdot \text{C))}$, then

$$r_{o\ \text{crit}} = \frac{20.8}{5.68}\ \text{cm} = 3.662\ \text{cm}$$

Note that the critical thickness is greater than the pipe thickness. Thus increasing the insulation will in fact increase the heat loss until the outside radius of the insulation is 3.662 cm. From that thickness on, the heat loss will be reduced.

Heat Source Systems

Tables 12-2 and 12-3 include one example each of a system with non-zero heat generation. These systems, referred to as **heat source systems**, need further discussion. If we limit our discussion to the steady-state situation, the general conduction equation has the form*

$$\nabla \cdot k\nabla T + \dot{q} = 0 \qquad (12\text{-}21)$$

Let us apply this to a cylindrical situation by assuming a very long circular cylinder—so long that heat lost from the ends may be neglected—with uniformly distributed heat sources ($\dot{q}$ is a constant). Further, let us allow the thermal conductivity, k, to be constant and the cylinder radius to be R. For these conditions Equation (12-21) reduces to

$$\frac{1}{r}\frac{d}{dr}\left(r\frac{dT}{dr}\right) + \frac{\dot{q}}{R} = 0 \qquad (12\text{-}22)$$

One integration of this equation yields

$$r\frac{dT}{dr} = \frac{-\dot{q}r^2}{2k} + C_1 \qquad (12\text{-}23)$$

and further

$$T = \frac{-\dot{q}r^2}{4k} + C_1 \ln(r) + C_2 \qquad (12\text{-}24)$$

---

*$\nabla \cdot k\nabla T$ is a shorthand notation for partial derivatives of Equation (12-17).

Note that Equation (12-24) has two constants of integration (Equation (12-22) is a second-order differential equation) which require two boundary conditions. Let us consider two cases.

Case I: First, let us assume the cylinder is solid with a specified surface temperature. Thus the boundary conditions for this case are:

(1) at $r = R$, $T = T_w$;   **(2)** at $r = 0$, $\dfrac{dT}{dr} = 0$ (symmetry condition)

Applying boundary condition (2), using Equation (12-23) yields

$$C_1 = 0$$

Using Equation (12-24) and boundary condition (1) yields

$$T_w = \frac{-\dot{q}r^2}{4k} + C_2 \qquad \text{or} \qquad C_2 = T_w + \frac{\dot{q}R^2}{4k}$$

Substitution of these values of the two boundary conditions into Equation (12-24) and arranging to a convenient form produces the following form of the temperature distribution (or profile):

$$T - T_w = \frac{\dot{q}}{4k}(R^2 - r^2) \tag{12-25}$$

Note from this equation that the maximum temperature occurs at $r = 0$, if $\dot{q}$ is positive, or

$$T_m = T_w + \frac{\dot{q}R^2}{4k} \tag{12-26}$$

Using the maximum temperature expression, $T(r = 0) = T_m$, a convenient nondimensional form of the temperature profile is obtained:

$$\frac{T - T_w}{T_m - T_w} = 1 - \left(\frac{r}{R}\right)^2 \tag{12-27}$$

Before going to Case II, an observation may help your intuition. The heat that is conducted to the surface equals the heat that was generated. That is,

$$-k(2\pi RL)\frac{dT}{dr}\bigg|_{r=R} = \dot{q}(\pi R^2 L)$$

where L is a length of the cylinder. Notice that this expression and boundary condition (2) represent equivalent conditions.

Case II: For this example let us again have uniformly distributed heat sources but allow the cylinder to be hollow. Further, let the inside and outside surface temperature be specified. Thus **(1)** at the inside surface, $r = r_i$, $T = T_i$ and **(2)** at the outside surface, $r = r_o$, $T = T_o$. Remember that Equation (12-24) is still the general solution—we must determine two new constants of integration. As before, to do this we apply the two boundary conditions, or

$$T_i = \frac{-\dot{q}r_i^2}{4k} + C_1 \ln r_i + C_2 \qquad T_o = \frac{-\dot{q}r_o^2}{4k} + C_1 \ln r_o + C_2$$

These two equations may be solved simultaneously for C_1 and C_2. Substitution of the expressions for C_1 and C_2 into Equation (12-24), yields the general solution

$$T - T_i = \frac{-\dot{q}}{4k}(r^2 - r_i^2) + \left[(T_o - T_i) + \frac{\dot{q}}{4k}(r_o^2 - r_i^2) \right] \frac{\ln(r/r_i)}{\ln(r_o/r_i)} \qquad \textbf{(12-28)}$$

Transient Heat Flow

Often it is necessary to know the heat transfer and temperature distribution under unsteady-state conditions (varying with time). As we have seen, the fundamental equation for all conduction in solids or fluids in which there is no substantial motion is the general conduction equation. Not much experience is required to realize that this is a very difficult equation to apply all the time. Therefore, convenient approximate procedures are needed. Just such a situation occurs when the thermal conductivity is large compared to hL. (L is a characteristic length of the body.)

For a well-stirred reservoir of fluid whose temperature is changing because of either a net rate of heat gain or loss, the applicable equation may be written

$$q_{net} = mc_v \frac{dT}{dt} \qquad k \gg hL \qquad \textbf{(12-29)}$$

where m is the mass of the body and c_v is the specific heat at constant volume of the body. For liquids and solids, the values of c_v and c_p are nearly equal and q_{net} may include all modes of heat transfer and is the difference between the rate of heat transfer to the body and the heat transfer away from the body (the first law). Equation (12-29) is simply an extension of the definition of the specific heat.

Consider a solid body that is suddenly immersed in a fluid at a lower temperature. If $k \gg hL$ such that the temperature of the body could be taken as uniform during the cooling process, the energy balance equation could be written as

$$q = hA(T - T_\infty) = -\rho cV \frac{dT}{dt} \qquad \textbf{(12-30)}$$

where A = surface area for convection
T = temperature of body at any time
T_∞ = fluid temperature
ρ = density of body

t = time

V = volume of body

If the fluid temperature remains constant, integration yields

$$T = T_\infty + (T_0 - T_\infty)e^{-(hA/\rho Vc)t} \tag{12-31}$$

where T_0 is the initial temperature of the body. This **lumped capacity** approach assumes a constant temperature throughout the solid that is equivalent to the condition that the internal resistance to the flow of heat is small compared to the surface convective resistance (that is, $k > hL$). The ratio of these resistances is referred to as the **Biot number** and designated Bi. Experience has shown that Equation (12-30) is accurate when the Biot number is less than 0.1, or

$$\text{Bi} = \frac{h(V/A)}{k} < 0.1 \tag{12-32}$$

before the approximate solution procedure of Equation (12-30) is applicable.

EXAMPLE 12-8

A small dam that may be idealized by a large slab 4 ft thick is to be poured completely in a short period of time. The hydration of the concrete results in the equivalent of a distributed heat source of constant strength of 7 Btu/hr-ft³. If both dam surfaces are at 60 F, determine the maximum temperature to which the concrete will be subjected, assuming steady-state conditions. The thermal conductivity of the wet concrete may be taken as 0.7 Btu/hr-ft-F.

SOLUTION:

$T(0) = 60°$ $T(4) = 60°$

14 Btu/hr ft²   14 Btu/hr ft²

$$\frac{d^2T}{dx^2} + \frac{\dot{q}}{k} = 0$$

$$T = -\frac{\dot{q}x^2}{2k} + C_1 x + C_2$$

At $x = 0$:

$$60 = \frac{-\dot{q}(0)}{2k} + C_1(0) + C_2$$

$$C_2 = 60$$

At x = 4':

$$60 = \frac{-\dot{q}(4)^2}{2k} + C_1(4) = 60$$

$$C_1 = \frac{(4)^2 \dot{q}}{(4)2k} = \frac{(4)7}{2(0.7)} = 20$$

$$T = -\frac{\dot{q}}{2k} x^2 + 20x + 60$$

$$= -\frac{7}{2(0.7)} x^2 + 20x + 60$$

$$= -5x^2 + 20x + 60$$

Check:

$$x = 0 \quad T = 60; \quad x = 4 \quad T = 60 \quad \text{(It checks.)}$$

$$\frac{dT}{dx} = -10x + 20 = 0 \quad \text{for max } T$$

yields

$$x = 2 \text{ ft} \quad \text{for } T_{max}$$

$$T_{max} = -5(2)^2 + 20(2) + 60 = -20 + 40 + 60 = 80 \text{ F}$$

EXAMPLE 12-9

A 4-mm-diameter copper wire is 1 m long and has 110 V across it. If the outer surface temperature of the wire is measured to be 100 C, what is the temperature of the wire in the center?

SOLUTION: Assume that $k_{copper} \approx 400$ W/(m · C). Then, from Equation (12-27),

$$T_0 = T_w + \frac{-\dot{q}R^2}{4k}$$

$$= 100C + \dot{q} \left(\frac{2 \text{ cm}}{2}\right)^2 \frac{1 \text{ m} \cdot C}{400 \text{ W}}$$

$$= 100C + \dot{q} \frac{10^{-7}}{8.4} \frac{m^3 \cdot C}{W}$$

To determine $\dot{q}$, apply the first law of thermodynamics:

$$\dot{q}(\pi r^2 L) = I^2 R \quad (R = \text{electrical resistance})$$

$$= \frac{\bar{V}^2}{R} \quad (\bar{V} = \text{voltage})$$

But $R = \bar{\rho}L/A$, where $\bar{\rho}$ is the resistivity of copper $(1.72 \times 10^{-8} \ \Omega \cdot m)$.
Thus

$$\dot{q} = \frac{\bar{V}^2}{R\pi r^2 L} = \frac{\bar{V}^2}{\bar{\rho}L} = \frac{\bar{V}^2}{\bar{\rho}L^2}$$

$$= \frac{(10 \ V)}{(1 \ m)^2 \ 1.72 \ (10^{-8}) \ \Omega \cdot m} = \frac{10^{10}}{1.72 \ m^3} \frac{V^2}{\Omega}$$

$$= 5.814 \ (10^3) \ MW/m^3$$

So

$$T_0 = \left(100 + 5.814 \ (10^9) \ \frac{W}{m^3} \frac{10^{-7}}{0.4} \frac{m^2}{W}\right) C$$

$$= 200 \ C$$

EXAMPLE 12-10

A cylinder of silver 1 in. in diameter by 1 in. long, initially at 480 F, is suddenly quenched in water at 212 F. If the convective heat transfer coefficient is 375 Btu/hr-ft²-F, determine the time (in seconds) required for the center of the silver cylinder to reach 250 F.

SOLUTION:

$$A = \pi DL + \left(\frac{\pi D^2}{4}\right)(2) = \frac{\pi}{144} + \frac{\pi}{2 \ (144)} = 0.0327 \ ft^2$$

$$V = \left(\frac{\pi D^2}{4}\right)(L) = \frac{\pi}{4 \ (144) \ (12)} = 0.000456 \ ft^3$$

$$Bi = \frac{hV}{kA} = \frac{375 \ (0.000456)}{237 \ (0.0327)}$$

$$= 0.022 < 0.1 \qquad \text{(use lumped approach)}$$

$$\frac{T - T_\infty}{T_0 - T_\infty} = e^{-\frac{hA}{(\rho v c)}t}$$

$$= \frac{250 - 212}{480 - 212} = e^{-\frac{375 \ (0.0327)}{657 \ (0.000456) \ (0.0559)}t}$$

$$0.1418 = e^{-732.2t}$$

$$-1.953 = -732.2t$$

$$t = 0.00267 \ hr = 9.6 \ sec$$

Extended Surfaces: Fins and Spines

Heat transfer from a surface may be increased by attaching an extended surface that increases the area available for the transfer. These compact,

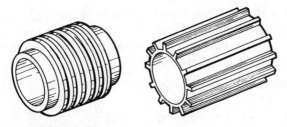

Figure 12-8. The two most common types of finned tubes: transverse fins and longitudinal fins.

tached to machinery maintain or control its temperature. Everyday examples of equipment employing extended surfaces are natural and forced-convection coils, shell-and-tube evaporators, motorcycles, transistors, and condensers. Fins are used inside tubes in condensers and dry expansion evaporators. (Figure 12-8).

Let us examine a simple fin by using a one-dimensional analysis. We will assume that a substantial temperature gradient occurs only in the x direction, an assumption that is satisfied if the fin is sufficiently thin. For most fins of practical interest the error introduced by this assumption is less than 1%. In practice, the overall accuracy of fin calculations is usually limited by uncertainties in values of the convection coefficient h. (The convection coefficient is seldom uniform over the entire surface.) Consider the constant cross-sectional area fin. Let us assume that k is constant and apply the first law of thermodynamics to a small control volume (see Figure 12-9).

With the aid of an energy balance on a differential volume (of width dx in Figure 12-9), the governing equation for the temperature may be obtained. Therefore

Energy generated in $(A\,dx)$ + energy in at $(x = x)$

$$= \text{energy out at } (x = x + dx)$$

$$+ \text{energy lost by convection from } (P\,dx)$$

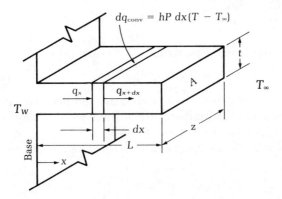

Figure 12-9. Straight fin of uniform cross section.

Care must be taken in applying Newton's law of cooling, Equation (12-2), to this incremental surface area of width dx and perimeter P. Letting A represent the constant cross sectional area of the fin, the terms of the word equation may be established:

$$\text{(Energy generated in } A\,dx) = \dot{q}A\,dx \tag{12-33}$$

$$\text{(Energy in at } x = x) = q_x = -kA\frac{dT}{dx} \tag{12-34}$$

$$\text{(Energy out at } x = x + dx) = q_{x+dx} \doteq qx + \frac{d\dot{q}_x}{dx}\,dx \tag{12-35}$$

$$\text{(Energy lost by convection)} = hP(T - T_\infty)\,dx \tag{12-36}$$

Substitution of Equations (12-33) to (12-36) into the word equation and cancelling terms yields

$$\dot{q}A = \frac{d}{dx}\left(-kA\frac{dT}{dx}\right) + hP(T - T_\infty)$$

Notice that this equation may be interpreted as a general one-dimensional fin equation applicable when $\dot{q}$, A, k, P and/or T_∞ are not constant. For our purposes A (and P) as well as k are constants. Thus since A is constant, this general fin equation reduces to

$$\frac{d}{dx}\left(k\frac{dT}{dx}\right) + \dot{q} = \frac{hP}{A}(T - T_\infty) \tag{12-37}$$

If, in addition, k is constant, the general fin equation is

$$\frac{d^2T}{dx^2} + \frac{\dot{q}}{k} = \frac{hP}{kA}(T - T_\infty) \tag{12-38}$$

Now let us consider a special case: no generation. With this restriction Equation (12-38) reduces to

$$\frac{d^2T}{dx^2} = \frac{hP}{kA}(T - T_\infty) = m^2(T - T_\infty) \tag{12-39}$$

where $m = \sqrt{hP/kA}$

The solution to this expression is, when T_∞ is constant

$$T - T_\infty = C_1 e^{mx} + C_2 e^{-mx} \tag{12-40}$$

The two constants of integration must be determined from two boundary conditions. The first is the temperature at the root of the fin (where it is attached to the prime surface of the device): $T = T_w$ at $x = 0$. Substitution of this into Equation (12-40) yields one condition on C_1 and C_2:

$$C_1 + C_2 = T_w - T_\infty$$

$$C_1 + C_2 = T_w - T_\infty$$

The other end of the fin may have a variety of conditions. As a matter of convenience let us assume the length of the fin is such that

$$x = L \qquad T = T_\infty \qquad \text{(where L is very large)}$$

This condition requires that $C_1 \to 0$. Therefore the solution is

$$\frac{T - T_\infty}{T_w - T_\infty} = e^{-mx} \tag{12-41}$$

The heat loss from the fin may be computed in two ways for steady state: count energy in or count energy out. The energy into the fin by conduction at the root is

$$q = -Ak\frac{dT}{dx}\bigg|_{x=0} \tag{12-42}$$

$$= -Ak(-m)(T_w - T_\infty)e^{-mx}\big|_{x=0}$$

$$= \sqrt{hPkA}\,(T_w - T_\infty)$$

The energy loss by convection from the fin surface is

$$q = \int_0^\infty h\,(T - T_\infty)\,dA_s \tag{12-43}$$

$$= hP\int_0^\infty (T - T_\infty)\,dx = hP\int_0^\infty (T_w - T_\infty)e^{-mx}\,dx$$

$$= \frac{hP}{m}(T_w - T_\infty) = \sqrt{hPkA}\,(T_w - T_\infty)$$

Table 12-4 presents two other commonly encountered conditions at $x = L$.

EXAMPLE 12-11

A 60-W soldering iron has a copper cylindrical tip, 0.32 cm in diameter, that extends 5 cm from the heating element. If the iron remains plugged in when not in use and the tip is surrounded by air at 27 C with a convective coefficient of 15 W/(m² · C), determine the base temperature (where tip meets heater). Assume the heating element is well insulated except for the heat path to the tip.

SOLUTION:
$$P = \pi(0.0032) = 0.010 \text{ m}$$

$$A = \frac{\pi d^2}{4} = 0.00000804 \text{ m}^2$$

$$L = 0.05 \text{ m}$$

Table 12-4 Some Extended-Surface Solutions

Extended Surface	Temperature Profile	Heat Flow

Straight fin or spine of uniform cross section

$$dq_{conv} = hP\,dx(T - T_\infty)$$

General: $T(x) = T_\infty + C_1 e^{mx} + C_2 e^{-mx}$

General:
$$q = -kA\frac{dT}{dx}\Big)_{x=0}$$
$$= \int_0^L hP(T - T_\infty)\,dx$$

Case I: End at $x = 0$ at temperature T_s; end at $x = L$ insulated.

$$\frac{T - T_\infty}{T_w - T_\infty} = \frac{\cosh m(L - x)}{\cosh mL}$$

Case I:
$$q_{rod} = -kA\frac{dT}{dx}\Big|_{x=0}$$
$$= \sqrt{PhkA}\,(T_w - T_\infty)\tanh mL$$

Case II: End at $x = 0$ at temperature T_s; end at $x = L$ exposed to fluid.

$$\frac{T - T_\infty}{T_w - T_\infty}$$
$$= \frac{\cosh m(L - x) + (h/km)\sinh m(L - x)}{\cosh mL + (h/mk)\sinh mL}$$

$$m = \sqrt{\frac{hP}{kA}} \approx \sqrt{\frac{2h}{kt}}$$

Case II:
$$q = \sqrt{hPkA}\,(T_w - T_\infty)$$
$$\times \frac{\sinh mL + (h/mk)\cosh mL}{\cosh mL + (h/mk)\sinh mL}$$

$$m = \sqrt{\frac{15\,(0.010)}{385\,(0.00000804)}} = 6.96$$

$$mL = 0.348$$

$$\frac{h}{mk} = 0.0056$$

$$q_{\text{fin, in}} = \sqrt{PhAk}\,(T_s - T_\infty)\,\frac{\sinh mL + (h/mk)\cosh mL}{\cosh mL + (h/mk)\sinh mL}$$

$$60 = \sqrt{(0.010)\,(15)\,(8.04 \times 10^{-6})\,(385)}$$

$$\times\,(T_s - 27)\,\frac{0.355 + 0.0059}{1.061 + 0.0020}$$

$$= 0.0215\,(T_s - 27)\,(0.3395) = 0.0073\,(T_s - 27)$$

$$T_s = 27 + 8200 = 8227\ \text{C}$$

Even with only 60 W of power, iron would burn up unless other heat paths exist.

As heat transfers from the root of a fin to its tip, there is a temperature drop due to the thermal resistance of the fin material. Therefore, the temperature difference between points of the fin and the surroundings varies—being greater at the root—which causes a corresponding variation in the heat flux. For this reason, increases in the length of a fin result in proportionately less additional heat transfer. To account for this effect, a factor called fin efficiency is introduced. **Fin efficiency** *is usually defined as the ratio of the actual heat transferred from the fin to the heat that would be transferred if the entire fin were at its root temperature:*

$$\text{Fin efficiency} = \frac{\text{actual heat transferred by fin}}{\substack{\text{heat transferred if entire fin area} \\ \text{were at the base temperature}}}$$

or

$$\eta_{\text{fin}} = \frac{\int h\,(T - T_\infty)\,dA}{\int h\,(T_w - T_\infty)\,dA} \tag{12-44}$$

where T_∞ is the temperature of the surrounding environment and T_w is the temperature at the fin root.

Thus the actual heat transfer from a fin is computed by

$$q = \eta_{fin} \int h\,(T_w - T_\infty)\,dA$$

$$= \eta_{fin}\,hA\,(T_w - T_\infty) \tag{12-45}$$

Determining the fin efficiency may be a difficult task if the fin does not have a constant cross-sectional area, so various charts are available for these situations.

In concluding let us review the assumptions made in this analysis:

1. The heat flow is steady; that is, the temperature at any point in the fin does not vary with time.

2. The fin material is homogeneous and the thermal conductivity is constant and uniform.

3. The coefficient of heat transfer for convection (h) is constant and uniform over the entire surface of the fin.

4. The temperature of the surrounding fluid is constant and uniform.

5. There are no temperature gradients within the fin other than along its length. Hence the fin's width and length are great when compared with its height.

6. There is no bond resistance to the flow of heat at the base of the fin.

7. The temperature at the base of the fin is uniform and constant.

EXAMPLE 12-12

The county engineer wishes to increase the heat transfer rate in the town's water tower by adding fins to only one side. The question is: On which side should the engineer add fins to increase the heat transfer? Assume that h (air side) = 2 W/(m² · C), η (air side) = 0.92, h (water side) = 45 W/(m² · C), and η (water side) = 0.39 (see the sketch). Notice that triangular fins are to be used for either option.

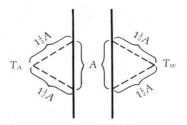

SOLUTION: As an approximation, assume that the problem can be modeled by a plane wall separating the water and air. Then carry out and approximate heat transfer analysis of the no-fin situation (as the tower now stands) and two analyses of the fin situations. In all cases, assume that the resistance to heat transfer of the steel wall is zero. Without fins:

$$q = \frac{A(T_A - T_w)}{1/h_A + 0 + 1/h_w} = 1.9149(T_A - T_w)A$$

Recall that with a fin

$$q_{fin} = hA\eta(T_w - T_\infty)$$

Hence

$$R_{fin} = \frac{1}{hA\eta}$$

With a fin on the air side only:

$$q = \frac{A(T_A - T_w)}{1/3\, h_A\eta_A + 0 + 1/h_w} = 4.9169(T_A - T_w)A$$

With a fin on water side only:

$$q = \frac{A(T_A - T_w)}{1/h_A + 0 + 1/3\, h_w\eta_w} = 1.9268(T_A - T_w)A$$

Thus the best heat transfer conditions occur when the fins are on the air side. Note that fins on both sides increase the heat transfer to only $4.9962(T_A - T_w)A$. This is about a 1.6% increase over the fin on the air side value.

Heat Exchangers

A **heat exchanger** *is a device that permits the transfer of heat from a warm fluid to a cooler fluid through an intermediate surface without mixing of the fluids.* In electronic equipment, a heat exchanger is often used to remove heat from a circulating liquid coolant that is applied directly to the components. Heat exchangers may be used to regulate temperature in a package containing electronic equipment or in a series of electronic packages. Simple heat exchangers are illustrated in Figures 12-10 and 12-11.

When heat is exchanged between two fluids flowing continuously through a heat exchanger, the local temperature difference ΔT varies along the flow path. To account for this variation, the heat transfer is calculated with the familiar rate equation:

$$q = UA\Delta T_m \tag{12-46}$$

where U is the **overall coefficient of heat transfer** from fluid to fluid, A is an area associated with the coefficient U, and ΔT_m is a mean temperature difference.

For parallel flow or counterflow heat exchangers and for any ex-

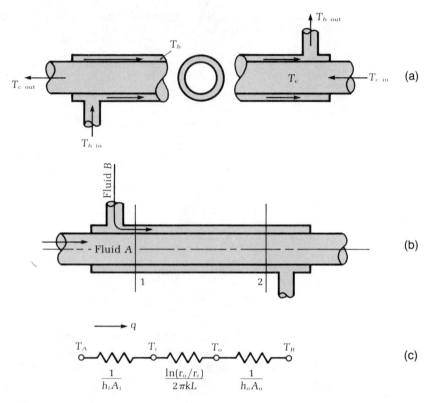

Figure 12-10. Double-pipe heat exchanger: (a) Simple counterflow exhanger; (b) parallel flow exchanger; (c) thermal resistance network for (a) and (b).

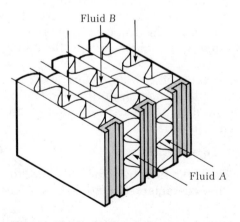

Figure 12-11. Plate-type heat exchanger.

changer in which the temperature of one of the fluids is substantially constant, this mean temperature difference is called the **log mean temperature difference (LMTD)** and is defined as

$$\Delta T_m = \frac{\Delta T_1 - \Delta T_2}{\ln (\Delta T_1/\Delta T_2)} \qquad (12\text{-}47)$$

where ΔT_1 and ΔT_2 are the differences in hot and cold fluids at the inlet and outlet, respectively.

The overall heat transfer coefficient is based on either the inside or outside surface areas of the inner tube. Thus for

$$U_i = \frac{1}{1/h_i + A_i \ln(r_o/r_i)/2\pi kL + (A_i/A_o)(1/h_o)}$$

$$\text{use } A_i \qquad (12\text{-}48)$$

and for
$$U_o = \frac{1}{(A_o/A_i)(1/h_i) + A_o \ln (r_o/r_i)/2\pi kL + 1/h_o}$$

$$\text{use } A_o \qquad (12\text{-}49)$$

The heat transfer surfaces for a heat exchanger, both inside and out, may become dirty, corroded, or coated with deposits from the fluids in the systems. From the heat transfer standpoint, these coatings are additional resistances to heat exchange and produce decreased performance. This effect is usually represented by **fouling factors** or **fouling resistances,** which must be included in the overall heat transfer coefficient. Thus

$$U_o = \frac{1}{1/h_o + R_o + R_i A_o/A_i + A_o \ln (r_o/r_i)/2\pi kL + A_o/h_i A_i} \qquad (12\text{-}50)$$

where U_o = design overall coefficient of heat transfer in Btu/hr-ft^2-F $(W/m^2 \cdot C)$ based on unit area of the outside tube surface

h_o = average unit-surface conductance of fluid on outside of tubing in Btu/hr-ft^2-F $(W/m^2 \cdot C)$

h_i = average unit-surface conductance of fluid inside tubing in Btu/hr-ft^2-F $(W/m^2 \cdot C)$

R_o = unit fouling resistance on outside of tubing in hr-ft^2-F/ Btu $(m^2 \cdot C/W)$

R_i = unit fouling resistance on inside of tubing in hr-ft^2-F/ Btu $(m^2 \cdot C/W)$

A_o/A_i = ratio of outside tube surface to inside tube surface

Typical values of this resistance are given for various fluids in Table 12-5. For preliminary engineering estimates of heat exchanger sizes and performance, it is often useful to use an order of magnitude value of the overall coefficient. Typical values of overall heat transfer coefficients for preliminary work are listed as Table 12-6.

Table 12-5 Fouling Factors

Fluid	Fouling Factor hr-ft^2-F/Btu	(m$^2 \cdot$ C)/W
Seawater below 125 F	0.0005	0.00009
City water or well water	0.001	0.0002
Treated boiler feedwater above 125 F	0.001	0.0002
River water	0.003	0.0006
Fuel oil	0.005	0.0009
Alcohol vapors	0.0005	0.00009
Steam (non-oil-bearing)	0.0005	0.00009
Industrial air	0.002	0.0004
Refrigerating liquid	0.001	0.0002

Table 12-6 Approximate Values for Overall Heat Transfer Coefficients

Application	Overall Coefficient Btu/hr-ft^2-F	W/(m$^2 \cdot$ C)
Steam to water		
Instantaneous heater	400–600	2267–3400
Storage tank heater	150–300	850–1700
Steam to oil		
Heavy fuel	10–30	57–170
Light fuel	30–70	170–397
Steam to gases	5–50	28–280
Water to compressed air	10–30	57–170
Water to water	150–275	850–1558
Water to lubricating oil	20–60	113–340
Water to condensing refrigerant-12	80–150	453–850
Water to condensing ammonia	150–250	850–1417
Water to organic solvents, alcohol	50–150	280–850
Water to boiling refrigerant-12	50–150	280–850
Water to gasoline	60–90	340–510
Water to brine	100–200	567–1133
Light organics to light organics	40–75	227–425
Medium organics to medium organics	20–60	113–340
Heavy organics to heavy organics	10–40	57–227
Heavy organics to light organics	10–60	57–340

EXAMPLE 12-13

Suppose that 20,000 lbm/hr of water is cooled from 190 to 150 F while passing through an exchanger. Assume that 30,000 lbm/hr of water entering at 90 F flows through the outer side of the heat exchanger. The overall heat transfer coefficient for the exchanger (U) is 300 Btu/hr-ft^2-F. Determine the areas required for parallel flow and for counterflow.

SOLUTION: $\quad q = 20{,}000\,(190 - 150) = 30{,}000\,(T_0 - 90) = 800{,}000$

$$T_0 = 116.7 \text{ F}$$

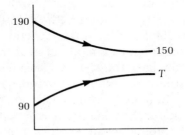

Parallel flow:

$$\Delta T_m = \frac{(190 - 90) - (150 - 116.7)}{\ln(100/33.3)} = 60.6 \text{ F}$$

$$A = \frac{q}{U\Delta T_m} = \frac{800{,}000}{(300)\,(60.6)} = 44 \text{ ft}^2$$

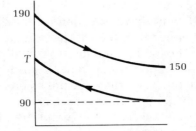

Counterflow:

$$\Delta T_m = \frac{(190 - 116.7) - (150 - 90)}{\ln(73.3/60)} = 66.6 \text{ F}$$

$$A = \frac{800{,}000}{(300)\,(66.6)} = 40 \text{ ft}^2$$

12–3 Convective Heat Transfer

Heat transfer by convection may be modeled in several steps. Assume a surface whose temperature is above that of a surrounding fluid. First, heat is conducted from the surface to immediately adjacent particles of fluid. This transfer will increase the temperature (and thus the internal energy) of these particles. The more energetic fluid particles will move to a region of lower temperature in the fluid and will mix with, and transfer energy to, other fluid particles. Notice that the energy is stored in the fluid particles and is transferred as a result of mass motion. This mode, therefore, does not depend merely on a temperature difference and does not strictly conform to the definition of heat transfer. The net heat transfer effect occurs in the direction of the temperature gradient and is thus classified as a mode of heat transfer.

Convection heat transfer is classified according to the mode of motivation: **free convection** or **forced convection.** When the mixing motion is due to density differences caused by temperature gradients, we speak of natural or free convection. When the mixing motion is caused by an external agent, such as a pump or blower, the process is called forced convection.

The effectiveness of heat transfer by convection depends largely on the mixing motion of the fluid. The effects of these mixing motions are included in the convection coefficient h of Newton's law of cooling:

$$\frac{q}{A} = h(T_h - T_c)$$

Hence the problem reduces to finding a convection coefficient h for the situation at hand. In some cases, generalized equations for the convection coefficients have been derived from fundamental principles. As far as we are concerned, the vast majority of cases will be obtained from correlations of experimental data. Normally, these correlations employ dimensionless numbers that have been determined by various means (dimensional analysis, analogy, and so on). Appendix B-9 lists some of the important dimensionless numbers.

In almost every case the properties for these numbers depend significantly on temperature. In the case of density ρ (and kinematic viscosity ν) there is also a pressure effect for gases. The temperature dependences mean that there may be a significant variation in the quantities through the region of gas next to the surface (the **thermal boundary layer**). The accuracy of the results of these theoretical relations and the dimensionless experimental correlations depends on the temperature chosen for evaluating the properties. In most cases, one of two average temperatures is used: bulk fluid temperature or mean film temperature. **Bulk fluid temperature** T_B is usually applied in the case of forced convection inside a closed duct or pipe. It is the energy-weighted average fluid temperature at a cross section—the temperature that would result if the fluid at a cross section were to be thoroughly mixed. **Mean film temperature** T_f is the straight arithmetic average of the surface temperature T_w and (in the case of flow external to the body) the free stream or undisturbed fluid temperature T_∞. Thus

$$T_f = \frac{T_w + T_\infty}{2}$$

In the case of internal flow, a mean film temperature that is the average of the surface temperature and the bulk temperature is sometimes used in place of the bulk temperature.

Table 12-7 outlines the general procedures for analyzing heat transfer by convection. Table 12-8 presents some order of magnitude values of h. Notice that Table 12-7 presents some new nondimensional numbers (Pr, Nu, Gr, Re). These numbers are discussed later in the chapter, but the table is presented at this point to make you aware of their influence on convective heat transfer.

Free Convection

Heat transfer involving motion in a fluid due to a difference in density and the action of gravity is called **free** or **natural convection**. Heat transfer coefficients for natural convection are generally much lower in

Table 12-7 Basic Procedure for Convection Calculations

Steps in determining q and ΔP:

1. Determine whether convection is free or forced.

2. Consider the geometry: internal or external flow.

3. Determine whether the flow is laminar or turbulent: Re for forced; Gr for free.

4. Evaluate Pr and Pe or Ra.

5. Using Re (or Gr) and Pr (or Pe or Ra) and system geometry, compute h (or Nu) and f from the proper predictive equation using fluid properties evaluated at proper temperature.

6. $q = hA(T_w - T_\infty)$ and $\Delta P = f(L/D_H)(\rho V^2/2g_c)$.

If the temperature difference between fluid and wall is not constant, use the log mean temperature difference:

$$\text{LMTD} = \frac{\Delta T_a - \Delta T_b}{\ln(\Delta T_a/\Delta T_b)} \qquad \text{where } a \text{ and } b \text{ are the two ends}$$

$$\Delta T_a = T_w - T_{\text{fluid}} \qquad \text{at end } a$$

$$\Delta T_b = T_w - T_{\text{fluid}} \qquad \text{at end } b$$

Table 12-8   Order of Magnitude of Convective Coefficients

Conditions	h, Btu/hr-ft^2-F	h, W/(m$^2 \cdot$ C)
Free convection		
Air	0.1–3	0.568–17.05
Water	1–10	5.684–56.85
Forced convection		
Air (superheated steam)	5–50	28.4–284
Water	50–2000	284 –11,370
Oil	10–300	56.8–1705
Boiling		
Water	500–10,000	2842–56,849
Refrigerant	200–1000	1137–5685
Condensation		
Steam	1000–20,000	5685–113,699
Refrigerant	200–2000	1137–11,370

magnitude than for forced convection. In fact, it is important not to ignore radiation in calculating the total heat loss or gain from a body. Radiant heat transfer may be of the same order of magnitude as natural convection, even at room temperatures, as evidenced by the fact that wall temperatures in a room can affect human comfort.

Natural or free convection is important in a wide variety of heating and cooling equipment. Examples are the evaporator and condenser of household refrigerators, baseboard radiators and convectors for space heating, and cooling panels for air conditioning. Natural convection is also involved in the heat losses or gains to the casings of equipment and in the cooling of electronic components. Let us consider in more detail the transfer by natural convection between a cold fluid and a hot surface. The physical description presented at the beginning of this chapter indicates that the heat transfer is influenced by three effects: (1) gravitational force due to thermal expansion, (2) viscous drag, and (3) thermal diffusion. Consequently, heat transfer may be expected to depend on the gravitational acceleration g, the coefficient of thermal expansion β and the kinematic viscosity ν ($= \mu/\rho$). The variables can be expressed in terms of a dimensionless number called the **Grashof number:**

$$\text{Gr} = \frac{\rho^2 g \beta (T - T_\infty) L^3}{\mu^2} \tag{12-51}$$

This dimensionless modulus represents the ratio of buoyant to viscous forces. Table 12-9 presents an empirical relation for predicting the convective coefficient for free convection. Note the direct dependence of h on the dimensionless convective heat transfer coefficient, called the **Nusselt number,** Nu. Equation (12-52) shows that the Nusselt number is a function of the Grashof number and the Prandtl number

Table 12-9 Free-Convection Heat Transfer Coefficients

General Relationship

$$\frac{hL}{k_f} = \mathrm{Nu}_f = C(\mathrm{Gr}_f\,\mathrm{Pr}_f)^n \tag{12-52}$$

Geometry	$\mathrm{Gr}_f\,\mathrm{Pr}_f$	C	n
Vertical planes and cylinders	10^4–10^9	0.59	1/4
	10^9–10^{13}	0.10	1/3
Horizontal cylinders	0–10^{-5}	0.40	0
	10^4–10^9	0.53	1/4
	10^9–10^{12}	0.13	1/3
Upper surface of heated plates or lower surface of cooled plates	2×10^4–8×10^6	0.54	1/4
	8×10^6–10^{11}	0.15	1/3
Lower surface of heated plates or upper surface of cooled plates	10^5–10^{11}	0.58	1/5
Wires (horizontal and vertical)	10^{-7}–1	1	0.1
Plates, short cylinders, spheres, blocks	10^4–10^9	0.60	1/4
Confined electronic components (relays, transformers, resistors, tubes)	$< 10^9$	1.45	0.23

Characteristic Dimension L

Vertical plates or pipes	L = height
Horizontal plates	L = length
Horizontal pipes	L = diameter
Horizontal or vertical wires	L = diameter
Horizontal flat plates	L = length of side for a square
	L = mean of the two dimensions for rectangle
	$L = 0.9 \times$ diameter for a circular disk
	$L = A/P$ for others
Spheres	L = radius
Short cylinders and blocks	$1/L = 1/L_{\mathrm{horiz}} + 1/L_{\mathrm{vert}}$
Miniature and subminiature tubes	L = tube height
Relays and transformers	L = vertical height
Horizontal resistors	$1/L = 1/\mathrm{diameter} + 1/\mathrm{length}$

$$\mathrm{Pr} = \frac{c_p\mu}{k}$$

and thus depends on the fluid properties, the temperature difference between the surface and the fluid ΔT, and the characteristic length of the surface L. The **Prandtl number** *is a measure of the momentum dissipation as compared to rate of heat diffusion in a fluid.* The constant C and the exponent n are experimentally correlated numbers that depend on the physical configuration and the nature of the flow. As may be seen from Table 12-9, the entire range of natural convection cannot be represented by a single value of the exponent n. The process must be divided into at least three regions: (1) the turbulent natural convection, (2) the laminar natural convection, and (3) a region having GrPr less than for laminar natural convection, for which the exponent n gradually diminishes from $\frac{1}{4}$ to lower values. For wires, the GrPr is likely to be very small so that the exponent n is only 0.1.

The first step in calculating the natural convection heat transfer coefficient is to calculate the quantity GrPr to determine whether the flow is laminar or turbulent. Then the appropriate values of C and n from Table 12-9 may be applied. Care should be taken that the correct characteristic length is used as indicated in the table. Turbulence occurs when the length or the temperature difference is large. Since the length of a pipe is generally larger than its diameter, the heat transfer coefficient is greater for vertical pipes than for horizontal pipes.

Convection from horizontal plates facing downward when heated, or upward when cooled, is a special case. Since the hot air is above the colder air, theoretically there is no reason for convection. Nevertheless, some convection is caused by secondary influences such as temperature differences on the edges of the plate. As a rough approximation, a coefficient of somewhat less than half the coefficient for a horizontal plate facing upward has been recommended.

Natural or free convection can affect the heat transfer coefficient in the presence of weak forced convection. As the forced-convection effect increases from zero, one passes through the mixed convection (or superimposed forced-on-free-convection) regime and then into the pure forced-convection regime. This may be seen in Figure 12-12, as the **Reynolds number,** Re, increases. This number, defined as

$$Re = \frac{\rho V x}{\mu}$$

represents a comparison of inertial forces and the viscous forces. Figure 12-12 indicates the approximate limits of these regimes for horizontal tubes.

Experimental results for free convection in enclosures are not always in agreement, but the general form of empirical equation normally used is

$$\frac{k_e}{k} = C\,(Gr_\delta Pr)^n \left(\frac{L}{\delta}\right)^m \tag{12-53}$$

Table 12-10 lists values of the constants C, n, and m for a number of physical circumstances. These values may be used for design purposes in the absence of specific data for the geometry or fluid being studied.

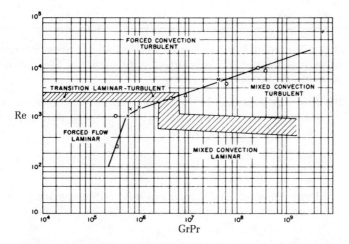

Figure 12-12. Regime of free, forced, and mixed convection for flow through horizontal tubes. *(Reprinted with permission from the 1977 Fundamentals Volume, ASHRAE Handbook and Product Directory.)*

Table 12-10 Free Convection in Enclosed spaces

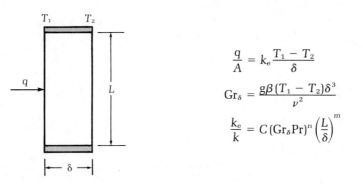

$$\frac{q}{A} = k_e \frac{T_1 - T_2}{\delta}$$

$$Gr_\delta = \frac{g\beta (T_1 - T_2)\delta^3}{\nu^2}$$

$$\frac{k_e}{k} = C (Gr_\delta Pr)^n \left(\frac{L}{\delta}\right)^m$$

Geometry	$Gr_\delta Pr$	C	n	m
Gas				
Vertical plate (isothermal)	< 2000	$k_e/k = 1.0$		
	6000–200,000	0.197	1/4	−1/9
	200,000–1.1 × 10⁷	0.073	1/3	−1/9
Horizontal plate (isothermal)	1700–7000	0.059	0.4	0
heated from below	7000–3.2 × 10⁵	0.212	1/4	0
	> 3.2 × 10⁵	0.061	1/3	0
Liquid				
Horizontal plate (isothermal)	1700–6000	0.012	0.6	0
heated from below	6000–37,000	0.375	0.2	0
	37,000–10⁸	0.13	0.3	0
	> 10⁸	0.057	1/3	0

EXAMPLE 12-14

A pipe, 15 cm in diameter, is positioned horizontally in a room whose temperature is maintained at 25 C. The surface temperature of the pipe is 275 C. What is the heat loss per length of pipe?

SOLUTION: From the statement of the problem we must assume that the heat loss is by free convection. It will be external flow and we must evaluate the Grashof number:

$$Gr = \frac{\rho^2 g \beta (T - T_\infty) L^3}{\mu^2}$$

From Appendix A-3-2 for air (notice that temperature conversion is necessary), for the temperatures of 150 C we get

$$Pr \doteq 0.71 \qquad \beta \doteq \frac{1}{T} = \frac{1}{423 \text{ K}} \qquad \nu = 2.84 \, (10^{-5}) \text{ m}^2/\text{s} = \frac{\mu}{\rho}$$

Hence

$$Gr = \frac{9.81 \text{ m/s}^2 \, (275 - 25) \text{ K} \, (0.15 \text{ m})^3}{423 \text{ K} \, (2.84 \text{ m}^2/\text{s})^2 \, 10^{-10}} = 2.426 \, (10^7)$$

and

$$GrPr \doteq 1.72 \, (10^7)$$

From Table 12-9:

$$h = Nu_f \left(\frac{k_f}{L} \right) = 0.53 \, \frac{k_f}{L} \, (GrPr)^{1/4}$$

$$\doteq 24.9 \text{ W/(m}^2 \cdot \text{K)}$$

Thus

$$q = hA \, (T_w - T_x) = 586.7 \text{ W}$$

EXAMPLE 12-15

A 1-in.-OD electrical transmission line carrying 5000 amp and having a resistance of 10^{-6} ohms per foot of length is placed horizontally in still air at 95 F. Determine the surface temperature of the line if the GrPr product has been determined to be 1.33×10^5.

SOLUTION: $\quad q = I^2 R = hA \, (T_s - T_x) = (5000)^2 \, (10^{-6}) \, 3.413 = 85.3 \text{ Btu/hr}$

$$h = \frac{k}{d} C \, (GrPr)^n \qquad C = 0.53 \qquad n = \tfrac{1}{4}$$

$$k \approx 0.018 \text{ Btu/hr-ft-F} \qquad d = \tfrac{1}{12} \text{ in.}$$

$$h = 0.018\,(12)\,(0.53)\,(1.33 \times 10^5)^{1/4} = 2.19 \text{ Btu/hr-ft}^2\text{-F}$$

$$85.3 = 2.19 \left[\frac{\pi\,(1)}{12}\right](T_s - 95)$$

$$T_s = 244 \text{ F}$$

Forced Convection

When a fluid flows over a flat plate, **a boundary layer** forms adjacent to the plate. As indicated schematically in Figure 12-13, the velocity of the fluid at the surface of the plate is zero and increases to its maximum free-stream value just at the edge of the boundary layer. It is important that you understand the mechanism of the forming of the boundary layer since the temperature change is concentrated in this region. Starting at the leading edge of the plate the boundary layer thickness is zero and, theoretically, the heat transfer coefficient is infinite. The flow within the boundary layer immediately downstream from the leading edge is laminar. As the flow proceeds along the plate, the laminar boundary layer increases in thickness until a critical value is reached. Then turbulent eddies develop within the bulk of the boundary layer. The exception is a thin laminar sublayer adjacent to the plate. The boundary layer beyond this critical region is referred to as a **turbulent boundary layer.** The region between the breakdown of the laminar boundary layer and the complete establishment of the turbulent boundary layer is called the **transition region.** Since the turbulent eddies greatly enhance the transport of heat into the main stream, the heat transfer coefficient begins to increase rapidly through the transition

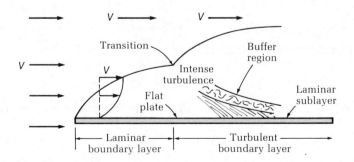

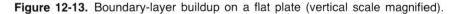

Figure 12-13. Boundary-layer buildup on a flat plate (vertical scale magnified).

region. For a flat plate with a smooth leading edge, the turbulent boundary layer starts at Reynolds numbers of about **500,000**, based on distance from the leading edge. For blunt-edged plates, it can begin at much smaller Reynolds numbers.

For flow in long tubes or channels of small hydraulic diameter (D_e = 4 · cross-sectional area for flow/total wetted perimeter), the laminar boundary layers on each wall increase until they meet, provided the velocity is sufficiently low. Beyond this point the velocity distribution does not change and no transition to turbulent flow takes place. The condition of no change in the velocity is referred to as **fully developed laminar flow.** At the other extreme—that is, for tubes of large diameter or at higher velocities—transition to turbulence takes place and fully developed turbulent flow is established as shown in Figure 12-14. Therefore, the length dimension that determines the critical Reynolds number is the hydraulic diameter of the channel. For smooth circular tubes, flow is laminar for Reynolds number below **2300** (approximate) and turbulent above.

The characteristic length L is the diameter of the tube, outside or inside, or the length of the plane plate. For other shapes, the hydraulic diameter D_e is used. This reduces to twice the distance between surfaces in the case of parallel plates or an annulus. Figure 12-15 gives values of the friction factor f for use with Equation (12-62).

The forced-convection correlations are presented as Tables 12-11 and 12-12. Note that the heat transfer is determined by the flow conditions—that is, the Reynolds number, the Prandtl number, and the fluid properties. The temperature indicated in parentheses in Tables 12-11 and 12-12 is the temperature to use in evaluating the physical properties.

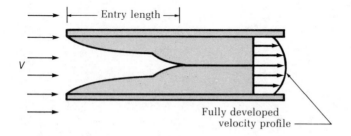
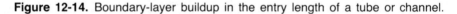

Figure 12-14. Boundary-layer buildup in the entry length of a tube or channel.

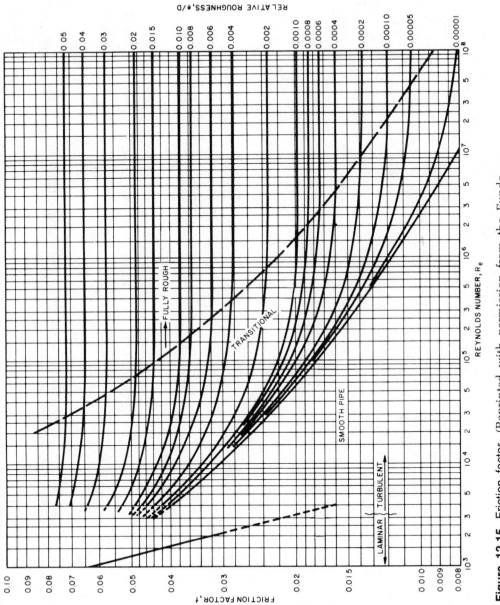

Figure 12-15. Friction factor. (Reprinted with permission from the *Funda-mentals* Volume, ASHRAE Handbook & Product Directory.)

Table 12-11 Forced-Convection Coefficients—External Flow

Flat plate (Critical Reynolds number = 500,000)

Laminar	Local:	$Nu_x = 0.332 Pr^{1/3} Re_x^{1/2}$	(12-54)
(T_f)	Average:	$\bar{h} = 2h_{x=L}$	
Turbulent	Local:	$St_x Pr^{2/3} = 0.0296 Re_x^{-1/5}$ $5 \times 10^5 < Re_x < 10^7$	(12-55)
(T_f)		$St_x Pr^{2/3} = 0.185(\log Re_x)^{-2.584}$ $10^7 < Re_x < 10^9$	(12-56)
		$St_x = h/\rho V c_p$	
Laminar and turbulent	Average:	$\overline{Nu_L} = \dfrac{hL}{k} = Pr^{1/3}(0.037\, Re_L^{0.8} - 850)$	(12-57)
(T_f)			

Cross flow

Cylinders

$$\frac{hd}{k_f} = C(Re_f)^n\, Pr^{1/3} \qquad (12\text{-}58)$$

Gases
(T_f)

Re	C	n
0.4–4	0.989	0.330
4–40	0.911	0.385
40–4000	0.683	0.466
4000–40,000	0.193	0.618
40,000–400,000	0.0266	0.805

Liquids	$Nu_f = (0.35 + 0.56\, Re_f^{0.52})\, Pr_f^{0.3}$ $10^{-1} < Re_f < 10^5$	(12-59)
(T_f)		
Spheres	$Nu = 2 + (0.4\, Re_d^{1/2} + 0.06\, Re_d^{2/3})\, Pr^{0.4} (\mu_\infty/\mu_w)^{1/4}$	(12-60)
(T_∞)	$3.5 < Re_d < 8 \times 10^4$ and $0.7 < Pr < 380$	

Table 12-12 Forced-Convection Coefficients—Internal Flow

Laminar (Critical Reynolds Number = 2300)

Smooth tubes (constant wall temperature)

$(T_{b,av})$

$$\mathrm{Nu}_d = 3.66 + \frac{0.0668\,(d/L)\,\mathrm{Re}_d\mathrm{Pr}}{1 + 0.04\left[(d/L)\,\mathrm{Re}_d\mathrm{Pr}\right]^{2/3}}$$
(12-61)

Rough tubes

$(T_{b,av} \text{ and } T_f)$

$$\mathrm{St}_b\mathrm{Pr}_f^{2/3} = \frac{f}{8}$$
(12-62)

Turbulent

Entrance region

$(T_{b,av})$

$$\mathrm{Nu}_d = 0.036\ \mathrm{Re}_d^{0.8}\,\mathrm{Pr}^{1/3}\left(\frac{d}{L}\right)^{0.055} \qquad \text{for } 10 < \frac{L}{d} < 400$$
(12-63)

Smooth tubes and ducts

1. $T_w - T_\infty \leq 100$ F for gases, except air
   ≤ 10 F for liquids

   (Use for air even if $T_w - T_\infty > 100$ F;
   but if > 100 F, use properties at T_f.)
2. $L/D > 60$
3. $2300 < \mathrm{Re} < 120{,}000$
4. $0.6 < \mathrm{Pr} < 100$
5. $(T_{b,av})$

$$\mathrm{Nu}_d = 0.023\ \mathrm{Re}_d^{0.8}\,\mathrm{Pr}^n \qquad n = \begin{cases} 0.4 \text{ for heating of the fluid} \\ 0.3 \text{ for cooling of the fluid} \end{cases}$$
(12-64)

1. $T_w - T_\infty > 100$ F for gases, except air
   > 10 F for liquids
2. $L/D > 60$
3. $2300 < \mathrm{Re} < 120{,}000$
4. $0.6 < \mathrm{Pr}$
5. $(T_{b,av} \text{ and } T_w)$

$$\mathrm{Nu}_d = 0.027\mathrm{Re}_d^{0.8}\,\mathrm{Pr}^{1/3}\left(\frac{\mu}{\mu_w}\right)^{0.14}$$
(12-65)

(table continues on next page)

Table 12-12 Forced-Convection Coefficients—Internal Flow (continued)

1. Constant heat flux
2. $L/D > 60$
3. $3.6 \times 10^3 < \mathrm{Re} < 9.05 \times 10^5$
4. $10^2 < \mathrm{Pe} < 10^4$
5. $\mathrm{Pr} < 0.1$
6. $(T_{b,av})$

$$\mathrm{Nu} = 482 + 0.0185 \mathrm{Pe}^{0.827} \qquad (12\text{-}66)$$

$$\mathrm{Pe} = DV\rho c_p / k$$

1. Constant wall temperature
2. $L/D > 60$
3. $\mathrm{Pe} > 100$
4. $\mathrm{Pr} < 0.1$
5. $(T_{b,av})$

$$\mathrm{Nu}_d = 5.0 + 0.025 \, (\mathrm{Re}_d \, \mathrm{Pr}_f)^{0.8} \qquad (12\text{-}67)$$

Rough tubes and ducts

$(T_{b,av}$ and $T_f)$

$$\mathrm{St}_b \, \mathrm{Pr}_f^{2/3} = \frac{f}{8} \qquad (12\text{-}62)$$

EXAMPLE 12-16

Water at the rate of 500 lbm/hr is to be cooled from 85 to 34 F while flowing through a $\frac{3}{4}$-inch, 16 BWG heat exchanger tube. The tube is maintained at 33 F by evaporating ammonia on the outside. Determine the length of tube required (in ft) and the corresponding pressure drop (psi). Use LMTD. (*Note:* For a $\frac{3}{4}$-in., 16 BWG tube: inside cross-sectional area = 0.0021 ft²; inside surface area per linear foot = 0.162 ft²; inside diameter = 0.62 in. For water at 60 F: c_p = 1.00 Btu/lbm-R; ρ = 62.34 lbm/ft³; μ = 2.71 lbm/ft-hr; k = 0.344 Btu/hr-ft²-F; Pr = 7.88.)

SOLUTION: $\dot{m} = \rho V A_c$

$$V = \frac{500}{62.34 \times 0.0021} = 3820 \text{ ft/hr}$$

$$\text{Re} = \frac{dV\rho}{\mu}$$

$$= \frac{0.62 \times 3820 \times 62.34}{12 \times 2.71} = 4540 \qquad \text{(turbulent since Re > 2300)}$$

$$\text{Nu} = \frac{hD}{k} = 0.023 \, \text{Re}^{0.8} \, \text{Pr}^{0.3} \qquad \text{(Table 12-12)}$$

$$= 0.023(4540)^{0.8} (7.88)^{0.3} = 35.9$$

$$h = \frac{0.344(35.9)(12)}{0.62} = 239 \text{ Btu/hr-ft}^2\text{-F}$$

$$Q = \dot{m}c_p(T_o - T_i) = 500(1.00)(85 - 34) = 25{,}500 \text{ Btu/hr}$$

$$= ha\Delta T_m$$

$$\Delta T_m = \frac{\Delta T_1 - \Delta T_2}{\ln(\Delta T_1/\Delta T_2)} = \frac{(85 - 33) - (34 - 33)}{\ln(52/1)}$$

$$= 12.9 \text{ F}$$

$$25{,}500 = 239(0.162)L(12.9)$$

$$L = 51 \text{ ft} \qquad \text{(Check: } L/D > 60)$$

$$\Delta p = f\frac{L}{D}\frac{\rho V^2}{2g_c} = 0.038 \frac{51 \times 12}{0.62} \times \frac{62.4(3800/3600)^2}{2 \times 32.2 \times 144}$$

$$= 0.284 \text{ psi}$$

EXAMPLE 12-17

Determine the heat transfer coefficient for water flowing in a 2.5-cm-diameter pipe at a rate of 1.5 kg/s. Assume that the bulk temperature is 45 C.

SOLUTION: We must assume in this case that the flow is forced convection. It is internal flow and we must evaluate the Reynolds number:

$$Re = \frac{\rho V d}{\mu}$$

Again from Appendix A-3-3 for water (temperature conversion is necessary) for $T = 318$ K.

Recall from continuity that

$$\dot{m} = \rho V A \quad \text{or} \quad V = \frac{\dot{m}}{\rho A}$$

Thus $V = 3.0$ m/s since

$$v \approx 0.001 \text{ m}^3/\text{kg}$$

and $Re \doteq 117{,}000$ since

$$\mu = 651(10^{-6}) \text{ kg/(m} \cdot \text{s)}$$

Now $117{,}000 > 2300$, implying that the flow is turbulent. So from Table 12-12 we get

$$Nu = 0.023 Re^{0.8} \ Pr^{0.4} \qquad \text{(cooling)}$$

Again, from the appendix we get

$$Pr \approx 4.3 \quad \text{and} \quad k \approx 6.32(10^{-4})\text{W/(m} \cdot \text{C)}$$

Thus $Nu = 495$ and

$$h = \frac{Nuk}{d} = 12.5 \text{ kW/(m}^2 \cdot \text{K)}$$

EXAMPLE 12-18

The flat roof of a building is 50 ft long in the wind direction. Find the average convective heat transfer coefficient, h, for a wind velocity of 15 mi/hr with the air at 50 F, 14.7 psia, $\rho = 0.078$ lb/ft^3.

SOLUTION: At 50 F:

$$\mu = 0.0427 \text{ lbm/ft/hr}$$

$$k = 0.0143 \text{ Btu/hr-ft-F}$$

$$c_p = 0.240 \text{ Btu/lbm-F}$$

$$Pr = 0.712$$

$$\rho = \frac{p}{RT} = \frac{(14.7)(144)}{(53.3)(510)} = 0.078$$

$$Re_{50'} = \frac{xV\rho}{\mu} = \frac{(50)(15)(5280)(0.078)}{0.0427}$$

$$= 7{,}230{,}000 \qquad \text{(turbulent since } Re > 500{,}000)$$

$$h = \frac{k}{L} \text{Pr}^{1/3} (0.037 \text{Re}_L^{0.8} - 850) \qquad \text{(Table 12-11)}$$

$$= \frac{0.0143}{50} (0.712)^{1/3} [0.037 (7.23 \times 10^6)^{0.8} - 850]$$

$$= 2.61 \text{ Btu/hr-ft}^2\text{-F}$$

EXAMPLE 12-19

A bridge deicing system uses hot air passed through tubes embedded in the concrete bridge floor. The tubes operate at 65 F when the mean temperature of the hot air is 300 F. Determine the heat transfer to the concrete per foot of 3-in.-ID tube if the mean air velocity is 36 ft/sec. (*Note:* Air properties: density, 0.050 lb/ft³; specific heat, 0.24 Btu/lbm-F; thermal conductivity, 0.018 Btu/hr-ft-F; viscosity, 1.5×10^{-5} lbm/ft-sec; Prandtl no., 0.71.)

SOLUTION:

$$\text{Re} = \frac{DV\rho}{\mu} = \frac{(3/12)(36)(0.050)}{1.5 \times 10^{-5}} = 30,000 \qquad \begin{array}{l}\text{(turbulent since}\\ \text{Re} > 2300)\end{array}$$

$$h = 0.023 \frac{k}{D} \text{Re}^{0.8} \text{Pr}^{0.3} = 0.023 \frac{0.018}{3/12} (30,000)^{0.8} (0.71)^{0.3} = 5.70$$

$$q = hA(T_b - T_w) = 5.70 \frac{3\pi}{12} (300 - 65) = 1052.0 \text{ Btu/hr-ft}$$

EXAMPLE 12-20

Air flows through a 12 in. × 18 in. air conditioning duct at average conditions of $T = 50$ F, $p = 14.7$ psia, $\rho = 0.078$ lbm/ft³, $V = 600$ ft/min, Re = 73,400. Determine: (1) the convective heat transfer coefficient h and (2) the pressure drop for a 10-ft length of duct.

SOLUTION: At 50 F:

$$\mu = 0.0459 \text{ lbm/ft-hr}$$

$$k = 0.0143 \text{ Btu/hr-ft R}$$

$$c_p = 0.240 \text{ Btu/lbm F}$$

$$\text{Pr} = 0.712 \qquad \text{(Appendix)}$$

$$\text{Re} = \frac{D_H V \rho}{\mu}$$

Since

$$D_H = \frac{4A}{P} = \frac{4(1)(1.5)}{1 + 1.5 + 1 + 1.5} = 1.2 \text{ ft}$$

we get

$$Re = \frac{(1.2)(600)(60)(0.078)}{0.0459} = 73,400 \qquad \text{(turbulent since } Re > 2300\text{)}$$

1. $h = 0.023\dfrac{k}{D}Re^{0.8}\,Pr^{0.4}$ \qquad (Table 12-12)

$$= 0.023\frac{0.0143}{1.2}(73,400)^{0.8}(0.712)^{0.4} = 1.87 \text{ Btu/hr-ft}^2\text{-F}$$

2. $\Delta p = \dfrac{f\rho(L/D)V^2}{2g_c}$ \qquad (from Table 12-7)

$f = 0.0192$ \qquad (from Figure 12-15)

$$\Delta P = 0.0192\,(0.078 \text{ lbm/ft}^3)\left(\frac{10}{1.2}\right)\left[\frac{(600/60)^2 \text{ ft}^2/\text{sec}^2}{(2)(32.2 \text{ lbm-ft/lbf-sec}^2)}\right]$$

$$= 0.0194 \text{ lbf/ft}^2$$

EXAMPLE 12-21

In a nuclear reactor, 2800 lbm/hr of sodium is heated from 390 to 510 F as it flows through a 10-ft heated length of $\frac{1}{2}$-in.-ID stainless steel tubing maintained at 700 F. Determine the actual convective heat transfer coefficient h between the liquid sodium and the tube.

SOLUTION: \qquad $q = \dot{m}c_p(T_o - T_i) = h(\pi dl)\Delta T = 2800(0.32)(510 - 390)$

$$= 107,520 \text{ Btu/hr}$$

$$\Delta T_m = \frac{(700 - 390) - (700 - 510)}{\ln(310/190)} = 245 \text{ F}$$

$$107,520 = h\left(\frac{1}{2 \times 12} \times 10\right)245$$

$$h = 335 \text{ Btu/hr-ft}^2\text{-F}$$

EXAMPLE 12-22

Prepare a wind chill chart. This chart lists the equivalent ambient temperature to account for the effect of the velocity of the wind. To make these estimates, neglect radiation and evaporation and assume that the human body is a vertical cylinder 0.5 m in diameter.

SOLUTION: From Table 12-11:

$$Nu = C\,Re^n\,Pr^{1/3}$$

or

$$h = C\frac{k}{d}\text{Re}^n \text{Pr}^{1/3}$$

The Reynolds number will, of course, be changing with the velocity. Since this is just an estimate, let us select

$$C = \frac{0.683 + 0.193}{2} = 0.438$$

and

$$n = \frac{0.466 + 0.618}{2} = 0.542$$

The heat loss by the body must be equal for the actual case and the equivalent (zero velocity) case. Thus

$$hA(37 - T) = (hA)_0(37 - T^*)$$

where T^* is the stagnation (zero velocity) reading, T is the actual temperature, and 37 is the approximate body temperature in degrees Celsius. In the quantities hA, only the velocity is changing; thus

$$V^{0.542}(37 - T) = V_0^{0.542}(37 - T^*)$$

We obviously cannot put $V_0 = 0$ into this equation or our model is lost. Suppose we let $V_0 = 1$ m/sec. Then

$$T^* = 37 - \left(\frac{V}{V_0}\right)^{0.542}(37 - T)$$

$$= 37 - V^{0.542}(37 - T)$$

where V has units of m/s. The following chart shows the results of using this equation:

T	4.467 m/s = 10 mi/hr	8.934 m/s = 20 mi/hr	13.8 m/s = 30 mi/hr
0	−46		
10	−26	−57	
20	− 1	−19	−34
30	21	14	8

Heat Transfer with Change in Phase

Our prior discussions of convection heat transfer have considered homogeneous single-phase fluids. Of equal importance are the convection processes associated with a change of phase of a fluid. The two most important examples are condensation and boiling, although heat transfer studies of solid–gas phase changes have recently become relevant.

In many types of power or refrigeration cycles one is interested in changing a vapor to a liquid, or a liquid to a vapor, depending on the part of the cycle under study. These changes are accomplished by boiling or condensation, and the engineer must understand the processes in order to design the appropriate heat-transfer equipment. High heat-transfer rates are obtained in boiling and condensation. This fact has led designers of compact heat exchangers to utilize the phenomena for heating or cooling purposes not necessarily associated with power cycles.

Boiling The flow of a two-phase mixture is characterized by various flow and thermal regimes. This is true whether vaporization takes place under natural convection or in forced flow. As in single-phase flow systems, the heat transfer coefficient of a two-phase mixture depends on the flow regime. The thermodynamic and transport properties of the vapor and the liquid, the conditions (roughness and wettability) of the heating surface, and other parameters influence the heat transfer coefficient in different ways in the different flow regimes. Consequently, to determine the heat transfer coefficient it becomes necessary to consider each flow and boiling regime separately.

The different regimes of pool boiling, first described by Nukiyama, are illustrated in Figure 12-16. **Pool boiling** *means boiling when the fluid has no net velocity (that is, no net motion—like a pool of water) and any motion is due to free convection and bubble movement.* When the temperature of the heating surface is near the fluid saturation temperature, heat is transferred by convection currents to the free surface where evaporation occurs (region I). Transition to nucleate boiling starts when the surface temperature exceeds saturation by a few degrees (region II). In **nucleate boiling** (region III), a thin layer of superheated liquid is formed adjacent to the heating surface. In this layer bubbles nucleate and grow from nucleation sites on the surface. The thermal resistance of this superheated liquid film is greatly reduced by the bubble-induced agitation and vaporization. An increase of the wall temperature is accompanied by a large increase in the number of bubbles that causes, in turn, a large increase of heat flux. As the heat flux, or the temperature difference, is further increased, more and more vapor is formed until a point is reached at which the flow of the liquid toward the surface is interrupted and a blanket of vapor tends to form. This gives rise to the maximum or peak heat flux in nucleate boiling (designated as point *a*). This flux is often referred to as the **burnout heat flux** because, for constant-power generating systems, an increase of the heat flux beyond this point results in a jump of the heater temperature (to point *b*), which is often beyond the melting point of a metal heating surface. In systems for which the surface temperature is beyond the value for peak heat flux (point *a*), a decrease of the heat flux density is observed. This condition is called the **transitional boiling regime** (region IV). In this regime liquid alternately falls onto the surface and is repulsed back by the explosive burst of vapor. At sufficiently high surface temperatures, a stable vapor film is formed at the heater surface, giving rise to the **film-boiling regime** (regions V and VI). Since the heat transfer is by conduction (and some radiation) across the vapor film, the

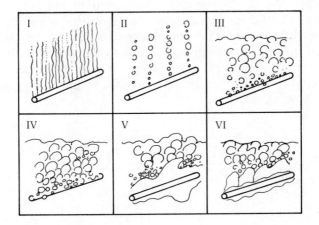

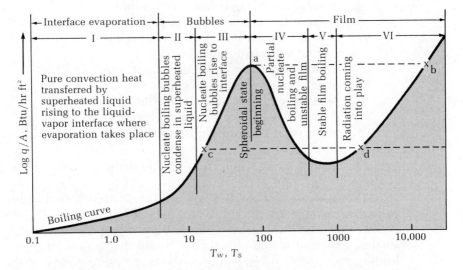

Figure 12-16. Regimes of pool boiling. (Reprinted by permission from the 1977 Fundamentals Volume ASHRAE Handbook and Product Directory.)

temperature of the heater is considerably higher than for comparable heat flux densities in the nucleate boiling regime.

The regime of nucleate boiling is of the greatest practical interest. Much information is now available for calculating boiling heat-transfer coefficients, but there is no universally reliable method for correlating all the data in this regime since the heat flux density is not a single-valued function of the temperature but depends also on the nucleating characteristics of the surface.

Experimental data that would lead to a method of describing, quantitatively, the nucleating characteristics have not yet been gathered. In the absence of these data, the only procedure that can be recommended for design purposes is the one proposed by Rohsenow. According to this procedure, the surface effects can be evaluated by performing a test at atmospheric pressure with a given surface liquid combination. Rohsenow correlated experimental data for nucleate pool boiling with the following relation:

$$\frac{c_1 \Delta T_x}{h_{fg} Pr_l^{1.7}} = C_{sf} \left[\frac{q/A}{\mu_l h_{fg}} \sqrt{\frac{g_c \sigma}{g(\rho_l - \rho_v)}} \right]^{0.33} \tag{12-68}$$

where

$\quad\quad c_l$ = specific heat of saturated liquid (Btu/lbm − F)
$\quad\Delta T_x$ = temperature excess = $T_w - T_{sat}$, (F)
$\quad\; h_{fg}$ = enthalpy of vaporization (Btu/lbm)
$\quad\; Pr_l$ = Prandtl number of saturated liquid
$\;\; q/A$ = heat flux per unit area (Btu/hr-ft²)
$\quad\;\; \mu_l$ = liquid viscosity (lbm/hr-ft)
$\quad\;\; \sigma$ = surface tension of liquid–vapor interface (lbf/ft)
$\quad\;\; g$ = gravitational acceleration
$\quad\;\; \rho_l$ = density of saturated liquid (lbm/ft³)
$\quad\;\; \rho_v$ = density of saturated vapor (lbm/ft³)
$\quad\; C_{sf}$ = constant determined from experimental data

Table 12-13 lists some values for C_{sf}. The surface tension for water is given in Table 12-14.

When a two-phase mixture of liquid and vapor flows inside a tube, a number of flow patterns can occur, depending on the mass fraction of liquid, the fluid properties of each phase, and the flow rate. In an evaporator tube the mass fraction of liquid decreases along the circuit leading to a series of changing gas–liquid flow patterns. If the fluid enters as a subcooled liquid, the first indications of vapor generation are bubbles forming at the heated tube wall (nucleation). Subsequently there can occur bubble, plug, churn (or semiannular), annular, spray-

Table 12-13   Coefficient C_{sf} for Various Liquid–Surface Combinations

Fluid–Surface Combination	C_{sf}
Water–copper	0.013
Water–platinum	0.013
Water–brass	0.0060
n-Butyl alcohol–copper	0.00305
Isopropyl alcohol–copper	0.00225
n-Pentane–chromium	0.015
Benzene–chromium	0.010
Ethyl alcohol–chromium	0.027
Water–emery–polished copper	0.0128
n-Pentane–emery–polished copper	0.0154
Carbon tetrachloride–emery–polished copper	0.0070
Water–ground and polished stainless steel	0.0080
Water–chemically etched stainless steel	0.0133
Water–mechanically polished stainless steel	0.0132
R-11–Inconel	0.0096
R-11–copper	0.0079

Table 12-14   Surface Tension for Water

Saturation temperature, F	Surface tension $(\sigma \times 10^4)$, lbf/ft
32	51.8
60	50.2
100	47.8
140	45.2
200	41.2
212	40.3
320	31.6
440	21.9
560	11.1
680	1.0
705.4	0

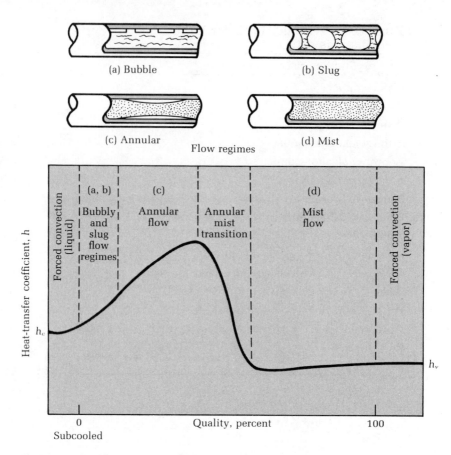

Figure 12-17. Flow regimes for forced-convection boiling.

annular, and mist flows as the vapor content increases. These flow patterns are illustrated (somewhat ideally) in Figure 12-17 for a horizontal tube evaporator.

EXAMPLE 12-23

While preparing water for tea, you observe the boiling water. You know the copper kettle being used is 20 cm in diameter and you estimate that the water temperature is 100 C and the boiling rate is 20 kg/hr. What is the kettle's temperature?

SOLUTION: Assume that you have nucleate boiling; thus the Rohsenow correlation may be used:

$$\frac{c_l(T_w - T_{sat})}{h_{fg}\,\Pr_l^{1.7}} = C_{sf}\left\{\frac{q/A}{\mu_l h_{fg}}\left[\frac{g_c\sigma}{g(\rho_l - \rho_{vap})}\right]^{1/2}\right\}^{0.33}$$

From the Appendix:

$$C_l = 1.2997 \text{ kJ/(kg} \cdot \text{C)}$$

$$T_{sat} = 100 \text{ C}$$

$$h_{fg} = 2256.92 \text{ kJ/kg}$$

$$C_{sf} = 0.013$$

$$\dot{q} = (2256.96 \text{ kJ/kg})(20 \text{ kg/hr}) = 45{,}139.2 \text{ kJ/hr}$$

$$\mu_{lg} = 1.02385 \text{ kg/(m} \cdot \text{s)}$$

$$\sigma = 59.97 \text{ kg/m}$$

$$\rho_l = 960.62 \text{ kg/m}^3$$

$$\rho_{vap} = 0.596 \text{ kg/m}^3$$

$$Pr_l = 1.756$$

$$A = 314.16 \text{ cm}^2$$

Hence

$$T_w = T_{sat} + \frac{C_{sf} h_{fg} \, Pr_l^{1.7}}{C_l} \left\{ \frac{q/A}{\mu_l h_{fg}} \left[\frac{g_c \sigma}{g(\rho_l - \rho_{vap})} \right]^{1/2} \right\}^{0.33}$$

$$= 100 \text{ C} + 20.3 \text{ C} = 120.3 \text{ C}$$

Condensing Condensation occurs when a saturated vapor comes in contact with a surface at a temperature lower than the saturation temperature of the vapor. Heat transfer from the vapor causes it to release its latent heat of vaporization and, hence, to condense on the surface. Normally a continuous flow of liquid is formed over the surface. The liquid, either in drops or as a film, offers a greater resistance to the removal of heat from the remaining vapor. The rate of heat flow depends on the thickness of the condensate film, which in turn depends on the rate at which vapor is condensed and the rate at which the condensate is removed from the condensing surface.

Two types of condensation have been observed in practice: film and drops. Film condensation occurs in most applications. The liquid condensate covers the condensing surface with a continuous film and normally flows off the surface by gravity. Film condensation inside a horizontal tube is a somewhat different phenomenon and requires more complex methods of analysis. Dropwise condensation occurs less frequently and has been observed for steam on highly polished surfaces or on surfaces contaminated with certain fatty acids.

Table 12-15 Approximate Values for Condensation Coefficients

Fluid	Geometry	Btu/hr-ft²-F	W/(m² · C)
Steam	Vertical surface	700–2000	3979–11,370
	Horizontal tubes	1700–4000	9664–22,740
Dowtherm A	Vertical surface	120–540	682–3070
Alcohols	Horizontal tube, 2.0-in. diameter	250–300	1421–1705
Refrigerant-12	Horizontal tubes	200–500	1137–2842

When condensation occurs on horizontal tubes and on short vertical plates, the motion of the condensate film is laminar. On vertical tubes and long vertical plates, the film may become turbulent. It is recommended that a Reynolds number of 1600 be used as the critical point at which the flow pattern changes from laminar to turbulent. This Reynolds number is based on the condensate flow rate in pounds per hour divided by the breadth, in feet, of the condensing surface. For a vertical tube, the breadth is the circumference of the tube in feet; for a horizontal tube, the breadth is twice the length of the tube in feet. The Reynolds number is $4\Gamma/\mu_f$, where Γ is the mass flow of condensate per unit of breadth (in lbm/hr-ft) and μ_f is the viscosity of the condensate (in lbm/hr-ft) at the film temperature T_f. In practice, condensation in shell-and-tube condensers with the vapor outside horizontal tubes is almost always laminar.

For laminar film condensation on horizontal tubes Nusselt obtained the relation

$$h = 0.725 \left[\frac{\rho(\rho - \rho_v)gh_{fg}k_f^3}{\mu_f d(T_g - T_w)} \right]^{1/4}$$

$$(12\text{-}69)$$

where d is the diameter of the tube. Table 12-15 presents approximate values of the convective coefficient when condensation is involved; that is, the table gives the normal ranges for the convective heat transfer coefficient.

12–4 Thermal Radiation

Thermal radiation heat transfer is one of the basic modes of the transfer of energy between regions of different temperature. It is distinguished from conduction and convection by the fact that it does not depend on an intermediate material as a carrier of the energy. On the contrary, thermal radiation heat transfer between two regions is impeded by the presence of a material in the intervening space. The radiation energy transfer may be considered to occur as a consequence of the action of either energy-carrying electromagnetic waves or energy-carrying corpuscles or quanta. This energy is emitted by atoms and molecules.

There are many types of electromagnetic radiation; thermal radiation is only one. Figure 12-18 illustrates part of the electromagnetic spectrum. Typically the 0.1 to 100 μm portion $(10^{-4} - 10^{-7}\,\text{m})$ of the spectrum is called the thermal radiation region. (Note that the human eye is sensitive to the 0.40 to 0.65 μm portion—the visible region.)

The amount and characteristics of the radiant energy emitted by a quantity of material depend on the nature of the material, its microscopic arrangement, and its absolute temperature. Although the rate of emission of energy is independent of the surroundings, the net energy transfer rate depends on the temperatures and spatial relationships of the various materials involved.

Blackbody Radiation

The rate of thermal radiation heat emission by a surface is very dependent on its absolute temperature and the wavelength. The relation between the heat emission rate and the temperature is very simple if the surface is black (a surface that absorbs all incident radiation and thus appears black to the eye). Planck in 1901 showed that the spectral distributional energy radiated by a **blackbody** at an absolute temperature T is given by

$$E_{b\lambda} = \frac{C_1 \lambda^{-5}}{e^{C_2/\lambda T} - 1} \tag{12-70}$$

where λ is the wavelength in centimeters and C_1 and C_2 are universal constants having the values 3.7413×10^{-5} erg-cm^2/sec and 1.4388 cm · K, respectively, and T is the absolute temperature in degrees Kelvin. A curve representative of Equation (12-70) is presented in Figure 12-19.

The symbol $E_{b\lambda}$ is used to denote the emitted flux per wavelength (monochromatic emissive power) and is defined as the energy emitted per unit surface area at wavelength λ per unit wavelength interval around λ. Equation (12-70), called **Planck's distribution law,** is the

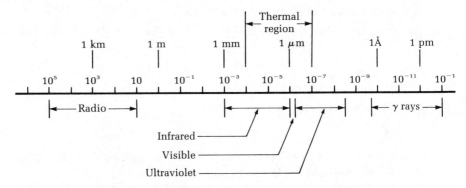

Figure 12-18. The electromagnetic spectrum (scale is in meters).

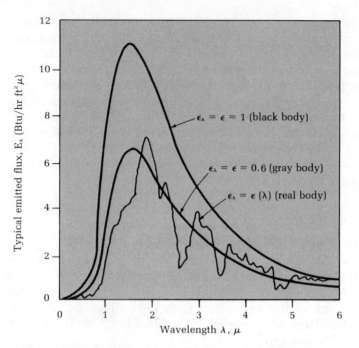

Figure 12-19. Emissive power of black and real surfaces.

basic equation of thermal radiation heat transfer. (See Appendix B-7 for further discussion.)

The rate of energy emission in the interval $d\lambda$ is equal to $E_{b\lambda}d\lambda$. Thus the total rate of heat emission per unit area of black surface, E_b, to the hemispherical region above it is given by the **Stefan–Boltzmann law** and may be obtained by direct integration of Planck's law:

$$E_b = \int_0^\infty E_{b\lambda}d\lambda = \sigma T^4 \qquad (12\text{-}71)$$

The constant σ has a magnitude of 5.669×10^{-9} W/(m$^2 \cdot$ R^4) or 0.1714×10^{-8} Btu/hr-ft^2-R^4. Notice that E_b is the heat flux quantity and is just the area under the Planck curve (Figure 12-19).

The peak of the distribution is determined by extremizing Planck's law (Equation 12-70). Thus

$$\frac{dE_{b\lambda}}{d\lambda} = 0 \qquad (12\text{-}72)$$

The result is

$$\lambda_{max}T = 5216 \; \mu \; \text{R} = 2898 \; \mu \; \text{K} \qquad (12\text{-}73)$$

where λ_{max} is in microns and T is in degrees absolute. Equation (12-73) is referred to as **Wien's displacement law**.

These are useful relationships to remember since they give a rough idea of the spectral energy distribution from a black surface. Values for the radiation functions are given in Appendix Table A-5-1. It may be noted from this table that 25% of the energy is radiated at wavelengths shorter than the maximum and 75% is radiated at wavelengths longer than the maximum.

EXAMPLE 12-24

1. Determine the amount of radiant energy emitted by a graybody of 0.73 emittance at a wavelength of 1.88 μ and a temperature of 1240 F (in Btu/hr-ft^2-μ).

2. For the graybody in part (1), determine the radiation emitted in the infrared region (in Btu/hr-ft^2). The infrared region is usually defined as 0.75 $\mu \leq \lambda \leq 300 \mu$.

SOLUTION: 1. $\lambda T = 1.88(1700) = 3196$ (see Appendix)

$$\frac{E_{b\lambda} \times 10^5}{\sigma T^5} = 6.312$$

$$E_{b\lambda} = \frac{6.312(0.1714 \times 10^{-8})(1700)^5}{10^5} \times 0.73$$

$$= 1121 \text{ Btu/hr-ft}^2 \mu$$

2. $\lambda_1 = 0.75$ $\lambda_1 T = 1275$ $\dfrac{E_b(0 - \lambda_1 T)}{\sigma T^4} = 0$

$\lambda_2 = 300$ $\lambda_2 T = 510{,}000$ $\dfrac{E_b(0 - \lambda_2 T)}{\sigma T^4} = 1.00$

Hence

100% or $E_{\text{infrared}} = \epsilon \sigma T^4 = 0.73(0.1714)(17)^4 = 10{,}450$ Btu/hr-ft^2

Actual Radiation

Thermophysical Properties Substances and surfaces of engineering interest show marked divergences from the Stefan–Boltzmann and Planck laws. But since E_b and $E_{b\lambda}$ are the maximum emissive powers for *any* given surface temperature, actual surfaces emit and absorb less readily and are called nonblack. The emissive power of a nonblack surface, at temperature T, radiating to the hemispherical region above it is

$$E = \epsilon E_b = \epsilon \sigma T^4 \tag{12-74}$$

where ϵ, called the **total hemispherical emittance,** is a function of the material, the condition of its surface, and the temperature of the surface

Table 12-16 Emittances of Coatings

Coating	Emittance
Enamel paint (all colors)	0.8–0.9
Lacquers (colors)	0.8–0.9
Heavy oil film	0.7–0.8
Aluminum paint	0.25–0.6*
Lacquer (clear) on clean stainless steel	0.6
Oxidized stainless steel	0.8
Special paints	0.2–0.3

* Depends on aluminum content.

(see Table 12-16).* The word *total* implies no wavelength dependence and is designated by no λ subscript (Figure 12-19).

To overcome these complexities for heat transfer calculations, gray surface characteristics (ϵ_λ = constant) are often assumed. Several important classes of surfaces actually do approximate this condition, at least in some regions of the spectrum. The resulting simplicity is convenient, but care must be exercised, especially for the high temperatures.

Figure 12-20 illustrates radiant energy falling on a surface where portions are absorbed, reflected, and transmitted through the material. Therefore, from the first law of thermodynamics,

$$\alpha + \tau + \rho = 1 \qquad (12\text{-}75)$$

where α = fraction of incident radiation absorbed (absorptance)
τ = fraction of incident radiation transmitted (transmittance)
ρ = fraction of incident radiation reflected (reflectance)

Since most of the materials encountered in engineering practice are opaque (solids) in the infrared region, $\tau = 0$ and $\alpha + \rho = 1$. Recall the definition of a black surface, which states that $\alpha = 1$, $\rho = 0$, and $\tau = 0$. Platinum black and gold black exhibit absorptances of about 98% in the infrared region—about as black as any surface found in nature. Any degree of blackness desired can be simulated by a small hole in a large enclosure. This can be seen by considering a ray of radiant energy entering the opening and undergoing many internal reflections before passing back out the opening. Certain commercial flat black paints also

---

* The term *emittance* is used here to conform to the practice in physical and electrical terminology of using the suffix "ance" to denote a property of a piece of material as it exists. The ending "ivity" is used to denote a property of the bulk material independent of geometry or surface condition. Thus emittance, reflectance, absorptance, and transmittance refer to actual pieces of material. Emissivity, reflectivity, absorptivity, and transmissivity refer to properties of materials that are optically smooth and thick enough to be opaque.

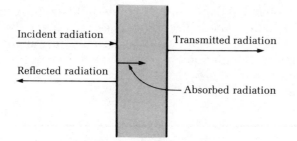

Figure 12-20. Radiative effects at surface.

exhibit emittances of 98% over a wide range of conditions. For the cost, they have more durable surfaces than gold or platinum black and are frequently used on radiation instruments and as standard reference in making emittance or reflectance measurements.

Kirchhoff's Law A useful relation between emittance and absorptance of any opaque surface may be developed directly from thermodynamic considerations. **Kirchhoff's law** *states that for any surface in thermo- dynamic equilibrium with its surroundings, the monochromatic emit- tance equals the monochromatic absorptance.*

$$\epsilon_\lambda = \alpha_\lambda \qquad \text{(12-76)}$$

If the surface is gray, or the incident radiation is from a black surface at the same temperature, then the relation of the total values is also true:

$$\epsilon = \alpha \qquad \text{(12-77)}$$

However, many surfaces are not gray. Moreover, care must be exercised in noting the temperatures of the surfaces under consideration. For many surfaces, absorptance for solar radiation is different from emit- tance for low-temperature radiation. This fact has been used to maintain thermal control of satellites with no internal temperature control.

Lambert's Law The foregoing discussion relates to total hemispherical radiation from surfaces. We have not made any assumptions regarding the distribution of energy in such a hemispherical region. From a heat transfer standpoint, the distribution of energy in this region has an important effect on the rates in various geometric arrangements.

Early in the study of radiation phenomena the spatial distribution of the emitted energy was investigated. A result of these investigations was the formulation of a law of radiant energy spatial distribution. **Lambert's law** *states that the flux distribution of radiant energy from an emitting surface varies as the cosine of the angle between the normal to the radiating surface and the line of observation.* Such radiation is called **diffuse radiation.** We will see later that the resulting Lambert intensity variation is constant.

Black surfaces obey Lambert's law. The law holds approximately for many actual radiation and reflection processes, especially those involving rough surfaces and nonmetallic materials. Most analyses of radiation heat-transfer calculations are based on the assumption of gray-diffuse radiation and reflection.

EXAMPLE 12-25

Using Equation (12-71) and Kirchhoff's law, relate the emittance ϵ and the monochromatic emittance ϵ_λ.

SOLUTION: By definition,

$$E_b = \int_0^\infty E_{b\lambda} d\lambda$$

Thus the total radiant energy emitted from a gray surface is

$$E = \epsilon E_b$$

The same definition can be used on a wavelength basis as well:

$$E_\lambda = \epsilon_\lambda E_{b\lambda}$$

Thus

$$\epsilon E_b = E = \int_0^\infty E_\lambda d\lambda = \int_0^\infty \epsilon_\lambda E_{b\lambda} d\lambda$$

or

$$\epsilon = \frac{1}{E_b} \int_0^\infty \epsilon_\lambda E_{b\lambda} d\lambda$$

$$= \frac{\int_0^\infty \epsilon_\lambda E_{b\lambda} d\lambda}{\int_0^\infty E_{b\lambda} d\lambda}$$

Thus this emittance ϵ is an energy-weighted average of ϵ_λ.

The next section deals with the techniques for estimating heat transfer rates between surfaces of different geometries, radiation characteristics, and orientations. The following assumptions are made: (1) all surfaces are either gray or black; (2) radiation and reflection processes are diffuse; (3) properties are uniform over the extent of the surfaces; (4) the material occupying the space between the radiating surfaces neither emits nor absorbs radiation. Although these assumptions are not strictly valid in many problems of practical importance, they are generally used as a matter of convenience. The considerable simplification they provide, however, means that we must consider the results to be approximate. In many cases, the development of new techniques without these assumptions would involve computations so lengthy and complex as to be impractical.

The Configuration Factor

The fraction of radiation leaving one surface, and being incident on a surfaces it irradiates, is indicated by a quantity variously called an **interception, view, configuration factor,** or **angle factor.** In terms of two surfaces, i and j, the configuration factor from surface i to surface j, F_{ij}, is defined as the fraction of diffuse radiant energy leaving surface i that falls *directly* on j. The configuration factor from j to i is similarly defined merely by interchanging the roles of i and j. This second configuration factor is not, in general, numerically equal to the first. However, under the assumptions noted previously,

$$F_{ij}A_i = F_{ji}A_j \qquad (12\text{-}78)$$

where A indicates surface area. Note that a concave surface may irradiate (or see) itself; thus, $F_{jj} \neq 0$. And if n surfaces form an enclosure, then

$$\sum_{j=1}^{n} F_{ij} = 1 \qquad (12\text{-}79)$$

The defining equation expressing the configuration factor, F_{ij} or F_{i-j}, between two surfaces is

$$F_{ij} = \frac{1}{A_i} \int_{A_i} \int_{A_j} \frac{\cos \phi_i \cos \phi_j}{\pi r^2} \, dA_i \, dA_j \qquad (12\text{-}80)$$

where dA_i and dA_j are elemental areas of the two surfaces, r is the distance between dA_i and dA_j, and ϕ_i and ϕ_j are the angles between the respective normals to dA_i and dA_j and the connecting line r. (See Appendix B-8 for details.) The solution of this equation in closed form is difficult, if not impossible, for all geometries. Numerical, graphic, and mechanical techniques have provided alternative methods, and numerical values of the configuration factor for many geometries encountered in engineering may be found in the literature. It must be emphasized that the expression for the angle factor is based on the assumption that the directional distribution of radiation leaving a surface is diffuse and uniformly distributed. These restrictions must be kept in mind when applying these factors for nonblack enclosures. The total flux leaving a surface must be constant along that surface so that the configuration factor is independent of the magnitude. Although it is unlikely that this condition is generally satisfied, the assumption is common for many engineering applications. Numerical values of configuration factors of some common geometries are given in Figures 12-21 through 12-24.

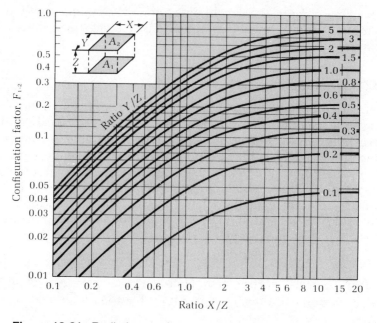

Figure 12-21. Radiation configuration factor for parallel rectangles.

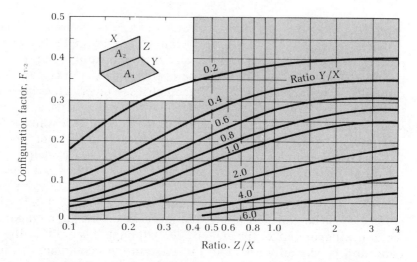

Figure 12-22. Radiation configuration factor for perpendicular rectangles with common edge.

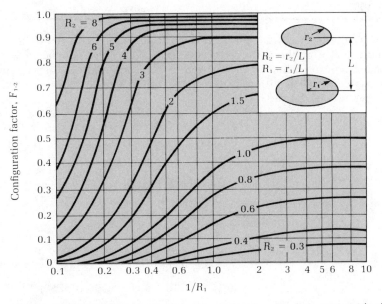

Figure 12-23. Radiation configuration factor for parallel, concentric disks.

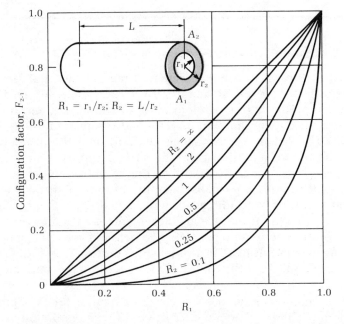

Figure 12-24. Radiation configuration factor for concentric cylinders.

EXAMPLE 12-26

The irradiation on a flat plate on earth is measured as 424 Btu/hr-ft². If the transmittance of the earth's atmosphere is 0.83, the diameter of the sun is 866,000 mi, the diameter of the earth is 8000 mi, and the distance from the center of the earth to the center of the sun is 93,000,000 mi, calculate the temperature of the sun.

SOLUTION: Apply the first law of thermodynamics:

$$\tau_{atm} F_{s-e} \sigma A_s T_s^4 = 424 A_{c,e} \qquad \text{(an energy balance)}$$

The sun, being a sphere, radiates uniformly in all directions. The flat plate is on a large enclosing sphere. Thus

$$\text{So} \quad F_{s-e} \doteq \frac{A(\text{disk})}{A(\text{enclosing sphere})}$$

$$0.83 \frac{(4000)^2}{4(93,000,000)^2}$$

$$\times (0.1717 \times 10^{-8}) 4\pi (433,000 \times 5280)^2 T_s^4$$

$$= 424\pi (4000 \times 5280)^2$$

$$T_s^4 = \frac{(424)(93 \times 10^6)^2}{0.83(0.1714 \times 10^{-8})(433,000)^2} = 1.3726(10^{16}) R^4$$

$$T = 10,824 \text{ R} = 10,364 \text{ F}$$

Radiant Exchange Between Surfaces

Of principal interest to the engineer in a radiant heat transfer calculation is the *net* rate of energy loss or gain at a surface. A surface radiates energy at a rate that is independent of its surroundings, and it absorbs and reflects the energy incident on it at a rate dependent on its surface condition. The net power exchange per unit area (flux), denoted by q_i/A_i, is the rate of emission from the surface minus the total rate of radiant energy absorbed at the surface due to all radiant effects by its surroundings (including perhaps the return of some of its own emission by reflection from surroundings). That is, q_i is the rate at which energy must be supplied to the surface material by other exchange processes if its temperature is to remain constant. To define q_i we must specify the total radiant surroundings. These total surroundings constitute in effect an enclosure and, from this point of view, all problems are enclosure problems.

Several methods have been devised for treating specific problems or classes of problems. A method for calculating the radiation exchange at each surface of an n opaque surface enclosure is presented here. The calculation. Two new terms need to be defined: **Irradiation** (G) *is the total radiant energy incident on a surface per unit time and per unit area;* **radiosity** *(R) is the total radiant energy that leaves a surface per unit time and per unit area.*

The method has one basic assumption: a "surface" of an enclosure is defined as a region where R and G are uniformly distributed. From this surface $(\tau = 0)$ the radiosity is the sum of the flux emitted and the flux reflected:

$$R = \epsilon E_b + \rho G \tag{12-81}$$

where ϵ (total hemispherical emittance) and ρ are constants (a gray body). From Equation (12-75) and Kirchhoff's law

$$\rho = 1 - \alpha = 1 - \epsilon \tag{12-82}$$

Thus

$$R = \epsilon E_b + (1 - \epsilon)G \tag{12-83}$$

The difference between the radiosity and the irradiation is defined as the net flux lost by the surface:

$$\frac{q}{A} = R - G = \epsilon E_b - \epsilon G \tag{12-84}$$

Eliminating G by using Equation (12-83) yields

$$q = \frac{E_b - R}{(1 - \epsilon)/\epsilon A} \tag{12-85}$$

Consider an enclosure made up of n isothermal surfaces with areas of $A_1, A_2, \ldots, A_n$, emittances of $\epsilon_1, \epsilon_2, \ldots, \epsilon_n$ and reflectances of $\rho_1, \rho_2, \ldots, \rho_n$. The total energy incident on surface i is the sum of the radiation incident on it coming from all n surfaces or

$$G_i A_i = \sum_{j=1}^{n} F_{ji} R_j A_j = \sum_{j=1}^{n} F_{ij} R_j A_i$$

or

$$G_i = \sum_{j=1}^{n} F_{ij} R_j \tag{12-86}$$

Notice that this is just a conservation of energy expression. Substitution of G_i into Equation (12-83) yields the following set of simultaneous equations:

$$R_i = \epsilon_i E_{bi} + (1 - \epsilon_i) \sum_{j=1}^{n} F_{ij} R_j \qquad i = 1, 2, \ldots, n \tag{12-87}$$

When each of the n surfaces is considered,

$$R_1 = \epsilon_1 \sigma T_1^4 + (1 - \epsilon_1) \sum_{j=1}^{n} F_{1j} R_j$$

$$R_2 = \epsilon_2 \sigma T_2^4 + (1 - \epsilon_2) \sum_{j=1}^{n} F_{2j} R_j$$

.

.

.

$$R_n = \epsilon_n \sigma T_n^4 + (1 - \epsilon_n) \sum_{j=1}^{n} F_{nj} R_j$$

Equation (12-87) can be solved manually for the unknown radiosities R_i if the number of surfaces is small. More complex enclosures require a computer.

Once the radiosities are known, the net radiant energy lost by each surface is determined from Equation (12-85):

$$q_i = \frac{E_{bi} - R_i}{(1 - \epsilon_i)/\epsilon_i A_i} \tag{12-88}$$

Sometimes it is desirable to determine the net radiant energy exchanged by only two of these surfaces. Under these circumstances,

$$q_{i-j} = A_i F_{ij} R_i - A_j F_{ji} R_j \tag{12-89}$$

If the surface is black, Equation (12-78) becomes indeterminate and we must use an alternative expression such as

$$q_i = \sum_{j=1}^{n} (R_i A_i F_{ij} - R_j A_j F_{ji}) \qquad F_{ij} A_i = F_{ji} A_j$$

$$= \sum_{j=1}^{n} F_{ij} A_i (R_i - R_j)$$

$$= \sum_{j=1}^{n} q_{i-j}$$

Many types of diffuse-radiation processes may be analyzed with the enclosure method discussed above by assigning, to surfaces having special characteristics, values of the thermophysical properties that are consistent. For example, an opening is treated as an equivalent surface area A_e with a reflectance of zero. If energy enters the enclosure through the opening, A_e is assigned the temperature seen through the opening; otherwise its temperature is taken as zero. If the loss through the opening is desired, q_e is found. A window in the enclosure is assigned its actual properties.

A surface in **thermal equilibrium** is one for which radiant emission is balanced by radiant absorption. This radiation balance is reached if the net radiant energy transfer is zero. Adiabatic surfaces, insulated surfaces for which q_{net} is zero, can be analyzed by applying the first law of thermodynamics.* The resulting equilibrium temperature of such a surface, subject to an incident flux Q/A, can be found from

$$T = \left(\frac{Q/A}{\sigma}\right)^{1/4}\left(\frac{\alpha}{\epsilon}\right)^{1/4}\left(\frac{A_{abs}}{A_{emit}}\right)^{1/4}$$ (12-90)

Use of the configuration factors and radiation properties, as defined, requires an assumption that each surface uniformly irradiates every other surface—that is, the surfaces are diffuse radiators. This is a good assumption for most nonmetal surfaces radiating in the infrared region. It is a poor assumption for highly polished metal surfaces. Subdividing the various surfaces and taking into account the variation of radiation properties with angle of incidence will improve the approximation, but the work required to obtain a solution increases rapidly with each subdivision. A balance must be found that is consistent with the importance of the problem.

EXAMPLE 12-27

Let us consider a "two-body problem"—that is, an enclosure with only two surfaces. Examples are two infinite planes or two infinite concentric cylinders. For this problem, imagine a room-shaped enclosure in which two opposing walls are surface 1 and all the other surfaces are 2 (see the sketch). Determine the net radiant flux exchange.

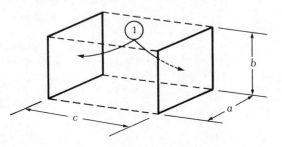

SOLUTION: Using the radiosity approach, we get

$$R_1 = \epsilon_1\sigma T_1^4 + \rho_2(F_{11}R_1 + F_{12}R_2)$$

$$R_2 = \epsilon_2\sigma T_2^4 + \rho_2(F_{21}R_1 + F_{22}R_2)$$

Solving these two equations simultaneously for R_1 and R_2 yields

---

* Since with this type of surface all radiation absorbed by the surface must leave, the surface may be treated in the radiosity equation as perfectly reflective. Alternately, for determining its temperature the surface may be treated as black; thus $T = (R/\sigma)^{\frac{1}{4}}$.

$$R_1 = \frac{\epsilon_1(1 - \rho_2 F_{22})\sigma T_1^4 + \rho_1 F_{12}\epsilon_2 \sigma T_2^4}{(1 - \rho_1 F_{11})(1 - \rho_2 F_{22}) - \rho_1 \rho_2 F_{12} F_{21}}$$

$$R_2 = \frac{\epsilon_2(1 - \rho_1 F_{11})\sigma T_2^4 + \rho_2 F_{21}\epsilon_1 \sigma T_1^4}{(1 - \rho_1 F_{11})(1 - \rho_2 F_{22}) - \rho_1 \rho_2 F_{12} F_{21}}$$

For this particular case

$$q_1 = q_{1-2} = -q_2$$

So

$$\frac{q}{A_1} = \frac{\sigma T_1^4 - R_1}{\rho_1/\epsilon_1}$$

$$= \frac{\sigma(T_1^4 - T_2^4)}{1/F_{12} + \rho_1/\epsilon_1 + (\rho_2/\epsilon_2)(A_1/A_2)} \tag{1}$$

Thus if $a = b = c = 3$ m, $\epsilon_1 = 0.5$, $\epsilon_2 = 0.8$, $T_1 = 1800$ C, and $T_2 = 1350$ C, we may determine the radiant exchange after we have used the charts to determine F_{12}. From Figure 12-22, we get

$$F_{11} = 0.2 \text{ (not zero)}$$

$$F_{12} = 1 - F_{11} = 0.8$$

Thus

$$\frac{q_1}{A_1} = \frac{(104 - 37)\text{W/cm}^2}{1/0.8 + 0.5/0.5 + (0.2/0.8)(2a^2/4a^2)} = \frac{67}{2.375} \frac{\text{W}}{\text{cm}^2}$$

$$= 28.21 \text{ W/cm}^2$$

Note the possibility of an electrical analog in Equation (1). In fact, perusal of Equations (12-88) and (12-89) would also indicate this possibility. That is,

$$q_i = \frac{E_{bi} R_i}{\rho_i/\epsilon_i A_i} \Rightarrow$$

and

$$q_{i-j} = \frac{R_i - R_j}{1/A_i F_{ij}} \Rightarrow$$

and Equation (1) $\Rightarrow$

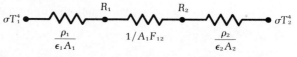

This type of electrical analog has been used for many years. It is unfortunate that this approach is quite unwieldy for a large number of surfaces.

EXAMPLE 12-28

A thermocouple is used to measure the temperature of a flame in a combustion chamber. If the thermocouple indicates 1400 F and the walls of the chamber are at 800 F, what is the error in the thermocouple reading? Assume all surfaces are black and the convective coefficient between the flame and the thermocouple is 20 Btu/hr-ft²-F.

SOLUTION:

$$q_{in_{conv}} = q_{out_{rad}}$$

$$hA_{tc}(T_f - T_{tc}) = \sigma F_{tc-w} A_{tc}(T_{tc}^4 - T_w^4) \qquad F_{tc-w} = 1$$

$$20(T_f - T_{tc}) = 0.1714[(18.6)^4 - (12.6)^4]$$

$$T_f - T_{tc} = 815 \text{ F error}$$

EXAMPLE 12-29

The surface temperature of insulation on a horizontal steam pipe passing through a 15 ft × 20 ft × 8 ft room is 145 F. The emittance of the insulation is 0.9. The room surfaces are black at 80 F. Determine the rate of heat loss by radiation per foot of pipe if the outside diameter of the insulation is 6.5 in.

SOLUTION:

$$R_1 = \epsilon \sigma T_1^4 + (1 - \epsilon_1)(F_{11}R_1 + F_{12}R_2)$$

$$= 0.9(0.1714)(6.05)^4 + 0.1(0 + 1R_2)$$

$$R_2 = \epsilon_2 \sigma T_2^4 + (1 - \epsilon_2)(F_{21}R_1 + F_{22}R_2)$$

$$= 0.1714(5.40)^4 = 145$$

$$R_1 = 0.9(229.6) + 14.5 = 221.2$$

$$q_1 = \frac{E_b - R_1}{(1 - \epsilon_1)/\epsilon_1 A_1} = \frac{229.6 - 221.2}{(1 - 0.9)/0.9[\pi(6.5/12)]}$$

$$= 128.6 \text{ Btu/hr-ft}$$

EXAMPLE 12-30

A 20-ft room, 10-ft high, is to be heated by means of electric resistance heating elements installed in the ceiling. The walls are well insulated. The floor temperature is 75 F. There is no reflected energy from the carpeting on the floor. If the painted ceiling (emittance = 0.92) is maintained at 105 F, determine the wattage required for the ceiling heaters (1 W = 3.413 Btu/hr).

SOLUTION: A_c = area of ceiling A_f = area of floor A_w = area of walls
$\qquad\qquad\quad$ = 400 ft² $\qquad\qquad\quad$ = 400 ft² $\qquad\qquad\quad$ = 800 ft²

$\qquad\qquad\quad \epsilon_c = 0.92 \qquad\qquad\qquad \epsilon_f = 1 \qquad\qquad\qquad \epsilon_w = 0$
$\qquad\qquad\qquad\qquad\qquad\qquad\qquad\qquad\qquad\qquad\qquad\qquad$ (perfectly reflecting)

$\qquad\qquad\quad T_c = 565 \qquad\qquad\qquad T_f = 535 \qquad\qquad\qquad T_w = ?$

$$F_{c-f} = 0.4 \qquad\qquad F_{f-c} = 0.4 \qquad\qquad F_{w-c} = (400/800)(0.6) = 0.3$$

$$F_{c-c} = 0 \qquad\qquad F_{f-f} = 0 \qquad\qquad F_{w-f} = 0.3$$

$$F_{c-w} = 0.6 \qquad\qquad F_{f-w} = 0.6 \qquad\qquad F_{w-w} = 0.4$$

$$R_c = \epsilon_c \sigma T_c^4 + (1 - \epsilon_c)(F_{c-c}R_c + F_{c-f}R_f + F_{c-w}R_w)$$

$$= 0.92(0.1714)(6.65)^4 + 0.08(0.4R_f + 0.6R_w)$$

$$R_f = (0.1714)(5.35)^4 + 0 = 140.4$$

$$R_w = 0 + (0.3R_c + 0.3R_f + 0.4R_w) \quad \text{or}$$

$$0.6R_w = 0.3R_c + 0.3(140.4)$$

$$R_w = 0.5R_c + 70.2$$

$$R_c = 0.92(174.66) + 0.08[0.4(140.4) + 0.6(0.5R_c + 70.2)]$$

$$= 160.69 + 4.49 + 0.02R_c + 3.37$$

$$= 171.99$$

$$q_c = \frac{E_{bc} - R_c}{(1 - \epsilon_c)/\epsilon_c A_c} = \frac{174.66 - 171.99}{0.08/0.92(400)}$$

$$= 12{,}282 \text{ Btu/hr} = 3598 \text{ W} \approx 3600 \text{ W}$$

EXAMPLE 12-31*

An industry has a 1000-ft pipeline with a constant-temperature fluid flowing through a standard 6-in. steel pipe (OD 6.625 in.) that is insulated with 2-in-thick calcium silicate insulation (see the sketch). The fluid temperature (which is approximately the temperature of the inside surface of the insulation) is 750 F, and the pipeline is inside a building that is at a temperature of 80 F. It is desired to find the total heat lost from the pipe to the building. Assume the emittance of the calcium silicate is 0.9 and its thermal conductivity is

$$k = 0.0284(1 + 0.00155T_{avg}) \; [=] \text{ Btu/hr-ft-F} \qquad\qquad (1)$$

$$T_{avg} = \frac{T_1 + T_2}{2} \; [=] \text{ F}$$

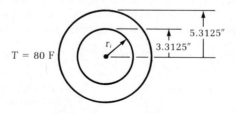

$T = 80$ F $\qquad r_i \qquad$ 3.3125" $\qquad$ 5.3125"

---

*From *Negligible Thermal Radiation*, by L. D. Simmons. Reprinted by permission from *Mechanical Engineering News*.

SOLUTION: Using Fourier's law, we find that the heat loss rate per foot by conduction (through insulation) is

$$q = \frac{2\pi k(T_i - T_o)}{\ln(r_o/r_i)} \tag{2}$$

The heat loss rate per foot from the outside surface by convection and radiation is

$$q = 2\pi h r_o(T_o - T_\infty) + 2\pi r_o \epsilon \sigma(T_o^4 - T_\infty^4) \tag{3}$$

where h for free convection is given by

$$h = \frac{0.227(T_o - T_\infty)^{1/4}}{r_o^{1/4}} \tag{4}$$

The radiation term of Equation (2) is written assuming that the pipe is small compared to the building. (Thus all radiation emitted by the pipe is absorbed by the building at 80 F.) Eliminating q by equating Equations (1) and (2) yields

$$\frac{k(T_i - T_o)}{\ln(r_o/r_i)} = h r_o(T_o - T_\infty) + r_o \epsilon \sigma(T_o^4 - T_\infty^4) \tag{5}$$

Assuming that r_o is fixed, Equations (1), (4), and (5) must be solved simultaneously for k, T_o, and h. This may be done quite easily using a Newton–Raphson iteration technique on a digital computer. The results then may be used in Equations (2) and (3) to determine the heat loss by each mode. The following table lists the results for various situations (including additional insulation):

Insulation	Heat Loss Rate, Btu/hr-ft	Outside Surface Temperature F	Loss by Convection	Loss by Radiation
2-in. Insulation (neglect radiation)	352	219	352	—
2-in. Insulation (include radiation)	385	150	156	229(59%)
4-in. Insulation	239	117	88	150(63%)
6-in. Insulation	185	104	63	122(66%)

Notice that the first case neglects radiation altogether ($\epsilon = 0$). Neglecting radiation raises the calculated surface temperature well above its true value (that is, the value accounting for radiation). This higher temperature increases h so that convection is increased, and the result is a calculated heat loss rate that is only modestly in error (low by 9%). The solution, including radiation, indicates that radiation accounts for a surprisingly high fraction of the total transmission from the surface. For the two cases with 4 in. and 6 in. of insulation, the difference between the surface temperature and the environment temperature is reduced. The fraction of the heat loss by radiation is *increased* due to the drop in h caused by T_o decreasing and r_o increasing.

Radiation Shields

The use of radiation shields is becoming more important due to the need to control the immediate environment at low cost. From our previous work, it should be clear that radiant heat transfer is reduced as the reflectance increases. But since this variable is usually beyond the control of the engineer, the shield is used between the hot (source) and cool (receiving) surfaces so that they do not add or subtract energy from the overall system. The radiation shield reduces the heat transfer from the hot surface by adding a resistance to the heat transfer.

To understand the use of these shields, consider a simple case of two infinite parallel surfaces with and without an intervening shield (Figure 12-25). Without the shield, the net radiant heat flux is

$$\frac{q}{A} = \frac{\sigma(T_1^4 - T_2^4)}{1/\epsilon_1 + 1/\epsilon_2 - 1} \qquad \text{(Recall Example 12-27)} \qquad \textbf{(12-91)}$$

If $\epsilon_1 = \epsilon_2$, then

$$\frac{q}{A} = \frac{\sigma(T_1^4 - T_2^4)}{2/\epsilon_1 - 1} \qquad \textbf{(12-92)}$$

Consistent with the first law of thermodynamics, the shield does not add or subtract heat from the system. Thus the heat transferred between surfaces 1 and 3 and between surfaces 3 and 2 is exactly equal. Therefore

$$\frac{q}{A} = \left(\frac{q}{A}\right)_{1-3} = \left(\frac{q}{A}\right)_{3-2} \qquad \textbf{(12-93)}$$

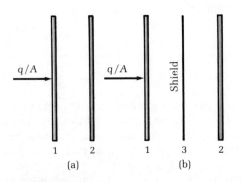

(a)   (b)

Figure 12-25. Schematic of infinite parallel surfaces (a) without and (b) with a radiation shield.

And using the radiosity approach

$$\frac{q}{A} = \frac{\sigma(T_1^4 - T_3^4)}{1/\epsilon_1 + 1/\epsilon_3 - 1} = \frac{\sigma(T_3^4 - T_2^4)}{1/\epsilon_3 + 1/\epsilon_2 - 1} \qquad (12\text{-}94)$$

The unknown in Equation (12-92) is T_3. For convenience only, let us assume that the emittances of all surfaces are equal $(\epsilon_1 = \epsilon_2 = \epsilon_3)$. Therefore

$$T_3^4 = \tfrac{1}{2}(T_1^4 + T_2^4) \qquad (12\text{-}95)$$

and the heat transfer is

$$\frac{q}{A} = \frac{\tfrac{1}{2}\sigma(T_1^4 - T_2^4)}{1/\epsilon_1 + 1/\epsilon_3 - 1} = \frac{\tfrac{1}{2}\sigma(T_1^4 - T_2^4)}{2/\epsilon_1 - 1} \qquad (12\text{-}96)$$

Thus the heat transfer is just one-half of that which would have occurred if there were no shield.

Multiple-shield problems may be treated in the same manner. The result for two infinite parallel surfaces with n shields, with all surfaces having the same emittance, is

$$\left. \frac{q}{A} \right)_{\substack{\text{with} \\ \text{shields}}} = \frac{1}{n+1} \left. \frac{q}{A} \right)_{\substack{\text{without} \\ \text{shields}}} \qquad (12\text{-}97)$$

if the temperatures of the heat transfer surfaces are maintained the same in both cases.

Solar Radiation

Solar Characteristics Solar energy approaches the earth as electromagnetic radiation extending from x-ray wavelengths (0.1 μ) to radio waves (100 m). Beyond the earth's atmosphere, distribution of energy detected would look very much like the designated $m = 0$ curve of Figure 12-26. The maximum spectral intensity occurs at approximately 0.48–0.50 μ in the green portion of the visible spectrum (the $m = 0$ curve). There are tables and charts of the sun's extraterrestrial spectral irradiance from 0.120 to 100 μm—the range where, for all practical purposes, all the sun's radiant energy is contained. The ultraviolet portion of the spectrum below 0.4 μm contains 8.73% of the total; another 38.15% is contained in the visible region between 0.40 and 0.70 μm; and the infrared region contains the remaining 53.12%. Thus 99% of the sun's radiant energy is contained between 0.28 and 4.96 μm.

In passing through the earth's atmosphere, some of the sun's direct radiation is scattered by nitrogen, oxygen, and other gases and particles (molecules small compared to the wavelength of the radiation). This scattered radiation causes the sky to be blue on clear days. Some of the direct radiation is scattered by aerosols, water droplets, dust, and the like (particles with diameters comparable to the wavelength).

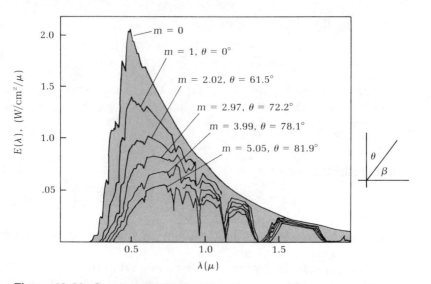

Figure 12-26. Spectral distribution of solar energy falling on a surface normal to the direction of irradiation for various zenith angles.

Attenuation of the solar rays is also caused by absorption—first by the ozone in the outer atmosphere, which causes a sharp cutoff at 0.29 μm of the ultraviolet radiation reaching the earth's surface. In the longer wavelengths, a series of absorption bands exists, caused by water vapor, carbon dioxide, and ozone. The total amount of attenuation at any given location is determined by the length of the atmospheric path that the rays traverse and by the composition of the atmosphere. The functional form of this transmittance is not known, but there are analytic models of the scattering and absorbing processes within the atmosphere. The path length is expressed in terms of the air mass m, which is the ratio of the mass of atmosphere in the earth–sun path to the mass that would exist if the sun were directly overhead at sea level ($m = 1.0$). For all practical purposes, $m = 1.0/\sin \beta$ at sea level. Beyond the earth's atmosphere, $m = 0$.

Numerous solar spectral distribution curves have been presented in the literature. These standard curves are based on both ground and high-altitude measurements. The important number for most engineering purposes is the solar constant (SC). *The **solar constant** is defined as the total flux of solar radiation on a surface normal to the sun's rays for $m = 0$.* The first successful measurement of the solar constant resulted in a value of 1323 W/m² and a solar spectral distribution, both of which were determined from ground measurements. Several other investigators have presented various values of the solar constant and solar spectral distributions; the 1954 reported value was 1428 W/m². These two values of the solar constant—1323 and 1428 W/m²—appear to be the extremes of all values published since 1940.

Even though ground-based measurements have continued to be made since 1954, no significant changes have been presented in the literature. The high-altitude measurements, beginning in about 1967, appear to be more reliable and yield solar constants from 1338 to 1370 W/m². The standard spectral distribution that has been adopted by ASTM for engineering use in the case of zero air mass ($m = 0$) is illustrated in Figure 12-26. The corresponding solar constant is 1353 W/m².

Solar Constant With the tools you have at your disposal, you can approximate the numbers given in the preceding section. Since the $m = 0$ curve looks very much like that of a blackbody, let us assume that the sun is a blackbody. Thus, from Wien's displacement law,

$$T \doteq \frac{2898 \; \mu K}{\lambda_{max}} \tag{12-73}$$

If λ_{max} is approximately 0.5 μ, then

$$T = 5796 \text{ K} = 10{,}465 \text{ R} \tag{12-98}$$

The temperature represents the surface temperature of a blackbody having a spectral distribution like the sun. Using the Stefan–Boltzmann law, we can calculate the flux at the surface of the sun. That is,

$$E_b = \sigma T^4$$

$$\doteq 22.5(10^6) \text{ Btu/hr-ft}^2 \tag{12-99}$$

$$\doteq 7.3 \text{ kW/cm}^2$$

Note that the total power leaving the sun (a blackbody) is approximately $1.3(10^{27})$ Btu/hr. Luckily not all of this power makes it to earth. To estimate how much gets to one solar distance (93×10^6 mi), refer to Equation (B-8-4) of Appendix B-8:

$$dE_{1 \to 2} = \cos \phi_1 \cos \phi_2 \frac{\sigma T_1^4 \, dA_1 \, dA_2}{\pi \; r^2} \tag{B-8-4}$$

Figure 12-27 illustrates the variables of this equation. Since we are interested only in the power at one solar radius (r), the solar constant could be defined as

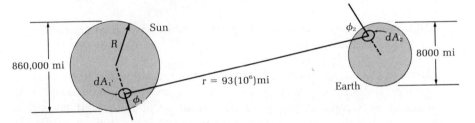

Figure 12-27. Geometric relationship of the earth and sun (not to scale).

$$SC = \int \frac{dE_{1 \to 2}}{\cos \phi_2 \, dA_2} = \frac{\sigma T_1^4}{\pi} \int \cos \phi_1 \, \frac{dA_1}{r^2} \tag{12-100}$$

But $dA_1 \doteq 2\pi R^2 \sin \phi_1 \, d\phi_1$; thus

$$SC \doteq 2\sigma T_1^4 \left(\frac{R^2}{r^2}\right) \int_0^{\pi/2} \cos \phi_1 \sin \phi_1 \, dA\phi_1$$

$$= \sigma T_1^4 \left(\frac{R^2}{r^2}\right)$$

$$= 22.5(10^6) \frac{\text{Btu}}{\text{hr-ft}^2} \left(\frac{4.3 \times 10^5}{9.3 \times 10^7}\right)^2$$

$$= 481.05 \text{ Btu/hr-ft}^2$$

$$= 1517.5 \text{ W/m}^2$$

Notice that this is not bad agreement with the measured values.

Earth Temperature Using this information and applying the first law of thermodynamics we can also estimate the average earth temperature. First we assume that the earth absorbs the power as a disk in a steady-state fashion and emits the power as a sphere. Thus, as per Equation (12-90),

$$T = \left(\frac{SC}{\sigma}\right)^{1/4} \left(\frac{\alpha}{\epsilon}\right)^{1/4} \left(\frac{A_{abs}}{A_{emit}}\right)^{1/4}$$

$$= \left(\frac{1517.5 \text{ W/m}^2}{5.669(10^{-8}) \text{ W/(m}^2 \cdot \text{K}^4)}\right)^{1/4} \left(\frac{1}{4}\right)^{1/4} \left(\frac{\alpha}{\epsilon}\right)^{1/4}$$

$$= 286.1 \left(\frac{\alpha}{\epsilon}\right)^{1/4} \text{K}$$

Recall that the absorptance is at the sun's temperature while the emittance is at the earth's temperature (T). This ratio has a large range of variation. For this example, we will allow it to be 1. Thus

$$T \doteq 278 \text{ K} = 13 \text{ C}$$

$$= 55 \text{ F} = 516 \text{ R}$$

Although this value seems too high, it is not bad when you consider the degree of approximation.

Closure This discussion of solar power is not intended to be comprehensive. Rather, our intent has been to inform you of the general properties of this radiant energy. For specific applications, it should be apparent that the power distribution from the sun can be considerably augmented. The classic example is the so-called **greenhouse effect.** Ordinary glass transmits solar radiation well up to 2.5 μ. Therefore, the majority of the solar power incident on the glass is transmitted. For the radiation whose wavelengths are greater than 3 μ, the transmittance is nearly zero. The majority of this long-wavelength power is reflected. Since the low-temperature sources inside the greenhouse emit radiation that cannot escape through the glass, the inside of the greenhouse warms up until other modes of heat transfer allow equilibrium to be established.

PROBLEMS

12-1 Give a reasonable value for each of the following:
a. Thermal conductivity of air (Btu/hr-ft^2-F/ft; W/(m · K))
b. Thermal conductivity of water (Btu/hr-ft^2-F/ft; W/(m · K))
c. Thermal conductivity of concrete (Btu/hr-ft^2-F/in.; W/(m · K))
d. Thermal conductivity of steel (Btu/hr-ft^2-F/ft; W/(m · K))
e. Specific heat of steel (Btu/lbm-F; J/(kg · K))
f. Specific heat of water (Btu/lbm-F; J/(kg · K))
g. Specific heat of air at constant pressure (Btu/lbm-F; J/(kg · K))

12-2 Determine the thermal conductivity in both English and SI units:
a. Plywood
b. Lightweight concrete (100 lbm)
c. Oak

12-3 Determine the thermal conductance in both Btu/hr-ft^2 and W/m^2:
a. $\frac{3}{4}$-in. plywood
b. 2-in. nominal thickness insulating roof deck
c. $\frac{1}{2}$-in. sand aggregate gypsum plaster

12-4 You have a 12 ft × 12 ft piece of aluminum 3 in. thick. How many Btu per hour will it lose with a temperature difference of 10 F impressed across the 3-in. width?

12-5 How thick would a wall of cinder concrete have to be to possess the same insulating properties as 1 in. of rock wool having a density of 10 lbm/ft^3?

12-6 A slab of thickness $L = 1$ ft has the instantaneous temperature distribution of $T = 200x^2 - 40x$. Determine the heat flux at $x = 0$, if $k = 0.02$ Btu/hr-ft-F. (Note: $x = L$ is the other surface.) Is the slab heating up or cooling down?

12-7 A wall of 0.8-ft thickness is to be constructed from material that has an average thermal conductivity of 0.75 Btu/hr-ft-F. The wall is to be insulated with material having an average thermal conductivity of 0.2 Btu/hr-ft-F so that the heat loss per square foot will not exceed 580 Btu/hr. The inner and outer surfaces of the composite wall are at 2400 F and 80 F; the insulation is located on the cooler side. Calculate: (a) the thickness of insulation required; (b) the values of the temperature gradient (dT/dx) at the interface between the wall and insulation; and (c) the temperature at this point.

12-8 A composite wall is formed of a 3-cm copper plate, a 3.2-mm layer of asbestos, and a 5-cm layer of fiberglass. The wall is subjected to an overall temperature difference of 460 C. Calculate the heat flow per unit area through the composite structure.

12-9 Calculate the heat loss per linear foot from a 3-in. steel schedule 40 pipe (3.07 in. ID; 3.50 in. OD; $k = 25$ Btu/hr-ft-F) covered with a $\frac{1}{2}$-in. thickness of asbestos insulation ($k = 0.11$ Btu/hr-ft-F). The pipe transports a fluid at 300 F with an inner-surface convective heat transfer coefficient of 20 Btu/hr-ft^2-F and is exposed to ambient air at 80 F with an average convective coefficient of 2.0 Btu/hr-ft^2-F.

12-10 An exterior wall of a building is composed of 6 in. of concrete backed by 2 in. of glass wool insulation followed by $\frac{3}{8}$-in. gypsum board. On a cold day, the outside air temperature is 5 F ($h_o = 6$ Btu/hr-ft^2-F) and you want to hold the room air temperature at 72 F ($h_i = 2$ Btu/hr-ft^2-F). Determine the heat that must be supplied by the furnace per square foot of wall.

12-11 A composite wall structure experiences a $- 10$ F surface temperature on the outside and 75 F surface temperature on the inside. The wall consists of 4-in.-thick outer face brick, a 2-in. ball of fiberglass insulation, and a $\frac{3}{8}$-in. sheet of gypsum board. Determine the U value and the heat flow rate per square foot. Plot the steady-state temperature profile across the wall.

12-12 A 1-in.-diameter thin copper tube carries refrigerant-12 at 35 F with an inside convective coefficient of 325 Btu/hr-ft^2-F. The tube is covered with $\frac{1}{2}$-in.-thick polyurethane foam insulation ($k = 0.015$ Btu/hr-ft-F). The outside of the insulation is exposed to air at 90 F with a convective coefficient of 4.0 Btu/hr-ft^2-F. The resistance of the copper tube is negligible. Determine the heat gain by the refrigerant in Btu/hr per foot of tube.

12-13 For a plane wall 4 in. thick, with faces maintained at 600 F and 300 F, determine the temperature 1 in. in from the hot surface if k of the material equals 0.6 Btu/hr-ft-F. Assume there is no internal generation. (a) Use the general conduction equation. (b) Use Fourier's law directly.

12-14 Starting with an elemental volume (see Figure 12-7), derive the general conduction equation in cylindrical coordinates.

12-15 During curing, a new concrete wall, 9 in. thick, has the following temperature distribution at a particular instant:

$$T = 85 + 3x - 0.2x^2$$

where T is in degrees Fahrenheit and x is in feet. At that particular time, determine: (a) the maximum temperature in the wall and (b) the heat flow per unit area out each face.

12-16 Calculate the maximum rate of heat transfer per foot from an insulated wire, $\frac{1}{8}$ in. in diameter, covered with $\frac{1}{16}$ in. insulation ($k = 0.8$ Btu/hr-ft-F) and suspended in still air at 80 F such that the convective heat transfer coefficient is 2 Btu/hr-ft^2-F. Assume that the maximum allowable temperature for the insulation is 200 F. If the wire has a resistance of 0.02 ohm/ft, what maximum voltage can be impressed per foot of length?

12-17 A nuclear fuel element in the form of large flat plate, 4 in. thick, has an internal generation rate of 80 Btu/hr-ft^3. One side of the plate is well insulated (no heat transfer past it); the other is kept at 300 F by transferring heat to a cooler fluid. Determine the maximum temperature in the plate if the material has a thermal conductivity of 20 Btu/hr-ft-F.

12-18 Determine the current-carrying capacity of the following insulated copper wire in still air at 80 F ($h = 2.0$ Btu/hr-ft^2-F). Wire: $\frac{1}{4}$-in. diameter; $k = 220$ Btu/hr-ft-F; $R = 0.003$ ohm/ft. Insulation: $\frac{1}{8}$-in.-thick plastic; $k = 0.56$ Btu/hr-ft-F; $c_p = 0.22$ Btu/lbm-F; density $= 54$ lb/ft^3; temperature limit $= 180$ F.

12-19 A copper rod $\frac{1}{2}$ in. in diameter and 3 ft long runs between two large bus bars. The rod is insulated on its lateral surface against the flow of heat and electric current. The bus bars will be at 60 F. What is the maximum current the rod may carry if its temperature is not to exceed 300 F at any point? Assume that the electrical resistivity of copper is constant at $1.72 \times 10^{-6} \, \Omega \cdot$ cm.

12-20 A wire whose resistance per foot of length is 2.3 ohms is embedded along the axis of a cylindrical cement tube of radii 0.02 in. and 0.4 in. A current of 5 amps is found to keep a steady difference of 225 F between the inner and outer surfaces. What is the thermal conductivity of the cement in Btu/hr-ft-F?

12-21 An aluminum sphere weighing 8 lb and initially at a temperature of 500 F is suddenly immersed in a fluid at 60 F. The convection heat transfer coefficient is 10 Btu/hr-ft^2-F. Estimate the time required to cool the aluminum to 200 F.

12-22 A 5-cm-diameter copper sphere is initially at a uniform temperature of 250 C. It is suddenly exposed to an environment at

30 C having a heat transfer coefficient $h = 28$ W/(m² · C). Calculate the time necessary for the sphere's temperature to reach 90 C.

12-23 A copper rod ($k = 215$ Btu/hr-ft-F) that is 4 ft long and $\frac{1}{2}$ in. in diameter runs between two walls maintained at temperatures T_1 and T_2. The rod is exposed to air at 60 F and the convective coefficient of heat transfer is 3.5 Btu/hr-ft²-F. The temperature distribution in the rod is given by

$$T = 60 + 1.9e^{1.25x} + 280e^{-1.25x}$$

where x is in feet and T is in degrees Fahrenheit. Determine: (a) T_1 and T_2, (b) the minimum temperature in the rod, and (c) the heat flow from the middle 2 ft of the rod to the air.

12-24 A $\frac{1}{2}$-in.-diameter copper rod ($k = 211$ Btu/hr-ft-F) is 18 in. long and is heated to 200 F at each end. If the fluid that surrounds the rod is 80 F and a film coefficient of 4.5 Btu/hr-ft²-F exists at the surface, determine the heat dissipated to the fluid by the first 4 in. of the rod.

12-25 The tip of a soldering iron consists of a $\frac{1}{4}$-in.-OD copper rod ($k = 221$ Btu/hr-ft-F) that is 3 in. long. If the tip must be 400 F, what is the temperature of the base and the wattage of the iron? Air temperature is 70 F and convective coefficient is 4.0 Btu/hr-ft²-F.

12-26 A copper rod, 2 ft long and $\frac{1}{2}$ in. in diameter, extends between two large steel plates. The temperature of one plate is maintained at 200 F. The measured temperature in the copper rod at the halfway (1 ft) location is 100 F when air at 80 F is blown across the rod with a convective heat transfer coefficient of 8 Btu/hr-ft²-F. Determine the temperature of the other steel plate.

12-27 Calculate the fin efficiency for a rectangular plate fin that is 1.6 in. high and 0.078 in. thick. The fin is made from aluminum and has an air-side heat transfer coefficient of 6.8 Btu/hr-ft²-F.

12-28 An iron pipe with a 5-cm OD is covered with 6.4-mm asbestos insulation ($k = 0.166$ W/(m · C)) followed by a 2.5-cm layer of fiberglass insulation ($k = 0.048$ W/(m · C)). The pipe's wall temperature is 315 C; the outside insulation temperature is 38 C. Calculate the interface temperature between the asbestos and fiberglass.

12-29 Water flows on the inside of a steel pipe ($k = 43$ W/(m · C)) with an ID of 2.5 cm. The wall thickness is 2 mm. The convection coefficient on the inside is 500 W/(m² · C); on the outside it is 12 W/(m² · C). Calculate the overall heat transfer coefficient.

12-30 An energy-conserving building has massive walls consisting of 4-in. building brick, 6-in. glass wool insulation, and 10-in. concrete (stone). Convective coefficients inside and outside may be taken as 1.46 and 6.00 Btu/hr-ft²-F, respectively. Determine the heat loss for winter design conditions of 72 F inside and 5 F outside in Btu/hr-ft² of the wall.

12-31 The uranium fuel elements in a nuclear reactor are in the form of hollow rods (tubes). The thermal conductivity of the uranium is 41 Btu/hr-ft²-F/ft. Each fuel element is 8 ft long; ID $= \frac{3}{4}$ in. and OD $= 1\frac{1}{2}$ in. The outside surface of the tubes is insulated. Coolant flows through the rods, maintaining the inside surface at 125 F. The reaction generates heat uniformly at the rate of 147 Btu/hr-ft³. For each fuel element determine: (a) the heat transfer to the coolant (Btu/hr) and (b) the maximum temperature in the fuel element (F).

12-32 Determine the temperature profile in a cylindrical geometry assuming constant thermal conductivity. (*Hint*: Recall Equation (12-11).)

12-33 Derive the temperature profiles and heat flow of the situations presented in Table 12-4.

12-34 Consider a furnace wall modeled as a composite surface consisting of four layers: (1) a 3.0-cm refractory brick; (2) an air gap; (3) 3 cm of insulation; and (4) 0.5 cm of plaster. The inside composite wall temperature is 1200 C while the room temperature is 25 C. What is the heat flux to the room? Use k (air) $= 0.0173$ W/(m · C) and the air gap dimension as 2 mm.

12-35 Determine the inside surface temperature for a wall having an indoor air temperature of 80 F and an outside air temperature of 15 F with an overall coefficient of heat transfer of 0.30 Btu/hr-ft²-F(wind velocity 7 mi/hr). h (inside surface) $= 1.47$ Btu/hr-ft²-F.

12-36 For a 50 ft × 2 ft wall with a U factor of 0.30 Btu/hr-ft²-F, determine the heat gain by the room if the outside air is 17 F warmer than the inside air.

12-37 Calculate the overall coefficient for a condenser tube ($k = 60$ Btu/hr-ft-F) having an inside and outside diameter of 0.902 in. and 1.00 in. Assume that the steam-side and water-side coefficients are 1000 Btu/hr-ft²-F and 750 Btu/hr-ft²-F, respectively.

12-38 A heat exchanger is to be built to heat water from 90 F at the rate of 30,000 lbm/hr while cooling 20,000 lbm/hr of water from 190 to 150 F. The specific heat of water may be taken as 1 Btu/lbm-F.

The overall coefficient U is 300 Btu/hr-ft²-F. Determine the area necessary for counterflow operation.

12-39 Hot gases cooling from 700 to 350 F are used to heat 200,000 lbm of oil per hour from 32 to 212 F in a parallel-flow heat exchanger whose gas-side surface area is 20,000 ft². If the oil has a specific heat of 0.60 Btu/lbm-F, determine: (a) the true mean temperature difference and (b) the overall coefficient of heat transfer from gas to oil based on the gas-side area.

12-40 A heat exchanger is to be designed to heat 30 gal/min of water from 60 to 150 F. Outline your method for designing this exchanger with steps, equations, heat sources, and similar details.

12-41 Air is used for cooling an electronics compartment. Atmospheric air enters at 60 F and the maximum allowable air temperature is 100 F. If the equipment in the compartment dissipates 2400 W of energy to the air, determine the necessary air flow rate in (a) lbm/hr and (b) ft³/min at inlet conditions.

12-42 A metal-cased vacuum tube, 1 in. in diameter by 3 in. high, dissipates 30 W when surrounded by air at 75 F. Determine the surface temperature of the case if all heat transfer is by free convection.

12-43 A double-plate glass window has a ½-in. air space separating the panes. The window is 4 ft × 6 ft high. Calculate the heat transfer through the air space for a temperature difference of 65 F.

12-44 Determine the rate of heat loss from the wall of a building in a 10 mi/hr wind blowing parallel to the wall. The wall is 100 ft long and 20 ft high; its surface temperature is 85 F and the temperature of the air is 35 F.

12-45 Determine the maximum power level (in kW) at which a solid copper bus bar, ¼ in. in diameter and 3 ft long, can operate if its surface temperature is not to exceed 400 F. Air at 100 F is blown across the rod at a velocity of 75 ft/sec.

12-46 A cylinder, 1 in. in diameter, is placed in a hot air duct at right angles to the direction of the air flow. The air and cylinder temperatures are 100 and 300 F, respectively. Calculate the value of the film coefficient for a mass flow velocity of 8000 lbm/hr-ft².

12-47 Air at atmospheric pressure is to be heated from 60 to 120 F in tubes of ¾-in. ID whose inside walls are maintained at 140 F. Find the tube length required for an average air velocity of 3 ft/sec.

12-48 In the Gemini space capsule a special fluid is used to transfer heat from the fuel cells, crew, and electronic equipment to the radiating surfaces and then to space. The fluid has the following properties: thermal conductivity, 0.215 Btu/hr-ft-F; specific heat, 0.39 Btu/lbm-F; density, 56 lbm/ft³; and viscosity, 1.4 lbm/ft-hr. If the fluid flows through ¼-in.-ID copper tubing at a velocity of 10 ft/sec, determine the convective heat transfer coefficient h between fluid and tube.

12-49 Forty pounds per minute of water is to be heated from 95 to 105 F as it flows through an electrically heated portion of a ½-in. thin-wall pipe. The heated length is 10 ft. Determine how much hotter the pipe must be than the water. (*Note*: Since the pipe is electrically heated, uniform heat flux will result.)

12-50 Water at the rate of 1.3 lbm/sec and mean velocity of 15.4 ft/sec is to be heated from 80 to 120 F while flowing through the heated section of a smooth ½-in.-diameter tube (cross-sectional area = 0.00136 ft²) maintained at 200 F. Determine the length of the heated section.

12-51 Atmospheric air at a velocity of 200 ft/sec and a temperature of 60 F enters a 2-ft-long square metal duct of 8 in. × 8 in. cross section. If the duct wall is at 300 F, determine the average unit-surface conductance (convective heat transfer coefficient, h).

12-52 Water flows through a 2-in.-diameter, thin-wall, copper tube with a mean velocity of 6 ft/sec. The mean water temperature is 140 F. The water is heated by condensing steam whose temperature is 220 F. The inside convective coefficient is 120 Btu/hr-ft²-F. Estimate the heat transfer per foot of pipe if the convective coefficient on the steam side is 1500 Btu/hr-ft²-F.

12-53 Water at the rate of 165 lbm/hr flows through a "solar-heated" garden hose of ⅝-in. ID and 75-ft length. The water enters at 62 F. As long as the hose remains at 110 F, what is the exit temperature of the water?

12-54 Liquid refrigerant-12 flows inside a 1.25-cm-diameter tube at a velocity of 3 m/s. Calculate the heat transfer coefficient for a bulk temperature of 10 C.

12-55 Air at 1 atm and 15 C flows through a long rectangular duct 7.5 cm × 15 cm. A 1.8-m section of the duct is maintained at 120 C, and the average air temperature at exit from this section is 65 C. Calculate the air flow rate and the total heat transfer.

12-56 You wish to heat 0.2 kg/s of water from 15.6 to 60 C while it is flowing through an electrically heated section of 2.50-cm-ID steel tubing. The tube temperature is not to exceed 149 C. Determine the heated length required (in meters).

12-57 Determine the heat loss (W/m) from steam at 112 C flowing inside a steel pipe that has an inside diameter of 2.54 cm and an outside diameter of 3.18 cm. Assume the pipe is covered with glass wool insulation that is 1.25 cm thick and the outside of the insulation is exposed to air at 23 C. The inside convective coefficient is 369 W/(m^2 · C); outside it is 17 W/(m^2 · C).

12-58 A power resistor, mounted vertically, is $\frac{1}{2}$ in. in diameter by 6 in. high and must operate at 310 F or less when surrounded by air at 90 F. Determine the maximum power rating (in watts) for the resistor if all heat transfer is by free convection with negligible heat loss from the ends.

12-59 Radiant flux of 300 Btu/hr-ft^2 strikes a surface where 200 Btu/hr-ft^2 is reflected and 60 Btu/hr-ft^2 is transmitted through the material. Compute the absorptance of the material.

12-60 A greenhouse is to be constructed of rigid polyethylene plastic. The plastic has the following properties: for wavelengths up to 0.75 μ, $\alpha = 0.05$ and $\rho = 0.05$; above 0.75μ, $\alpha = 0.30$ and $\rho = 0.55$. If 224 Btu/hr of radiant energy from a blackbody source at 10,000 F falls on a square foot of the plastic, determine the amount of radiant energy that passes through the plastic (in Btu/hr-ft^2).

12-61 Find the maximum monochromatic emissive power for a blackbody at 600 F and for a graybody at 600 F having an emittance of 0.43. At what wavelength does the maximum emission occur? How much is emitted at a wavelength of 2.5 μ?

12-62 Find the amount of radiant energy emitted between wavelengths of 0.20 and 0.83 μ for a 6000 F graybody having an emittance of 0.71.

12-63 How much of the sun's radiation is in the visible range?

12-64 **a.** Determine the amount of radiant energy emitted by a graybody of 0.70 emittance at a wavelength of 0.75 and a temperature of 1140 F (in Btu/hr-ft^2).

b. For a blackbody surface at 3000 F, determine the radiant energy emitted in the infrared region (0.7 to 1000 μ).

c. For a blackbody at 300 R, determine the wavelength at which maximum monochromatic emission occurs.

12-65 Determine the configuration factor between the sun and the earth and the earth and the sun (see the sketch). (*Hint:* You will have to apply the basic definition.)

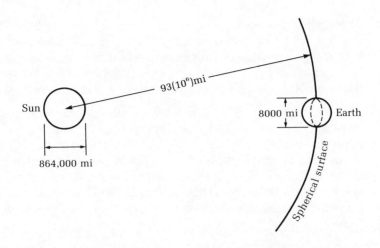

12-66 If the sun can be considered a blackbody at 10,800 R, determine the normal incident radiation on a square foot of the earth's surface (no atmosphere).

12-67 A bank of thermoelectric modules is used for auxiliary cooling on a spacecraft. The modules form a plate, 1 ft × 3 ft, from which heat must be radiated to space. The plate surface has an emittance of 0.7 and is at 70 F. The plate "sees" space (which may be treated as a blackbody at 0 R) and nothing else. Determine the rate (in Btu/hr) at which heat is rejected from the thermoelectric units.

12-68 A copper sphere (area, 0.5 ft²; emittance, 0.3) is placed in a large room. The temperature of the sphere is maintained at 300 F while the room is held at 100 F. The walls of the room are black. Determine the net rate of heat loss from the copper sphere. Derive or explain all the relations you use.

12-69 A sphere is heated internally to a temperature of 140 F. The emittance of the surface is 0.1. For a surrounding temperature of 40 F, the heat loss due to radiation is 370 Btu/hr. If the temperature of the sphere is increased to 540 F, compute the heat loss to the surroundings.

12-70 The emittance of opaque coatings (such as paint) is determined by applying the coating to the exterior of a 6-in.-diameter hollow sphere that contains an electric heating element. The sphere is then placed in an evacuated container that has black walls. In one such test, the heating element dissipated 100 W of power and the exterior surface of the sphere stabilized at 400 F when the container walls were kept at 80 F. What is the reflectance of the coating?

12-71 The surface temperature of insulation on a horizontal pipe pass-ing through a large room is 120 F. The emittance of the insulation is 0.90. The temperature of the surrounding air and room is 80 F. Determine the rate of heat loss per foot of pipe when the outside diameter is 8.5 in. Include both radiation and convection.

12-72 A black sphere, 1 in. in diameter, is placed in a large infrared heating oven whose walls are maintained at 700 F. The tem-perature of the air in the oven is 200 F and the heat transfer coefficient for convection between the surface of the sphere and the air is 5 Btu/hr-ft²-F. (a) Estimate the net rate of heat flow to the sphere when its temperature is 100 F. (b) Determine the equi-librium temperature of the sphere.

12-73 A large slab of steel 4 in. thick has a 4-in-diameter hole with axis normal to the surface. The sides of the hole have an emittance of 0.2. The surroundings are black. Determine the net rate of heat loss from the hole if the plate is at 1000 F and the surroundings are at 80 F. Work this problem in two ways: (a) network and (b) radiosity equations.

12-74 A 20 ft × 20 ft room with 8-ft ceiling is to be heated by means of a panel-heating installation in the ceiling. There are three inte-rior walls that can be considered as nonconducting but re-radiating (q = 0). The exterior wall has an average emittance of 0.90 and loses 1005 Btu/hr to the outside when the exterior air temperature is 20 F. The floor temperature is to be kept at 75 F for comfort. There is no reflected energy from the carpeting on the floor. If the painted ceiling (emittance of 0.92) is maintained at 105 F, determine: (a) the net heat loss or gain by each surface; (b) the temperature of each wall; and (c) the heating cost per hour at 5¢/kW-hr.

12-75 A 10 ft × 10 ft × 8 ft high room is to be heated by means of a panel-heating installation in the ceiling. The ceiling is painted with a white enamel and is maintained at 110 F. The walls are well insulated. The floor may be treated as a blackbody at 70 F. Neglecting convection, determine: (a) the net heat loss or gain by each surface; (b) the heating cost per hour at 5¢/kW-hr; and (c) the temperature of the walls.

12-76 A polished aluminum sphere (emittance of 0.05) is suspended by a fine wire in the center of a large treating chamber that is 10 ft × 10 ft × 10 ft. The ceiling of the chamber (surface 1) has an emittance of 0.85 and is at 125 F. The floor (surface 2) is black and at 70 F. The walls (surface 3) are also black but at 75 F. The sphere (surface 4) has a surface area of 1.0 ft². Neglecting con-vection, find the temperature of the sphere.

12-77 Taking the moon as a disk 2000 mi in diameter and 93,000,000 mi from the sun, and assuming that its emittance equals its absorptance for all wavelengths, compute its mean surface temperature due to the sun's heat only. Assume that it receives heat as a disk but radiates to empty space as a sphere.

12-78 What is the direct radiant interchange between a black heated concrete floor slab, 12 ft × 16 ft, and the black plaster wall adjoining the 16-ft side if the slab is at 120 F and the wall is at 65 F?

12-79 What is the net radiant interchange between 1 mi^2 of cloudy sky and 1 mi^2 of open water beneath the clouds if the water is at 58 F and the clouds are at 70 F? How many pounds per hour of water will this energy rate condense from saturated vapor conditions at 70 F? Assume that the cloud and water layers are black. If the upper layer of cloud is receiving 190 Btu/hr-ft^2 from the sun, will it rain or clear?

12-80 Considering the emittance of the earth as 0.9 and its absorptance to solar radiation as 0.83, estimate the average temperature of the earth. Assume that it receives heat as a disk and radiates as a sphere.

12-81 Estimate the equilibrium temperature of the planet Mars, which has a diameter of 4300 mi and revolves around the sun at a distance of 149 × 10^6 mi. The sun acts as a blackbody with a temperature of 10,000 F and has a radius of 432,000 mi. The albedo of Mars (the fraction of the incoming radiation that is reflected to space) is 0.15. (*Hint*: Determine a configuration factor from basic concepts.)

12-82 A solar house has a roof area of 400 ft^2. Assuming that its absorptance to sunlight is 0.95 and its emittance to space is 0.5, what is the heat available for the house if the collector temperature is 150 F? If it is 250 F? What would be the best collection temperature from the standpoint of maximum energy going into the house? (Assume that the roof is a flat surface facing the sun at a distance of 93,000,000 mi, and ignore atmospheric losses.)

12-83 The earth's atmosphere has a transmittance to solar radiation of 0.54. Determine the maximum amount of water that can be heated from 60 to 140 F (lbm/hr) by a 3 ft × 6 ft solar collector having a collection efficiency of 45%. Take the sun as a blackbody at 10,800 R and use the solar system dimensions shown in the sketch of Problem 12-65.

12-84 A 3 m × 3 m × 2.5 m high room is to be heated by means of a panel-heating installation in the ceiling. The ceiling is painted with a white enamel and is maintained at 43 C. The walls are well insulated. The floor is maintained at 21 C. Neglecting convection determine: (a) the net heat loss or gain by each surface; (b) the heating cost per hour at 6¢/kW-hr; and (c) the temperature of the walls (degrees Celsius).

12-85 What is the direct radiant interchange between a black heated concrete floor slab, 4 m × 5 m, and a black plaster wall adjoining the 5-m side if the slab is at 50 C and the wall is at 19 C? The other dimension of the wall is $2\frac{1}{2}$ m.

Appendices

Appendix A-1: Steam Tables

Table A-1-1 Saturated Steam: Temperature Table (English)

Temp Fahr. T	Abs. Press lbf/in.² p	Specific Volume, ft³/lbm Sat. Liquid v_f	Evap. v_{fg}	Sat. Vapor v_g	Enthalpy, Btu/lbm Sat. Liquid h_f	Evap. h_{fg}	Sat. Vapor h_g	Entropy, Btu/lbm R Sat. Liquid s_f	Evap. s_{fg}	Sat. Vapor s_g	Temp Fahr. T
32.0†	0.08859	0.016022	3304.7	3304.7	−0.0179	1075.5	1075.5	0.0000	2.1873	2.1873	32.0
34.0	0.09600	0.016021	3061.9	3061.9	1.996	1074.4	1076.4	0.0041	2.1762	2.1802	34.0
36.0	0.10395	0.016020	2839.0	2839.0	4.008	1073.2	1077.2	0.0081	2.1651	2.1732	36.0
38.0	0.11249	0.016019	2634.1	2634.2	6.018	1072.1	1078.1	0.0122	2.1541	2.1663	38.0
40.0	0.12163	0.016019	2445.8	2445.8	8.027	1071.0	1079.0	0.0162	2.1432	2.1594	40.0
42.0	0.13143	0.016019	2272.4	2272.4	10.035	1069.8	1079.9	0.0202	2.1325	2.1527	42.0
44.0	0.14192	0.016019	2112.8	2112.8	12.041	1068.7	1080.7	0.0242	2.1217	2.1459	44.0
46.0	0.15314	0.016020	1965.7	1965.7	14.047	1067.6	1081.6	0.0282	2.1111	2.1393	46.0
48.0	0.16514	0.016021	1830.0	1830.0	16.051	1066.4	1082.5	0.0321	2.1006	2.1327	48.0
50.0	0.17796	0.016023	1704.8	1704.8	18.054	1065.3	1083.4	0.0361	2.0901	2.1262	50.0
52.0	0.19165	0.016024	1589.2	1589.2	20.057	1064.2	1084.2	0.0400	2.0798	2.1197	52.0
54.0	0.20625	0.016026	1482.4	1482.4	22.058	1063.1	1085.1	0.0439	2.0695	2.1134	54.0
56.0	0.22183	0.016028	1383.6	1383.6	24.059	1061.9	1086.0	0.0478	2.0593	2.1070	56.0
58.0	0.23843	0.016031	1292.2	1292.2	26.060	1060.8	1086.9	0.0516	2.0491	2.1008	58.0
60.0	0.25611	0.016033	1207.6	1207.6	28.060	1059.7	1087.7	0.0555	2.0391	2.0946	60.0
62.0	0.27494	0.016036	1129.2	1129.2	30.059	1058.5	1088.6	0.0593	2.0291	2.0885	62.0
64.0	0.29497	0.016039	1056.5	1056.5	32.058	1057.4	1089.5	0.0632	2.0192	2.0824	64.0
66.0	0.31626	0.016043	989.0	989.1	34.056	1056.3	1090.4	0.0670	2.0094	2.0764	66.0
68.0	0.33889	0.016046	926.5	926.5	36.054	1055.2	1091.2	0.0708	1.9996	2.0704	68.0

† Approximate Triple Point

Source: Reprinted with permission from ASME.

Table A-1-1 Saturated Steam: Temperature Table (English) (cont.)

Temp Fahr. T	Abs. Press lbf/in.² p	Specific Volume, ft³/lbm Sat. Liquid v_f	Evap. v_fg	Sat. Vapor v_g	Enthalpy, Btu/lbm Sat. Liquid h_f	Evap. h_fg	Sat. Vapor h_g	Entropy, Btu/lbm R Sat. Liquid s_f	Evap. s_fg	Sat. Vapor s_g	Temp Fahr. T
70.0	0.36292	0.016050	868.3	868.4	38.052	1054.0	1092.1	0.0745	1.9900	2.0645	70.0
72.0	0.38844	0.016054	814.3	814.3	40.049	1052.9	1093.0	0.0783	1.9804	2.0587	72.0
74.0	0.41550	0.016058	764.1	764.1	42.046	1051.8	1093.8	0.0821	1.9708	2.0529	74.0
76.0	0.44420	0.016063	717.4	717.4	44.043	1050.7	1094.7	0.0858	1.9614	2.0472	76.0
78.0	0.47461	0.016067	673.8	673.9	46.040	1049.5	1095.6	0.0895	1.9520	2.0415	78.0
80.0	0.50683	0.016072	633.3	633.3	48.037	1048.4	1096.4	0.0932	1.9426	2.0359	80.0
82.0	0.54093	0.016077	595.5	595.5	50.033	1047.3	1097.3	0.0969	1.9334	2.0303	82.0
84.0	0.57702	0.016082	560.3	560.3	52.029	1046.1	1098.2	0.1006	1.9242	2.0248	84.0
86.0	0.61518	0.016087	527.5	527.5	54.026	1045.0	1099.0	0.1043	1.9151	2.0193	86.0
88.0	0.65551	0.016093	496.8	496.8	56.022	1043.9	1099.9	0.1079	1.9060	2.0139	88.0
90.0	0.69813	0.016099	468.1	468.1	58.018	1042.7	1100.8	0.1115	1.8970	2.0086	90.0
92.0	0.74313	0.016105	441.3	441.3	60.014	1041.6	1101.6	0.1152	1.8881	2.0033	92.0
94.0	0.79062	0.016111	416.3	416.3	62.010	1040.5	1102.5	0.1188	1.8792	1.9980	94.0
96.0	0.84072	0.016117	392.8	392.9	64.006	1039.3	1103.3	0.1224	1.8704	1.9928	96.0
98.0	0.89356	0.016123	370.9	370.9	66.003	1038.2	1104.2	0.1260	1.8617	1.9876	98.0
100.0	0.94924	0.016130	350.4	350.4	67.999	1037.1	1105.1	0.1295	1.8530	1.9825	100.0
102.0	1.00789	0.016137	331.1	331.1	69.995	1035.9	1105.9	0.1331	1.8444	1.9775	102.0
104.0	1.06965	0.016144	313.1	313.1	71.992	1034.8	1106.8	0.1366	1.8358	1.9725	104.0
106.0	1.1347	0.016151	296.16	296.18	73.99	1033.6	1107.6	0.1402	1.8273	1.9675	106.0
108.0	1.2030	0.016158	280.28	280.30	75.98	1032.5	1108.5	0.1437	1.8188	1.9626	108.0
110.0	1.2750	0.016165	265.37	265.39	77.98	1031.4	1109.3	0.1472	1.8105	1.9577	110.0
112.0	1.3505	0.016173	251.37	251.38	79.98	1030.2	1110.2	0.1507	1.8021	1.9528	112.0
114.0	1.4299	0.016180	238.21	238.22	81.97	1029.1	1111.1	0.1542	1.7938	1.9480	114.0
116.0	1.5133	0.016188	225.84	225.85	83.97	1027.9	1111.9	0.1577	1.7856	1.9433	116.0
118.0	1.6009	0.016196	214.20	214.21	85.97	1026.8	1112.7	0.1611	1.7774	1.9386	118.0
120.0	1.6927	0.016204	203.25	203.26	87.97	1025.6	1113.6	0.1646	1.7693	1.9339	120.0
122.0	1.7891	0.016213	192.94	192.95	89.96	1024.5	1114.4	0.1680	1.7613	1.9293	122.0
124.0	1.8901	0.016221	183.23	183.24	91.96	1023.3	1115.3	0.1715	1.7533	1.9247	124.0
126.0	1.9959	0.016229	174.08	174.09	93.96	1022.2	1116.1	0.1749	1.7453	1.9202	126.0
128.0	2.1068	0.016238	165.45	165.47	95.96	1021.0	1117.0	0.1783	1.7374	1.9157	128.0

Temp	Press	v_f			h_f	h_fg	h_g	s_f	s_fg	s_g	Temp
130.0	2.2230	0.016247	157.33	157.32	97.96	1019.8	1117.8	0.1817	1.7295	1.9112	130.0
132.0	2.3445	0.016256	149.66	149.64	99.95	1018.7	1118.6	0.1851	1.7217	1.9068	132.0
134.0	2.4717	0.016265	142.41	142.40	101.95	1017.5	1119.5	0.1884	1.7140	1.9024	134.0
136.0	2.6047	0.016274	135.57	135.55	103.95	1016.4	1120.3	0.1918	1.7063	1.8980	136.0
138.0	2.7438	0.016284	129.11	129.09	105.95	1015.2	1121.1	0.1951	1.6986	1.8937	138.0
140.0	2.8892	0.016293	123.00	122.98	107.95	1014.0	1122.0	0.1985	1.6910	1.8895	140.0
142.0	3.0411	0.016303	117.22	117.21	109.95	1012.9	1122.8	0.2018	1.6834	1.8852	142.0
144.0	3.1997	0.016312	111.76	111.74	111.95	1011.7	1123.6	0.2051	1.6759	1.8810	144.0
146.0	3.3653	0.016322	106.59	106.58	113.95	1010.5	1124.5	0.2084	1.6684	1.8769	146.0
148.0	3.5381	0.016332	101.70	101.68	115.95	1009.3	1125.3	0.2117	1.6610	1.8727	148.0
150.0	3.7184	0.016343	97.07	97.05	117.95	1008.2	1126.1	0.2150	1.6536	1.8686	150.0
152.0	3.9065	0.016353	92.68	92.66	119.95	1007.0	1126.9	0.2183	1.6463	1.8646	152.0
154.0	4.1025	0.016363	88.52	88.50	121.95	1005.8	1127.7	0.2216	1.6390	1.8606	154.0
156.0	4.3068	0.016374	84.57	84.56	123.95	1004.6	1128.6	0.2248	1.6318	1.8566	156.0
158.0	4.5197	0.016384	80.83	80.82	125.96	1003.4	1129.4	0.2281	1.6245	1.8526	158.0
160.0	4.7414	0.016395	77.29	77.27	127.96	1002.2	1130.2	0.2313	1.6174	1.8487	160.0
162.0	4.9722	0.016406	73.92	73.90	129.96	1001.0	1131.0	0.2345	1.6103	1.8448	162.0
164.0	5.2124	0.016417	70.72	70.70	131.96	999.8	1131.8	0.2377	1.6032	1.8409	164.0
166.0	5.4623	0.016428	67.68	67.67	133.97	998.6	1132.6	0.2409	1.5961	1.8371	166.0
168.0	5.7223	0.016440	64.80	64.78	135.97	997.4	1133.4	0.2441	1.5892	1.8333	168.0
170.0	5.9926	0.016451	62.06	62.04	137.97	996.2	1134.2	0.2473	1.5822	1.8295	170.0
172.0	6.2736	0.016463	59.45	59.43	139.98	995.0	1135.0	0.2505	1.5753	1.8258	172.0
174.0	6.5656	0.016474	56.97	56.95	141.98	993.8	1135.8	0.2537	1.5684	1.8221	174.0
176.0	6.8690	0.016486	54.61	54.59	143.99	992.6	1136.6	0.2568	1.5616	1.8184	176.0
178.0	7.1840	0.016498	52.36	52.35	145.99	991.4	1137.4	0.2600	1.5548	1.8147	178.0
180.0	7.5110	0.016510	50.22	50.21	148.00	990.2	1138.2	0.2631	1.5480	1.8111	180.0
182.0	7.850	0.016522	48.189	48.172	150.01	989.0	1139.0	0.2662	1.5413	1.8075	182.0
184.0	8.203	0.016534	46.249	46.232	152.01	987.8	1139.8	0.2694	1.5346	1.8040	184.0
186.0	8.568	0.016547	44.400	44.383	154.02	986.5	1140.5	0.2725	1.5279	1.8004	186.0
188.0	8.947	0.016559	42.638	42.621	156.03	985.3	1141.3	0.2756	1.5213	1.7969	188.0
190.0	9.340	0.016572	40.957	40.941	158.04	984.1	1142.1	0.2787	1.5148	1.7934	190.0
192.0	9.747	0.016585	39.354	39.337	160.05	982.8	1142.9	0.2818	1.5082	1.7900	192.0
194.0	10.168	0.016598	37.824	37.808	162.05	981.6	1143.7	0.2848	1.5017	1.7865	194.0
196.0	10.605	0.016611	36.364	36.348	164.06	980.4	1144.4	0.2879	1.4952	1.7831	196.0
198.0	11.058	0.016624	34.970	34.954	166.08	979.1	1145.2	0.2910	1.4888	1.7798	198.0

Table A-1-1 Saturated Steam: Temperature Table (English) (cont.)

Temp Fahr. T	Abs. Press lbf/in.² p	Specific Volume, ft³/lbm			Enthalpy, Btu/lbm			Entropy, Btu/lbm R			Temp Fahr. T
		Sat. Liquid v_f	Evap. v_{fg}	Sat. Vapor v_g	Sat. Liquid h_f	Evap. h_{fg}	Sat. Vapor h_g	Sat. Liquid s_f	Evap. s_{fg}	Sat. Vapor s_g	
200.0	11.526	0.016637	33.622	33.639	168.09	977.9	1146.0	0.2940	1.4824	1.7764	200.0
204.0	12.512	0.016664	31.135	31.151	172.11	975.4	1147.5	0.3001	1.4697	1.7698	204.0
208.0	13.568	0.016691	28.862	28.878	176.14	972.8	1149.0	0.3061	1.4571	1.7632	208.0
212.0	14.696	0.016719	26.782	26.799	180.17	970.3	1150.5	0.3121	1.4447	1.7568	212.0
216.0	15.901	0.016747	24.878	24.894	184.20	967.8	1152.0	0.3181	1.4323	1.7505	216.0
220.0	17.186	0.016775	23.131	23.148	188.23	965.2	1153.4	0.3241	1.4201	1.7442	220.0
224.0	18.556	0.016805	21.529	21.545	192.27	962.6	1154.9	0.3300	1.4081	1.7380	224.0
228.0	20.015	0.016834	20.056	20.073	196.31	960.0	1156.3	0.3359	1.3961	1.7320	228.0
232.0	21.567	0.016864	18.701	18.718	200.35	957.4	1157.8	0.3417	1.3842	1.7260	232.0
236.0	23.216	0.016895	17.454	17.471	204.40	954.8	1159.2	0.3476	1.3725	1.7201	236.0
240.0	24.968	0.016926	16.304	16.321	208.45	952.1	1160.6	0.3533	1.3609	1.7142	240.0
244.0	26.826	0.016958	15.243	15.260	212.50	949.5	1162.0	0.3591	1.3494	1.7085	244.0
248.0	28.796	0.016990	14.264	14.281	216.56	946.8	1163.4	0.3649	1.3379	1.7028	248.0
252.0	30.883	0.017022	13.358	13.375	220.62	944.1	1164.7	0.3706	1.3266	1.6972	252.0
256.0	33.091	0.017055	12.520	12.538	224.69	941.4	1166.1	0.3763	1.3154	1.6917	256.0
260.0	35.427	0.017089	11.745	11.762	228.76	938.6	1167.4	0.3819	1.3043	1.6862	260.0
264.0	37.894	0.017123	11.025	11.042	232.83	935.9	1168.7	0.3876	1.2933	1.6808	264.0
268.0	40.500	0.017157	10.358	10.375	236.91	933.1	1170.0	0.3932	1.2823	1.6755	268.0
272.0	43.249	0.017193	9.738	9.755	240.99	930.3	1171.3	0.3987	1.2715	1.6702	272.0
276.0	46.147	0.017228	9.162	9.180	245.08	927.5	1172.5	0.4043	1.2607	1.6650	276.0
280.0	49.200	0.017264	8.627	8.644	249.17	924.6	1173.8	0.4098	1.2501	1.6599	280.0
284.0	52.414	0.01730	8.1280	8.1453	253.3	921.7	1175.0	0.4154	1.2395	1.6548	284.0
288.0	55.795	0.01734	7.6634	7.6807	257.4	918.8	1176.2	0.4208	1.2290	1.6498	288.0
292.0	59.350	0.01738	7.2301	7.2475	261.5	915.9	1177.4	0.4263	1.2186	1.6449	292.0
296.0	63.084	0.01741	6.8259	6.8433	265.6	913.0	1178.6	0.4317	1.2082	1.6400	296.0
300.0	67.005	0.01745	6.4483	6.4658	269.7	910.0	1179.7	0.4372	1.1979	1.6351	300.0
304.0	71.119	0.01749	6.0955	6.1130	273.8	907.0	1180.9	0.4426	1.1877	1.6303	304.0
308.0	75.433	0.01753	5.7655	5.7830	278.0	904.0	1182.0	0.4479	1.1776	1.6256	308.0
312.0	79.953	0.01757	5.4566	5.4742	282.1	901.0	1183.1	0.4533	1.1676	1.6209	312.0
316.0	84.688	0.01761	5.1673	5.1849	286.3	897.9	1184.1	0.4586	1.1576	1.6162	316.0

Temp											Temp
320.0	89.643	0.01766	4.8961	4.9138	290.4	894.8	1185.2	0.4640	1.1477	1.6116	320.0
324.0	94.826	0.01770	4.6418	4.6595	294.6	891.6	1186.2	0.4692	1.1378	1.6071	324.0
328.0	100.245	0.01774	4.4030	4.4208	298.7	888.5	1187.2	0.4745	1.1280	1.6025	328.0
332.0	105.907	0.01779	4.1788	4.1966	302.9	885.3	1188.2	0.4798	1.1183	1.5981	332.0
336.0	111.820	0.01783	3.9681	3.9859	307.1	882.1	1189.1	0.4850	1.1086	1.5936	336.0
340.0	117.992	0.01787	3.7699	3.7878	311.3	878.8	1190.1	0.4902	1.0990	1.5892	340.0
344.0	124.430	0.01792	3.5834	3.6013	315.5	875.5	1191.0	0.4954	1.0894	1.5849	344.0
348.0	131.142	0.01797	3.4078	3.4258	319.7	872.2	1191.1	0.5006	1.0799	1.5806	348.0
352.0	138.138	0.01801	3.2423	3.2603	323.9	868.9	1192.7	0.5058	1.0705	1.5763	352.0
356.0	145.424	0.01806	3.0863	3.1044	328.1	865.5	1193.6	0.5110	1.0611	1.5721	356.0
360.0	153.010	0.01811	2.9392	2.9573	332.3	862.1	1194.4	0.5161	1.0517	1.5678	360.0
364.0	160.903	0.01816	2.8002	2.8184	336.5	858.6	1195.2	0.5212	1.0424	1.5637	364.0
368.0	169.113	0.01821	2.6691	2.6873	340.8	855.1	1195.9	0.5263	1.0332	1.5595	368.0
372.0	177.648	0.01826	2.5451	2.5633	345.0	851.6	1196.7	0.5314	1.0240	1.5554	372.0
376.0	186.517	0.01831	2.4279	2.4462	349.3	848.1	1197.4	0.5365	1.0148	1.5513	376.0
380.0	195.729	0.01836	2.3170	2.3353	353.6	844.5	1198.0	0.5416	1.0057	1.5473	380.0
384.0	205.294	0.01842	2.2120	2.2304	357.9	840.8	1198.7	0.5466	0.9966	1.5432	384.0
388.0	215.220	0.01847	2.1126	2.1311	362.2	837.2	1199.3	0.5516	0.9876	1.5392	388.0
392.0	225.516	0.01853	2.0184	2.0369	366.5	833.4	1199.9	0.5567	0.9786	1.5352	392.0
396.0	236.193	0.01858	1.9291	1.9477	370.8	829.7	1200.4	0.5617	0.9696	1.5313	396.0
400.0	247.259	0.01864	1.8444	1.8630	375.1	825.9	1201.0	0.5667	0.9607	1.5274	400.0
404.0	258.725	0.01870	1.7640	1.7827	379.4	822.0	1201.5	0.5717	0.9518	1.5234	404.0
408.0	270.600	0.01875	1.6877	1.7064	383.8	818.2	1201.9	0.5766	0.9429	1.5195	408.0
412.0	282.894	0.01881	1.6152	1.6340	388.1	814.2	1202.4	0.5816	0.9341	1.5157	412.0
416.0	295.617	0.01887	1.5463	1.5651	392.5	810.2	1202.8	0.5866	0.9253	1.5118	416.0
420.0	308.780	0.01894	1.4808	1.4997	396.9	806.2	1203.1	0.5915	0.9165	1.5080	420.0
424.0	322.391	0.01900	1.4184	1.4374	401.3	802.2	1203.5	0.5964	0.9077	1.5042	424.0
428.0	336.463	0.01906	1.3591	1.3782	405.7	798.0	1203.7	0.6014	0.8990	1.5004	428.0
432.0	351.00	0.01913	1.30266	1.32179	410.1	793.9	1204.0	0.6063	0.8903	1.4966	432.0
436.0	366.03	0.01919	1.24887	1.26806	414.6	789.7	1204.2	0.6112	0.8816	1.4928	436.0
440.0	381.54	0.01926	1.19761	1.21687	419.0	785.4	1204.4	0.6161	0.8729	1.4890	440.0
444.0	397.56	0.01933	1.14874	1.16806	423.5	781.1	1204.6	0.6210	0.8643	1.4853	444.0
448.0	414.09	0.01940	1.10212	1.12152	428.0	776.7	1204.7	0.6259	0.8557	1.4815	448.0
452.0	431.14	0.01947	1.05764	1.07711	432.5	772.3	1204.8	0.6308	0.8471	1.4778	452.0
456.0	448.73	0.01954	1.01518	1.03472	437.0	767.8	1204.8	0.6356	0.8385	1.4741	456.0

Table A-1-1 Saturated Steam: Temperature Table (English) (cont.)

Temp Fahr. T	Abs. Press lbf/in.² p	Specific Volume, ft³/lbm			Enthalpy, Btu/lbm			Entropy, Btu/lbm R			Temp Fahr. T
		Sat. Liquid v_f	Evap. v_{fg}	Sat. Vapor v_g	Sat. Liquid h_f	Evap. h_{fg}	Sat. Vapor h_g	Sat. Liquid s_f	Evap. s_{fg}	Sat. Vapor s_g	
460.0	466.87	0.01961	0.97463	0.99424	441.5	763.2	1204.8	0.6405	0.8299	1.4704	460.0
464.0	485.56	0.01969	0.93588	0.95557	446.1	758.6	1204.7	0.6454	0.8213	1.4667	464.0
468.0	504.83	0.01976	0.89885	0.91862	450.7	754.0	1204.6	0.6502	0.8127	1.4629	468.0
472.0	524.67	0.01984	0.86345	0.88329	455.2	749.3	1204.5	0.6551	0.8042	1.4592	472.0
476.0	545.11	0.01992	0.82958	0.84950	459.9	744.5	1204.3	0.6599	0.7956	1.4555	476.0
480.0	566.15	0.02000	0.79716	0.81717	464.5	739.6	1204.1	0.6648	0.7871	1.4518	480.0
484.0	587.81	0.02009	0.76613	0.78622	469.1	734.7	1203.8	0.6696	0.7785	1.4481	484.0
488.0	610.10	0.02017	0.73641	0.75658	473.8	729.7	1203.5	0.6745	0.7700	1.4444	488.0
492.0	633.03	0.02026	0.70794	0.72820	478.5	724.6	1203.1	0.6793	0.7614	1.4407	492.0
496.0	656.61	0.02034	0.68065	0.70100	483.2	719.5	1202.7	0.6842	0.7528	1.4370	496.0
500.0	680.86	0.02043	0.65448	0.67492	487.9	714.3	1202.2	0.6890	0.7443	1.4333	500.0
504.0	705.78	0.02053	0.62938	0.64991	492.7	709.0	1201.7	0.6939	0.7357	1.4296	504.0
508.0	731.40	0.02062	0.60530	0.62592	497.5	703.7	1201.1	0.6987	0.7271	1.4258	508.0
512.0	757.72	0.02072	0.58218	0.60289	502.3	698.2	1200.5	0.7036	0.7185	1.4221	512.0
516.0	784.76	0.02081	0.55997	0.58079	507.1	692.7	1199.8	0.7085	0.7099	1.4183	516.0
520.0	812.53	0.02091	0.53864	0.55956	512.0	687.0	1199.0	0.7133	0.7013	1.4146	520.0
524.0	841.04	0.02102	0.51814	0.53916	516.9	681.3	1198.2	0.7182	0.6926	1.4108	524.0
528.0	870.31	0.02112	0.49843	0.51955	521.8	675.5	1197.3	0.7231	0.6839	1.4070	528.0
532.0	900.34	0.02123	0.47947	0.50070	526.8	669.6	1196.4	0.7280	0.6752	1.4032	532.0
536.0	931.17	0.02134	0.46123	0.48257	531.7	663.6	1195.4	0.7329	0.6665	1.3993	536.0
540.0	962.79	0.02146	0.44367	0.46513	536.8	657.5	1194.3	0.7378	0.6577	1.3954	540.0
544.0	995.22	0.02157	0.42677	0.44834	541.8	651.3	1193.1	0.7427	0.6489	1.3915	544.0
548.0	1028.49	0.02169	0.41048	0.43217	546.9	645.0	1191.9	0.7476	0.6400	1.3876	548.0
552.0	1062.59	0.02182	0.39479	0.41660	552.0	638.5	1190.6	0.7525	0.6311	1.3837	552.0
556.0	1097.55	0.02194	0.37966	0.40160	557.2	632.0	1189.2	0.7575	0.6222	1.3797	556.0
560.0	1133.38	0.02207	0.36507	0.38714	562.4	625.3	1187.7	0.7625	0.6132	1.3757	560.0
564.0	1170.10	0.02221	0.35099	0.37320	567.6	618.5	1186.1	0.7674	0.6041	1.3716	564.0
568.0	1207.72	0.02235	0.33741	0.35975	572.9	611.5	1184.5	0.7725	0.5950	1.3675	568.0
572.0	1246.26	0.02249	0.32429	0.34678	578.3	604.5	1182.7	0.7775	0.5859	1.3634	572.0
576.0	1285.74	0.02264	0.31162	0.33426	583.7	597.2	1180.9	0.7825	0.5766	1.3592	576.0

Temp											Temp
580.0	1326.17	0.02279	0.29937	0.32216	589.1	589.9	1179.0	0.7876	0.5673	1.3550	580.0
584.0	1367.7	0.02295	0.28753	0.31048	594.6	582.4	1176.9	0.7927	0.5580	1.3507	584.0
588.0	1410.0	0.02311	0.27608	0.29919	600.1	574.7	1174.8	0.7978	0.5485	1.3464	588.0
592.0	1453.3	0.02328	0.26499	0.28827	605.7	566.8	1172.6	0.8030	0.5390	1.3420	592.0
596.0	1497.8	0.02345	0.25425	0.27770	611.4	558.8	1170.2	0.8082	0.5293	1.3375	596.0
600.0	1543.2	0.02364	0.24384	0.26747	617.1	550.6	1167.7	0.8134	0.5196	1.3330	600.0
604.0	1589.7	0.02382	0.23374	0.25757	622.9	542.2	1165.1	0.8187	0.5097	1.3284	604.0
608.0	1637.3	0.02402	0.22394	0.24796	628.8	533.6	1162.4	0.8240	0.4997	1.3238	608.0
612.0	1686.1	0.02422	0.21442	0.23865	634.8	524.7	1159.5	0.8294	0.4896	1.3190	612.0
616.6	1735.9	0.02444	0.20516	0.22960	640.8	515.6	1156.4	0.8348	0.4794	1.3141	616.6
620.0	1786.9	0.02466	0.19615	0.22081	646.9	506.3	1153.2	0.8403	0.4689	1.3092	620.0
624.0	1839.0	0.02489	0.18737	0.21226	653.1	496.6	1149.8	0.8458	0.4583	1.3041	624.0
628.0	1892.4	0.02514	0.17880	0.20394	659.5	486.7	1146.1	0.8514	0.4474	1.2988	628.0
632.0	1947.0	0.02539	0.17044	0.19583	665.9	476.4	1142.2	0.8571	0.4364	1.2934	632.0
636.0	2002.8	0.02566	0.16226	0.18792	672.4	465.7	1138.1	0.8628	0.4251	1.2879	636.0
640.0	2059.9	0.02595	0.15427	0.18021	679.1	454.6	1133.7	0.8686	0.4134	1.2821	640.0
644.0	2118.3	0.02625	0.14644	0.17269	685.9	443.1	1129.0	0.8746	0.4015	1.2761	644.0
648.0	2178.1	0.02657	0.13876	0.16534	692.9	431.1	1124.0	0.8806	0.3893	1.2699	648.0
652.0	2239.2	0.02691	0.13124	0.15816	700.0	418.7	1118.7	0.8868	0.3767	1.2634	652.0
656.0	2301.7	0.02728	0.12387	0.15115	707.4	405.7	1113.1	0.8931	0.3637	1.2567	656.0
660.0	2365.7	0.02768	0.11663	0.14431	714.9	392.1	1107.0	0.8995	0.3502	1.2498	660.0
664.0	2431.1	0.02811	0.10947	0.13757	722.9	377.7	1100.6	0.9064	0.3361	1.2425	664.0
668.0	2498.1	0.02858	0.10229	0.13087	731.5	362.1	1093.5	0.9137	0.3210	1.2347	668.0
672.0	2566.6	0.02911	0.09514	0.12424	740.2	345.7	1085.9	0.9212	0.3054	1.2266	672.0
676.0	2636.8	0.02970	0.08799	0.11769	749.2	328.5	1077.6	0.9287	0.2892	1.2179	676.0
680.0	2708.6	0.03037	0.08080	0.11117	758.5	310.1	1068.5	0.9365	0.2720	1.2086	680.0
684.0	2782.1	0.03114	0.07349	0.10463	768.2	290.2	1058.4	0.9447	0.2537	1.1984	684.0
688.0	2857.4	0.03204	0.06595	0.09799	778.8	268.2	1047.0	0.9535	0.2337	1.1872	688.0
692.0	2934.5	0.03313	0.05797	0.09110	790.5	243.1	1033.6	0.9634	0.2110	1.1744	692.0
696.0	3013.4	0.03455	0.04916	0.08371	804.4	212.8	1017.2	0.9749	0.1841	1.1591	696.0
700.0	3094.3	0.03662	0.03857	0.07519	822.4	172.7	995.2	0.9901	0.1490	1.1390	700.0
702.0	3135.5	0.03824	0.03173	0.06997	835.0	144.7	979.7	1.0006	0.1246	1.1252	702.0
704.0	3177.2	0.04108	0.02192	0.06330	854.2	102.0	956.2	1.0169	0.0877	1.1046	704.0
705.0	3198.3	0.04427	0.01304	0.05730	873.0	61.4	934.4	1.0329	0.0527	1.0856	705.0
705.47*	3208.2	0.05078	0.00000	0.05078	906.0	0.0	906.0	1.0612	0.0000	1.0612	705.47*

CRITICAL TEMPERATURE

CRITICAL PRESSURE

Table A-1-2 Saturated Steam: Pressure Table (English)

Abs. Press. lbf/in.² p	Temp Fahr. T	Specific Volume, ft³/lbm Sat. Liquid v_f	Evap. v_{fg}	Sat. Vapor v_g	Enthalpy, Btu/lbm Sat. Liquid h_f	Evap. h_{fg}	Sat. Vapor h_g	Entropy, Btu/lbm R Sat. Liquid s_f	Evap. s_{fg}	Sat. Vapor s_g	Abs. Press. lbf/in.² p
0.08865	32.018	0.016022	3302.4	3302.4	0.0003	1075.5	1075.5	0.0000	2.1872	2.1872	0.08865
0.50	79.586	0.016071	641.5	641.5	47.623	1048.6	1096.2	0.0925	1.9446	2.0370	0.50
1.0	101.74	0.016136	333.59	333.60	69.73	1036.1	1105.8	0.1326	1.8455	1.9781	1.0
5.0	162.24	0.016407	73.515	73.532	130.20	1000.9	1131.1	0.2349	1.6094	1.8443	5.0
10.0	193.21	0.016592	38.404	38.420	161.26	982.1	1143.3	0.2836	1.5043	1.7879	10.0
14.696	212.00	0.016719	26.782	26.799	180.17	970.3	1150.5	0.3121	1.4447	1.7568	14.696
15.0	213.03	0.016726	26.274	26.290	181.21	969.7	1150.9	0.3137	1.4415	1.7552	15.0
20.0	227.96	0.016834	20.070	20.087	196.27	960.1	1156.3	0.3358	1.3962	1.7320	20.0
30.0	250.34	0.017009	13.7266	13.7436	218.9	945.2	1164.1	0.3682	1.3313	1.6995	30.0
40.0	267.25	0.017151	10.4794	10.4965	236.1	933.6	1169.8	0.3921	1.2844	1.6765	40.0
50.0	281.02	0.017274	8.4967	8.5140	250.2	923.9	1174.1	0.4112	1.2474	1.6586	50.0
60.0	292.71	0.017383	7.1562	7.1736	262.2	915.4	1177.6	0.4273	1.2167	1.6440	60.0
70.0	302.93	0.017482	6.1875	6.2050	272.7	907.8	1180.6	0.4411	1.1905	1.6316	70.0
80.0	312.04	0.017573	5.4536	5.4711	282.1	900.9	1183.1	0.4534	1.1675	1.6208	80.0
90.0	320.28	0.017659	4.8779	4.8953	290.7	894.6	1185.3	0.4643	1.1470	1.6113	90.0
100.0	327.82	0.017740	4.4133	4.4310	298.5	888.6	1187.2	0.4743	1.1284	1.6027	100.0
110.3	334.79	0.01782	4.0306	4.0484	305.8	883.1	1188.9	0.4834	1.1115	1.5950	110.3
120.0	341.27	0.01789	3.7097	3.7275	312.6	877.8	1190.4	0.4919	1.0960	1.5879	120.0
130.0	347.33	0.01796	3.4364	3.4544	319.0	872.8	1191.7	0.4998	1.0815	1.5813	130.0
140.0	353.04	0.01803	3.2010	3.2190	325.0	868.0	1193.0	0.5071	1.0681	1.5752	140.0
150.0	358.43	0.01809	2.9958	3.0139	330.6	863.4	1194.1	0.5141	1.0554	1.5695	150.0
160.0	363.55	0.01815	2.8155	2.8336	336.1	859.0	1195.1	0.5206	1.0435	1.5641	160.0
170.0	368.42	0.01821	2.6556	2.6738	341.2	854.8	1196.0	0.5269	1.0322	1.5591	170.0
180.0	373.08	0.01827	2.5129	2.5312	346.2	850.7	1196.9	0.5328	1.0215	1.5543	180.0
190.0	377.53	0.01833	2.3847	2.4030	350.9	846.7	1197.6	0.5384	1.0113	1.5498	190.0
200.0	381.80	0.01839	2.2689	2.2873	355.5	842.8	1198.3	0.5438	1.0016	1.5454	200.0
210.0	385.91	0.01844	2.16373	2.18217	359.9	839.1	1199.0	0.5490	0.9923	1.5413	210.0
220.0	389.88	0.01850	2.06779	2.08629	364.2	835.4	1199.6	0.5540	0.9834	1.5374	220.0
230.0	393.70	0.01855	1.97991	1.99846	368.3	831.8	1200.1	0.5588	0.9748	1.5336	230.0
240.0	397.39	0.01860	1.89909	1.91769	372.3	828.4	1200.6	0.5634	0.9665	1.5299	240.0
250.0	400.97	0.01865	1.82452	1.84317	376.1	825.0	1201.1	0.5679	0.9585	1.5264	250.0

P	T	v_f	v_{fg}	v_g	h_f	h_{fg}	h_g	s_f	s_{fg}	s_g
260.0	404.44	0.01870	1.75548	1.77418	379.9	821.6	1201.5	0.5722	0.9508	1.5230
270.0	407.80	0.01875	1.69137	1.71013	383.6	818.3	1201.9	0.5764	0.9433	1.5197
280.0	411.07	0.01880	1.63169	1.65049	387.1	815.1	1202.3	0.5805	0.9361	1.5166
290.0	414.25	0.01885	1.57597	1.59482	390.6	812.0	1202.6	0.5844	0.9291	1.5135
300.0	417.35	0.01889	1.52384	1.54274	394.0	808.9	1202.9	0.5882	0.9223	1.5105
350.0	431.73	0.01912	1.30642	1.32554	409.8	794.2	1204.0	0.6059	0.8909	1.4968
400.0	444.60	0.01934	1.14162	1.16095	424.2	780.4	1204.6	0.6217	0.8630	1.4847
450.0	456.28	0.01954	1.01224	1.03179	437.3	767.5	1204.8	0.6360	0.8378	1.4738
500.0	467.01	0.01975	0.90787	0.92762	449.5	755.1	1204.7	0.6490	0.8148	1.4639
550.0	476.94	0.01994	0.82183	0.84177	460.9	743.3	1204.3	0.6611	0.7936	1.4547
600.0	486.20	0.02013	0.74962	0.76975	471.7	732.0	1203.7	0.6723	0.7738	1.4461
650.0	494.89	0.02032	0.68811	0.70843	481.9	720.9	1202.8	0.6828	0.7552	1.4381
700.0	503.08	0.02050	0.63505	0.65556	491.6	710.2	1201.8	0.6928	0.7377	1.4304
750.0	510.84	0.02069	0.58880	0.60949	500.9	699.8	1200.7	0.7022	0.7210	1.4232
800.0	518.21	0.02087	0.54809	0.56896	509.8	689.6	1199.4	0.7111	0.7051	1.4163
850.0	525.24	0.02105	0.51197	0.53302	518.4	679.5	1198.0	0.7197	0.6899	1.4096
900.0	531.95	0.02123	0.47968	0.50091	526.7	669.7	1196.4	0.7279	0.6753	1.4032
950.0	538.39	0.02141	0.45064	0.47205	534.7	660.0	1194.7	0.7358	0.6612	1.3970
1000.0	544.58	0.02159	0.42436	0.44596	542.6	650.4	1192.9	0.7434	0.6476	1.3910
1050.0	550.53	0.02177	0.40047	0.42224	550.1	640.9	1191.0	0.7507	0.6344	1.3851
1100.0	556.28	0.02195	0.37863	0.40058	557.5	631.5	1189.1	0.7578	0.6216	1.3794
1150.0	561.82	0.02214	0.35859	0.38245	564.8	622.2	1187.0	0.7647	0.6091	1.3738
1200.0	567.19	0.02232	0.34013	0.36245	571.9	613.0	1184.8	0.7714	0.5969	1.3683
1250.0	572.38	0.02250	0.32306	0.34556	578.8	603.8	1182.6	0.7780	0.5850	1.3630
1300.0	577.42	0.02269	0.30722	0.32991	585.6	594.6	1180.2	0.7843	0.5733	1.3577
1350.0	582.32	0.02288	0.29250	0.31537	592.3	585.4	1177.8	0.7906	0.5620	1.3525
1400.0	587.07	0.02307	0.27871	0.30178	598.8	576.5	1175.3	0.7966	0.5507	1.3474
1450.0	591.70	0.02327	0.26584	0.28911	605.3	567.4	1172.8	0.8026	0.5397	1.3423
1500.0	596.20	0.02346	0.25372	0.27719	611.7	558.4	1170.1	0.8085	0.5288	1.3373
1550.0	600.59	0.02366	0.24235	0.26601	618.0	549.4	1167.4	0.8142	0.5182	1.3324
1600.0	604.87	0.02387	0.23159	0.25545	624.2	540.3	1164.5	0.8199	0.5076	1.3274
1650.0	609.05	0.02407	0.22143	0.24551	630.4	531.3	1161.6	0.8254	0.4971	1.3225
1700.0	613.13	0.02428	0.21178	0.23607	636.5	522.2	1158.6	0.8309	0.4867	1.3176
1750.0	617.12	0.02450	0.20263	0.22713	642.5	513.1	1155.6	0.8363	0.4765	1.3128
1800.0	621.02	0.02472	0.19390	0.21861	648.5	503.8	1152.3	0.8417	0.4662	1.3079

Table A-1-2 Saturated Steam: Pressure Table (English) (cont.)

Abs. Press. lbf/in.² p	Temp Fahr. T	Specific Volume, ft³/lbm			Enthalpy, Btu/lbm			Entropy, Btu/lbm R			Abs. Press. lbf/in.² p
		Sat. Liquid v_f	Evap. v_{fg}	Sat. Vapor v_g	Sat. Liquid h_f	Evap. h_{fg}	Sat. Vapor h_g	Sat. Liquid s_f	Evap. s_{fg}	Sat. Vapor s_g	
1850.0	624.83	0.02495	0.18558	0.21052	654.5	494.6	1149.0	0.8470	0.4561	1.3030	**1850.0**
1900.0	628.56	0.02517	0.17761	0.20278	660.4	485.2	1145.6	0.8522	0.4459	1.2981	**1900.0**
1950.0	632.22	0.02541	0.16999	0.19540	666.3	475.8	1142.0	0.8574	0.4358	1.2931	**1950.0**
2000.0	635.80	0.02565	0.16266	0.18831	672.1	466.2	1138.3	0.8625	0.4256	1.2881	**2000.0**
2100.0	642.76	0.02615	0.14885	0.17501	683.8	446.7	1130.5	0.8727	0.4053	1.2780	**2100.0**
2200.0	649.45	0.02669	0.13603	0.16272	695.5	426.7	1122.2	0.8828	0.3848	1.2676	**2200.0**
2300.0	655.89	0.02727	0.12406	0.15133	707.2	406.0	1113.2	0.8929	0.3640	1.2569	**2300.0**
2400.0	662.11	0.02790	0.11287	0.14076	719.0	384.8	1103.7	0.9031	0.3430	1.2460	**2400.0**
2500.0	668.11	0.02859	0.10209	0.13068	731.7	361.6	1093.3	0.9139	0.3206	1.2345	**2500.0**
2600.0	673.91	0.02938	0.09172	0.12110	744.5	337.6	1082.0	0.9247	0.2977	1.2225	**2600.0**
2700.0	679.53	0.03029	0.08165	0.11194	757.3	312.3	1069.7	0.9356	0.2741	1.2097	**2700.0**
2800.0	684.96	0.03134	0.07171	0.10305	770.7	285.1	1055.8	0.9468	0.2491	1.1958	**2800.0**
2900.0	690.22	0.03262	0.06158	0.09420	785.1	254.7	1039.8	0.9588	0.2215	1.1803	**2900.0**
3000.0	695.33	0.03428	0.05073	0.08500	801.8	218.4	1020.3	0.9728	0.1891	1.1619	**3000.0**
3100.0	700.28	0.03681	0.03771	0.07452	824.0	169.3	993.3	0.9914	0.1460	1.1373	**3100.0**
3200.0	705.08	0.04472	0.01191	0.05663	875.5	56.1	931.6	1.0351	0.0482	1.0832	**3200.0**
3208.2*	705.47	0.05078	0.00000	0.05078	906.0	0.0	906.0	1.0612	0.0000	1.0612	**3208.2***

Source: Reprinted with permission from ASME.

Table A-1-3 Superheated Steam (No Liquid Present) (English)

Abs. Press. lbf/in.² (Sat. Temp. F)		Sat. Liq.	Sat. Vap.	200	250	300	350	400	450	500	600	700	800	900	1000	1100	1200
										Temperature—F							
1 (101.74)	v	0.01614	333.6	392.5	422.4	452.3	482.1	511.9	541.7	571.5	631.1	690.7	750.3	809.8	869.4	929.0	988.6
	h	69.73	1105.8	1150.2	1172.9	1195.7	1218.7	1241.8	1265.1	1288.6	1336.1	1384.5	1433.7	1483.8	1534.9	1586.8	1639.7
	s	0.1326	1.9781	2.0509	2.0841	2.1152	2.1445	2.1722	2.1985	2.2237	2.2708	2.3144	2.3551	2.3934	2.4296	2.4640	2.4969
5 (162.24)	v	0.01641	73.53	78.14	84.21	90.24	96.25	102.24	108.23	114.21	126.15	138.08	150.01	161.94	173.86	185.78	197.70
	h	130.20	1131.1	1148.6	1171.7	1194.8	1218.0	1241.3	1264.7	1288.2	1335.9	1384.3	1433.6	1483.7	1534.7	1586.7	1639.6
	s	0.2349	1.8443	1.8716	1.9054	1.9369	1.9664	1.9943	2.0208	2.0460	2.0932	2.1369	2.1776	2.2159	2.2521	2.2866	2.3194
10 (193.21)	v	0.01659	38.42	38.84	41.93	44.98	48.02	51.03	54.04	57.04	63.03	69.00	74.98	80.94	86.91	92.87	98.84
	h	161.26	1143.3	1146.6	1170.2	1193.7	1217.1	1240.6	1264.1	1287.8	1335.5	1384.0	1433.4	1483.5	1534.6	1586.6	1639.5
	s	0.2836	1.7879	1.7928	1.8273	1.8593	1.8892	1.9173	1.9439	1.9692	2.0166	2.0603	2.1011	2.1394	2.1757	2.2101	2.2430
14.696 (212.00)	v	.0167	26.799		28.42	30.52	32.60	34.67	36.72	38.77	42.86	46.93	51.00	55.06	59.13	63.19	67.25
	h	180.17	1150.5		1168.8	1192.6	1216.3	1239.9	1263.6	1287.4	1335.2	1383.8	1433.2	1483.4	1534.5	1586.5	1639.4
	s	.3121	1.7568		1.7833	1.8158	1.8459	1.8743	1.9010	1.9265	1.9739	2.0177	2.0585	2.0969	2.1332	2.1676	2.2005
15 (213.03)	v	0.01673	26.290		27.837	29.899	31.939	33.963	35.977	37.985	41.986	45.978	49.964	53.946	57.926	61.905	65.882
	h	181.21	1150.9		1168.7	1192.5	1216.2	1239.9	1263.6	1287.3	1335.2	1383.8	1433.2	1483.4	1534.5	1586.5	1639.4
	s	0.3137	1.7552		1.7809	1.8134	1.8437	1.8720	1.8988	1.9242	1.9717	2.0155	2.0563	2.0946	2.1309	2.1653	2.1982
20 (227.96)	v	0.01683	20.087		20.788	22.356	23.900	25.428	26.946	28.457	31.466	34.465	37.458	40.447	43.435	46.420	49.405
	h	196.27	1156.3		1167.1	1191.4	1215.4	1239.2	1263.0	1286.9	1334.9	1383.5	1432.9	1483.2	1534.3	1586.3	1639.3
	s	0.3358	1.7320		1.7475	1.7805	1.8111	1.8397	1.8666	1.8921	1.9397	1.9836	2.0244	2.0628	2.0991	2.1336	2.1665
25 (240.07)	v	0.01693	16.301		16.558	17.829	19.076	20.307	21.527	22.740	25.153	27.557	29.954	32.348	34.740	37.130	39.518
	h	208.52	1160.6		1165.6	1190.2	1214.5	1238.5	1262.5	1286.4	1334.6	1383.3	1432.7	1483.0	1534.2	1586.2	1639.2
	s	0.3535	1.7141		1.7212	1.7547	1.7856	1.8145	1.8415	1.8672	1.9149	1.9588	1.9997	2.0381	2.0744	2.1089	2.1418
30 (250.34)	v	0.01701	13.744			14.810	15.859	16.892	17.914	18.929	20.945	22.951	24.952	26.949	28.943	30.936	32.927
	h	218.93	1164.1			1189.0	1213.6	1237.8	1261.9	1286.0	1334.2	1383.0	1432.5	1482.8	1534.0	1586.1	1639.0
	s	0.3682	1.6995			1.7334	1.7647	1.7937	1.8210	1.8467	1.8946	1.9386	1.9795	2.0179	2.0543	2.0888	2.1217

Table A-1-3 Superheated Steam (No Liquid Present) (English) (cont.)

Abs. Press. lbf/in.² (Sat. Temp, F)		Sat. Liq.	Sat. Vap.	200	250	300	350	400	450	500	600	700	800	900	1000	1100	1200
35 (259.29)	v	0.01708	11.896			12.654	13.562	14.453	15.334	16.207	17.939	19.662	21.379	23.092	24.803	26.512	28.220
	h	228.03	1167.1			1187.8	1212.7	1237.1	1261.3	1285.5	1333.9	1382.8	1432.3	1482.7	1533.9	1586.0	1638.9
	s	0.3809	1.6872			1.7152	1.7468	1.7761	1.8035	1.8294	1.8774	1.9214	1.9624	2.0009	2.0372	2.0717	2.1046
40 (267.25)	v	0.01715	10.497			11.036	11.838	12.624	13.398	14.165	15.685	17.195	18.699	20.199	21.697	23.194	24.689
	h	236.14	1169.8			1186.6	1211.7	1236.4	1260.8	1285.0	1333.6	1382.5	1432.1	1482.5	1533.7	1585.8	1638.8
	s	0.3921	1.6765			1.6992	1.7312	1.7608	1.7883	1.8143	1.8624	1.9065	1.9476	1.9860	2.0224	2.0569	2.0899
45 (274.44)	v	0.01721	9.399			9.777	10.497	11.201	11.892	12.577	13.932	15.276	16.614	17.950	19.282	20.613	21.943
	h	243.49	1172.1			1185.4	1210.4	1235.7	1260.2	1284.6	1333.3	1382.3	1431.9	1482.3	1533.6	1585.7	1638.7
	s	0.4021	1.6671			1.6849	1.7173	1.7471	1.7748	1.8010	1.8492	1.8934	1.9345	1.9730	2.0093	2.0439	2.0768
50 (281.02)	v	0.01727	8.514			8.769	9.424	10.062	10.688	11.306	12.529	13.741	14.947	16.150	17.350	18.549	19.746
	h	250.21	1174.1			1184.1	1209.9	1234.9	1259.6	1284.1	1332.9	1382.0	1431.7	1482.2	1533.4	1585.6	1638.6
	s	0.4112	1.6586			1.6720	1.7048	1.7349	1.7628	1.7890	1.8374	1.8816	1.9227	1.9613	1.9977	2.0322	2.0652
55 (287.07)	v	0.01733				7.945	8.546	9.130	9.702	10.267	11.381	12.485	13.583	14.677	15.769	16.859	17.948
	h	256.43				1182.9	1208.9	1234.2	1259.1	1283.6	1332.6	1381.8	1431.5	1482.0	1533.3	1585.5	1638.5
	s	0.4196				1.6601	1.6933	1.7237	1.7518	1.7781	1.8266	1.8710	1.9121	1.9507	1.987	2.022	2.055
60 (292.71)	v	0.01738	7.174			7.257	7.815	8.354	8.881	9.400	10.425	11.438	12.446	13.450	14.452	15.452	16.450
	h	262.21	1177.6			1181.6	1208.0	1233.5	1258.5	1283.2	1332.3	1381.5	1431.3	1481.8	1533.2	1585.3	1638.4
	s	0.4273	1.6440			1.6492	1.6934	1.7134	1.7417	1.7681	1.8168	1.8612	1.9024	1.9410	1.9774	2.0120	2.0450
65 (297.98)	v	0.01743	6.653			6.675	7.195	7.697	8.186	8.667	9.615	10.552	11.484	12.412	13.337	14.261	15.183
	h	267.63	1179.1			1180.3	1207.0	1232.7	1257.9	1282.7	1331.9	1381.3	1431.1	1481.6	1533.0	1585.2	1638.3
	s	0.4344	1.6375			1.6390	1.6731	1.7040	1.7324	1.7590	1.8077	1.8522	1.8935	1.9321	1.9685	2.0031	2.0361
70 (302.93)	v	0.01748	6.205				6.664	7.133	7.590	8.039	8.922	9.793	10.659	11.522	12.382	13.240	14.097
	h	272.74	1180.6				1206.0	1232.0	1257.3	1282.2	1331.6	1381.0	1430.9	1481.5	1532.9	1585.1	1638.2
	s	0.4411	1.6316				1.6640	1.6951	1.7237	1.7504	1.7993	1.8439	1.8852	1.9238	1.9603	1.9949	2.0279
75 (307.61)	v	0.01753	5.814				6.204	6.645	7.074	7.494	8.320	9.135	9.945	10.750	11.553	12.355	13.155
	h	277.56	1181.9				1205.0	1231.2	1256.7	1281.7	1331.3	1380.7	1430.7	1481.3	1532.7	1585.0	1638.1
	s	0.4474	1.6260				1.6554	1.6868	1.7156	1.7424	1.7915	1.8361	1.8774	1.9161	1.9526	1.9872	2.0202

Temperature—F

Temperature—Fahr.

Abs. Press. Lb./Sq. In. (Sat. Temp.)		Sat. Liq.	Sat. Vap.	350	400	450	500	550	600	700	800	900	1000	1100	1200	1300	1400
80 (312.04)	v	0.01757	5.471	5.801	6.218	6.622	7.018	7.408	7.794	8.560	9.319	10.075	10.829	11.581	12.331	13.081	13.829
	h	282.15	1183.1	1204.0	1230.5	1256.1	1281.3	1306.2	1330.9	1380.5	1430.5	1481.1	1532.6	1584.9	1638.0	1692.0	1746.8
	s	0.4534	1.6208	1.6473	1.6790	1.7080	1.7349	1.7602	1.7842	1.8289	1.8702	1.9089	1.9454	1.9800	2.0131	2.0446	2.0750
85 (316.26)	v	0.01762	5.167	5.445	5.840	6.223	6.597	6.966	7.330	8.052	8.768	9.480	10.190	10.898	11.604	12.310	13.014
	h	286.52	1184.2	1203.0	1229.7	1255.5	1280.8	1305.8	1330.6	1380.2	1430.3	1481.0	1532.4	1584.7	1637.9	1691.9	1746.8
	s	0.4590	1.6159	1.6396	1.6716	1.7008	1.7279	1.7532	1.7772	1.8220	1.8634	1.9021	1.9386	1.9733	2.0063	2.0379	2.0682
90 (320.28)	v	0.01766	4.895	5.128	5.505	5.869	6.223	6.572	6.917	7.600	8.277	8.950	9.621	10.290	10.958	11.625	12.290
	h	290.69	1185.3	1202.0	1228.9	1254.9	1280.3	1305.4	1330.2	1380.0	1430.1	1480.8	1532.3	1584.6	1637.8	1691.8	1746.7
	s	0.4643	1.6113	1.6323	1.6646	1.6940	1.7212	1.7467	1.7707	1.8156	1.8570	1.8957	1.9323	1.9669	2.0000	2.0316	2.0619
95 (324.13)	v	0.01770	4.651	4.845	5.205	5.551	5.889	6.221	6.548	7.196	7.838	8.477	9.113	9.747	10.380	11.012	11.643
	h	294.70	1186.2	1200.9	1228.1	1254.3	1279.8	1305.0	1329.9	1379.7	1429.9	1480.6	1532.1	1584.5	1637.7	1691.7	1746.6
	s	0.4694	1.6069	1.6253	1.6580	1.6876	1.7149	1.7404	1.7645	1.8094	1.8509	1.8897	1.9262	1.9609	1.9940	2.0256	2.0559
100 (327.82)	v	0.01774	4.431	4.590	4.935	5.266	5.588	5.904	6.216	6.833	7.443	8.050	8.655	9.258	9.860	10.460	11.060
	h	298.54	1187.2	1199.9	1227.4	1253.7	1279.3	1304.6	1329.6	1379.5	1429.7	1480.4	1532.0	1584.4	1637.6	1691.6	1746.5
	s	0.4743	1.6027	1.6187	1.6516	1.6814	1.7088	1.7344	1.7586	1.8036	1.8451	1.8839	1.9205	1.9552	1.9883	2.0199	2.0502
105 (331.37)	v	0.01778	4.231	4.359	4.690	5.007	5.315	5.617	5.915	6.504	7.086	7.665	8.241	8.816	9.389	9.961	10.532
	h	302.24	1188.0	1198.8	1226.6	1253.1	1278.8	1304.2	1329.2	1379.2	1429.4	1480.3	1531.8	1584.2	1637.5	1691.5	1746.4
	s	0.4790	1.5988	1.6122	1.6455	1.6755	1.7031	1.7288	1.7530	1.7981	1.8396	1.8785	1.9151	1.9498	1.9828	2.0145	2.0448
110 (334.79)	v	0.01782	4.048	4.149	4.468	4.772	5.068	5.357	5.642	6.205	6.761	7.314	7.865	8.413	8.961	9.507	10.053
	h	305.80	1188.9	1197.7	1225.8	1252.5	1278.3	1303.8	1328.8	1379.0	1429.2	1480.1	1531.7	1584.1	1637.4	1691.4	1746.4
	s	0.4834	1.5950	1.6061	1.6396	1.6698	1.6975	1.7233	1.7476	1.7928	1.8344	1.8732	1.9099	1.9446	1.9777	2.0093	2.0397
115 (338.08)	v	0.01785	3.881	3.957	4.265	4.558	4.841	5.119	5.392	5.932	6.465	6.994	7.521	8.046	8.570	9.093	9.615
	h	309.25	1189.6	1196.7	1225.0	1251.8	1277.9	1303.3	1328.6	1378.7	1429.0	1479.9	1531.6	1584.0	1637.2	1691.4	1746.3
	s	0.4877	1.5913	1.6001	1.6340	1.6644	1.6922	1.7181	1.7425	1.7877	1.8294	1.8682	1.9049	1.9396	1.9727	2.0044	2.0347
120 (341.27)	v	0.01789	3.7275	3.7815	4.0786	4.3610	4.6341	4.9009	5.1637	5.6813	6.1928	6.7006	7.2060	7.7096	8.2119	8.7130	9.2134
	h	312.58	1190.4	1195.6	1224.1	1251.2	1277.4	1302.9	1328.2	1378.4	1428.8	1479.8	1531.4	1583.9	1637.1	1691.3	1746.2
	s	0.4919	1.5879	1.5943	1.6286	1.6592	1.6872	1.7132	1.7376	1.7829	1.8246	1.8635	1.9001	1.9349	1.9680	1.9996	2.0300
130 (347.33)	v	0.01796	3.4544	3.4699	3.7489	4.0129	4.2672	4.5151	4.7589	5.2384	5.7118	6.1814	6.6486	7.1140	7.5781	8.0411	8.5033
	h	318.95	1191.7	1193.4	1222.5	1249.9	1276.4	1302.1	1327.5	1377.9	1428.4	1479.4	1531.1	1583.6	1636.9	1691.1	1746.1
	s	0.4998	1.5813	1.5833	1.6182	1.6493	1.6775	1.7037	1.7283	1.7737	1.8155	1.8545	1.8911	1.9259	1.9591	1.9907	2.0211

Table A-1-3 Superheated Steam (No Liquid Present) (English) (*cont.*)

Abs. Press. lbf/in.^{2} (Sat. Temp, F)		Sat. Liq.	Sat. Vap.	350	400	450	500	550	600	700	800	900	1000	1100	1200	1300	1400
140 (353.04)	v	0.01803	3.2190		3.4661	3.7143	3.9526	4.1844	4.4119	4.8588	5.2995	5.7364	6.1709	6.6036	7.0349	7.4652	7.8946
	h	324.96	1193.0		1220.8	1248.7	1275.3	1301.3	1326.8	1377.4	1428.0	1479.1	1530.8	1583.4	1636.7	1690.9	1745.9
	s	0.5071	1.5752		1.6085	1.6400	1.6686	1.6949	1.7196	1.7652	1.8071	1.8461	1.8828	1.9176	1.9508	1.9825	2.0129
150 (358.43)	v	0.01809	3.0139		3.2208	3.4555	3.6799	3.8978	4.1112	4.5298	4.9421	5.3507	5.7568	6.1612	6.5642	6.9661	7.3671
	h	330.65	1194.1		1219.1	1247.4	1274.3	1300.5	1326.1	1376.9	1427.6	1478.7	1530.5	1583.1	1636.5	1690.7	1745.7
	s	0.5141	1.5695		1.5993	1.6313	1.6602	1.6867	1.7115	1.7573	1.7992	1.8383	1.8751	1.9099	1.9431	1.9748	2.0052
160 (363.55)	v	0.01815	2.8336		3.0060	3.2288	3.4413	3.6469	3.8480	4.2420	4.6295	5.0132	5.3945	5.7741	6.1522	6.5293	6.9055
	h	336.07	1195.1		1217.4	1246.0	1273.3	1299.6	1325.4	1376.4	1427.2	1478.4	1530.3	1582.9	1636.3	1690.5	1745.6
	s	0.5206	1.5641		1.5906	1.6231	1.6522	1.6790	1.7039	1.7499	1.7919	1.8310	1.8678	1.9027	1.9359	1.9676	1.9980
170 (368.42)	v	0.01821	2.6738		2.8162	3.0288	3.2306	3.4255	3.6158	3.9879	4.3536	4.7155	5.0749	5.4325	5.7888	6.1440	6.4983
	h	341.24	1196.0		1215.6	1244.7	1272.2	1298.8	1324.7	1375.8	1426.8	1478.0	1530.0	1582.6	1636.1	1690.4	1745.4
	s	0.5269	1.5591		1.5823	1.6152	1.6447	1.6717	1.6968	1.7428	1.7850	1.8241	1.8610	1.8959	1.9291	1.9608	1.9913
180 (373.08)	v	0.01827	2.5312		2.6474	2.8508	3.0433	3.2286	3.4093	3.7621	4.1084	4.4508	4.7907	5.1289	5.4657	5.8014	6.1363
	h	346.19	1196.9		1213.8	1243.4	1271.2	1297.9	1324.0	1375.3	1426.3	1477.7	1529.7	1582.4	1635.9	1690.2	1745.3
	s	0.5328	1.5543		1.5743	1.6078	1.6376	1.6647	1.6900	1.7362	1.7784	1.8176	1.8545	1.8894	1.9227	1.9545	1.9849
190 (377.53)	v	0.01833	2.4030		2.4961	2.6915	2.8756	3.0525	3.2246	3.5601	3.8889	4.2140	4.5365	4.8572	5.1766	5.4949	5.8124
	h	350.94	1197.6		1212.0	1242.0	1270.1	1297.1	1323.3	1374.8	1425.9	1477.4	1529.4	1582.1	1635.7	1690.0	1745.1
	s	0.5384	1.5498		1.5667	1.6006	1.6307	1.6581	1.6835	1.7299	1.7722	1.8115	1.8484	1.8834	1.9166	1.9484	1.9789
200 (381.80)	v	0.01839	2.2873		2.3598	2.5480	2.7247	2.8939	3.0583	3.3783	3.6915	4.0008	4.3077	4.6128	4.9165	5.2191	5.5209
	h	355.51	1198.3		1210.1	1240.6	1269.0	1296.2	1322.6	1374.3	1425.5	1477.0	1529.1	1581.9	1635.4	1689.8	1745.0
	s	0.5438	1.5454		1.5593	1.5938	1.6242	1.6518	1.6773	1.7239	1.7663	1.8057	1.8426	1.8776	1.9109	1.9427	1.9732
				400	450	500	550	600	700	800	900	1000	1100	1200	1300	1400	1500
210 (385.91)	v	0.01844	2.1822	2.2364	2.4181	2.5880	2.7504	2.9078	3.2137	3.5128	3.8080	4.1007	4.3915	4.6811	4.9695	5.2571	5.5440
	h	359.91	1199.0	1208.02	1239.2	1268.0	1295.3	1321.9	1373.7	1425.1	1476.7	1528.8	1581.6	1635.2	1689.6	1744.8	1800.8
	s	0.5490	1.5413	1.5522	1.5872	1.6180	1.6458	1.6715	1.7182	1.7607	1.8001	1.8371	1.8721	1.9054	1.9372	1.9677	1.9970

Press. (Sat. Temp)																	
220 (389.88)	v	0.01850	2.0863	2.1240	2.2999	2.4638	2.6199	2.7710	3.0642	3.3504	3.6327	3.9125	4.1905	4.4671	4.7426	5.0173	5.2913
	h	364.17	1199.6	1206.3	1237.8	1266.9	1294.5	1321.2	1373.2	1424.7	1476.3	1528.5	1581.4	1635.0	1689.4	1744.7	1800.6
	s	0.5540	1.5374	1.5453	1.5808	1.6120	1.6400	1.6658	1.7128	1.7553	1.7948	1.8318	1.8668	1.9002	1.9320	1.9625	1.9919
230 (393.70)	v	0.01855	1.9985	2.0212	2.1919	2.3503	2.5008	2.6461	2.9276	3.2020	3.4726	3.7406	4.0068	4.2717	4.5355	4.7984	5.0606
	h	368.28	1200.1	1204.4	1236.3	1265.7	1293.6	1320.4	1372.7	1424.2	1476.0	1528.2	1581.1	1634.8	1689.3	1744.5	1800.5
	s	0.5588	1.5336	1.5385	1.5747	1.6062	1.6344	1.6604	1.7075	1.7502	1.7897	1.8268	1.8618	1.8952	1.9270	1.9576	1.9869
240 (397.39)	v	0.01860	1.9177	1.9268	2.0928	2.2462	2.3915	2.5316	2.8024	3.0661	3.3259	3.5831	3.8385	4.0926	4.3456	4.5977	4.8492
	h	372.27	1200.6	1202.4	1234.9	1264.6	1292.7	1319.7	1372.1	1423.8	1475.6	1527.9	1580.9	1634.6	1689.1	1744.3	1800.4
	s	0.5634	1.5299	1.5320	1.5687	1.6006	1.6291	1.6552	1.7025	1.7452	1.7848	1.8219	1.8570	1.8904	1.9223	1.9528	1.9822
250 (400.97)	v	0.01865	1.8432		2.0016	2.1504	2.2909	2.4262	2.6872	2.9410	3.1909	3.4382	3.6837	3.9278	4.1709	4.4131	4.6546
	h	376.14	1201.1		1233.4	1263.5	1291.8	1319.0	1371.6	1423.4	1475.3	1527.6	1580.6	1634.4	1688.9	1744.2	1800.2
	s	0.5679	1.5264		1.5629	1.5951	1.6239	1.6502	1.6976	1.7405	1.7801	1.8173	1.8524	1.8858	1.9177	1.9482	1.9776
260 (404.44)	v	0.01870	1.7742		1.9173	2.0619	2.1981	2.3289	2.5808	2.8256	3.0663	3.3044	3.5408	3.7758	4.0097	4.2427	4.4750
	h	379.90	1201.5		1231.9	1262.4	1290.9	1318.2	1371.1	1423.0	1474.9	1527.3	1580.4	1634.2	1688.7	1744.0	1800.1
	s	0.5722	1.5230		1.5573	1.5899	1.6189	1.6453	1.6930	1.7359	1.7756	1.8128	1.8480	1.8814	1.9133	1.9439	1.9732
270 (407.80)	v	0.01875	1.7101		1.8391	1.9799	2.1121	2.2388	2.4824	2.7186	2.9509	3.1806	3.4084	3.6349	3.8603	4.0849	4.3087
	h	383.56	1201.9		1230.4	1261.2	1290.0	1317.5	1370.5	1422.6	1474.6	1527.1	1580.1	1634.0	1688.5	1743.9	1800.0
	s	0.5764	1.5197		1.5518	1.5848	1.6140	1.6406	1.6885	1.7315	1.7713	1.8085	1.8437	1.8771	1.9090	1.9396	1.9690
280 (411.07)	v	0.01880	1.6505		1.7665	1.9037	2.0322	2.1551	2.3909	2.6194	2.8437	3.0655	3.2855	3.5042	3.7217	3.9384	4.1543
	h	387.12	1202.3		1228.8	1260.0	1289.1	1316.8	1370.0	1422.1	1474.2	1526.8	1579.9	1633.8	1688.4	1743.7	1799.8
	s	0.5805	1.5166		1.5464	1.5798	1.6093	1.6361	1.6841	1.7273	1.7671	1.8043	1.8395	1.8730	1.9050	1.9356	1.9649
290 (414.25)	v	0.01885	1.5948		1.6988	1.8327	1.9578	2.0772	2.3058	2.5269	2.7440	2.9585	3.1711	3.3824	3.5926	3.8019	4.0106
	h	390.60	1202.6		1227.3	1258.8	1288.1	1316.0	1369.5	1421.7	1473.9	1526.5	1579.6	1633.5	1688.2	1743.6	1799.7
	s	0.5844	1.5135		1.5412	1.5750	1.6048	1.6317	1.6799	1.7232	1.7630	1.8003	1.8356	1.8690	1.9010	1.9316	1.9610
300 (417.35)	v	0.01889	1.5427		1.6356	1.7665	1.8883	2.0044	2.2263	2.4407	2.6509	2.8585	3.0643	3.2688	3.4721	3.6746	3.8764
	h	393.99	1202.9		1225.7	1257.7	1287.2	1315.2	1368.9	1421.3	1473.6	1526.2	1579.4	1633.3	1688.0	1743.4	1799.6
	s	0.5882	1.5105		1.5351	1.5703	1.6003	1.6274	1.6758	1.7192	1.7591	1.7964	1.8317	1.8652	1.8972	1.9278	1.9572
310 (420.36)	v	0.01894	1.4939		1.5763	1.7044	1.8233	1.9363	2.1520	2.3600	2.5638	2.7650	2.9644	3.1625	3.3594	3.5555	3.7509
	h	397.30	1203.2		1224.1	1256.5	1286.3	1314.5	1368.4	1420.9	1473.2	1525.9	1579.2	1633.1	1687.8	1743.3	1799.4
	s	0.5920	1.5076		1.5311	1.5657	1.5960	1.6233	1.6719	1.7153	1.7553	1.7927	1.8280	1.8615	1.8935	1.9241	1.9536
320 (423.31)	v	0.01899	1.4480		1.5207	1.6462	1.7623	1.8725	2.0823	2.2843	2.4821	2.6774	2.8708	3.0628	3.2538	3.4438	3.6332
	h	400.53	1203.4		1222.5	1255.2	1285.3	1313.7	1367.8	1420.5	1472.9	1525.6	1578.9	1632.9	1687.6	1743.1	1799.3
	s	0.5956	1.5048		1.5261	1.5612	1.5918	1.6192	1.6680	1.7116	1.7516	1.7890	1.8243	1.8579	1.8899	1.9206	1.9500

Table A-1-3 Superheated Steam (No Liquid Present) (English) (cont.)

Abs. Press. lbf/in.² (Sat. Temp, F)		Sat. Liq.	Sat. Vap.	|	400	450	500	550	600	650	700	800	900	1000	1100	1200	1300	1400	1500
330 (426.18)	v	0.01903	1.4048			1.4684	1.5915	1.7050	1.8125		2.0168	2.2132	2.4054	2.5950	2.7828	2.9692	3.1545	3.3389	3.5227
	h	403.70	1203.6			1220.9	1254.0	1284.4	1313.0		1367.3	1420.0	1472.5	1525.3	1578.7	1632.7	1687.5	1742.9	1799.2
	s	0.5991	1.5021			1.5213	1.5568	1.5876	1.6153		1.6643	1.7079	1.7480	1.7855	1.8208	1.8544	1.8864	1.9171	1.9466
340 (428.99)	v	0.01908	1.3640			1.4191	1.5399	1.6511	1.7561		1.9552	2.1463	2.3333	2.5175	2.7000	2.8811	3.0611	3.2402	3.4186
	h	406.80	1203.8			1219.2	1252.8	1283.4	1312.2		1366.7	1419.6	1472.2	1525.0	1578.4	1632.5	1687.3	1742.8	1799.0
	s	0.6026	1.4994			1.5165	1.5525	1.5836	1.6114		1.6606	1.7044	1.7445	1.7820	1.8174	1.8510	1.8831	1.9138	1.9432
350 (431.73)	v	0.01912	1.3255			1.3725	1.4913	1.6002	1.7028		1.8970	2.0832	2.2652	2.4445	2.6219	2.7980	2.9730	3.1471	3.3205
	h	409.83	1204.0			1217.5	1251.5	1282.4	1311.4		1366.2	1419.2	1471.8	1524.7	1578.2	1632.3	1687.1	1742.6	1798.9
	s	0.6059	1.4968			1.5119	1.5483	1.5797	1.6077		1.6571	1.7009	1.7411	1.7787	1.8141	1.8477	1.8798	1.9105	1.9400
360 (434.41)	v	0.01917	1.2891			1.3285	1.4454	1.5521	1.6525		1.8421	2.0237	2.2009	2.3755	2.5482	2.7196	2.8898	3.0592	3.2279
	h	412.81	1204.1			1215.8	1250.3	1281.5	1310.6		1365.6	1418.7	1471.5	1524.4	1577.9	1632.1	1686.9	1742.5	1798.8
	s	0.6092	1.4943			1.5073	1.5441	1.5758	1.6040		1.6536	1.6976	1.7379	1.7754	1.8109	1.8445	1.8766	1.9073	1.9368
380 (439.61)	v	0.01925	1.2218			1.2472	1.3606	1.4635	1.5598		1.7410	1.9139	2.0825	2.2484	2.4124	2.5750	2.7366	2.8973	3.0572
	h	418.59	1204.4			1212.4	1247.7	1279.5	1309.0		1364.5	1417.9	1470.8	1523.8	1577.4	1631.6	1686.5	1742.2	1798.5
	s	0.6156	1.4894			1.4982	1.5360	1.5683	1.5969		1.6470	1.6911	1.7315	1.7692	1.8047	1.8384	1.8705	1.9012	1.9307
400 (444.60)	v	0.01934	1.1610			1.1738	1.2841	1.3836	1.4763	1.5646	1.6499	1.8151	1.9759	2.1339	2.2901	2.4450	2.5987	2.7515	2.9037
	h	424.17	1204.6			1208.8	1245.1	1277.5	1307.4	1335.9	1363.4	1417.0	1470.1	1523.3	1576.9	1631.2	1686.2	1741.9	1798.2
	s	0.6217	1.4847			1.4894	1.5282	1.5611	1.5901	1.6163	1.6406	1.6850	1.7255	1.7632	1.7988	1.8325	1.8647	1.8955	1.9250
420 (449.40)	v	0.01942	1.1057			1.1071	1.2148	1.3113	1.4007	1.4856	1.5676	1.7258	1.8795	2.0304	2.1795	2.3273	2.4739	2.6196	2.7647
	h	429.56	1204.7			1205.2	1242.4	1275.4	1305.8	1334.5	1362.3	1416.2	1469.4	1522.7	1576.4	1630.8	1685.8	1741.6	1798.0
	s	0.6276	1.4802			1.4808	1.5206	1.5542	1.5835	1.6100	1.6345	1.6791	1.7197	1.7575	1.7932	1.8269	1.8591	1.8899	1.9195
440 (454.03)	v	0.01950	1.0554				1.1517	1.2454	1.3319	1.4138	1.4926	1.6445	1.7918	1.9363	2.0790	2.2203	2.3605	2.4998	2.6384
	h	434.77	1204.8				1239.7	1273.4	1304.2	1333.2	1361.1	1415.3	1468.7	1522.1	1575.9	1630.4	1685.5	1741.2	1797.7
	s	0.6332	1.4759				1.5132	1.5474	1.5772	1.6040	1.6286	1.6734	1.7142	1.7521	1.7878	1.8216	1.8538	1.8847	1.9143
460 (458.50)	v	0.01959	1.0092				1.0939	1.1852	1.2691	1.3482	1.4242	1.5703	1.7117	1.8504	1.9872	2.1226	2.2569	2.3903	2.5230
	h	439.83	1204.8				1236.9	1271.3	1302.5	1331.8	1360.0	1414.4	1468.0	1521.5	1575.4	1629.9	1685.1	1740.9	1797.4
	s	0.6387	1.4718				1.5060	1.5409	1.5711	1.5982	1.6230	1.6680	1.7089	1.7469	1.7826	1.8165	1.8488	1.8797	1.9093

Temperature—F

Press. (sat. temp)																
480 (462.82)	v	0.01967	0.9668	1.0409	1.1300	1.2115	1.2881	1.3615	1.5023	1.6384	1.7716	1.9030	2.0330	2.1619	2.2900	2.4173
	h	444.75	1204.8	1234.1	1269.1	1300.8	1330.5	1358.8	1413.6	1467.3	1520.9	1574.9	1629.5	1684.7	1740.6	1797.2
	s	0.6439	1.4677	1.4990	1.5346	1.5652	1.5925	1.6176	1.6628	1.7038	1.7419	1.7777	1.8116	1.8439	1.8748	1.9045
500 (467.01)	v	0.01975	0.9276	0.9919	1.0791	1.1584	1.2327	1.3037	1.4397	1.5708	1.6992	1.8256	1.9507	2.0746	2.1977	2.3200
	h	449.52	1204.7	1231.2	1267.0	1299.1	1329.1	1357.7	1412.7	1466.6	1520.3	1574.4	1629.1	1684.4	1740.3	1796.9
	s	0.6490	1.4639	1.4921	1.5284	1.5595	1.5871	1.6123	1.6578	1.6990	1.7371	1.7730	1.8069	1.8393	1.8702	1.8998
520 (471.07)	v	0.01982	0.8914	0.9466	1.0321	1.1094	1.1816	1.2504	1.3819	1.5085	1.6323	1.7542	1.8746	1.9940	2.1125	2.2302
	h	454.18	1204.5	1228.3	1264.8	1297.4	1327.7	1356.5	1411.8	1465.9	1519.7	1573.9	1628.7	1684.0	1740.0	1796.7
	s	0.6540	1.4601	1.4853	1.5223	1.5539	1.5818	1.6072	1.6530	1.6943	1.7325	1.7684	1.8024	1.8348	1.8657	1.8954
540 (475.01)	v	0.01990	0.8577	0.9045	0.9884	1.0640	1.1342	1.2010	1.3284	1.4508	1.5704	1.6880	1.8042	1.9193	2.0336	2.1471
	h	458.71	1204.4	1225.3	1262.5	1295.7	1326.3	1355.3	1410.9	1465.1	1519.1	1573.4	1628.2	1683.6	1739.7	1796.4
	s	0.6587	1.4565	1.4786	1.5164	1.5485	1.5767	1.6023	1.6483	1.6897	1.7280	1.7640	1.7981	1.8305	1.8615	1.8911
560 (478.84)	v	0.01998	0.8264	0.8653	0.9479	1.0217	1.0902	1.1552	1.2787	1.3972	1.5129	1.6266	1.7388	1.8500	1.9603	2.0699
	h	463.14	1204.2	1222.2	1260.3	1293.9	1324.9	1354.2	1410.0	1464.4	1518.6	1572.9	1627.8	1683.3	1739.4	1796.1
	s	0.6634	1.4529	1.4720	1.5106	1.5431	1.5717	1.5975	1.6438	1.6853	1.7237	1.7598	1.7939	1.8263	1.8573	1.8870
580 (482.57)	v	0.02006	0.7971	0.8287	0.9100	0.9824	1.0492	1.1125	1.2324	1.3473	1.4593	1.5693	1.6780	1.7855	1.8921	1.9980
	h	467.47	1203.9	1219.1	1258.0	1292.1	1323.4	1353.0	1409.2	1463.7	1518.0	1572.4	1627.4	1682.9	1739.1	1795.9
	s	0.6679	1.4495	1.4654	1.5049	1.5380	1.5668	1.5929	1.6394	1.6811	1.7196	1.7556	1.7898	1.8223	1.8533	1.8831
600 (486.20)	v	0.02013	0.7697	0.7944	0.8746	0.9456	1.0109	1.0726	1.1892	1.3008	1.4093	1.5160	1.6211	1.7252	1.8284	1.9309
	h	471.70	1203.7	1215.9	1255.6	1290.3	1322.0	1351.8	1408.3	1463.0	1517.4	1571.9	1627.0	1682.6	1738.8	1795.6
	s	0.6723	1.4461	1.4590	1.4993	1.5329	1.5621	1.5884	1.6351	1.6769	1.7155	1.7517	1.7859	1.8184	1.8494	1.8792
650 (494.89)	v	0.02032	0.7084	0.7173	0.7954	0.8634	0.9254	0.9835	1.0929	1.1969	1.2979	1.3969	1.4944	1.5909	1.6864	1.7813
	h	481.89	1202.8	1207.6	1249.6	1285.7	1318.3	1348.7	1406.0	1461.2	1515.9	1570.7	1625.9	1681.6	1738.0	1794.9
	s	1.6828	1.4381	1.4430	1.4858	1.5207	1.5507	1.5775	1.6249	1.6671	1.7059	1.7422	1.7765	1.8092	1.8403	1.8701
700 (503.08)	v	0.02050	0.6556		0.7271	0.7928	0.8520	0.9072	1.0102	1.1078	1.2023	1.2948	1.3858	1.4757	1.5647	1.6530
	h	491.60	1201.8		1243.4	1281.0	1314.6	1345.6	1403.7	1459.4	1514.4	1569.4	1624.8	1680.7	1737.2	1794.3
	s	0.6928	1.4304		1.4726	1.5090	1.5399	1.5673	1.6154	1.6580	1.6970	1.7335	1.7679	1.8006	1.8318	1.8617
750 (510.84)	v	0.02069	0.6095		0.6676	0.7313	0.7882	0.8409	0.9386	1.0306	1.1195	1.2063	1.2916	1.3759	1.4592	1.5419
	h	500.89	1200.7		1236.9	1276.1	1310.7	1342.5	1401.5	1457.6	1512.9	1568.2	1623.8	1679.8	1736.4	1793.6
	s	0.7022	1.4232		1.4598	1.4977	1.5296	1.5577	1.6065	1.6494	1.6886	1.7252	1.7598	1.7926	1.8239	1.8538

Table A-1-3 Superheated Steam (No Liquid Present) (English) (cont.)

Abs. Press. lbf/in.² (Sat. Temp, F)		Sat. Liq.	Sat. Vap.	450	500	550	600	650	700	800	900	1000	1100	1200	1300	1400	1500
800 (518.21)	v	0.02087	0.5690			0.6151	0.6774	0.7323	0.7828	0.8759	0.9631	1.0470	1.1289	1.2093	1.2885	1.3669	1.4446
	h	509.81	1199.4			1230.1	1271.1	1306.8	1339.3	1399.1	1455.8	1511.4	1566.9	1622.7	1678.9	1735.7	1792.9
	s	0.7111	1.4163			1.4472	1.4869	1.5198	1.5484	1.5980	1.6413	1.6807	1.7175	1.7522	1.7851	1.8164	1.8464
850 (525.24)	v	0.02105	0.5330			0.5683	0.6296	0.6829	0.7315	0.8205	0.9034	0.9830	1.0606	1.1366	1.2115	1.2855	1.3588
	h	518.40	1198.0			1223.0	1265.9	1302.0	1336.0	1396.8	1454.0	1510.0	1565.7	1621.6	1678.0	1734.9	1792.3
	s	0.7197	1.4096			1.4347	1.4763	1.5102	1.5396	1.5899	1.6336	1.6733	1.7102	1.7450	1.7780	1.8094	1.8395
900 (531.95)	v	0.02123	0.5009			0.5263	0.5869	0.6388	0.6858	0.7713	0.8504	0.9262	0.9998	1.0720	1.1430	1.2131	1.2825
	h	526.70	1196.4			1215.5	1260.6	1298.6	1332.7	1394.4	1452.2	1508.5	1564.4	1620.6	1677.1	1734.1	1791.6
	s	0.7279	1.4032			1.4223	1.4659	1.5010	1.5311	1.5822	1.6263	1.6662	1.7033	1.7382	1.7713	1.8028	1.8329

Temperature—F

Abs. Press. lbf/in.² (Sat. Temp, F)		Sat. Liq.	Sat. Vap.	550	600	650	700	750	800	850	900	1000	1100	1200	1300	1400	1500
950 (538.39)	v	0.02141	0.4721	0.4883	0.5485	0.5993	0.6449	0.6871	0.7272	0.7656	0.8030	0.8753	0.9455	1.0142	1.0817	1.1484	1.2143
	h	534.74	1194.7	1207.6	1255.1	1294.4	1329.3	1361.5	1392.0	1421.5	1450.3	1507.0	1563.2	1619.5	1676.2	1733.3	1791.0
	s	0.7358	1.3970	1.4098	1.4557	1.4921	1.5228	1.5500	1.5748	1.5977	1.6193	1.6595	1.6967	1.7317	1.7649	1.7965	1.8267
1000 (544.58)	v	0.02159	0.4460	0.4535	0.5137	0.5636	0.6080	0.6489	0.6875	0.7245	0.7603	0.8295	0.8966	0.9622	1.0266	1.0901	1.1529
	h	542.55	1192.9	1199.3	1249.3	1290.1	1325.9	1358.7	1389.6	1419.4	1448.5	1505.4	1561.9	1618.4	1675.3	1732.5	1790.3
	s	0.7434	1.3910	1.3973	1.4457	1.4833	1.5149	1.5426	1.5677	1.5908	1.6126	1.6530	1.6905	1.7256	1.7589	1.7905	1.8207
1050 (550.53)	v	0.02177	0.4222		0.4821	0.5312	0.5745	0.6142	0.6515	0.6872	0.7216	0.7881	0.8524	0.9151	0.9767	1.0373	1.0973
	h	550.15	1191.0		1243.4	1285.7	1322.4	1355.8	1387.2	1417.3	1446.6	1503.9	1560.7	1617.4	1674.4	1731.8	1789.6
	s	0.7507	1.3851		1.4358	1.4748	1.5072	1.5354	1.5608	1.5842	1.6062	1.6469	1.6845	1.7197	1.7531	1.7848	1.8151
1100 (556.28)	v	0.02195	0.4006		0.4531	0.5017	0.5440	0.5826	0.6188	0.6533	0.6865	0.7505	0.8121	0.8723	0.9313	0.9894	1.0468
	h	557.55	1189.1		1237.3	1281.2	1318.8	1352.9	1384.7	1415.2	1444.7	1502.4	1559.4	1616.3	1673.5	1731.0	1789.0
	s	0.7578	1.3794		1.4259	1.4664	1.4996	1.5284	1.5542	1.5779	1.6000	1.6410	1.6787	1.7141	1.7475	1.7793	1.8097
1150 (561.82)	v	0.02214	0.3807		0.4263	0.4746	0.5162	0.5538	0.5889	0.6223	0.6544	0.7161	0.7754	0.8332	0.8899	0.9456	1.0007
	h	564.78	1187.0		1230.9	1276.6	1315.2	1349.9	1382.2	1413.0	1442.8	1500.9	1558.1	1615.2	1672.6	1730.2	1788.3
	s	0.7647	1.3738		1.4160	1.4582	1.4923	1.5216	1.5478	1.5717	1.5941	1.6353	1.6732	1.7087	1.7422	1.7741	1.8045
1200 (567.19)	v	0.02232	0.3624		0.4016	0.4497	0.4905	0.5273	0.5615	0.5939	0.6250	0.6845	0.7418	0.7974	0.8519	0.9055	0.9584
	h	571.85	1184.8		1224.2	1271.8	1311.5	1346.5	1379.7	1410.8	1440.9	1499.4	1556.9	1614.2	1671.6	1729.4	1787.6
	s	0.7714	1.3683		1.4061	1.4501	1.4851	1.5150	1.5415	1.5658	1.5883	1.6298	1.6679	1.7035	1.7371	1.7691	1.7996

Abs. Press. Lbs. Sq. In. (Sat. Temp.)		Sat.														
1300 (577.42)	v	0.02269	0.3299	0.3570	0.4052	0.4451	0.4804	0.5129	0.5436	0.5729	0.6287	0.6822	0.7341	0.7847	0.8345	0.8836
	h	585.58	1180.2	1209.9	1261.9	1303.9	1340.8	1374.6	1406.4	1437.1	1496.3	1554.3	1612.0	1669.8	1727.9	1786.3
	s	0.7843	1.3577	1.3860	1.4340	1.4711	1.5022	1.5296	1.5544	1.5773	1.6194	1.6578	1.6937	1.7275	1.7596	1.7902
1400 (587.07)	v	0.02307	0.3018	0.3176	0.3667	0.4059	0.4400	0.4712	0.5004	0.5282	0.5809	0.6311	0.6798	0.7272	0.7737	0.8195
	h	598.83	1175.3	1194.1	1251.4	1296.1	1334.5	1369.3	1402.0	1433.2	1493.2	1551.8	1609.9	1668.0	1726.3	1785.0
	s	0.7966	1.3474	1.3652	1.4181	1.4575	1.4900	1.5182	1.5436	1.5670	1.6096	1.6484	1.6845	1.7185	1.7508	1.7815
1500 (596.20)	v	0.02346	0.2772	0.2820	0.3328	0.3717	0.4049	0.4350	0.4629	0.4894	0.5394	0.5869	0.6327	0.6773	0.7210	0.7639
	h	611.68	1170.1	1176.3	1240.2	1287.9	1328.0	1364.0	1397.4	1429.2	1490.1	1549.2	1607.7	1666.2	1724.8	1783.7
	s	0.8085	1.3373	1.3431	1.4022	1.4443	1.4782	1.5073	1.5333	1.5572	1.6004	1.6395	1.6759	1.7101	1.7425	1.7734
1600 (604.87)	v	0.02387	0.2555		0.3026	0.3415	0.3741	0.4032	0.4301	0.4555	0.5031	0.5482	0.5915	0.6336	0.6748	0.7153
	h	624.20	1164.5		1228.3	1279.4	1321.4	1358.5	1392.8	1425.2	1486.9	1546.6	1605.6	1664.3	1723.2	1782.3
	s	0.8199	1.3274		1.3861	1.4312	1.4667	1.4968	1.5235	1.5478	1.5916	1.6312	1.6678	1.7022	1.7347	1.7657
1700 (613.13)	v	0.02428	0.2361		0.2754	0.3147	0.3468	0.3751	0.4011	0.4255	0.4711	0.5140	0.5552	0.5951	0.6341	0.6724
	h	636.45	1158.6		1215.3	1270.5	1314.5	1352.9	1388.1	1421.2	1483.8	1544.0	1603.4	1662.5	1721.7	1781.0
	s	0.8309	1.3176		1.3697	1.4183	1.4555	1.4867	1.5140	1.5388	1.5833	1.6232	1.6601	1.6947	1.7274	1.7585
1800 (621.02)	v	0.02472	0.2186		0.2505	0.2906	0.3223	0.3500	0.3752	0.3988	0.4426	0.4836	0.5229	0.5609	0.5980	0.6343
	h	648.49	1152.3		1201.2	1261.1	1307.4	1347.2	1383.3	1417.1	1480.6	1541.4	1601.2	1660.7	1720.1	1779.7
	s	0.8417	1.3079		1.3526	1.4054	1.4446	1.4768	1.5049	1.5302	1.5753	1.6156	1.6528	1.6876	1.7204	1.7516
1900 (628.56)	v	0.02517	0.2028		0.2274	0.2687	0.3004	0.3275	0.3521	0.3749	0.4171	0.4565	0.4940	0.5303	0.5656	0.6002
	h	660.36	1145.6		1185.7	1251.3	1300.2	1341.4	1378.4	1412.9	1477.4	1538.8	1599.1	1658.8	1718.6	1778.4
	s	0.8522	1.2981		1.3346	1.3925	1.4338	1.4672	1.4960	1.5219	1.5677	1.6084	1.6458	1.6808	1.7138	1.7451
2000 (635.80)	v	0.02565	0.1883		0.2056	0.2488	0.2805	0.3072	0.3312	0.3534	0.3942	0.4320	0.4680	0.5027	0.5365	0.5695
	h	672.11	1138.3		1168.3	1240.9	1292.6	1335.4	1373.5	1408.7	1474.1	1536.2	1596.9	1657.0	1717.0	1777.1
	s	0.8625	1.2881		1.3154	1.3794	1.4231	1.4578	1.4874	1.5138	1.5603	1.6014	1.6391	1.6743	1.7075	1.7389
2100 (642.76)	v	0.02615	0.1750		0.1847	0.2304	0.2624	0.2888	0.3123	0.3339	0.3734	0.4099	0.4445	0.4778	0.5101	0.5418
	h	683.79	1130.5		1148.5	1229.8	1284.9	1329.3	1368.4	1404.4	1470.9	1533.6	1594.7	1655.2	1715.4	1775.7
	s	0.8727	1.2780		1.2942	1.3661	1.4125	1.4486	1.4790	1.5060	1.5532	1.5948	1.6327	1.6681	1.7014	1.7330
2200 (649.45)	v	0.02669	0.1627		0.1636	0.2134	0.2458	0.2720	0.2950	0.3161	0.3545	0.3897	0.4231	0.4551	0.4862	0.5165
	h	695.46	1122.2		1123.9	1218.0	1276.8	1323.1	1363.3	1400.0	1467.6	1530.9	1592.5	1653.3	1713.9	1774.4
	s	0.8828	1.2676		1.2691	1.3523	1.4020	1.4395	1.4708	1.4984	1.5463	1.5883	1.6266	1.6622	1.6956	1.7273

Table A-1-3 Superheated Steam (No Liquid Present) (English)

Temperature—F

Abs. Press. lbf/in.² (Sat. Temp, F)		Sat. Liq.	Sat. Vap.	550	600	650	700	750	800	850	900	950	1000	1050	1100	1150	1200	1300	1400	1500
2300 (655.89)	v	0.02727	0.1513				0.1975	0.2305	0.2566	0.2793	0.2999		0.3372		0.3714		0.4035	0.4344	0.4643	0.4935
	h	707.18	1113.2				1205.3	1268.4	1316.7	1358.1	1395.7		1464.2		1528.3		1590.3	1651.5	1712.3	1773.1
	s	0.8929	1.2569				1.3381	1.3914	1.4305	1.4628	1.4910		1.5397		1.5821		1.6207	1.6565	1.6901	1.7219
2400 (662.11)	v	0.02790	0.1408				0.1824	0.2164	0.2424	0.2648	0.2850	0.3037	0.3214	0.3382	0.3545	0.3703	0.3856	0.4155	0.4443	0.4724
	h	718.95	1103.7				1191.6	1259.7	1310.1	1352.8	1391.2	1426.9	1460.9	1493.7	1525.6	1557.0	1588.1	1649.6	1710.8	1771.8
	s	0.9031	1.2460				1.3232	1.3808	1.4217	1.4549	1.4837	1.5095	1.5332	1.5553	1.5761	1.5959	1.6149	1.6509	1.6847	1.7167
2500 (668.11)	v	0.02859	0.1307				0.1681	0.2032	0.2293	0.2514	0.2712	0.2896	0.3068	0.3232	0.3390	0.3543	0.3692	0.3980	0.4259	0.4529
	h	731.71	1093.3				1176.7	1250.6	1303.4	1347.4	1386.7	1423.1	1457.5	1490.7	1522.9	1554.6	1585.9	1647.8	1709.2	1770.4
	s	0.9139	1.2345				1.3076	1.3701	1.4129	1.4472	1.4766	1.5029	1.5269	1.5492	1.5703	1.5903	1.6094	1.6456	1.6796	1.7116
2600 (673.91)	v	0.02938	0.1211				0.1544	0.1909	0.2171	0.2390	0.2585	0.2765	0.2933	0.3093	0.3247	0.3395	0.3540	0.3819	0.4088	0.4350
	h	744.47	1082.0				1160.2	1241.1	1296.5	1341.9	1382.1	1419.2	1454.1	1487.7	1520.2	1552.2	1583.7	1646.0	1707.7	1769.1
	s	0.9247	1.2225				1.2908	1.3592	1.4042	1.4395	1.4696	1.4964	1.5208	1.5434	1.5646	1.5848	1.6040	1.6405	1.6746	1.7068
2700 (679.53)	v	0.03029	0.1119				0.1411	0.1794	0.2058	0.2275	0.2468	0.2644	0.2809	0.2965	0.3114	0.3259	0.3399	0.3670	0.3931	0.4184
	h	757.34	1069.7				1142.0	1231.1	1289.5	1336.3	1377.5	1415.2	1450.7	1484.6	1517.5	1549.8	1581.5	1644.1	1706.1	1767.8
	s	0.9356	1.2097				1.2727	1.3481	1.3954	1.4319	1.4628	1.4900	1.5148	1.5376	1.5591	1.5794	1.5988	1.6355	1.6697	1.7021
2800 (684.96)	v	0.03134	0.1030				0.1278	0.1685	0.1952	0.2168	0.2358	0.2531	0.2693	0.2845	0.2991	0.3132	0.3268	0.3532	0.3785	0.4030
	h	770.69	1055.8				1121.2	1220.6	1282.2	1330.7	1372.8	1411.2	1447.2	1481.6	1514.8	1547.3	1579.3	1642.2	1704.5	1766.5
	s	0.9468	1.1958				1.2527	1.3368	1.3867	1.4245	1.4561	1.4838	1.5089	1.5321	1.5537	1.5742	1.5938	1.6306	1.6651	1.6975
2900 (690.22)	v	0.03262	0.0942				0.1138	0.1581	0.1853	0.2068	0.2256	0.2427	0.2585	0.2734	0.2877	0.3014	0.3147	0.3403	0.3649	0.3887
	h	785.13	1039.8				1095.3	1209.6	1274.7	1324.9	1368.0	1407.2	1443.7	1478.5	1512.1	1544.9	1577.0	1640.4	1703.0	1765.2
	s	0.9588	1.1803				1.2283	1.3251	1.3780	1.4171	1.4494	1.4777	1.5032	1.5266	1.5485	1.5692	1.5889	1.6259	1.6605	1.6931
3000 (695.33)	v	0.03428	0.0850				0.0982	0.1483	0.1759	0.1975	0.2161	0.2329	0.2484	0.2630	0.2770	0.2904	0.3033	0.3282	0.3522	0.3753
	h	801.84	1020.3				1060.5	1197.9	1267.0	1319.0	1363.2	1403.1	1440.2	1475.4	1509.4	1542.4	1574.8	1638.5	1701.4	1763.8
	s	0.9728	1.1619				1.1966	1.3131	1.3692	1.4097	1.4429	1.4717	1.4976	1.5213	1.5434	1.5642	1.5841	1.6214	1.6561	1.6888
3100 (700.28)	v	0.03681	0.0745					0.1389	0.1671	0.1887	0.2071	0.2237	0.2390	0.2533	0.2670	0.2800	0.2927	0.3170	0.3403	0.3628
	h	823.97	993.3					1185.4	1259.1	1313.0	1358.4	1399.0	1436.7	1472.3	1506.6	1539.9	1572.6	1636.7	1699.8	1762.5
	s	0.9914	1.1373					1.3007	1.3604	1.4024	1.4364	1.4658	1.4920	1.5161	1.5384	1.5594	1.5794	1.6169	1.6518	1.6847

P																
3200 (705.08)	v	0.04472	0.0566	0.1300	0.1588	0.1804	0.1987	0.2151	0.2301	0.2442	0.2576	0.2704	0.2827	0.3065	0.3291	0.3510
	h	875.54	931.6	1172.3	1250.9	1306.9	1353.4	1394.9	1433.1	1469.2	1503.8	1537.4	1570.3	1634.8	1698.3	1761.2
	s	1.0351	1.0832	1.2877	1.3515	1.3951	1.4300	1.4600	1.4866	1.5110	1.5335	1.5547	1.5749	1.6126	1.6477	1.6806
3300	v			0.1213	0.1510	0.1727	0.1908	0.2070	0.2218	0.2357	0.2488	0.2613	0.2734	0.2966	0.3187	0.3400
	h			1158.2	1242.5	1300.7	1348.4	1390.7	1429.5	1466.1	1501.0	1534.9	1568.1	1632.9	1696.7	1759.9
	s			1.2742	1.3425	1.3879	1.4237	1.4542	1.4813	1.5059	1.5287	1.5501	1.5704	1.6084	1.6436	1.6767
3400	v			0.1129	0.1435	0.1653	0.1834	0.1994	0.2140	0.2276	0.2405	0.2528	0.2646	0.2872	0.3088	0.3296
	h			1143.2	1233.7	1294.3	1343.4	1386.4	1425.9	1462.9	1498.3	1532.4	1565.8	1631.1	1695.1	1758.5
	s			1.2600	1.3334	1.3807	1.4174	1.4486	1.4761	1.5010	1.5240	1.5456	1.5660	1.6042	1.6396	1.6728
3500	v			0.1048	0.1364	0.1583	0.1764	0.1922	0.2066	0.2200	0.2326	0.2447	0.2563	0.2784	0.2995	0.3198
	h			1127.1	1224.6	1287.8	1338.2	1382.2	1422.2	1459.7	1495.5	1529.9	1563.6	1629.2	1693.6	1757.2
	s			1.2450	1.3242	1.3734	1.4112	1.4430	1.4709	1.4962	1.5194	1.5412	1.5618	1.6002	1.6358	1.6691
3600	v			0.0966	0.1296	0.1517	0.1697	0.1854	0.1996	0.2128	0.2252	0.2371	0.2485	0.2702	0.2908	0.3106
	h			1108.6	1215.3	1281.2	1333.0	1377.9	1418.6	1456.5	1492.6	1527.4	1561.3	1627.3	1692.0	1755.9
	s			1.2281	1.3148	1.3662	1.4050	1.4374	1.4658	1.4914	1.5149	1.5369	1.5576	1.5962	1.6320	1.6654
3800	v			0.0799	0.1169	0.1395	0.1574	0.1729	0.1868	0.1996	0.2116	0.2231	0.2340	0.2549	0.2746	0.2936
	h			1064.2	1195.5	1267.6	1322.4	1369.1	1411.2	1450.1	1487.0	1522.4	1556.8	1623.6	1688.9	1753.2
	s			1.1888	1.2955	1.3517	1.3928	1.4265	1.4558	1.4821	1.5061	1.5284	1.5495	1.5886	1.6247	1.6584
4000	v			0.0631	0.1052	0.1284	0.1463	0.1616	0.1752	0.1877	0.1994	0.2105	0.2210	0.2411	0.2601	0.2783
	h			1007.4	1174.3	1253.4	1311.6	1360.2	1403.6	1443.6	1481.3	1517.3	1552.2	1619.8	1685.7	1750.6
	s			1.1396	1.2754	1.3371	1.3807	1.4158	1.4461	1.4730	1.4976	1.5203	1.5417	1.5812	1.6177	1.6516
4200	v			0.0498	0.0945	0.1183	0.1362	0.1513	0.1647	0.1769	0.1883	0.1991	0.2093	0.2287	0.2470	0.2645
	h			950.1	1151.6	1238.6	1300.4	1351.2	1396.0	1437.1	1475.5	1512.2	1547.6	1616.1	1682.6	1748.0
	s			1.0905	1.2544	1.3223	1.3686	1.4053	1.4366	1.4642	1.4893	1.5124	1.5341	1.5742	1.6109	1.6452
4400	v			0.0421	0.0846	0.1090	0.1270	0.1420	0.1552	0.1671	0.1782	0.1887	0.1986	0.2174	0.2351	0.2519
	h			909.5	1127.3	1223.3	1289.0	1342.0	1388.3	1430.4	1469.7	1507.1	1543.0	1612.3	1679.4	1745.3
	s			1.0556	1.2325	1.3073	1.3566	1.3949	1.4272	1.4556	1.4812	1.5048	1.5268	1.5673	1.6044	1.6389

v = specific volume, ft³/lbm
h = enthalpy, Btu/lbm
s = entropy, Btu/lbm R

Source: Reprinted with permission from ASME.

Table A-1-4 Compressed Liquid (English)

T	v	u	h	s	v	u	h	s	v	u	h	s
	$p = 500(467.13)$				$p = 1000(544.75)$				$p = 1500(596.39)$			
Sat	0.019748	447.70	449.53	0.64904	0.021591	538.39	542.38	0.74320	0.023461	604.97	611.48	0.80824
32	0.015994	0.00	1.49	0.00000	0.015967	0.03	2.99	0.00005	0.015939	0.05	4.47	0.00007
50	0.015998	18.02	19.50	0.03599	0.015972	17.99	20.94	0.03592	0.015946	17.95	22.38	0.03584
100	0.016106	67.87	69.36	0.12932	0.016082	67.70	70.68	0.12901	0.016058	67.53	71.99	0.12870
150	0.016318	117.66	119.17	0.21457	0.016293	117.38	120.40	0.21410	0.016268	117.10	121.62	0.21364
200	0.016608	167.65	169.19	0.29341	0.016580	167.26	170.32	0.29281	0.016554	166.87	171.46	0.29221
250	0.016972	217.99	219.56	0.36702	0.016941	217.47	220.61	0.36628	0.016910	216.96	221.65	0.36554
300	0.017416	268.92	270.53	0.43641	0.017379	268.24	271.46	0.43552	0.017343	267.58	272.39	0.43463
350	0.017954	320.71	322.37	0.50249	0.017909	319.83	323.15	0.50140	0.017865	318.98	323.94	0.50034
400	0.018608	373.68	375.40	0.56604	0.018550	372.55	375.98	0.56472	0.018493	371.45	376.59	0.56343
450	0.019420	428.40	430.19	0.62798	0.019340	426.89	430.47	0.62632	0.019264	425.44	430.79	0.62470
500					0.02036	483.8	487.5	0.6874	0.02024	481.8	487.4	0.6853
550									0.02158	542.1	548.1	0.7469

	p = 2000 (636.00)				p = 3000 (695.52)				p = 5000			
	v	u	h	s	v	u	h	s	v	u	h	s
Sat	0.025649	662.40	671.89	0.86227	0.034310	783.45	802.50	0.97320				
32	0.015912	0.06	5.95	0.00008	0.015859	0.09	8.90	0.00009	0.015755	0.11	14.70	−0.00001
50	0.015920	17.91	23.81	0.03575	0.015870	17.84	26.65	0.03555	0.015773	17.67	32.26	0.03508
100	0.016034	67.37	73.30	0.12839	0.015987	67.04	75.91	0.12777	0.015897	66.40	81.11	0.12651
200	0.016527	166.49	172.60	0.29162	0.016476	165.74	174.89	0.29046	0.016376	164.32	179.47	0.28818
300	0.017308	266.93	273.33	0.43376	0.017240	265.66	275.23	0.43205	0.017110	263.25	279.08	0.42875
400	0.018439	370.38	377.21	0.56216	0.018334	368.32	378.50	0.55970	0.018141	364.47	381.25	0.55506
450	0.019191	424.04	431.14	0.62313	0.019053	421.36	431.93	0.62011	0.018803	416.44	433.84	0.61451
500	0.02014	479.8	487.3	0.6832	0.019944	476.2	487.3	0.6794	0.019603	469.8	487.9	0.6724
560	0.02172	551.8	559.8	0.7565	0.021382	546.2	558.0	0.7508	0.020835	536.7	556.0	0.7411
600	0.02330	605.4	614.0	0.8086	0.02274	597.0	609.6	0.8004	0.02191	584.0	604.2	0.7876
640					0.02475	654.3	668.0	0.8545	0.02334	634.6	656.2	0.8357
680					0.02879	728.4	744.3	0.9226	0.02535	690.6	714.1	0.8873
700									0.02676	721.8	746.6	0.9156

From p. 662 Fundamentals of Classical Thermodynamics, by Van Wylen and Sonntag, Second Edition, John Wiley and Sons, Inc. 1973.

v = specific volume, ft³/lbm; u = internal energy, Btu/lbm; h = enthalpy, Btu/lbm; s = entropy, Btu/lbm R; T, F; p, psia

Table A-1-5 Saturated Steam: Temperature Table (SI)

Temp C T	Abs. Press. kPa p	Specific Volume, m³/kg Sat. Liq. v_f	Evap. v_{fg}	Sat. Vapor v_g	Enthalpy, kJ/kg Sat. Liquid h_f	Evap. h_{fg}	Sat. Vapor h_g	Entropy, kJ/kg K Sat. Liquid s_f	Evap. s_{fg}	Sat. Vapor s_g
0.01	0.6112+	0.0010002	206.16	206.16	0.00	2501.6	2501.6	0.0000	9.1575	9.1575
1.0	0.6566	0.0010001	192.61	192.61	4.17	2499.2	2503.4	0.0153	9.1158	9.1311
2.0	0.7055	0.0010001	179.92	179.92	8.39	2496.8	2505.2	0.0306	9.0741	9.1047
3.0	0.7575	0.0010001	168.17	168.17	12.60	2494.5	2507.1	0.0459	9.0326	9.0785
4.0	0.8129	0.0010000	157.27	157.27	16.80	2492.1	2508.9	0.0611	8.9915	9.0526
5.0	0.8718	0.0010000	147.16	147.16	21.01	2489.7	2510.7	0.0762	8.9507	9.0269
6.0	0.9345	0.0010000	137.78	137.78	25.21	2487.4	2512.6	0.0193	8.9102	9.0015
7.0	1.0012	0.0010001	129.06	129.06	29.41	2485.0	2514.4	0. 3	8.8699	8.9762
8.0	1.0720	0.0010001	120.96	120.97	33.60	2482.6	2516.2	0.1213	8.8300	8.9513
9.0	1.1472	0.0010002	113.43	113.44	37.80	2480.3	2518.1	0.1362	8.7903	.9265
10.0	1.2270	0.0010003	106.43	106.43	41.99	2477.9	2519.9	0.1510	8.7510	8.9020
12.0	1.4014	0.0010004	93.83	93.84	50.38	2473.2	2523.6	0.1805	8.6731	8.8536
14.0	1.5973	0.0010007	82.90	82.90	58.75	2468.5	2527.2	0.2098	8.5963	8.8060
16.0	1.8168	0.0010010	73.38	73.38	67.13	2463.8	2530.9	0.2388	8.5205	8.7593
18.0	2.0624	0.0010013	65.09	65.09	75.50	2459.0	2534.5	0.2677	8.4458	8.7135
20.0	2.337	0.0010017	57.84	57.84	83.86	2454.3	2538.2	0.2963	8.3721	8.6684
22.0	2.642	0.0010022	51.49	51.49	92.23	2449.6	2541.8	0.3247	8.2994	8.6241
24.0	2.982	0.0010026	45.92	45.93	100.59	2444.9	2545.5	0.3530	8.2277	8.5806
26.0	3.360	0.0010032	41.03	41.03	108.95	2440.2	2549.1	0.3810	8.1569	8.5379
28.0	3.778	0.0010037	36.73	36.73	117.31	2435.4	2552.7	0.4088	8.0870	8.4959
30.0	4.241	0.0010043	32.93	32.93	125.66	2430.7	2556.4	0.4365	8.0181	8.4546
32.0	4.753	0.0010049	29.57	29.57	134.02	2425.9	2560.0	0.4640	7.9500	8.4140
34.0	5.318	0.0010056	26.60	26.60	142.38	2421.2	2563.6	0.4913	7.8828	8.3740
36.0	5.940	0.0010063	23.97	23.97	150.74	2416.4	2567.2	0.5184	7.8164	8.3348
38.0	6.624	0.0010070	21.63	21.63	159.06	2411.7	2570.8	0.5453	7.7509	8.2962
40.0	7.375	0.0010078	19.545	19.546	167.45	2406.9	2574.4	0.5721	7.6861	8.2583
42.0	8.198	0.0010086	17.691	17.692	175.81	2402.1	2577.9	0.5987	7.6222	8.2209

44.0	9.100	0.0010094	16.035	16.036	184.17	2397.3	2581.5	0.6252	7.5590	8.1842
46.0	10.086	0.0010103	14.556	14.557	192.53	2392.5	2585.1	0.6514	7.4966	8.1481
48.0	11.162	0.0010112	13.232	13.233	200.89	2387.7	2588.6	0.6776	7.4350	8.1125
50.0	12.335	0.0010121	12.045	12.046	209.26	2382.9	2592.2	0.7035	7.3741	8.0776
52.0	13.613	0.0010131	10.979	10.980	217.62	2378.1	2595.7	0.7293	7.3138	8.0432
54.0	15.002	0.0010140	10.021	10.022	225.99	2373.2	2599.2	0.7550	7.2543	8.0093
56.0	16.511	0.0010150	9.158	9.159	234.35	2368.4	2602.7	0.7804	7.1955	7.9759
58.0	18.147	0.0010161	8.380	8.381	242.72	2363.5	2606.2	0.8058	7.1373	7.9431
60.0	19.920	0.0010171	7.678	7.679	251.09	2358.6	2609.7	0.8310	7.0798	7.9108
62.0	21.838	0.0010182	7.043	7.044	259.46	2353.7	2613.2	0.8560	7.0230	7.8790
64.0	23.912	0.0010193	6.468	6.469	267.84	2348.8	2616.6	0.8809	6.9667	7.8477
66.0	26.150	0.0010205	5.947	5.948	276.21	2343.9	2610.1	0.9057	6.9111	7.8168
68.0	28.563	0.0010217	5.475	5.476	284.59	2338.9	2623.6	0.9303	6.8561	7.7864
70.0	31.16	0.0010288	5.045	5.046	292.97	2334.0	2626.9	0.9548	6.8017	7.7565
72.0	33.96	0.0010241	4.655	4.656	301.36	2329.0	2630.3	0.9792	6.7478	7.7270
74.0	36.96	0.0010253	4.299	4.300	309.74	2324.0	2633.7	1.0034	6.6945	7.6979
76.0	40.10	0.0010266	3.976	3.976	318.13	2318.9	2637.1	1.0275	6.6418	7.6693
78.0	43.65	0.0010279	3.679	3.680	326.52	2313.9	2640.4	1.0514	6.5896	7.6410
80.0	47.36	0.0010292	3.408	3.409	334.92	2308.8	2643.8	1.0753	6.5380	7.6132
82.0	51.33	0.0010305	3.161	3.162	343.31	2303.8	2647.1	1.0990	6.4868	7.5858
84.0	55.57	0.0010319	2.934	2.935	351.71	2298.6	2650.4	1.1225	6.4362	7.5588
86.0	60.11	0.0010333	2.726	2.727	360.12	2293.5	2653.6	1.1460	6.3861	7.5351
88.0	64.95	0.0010347	2.535	2.536	368.53	2288.4	2656.9	1.1693	6.3365	7.5058
90.0	70.11	0.0010361	2.3603	2.3613	376.94	2283.2	2660.1	1.1925	6.2873	7.4799
92.0	75.61	0.0010376	2.1992	2.2002	385.36	2278.0	2663.4	1.2156	6.2387	7.4543
94.0	81.46	0.0010391	2.0509	2.0519	393.78	2272.9	2666.6	1.2386	6.1905	7.4291
96.0	87.69	0.0010406	1.9143	1.9153	402.20	2267.5	2669.7	1.2615	6.1427	7.4042
98.0	94.30	0.0010421	1.7883	1.7893	410.63	2262.2	2672.9	1.2842	6.0954	7.3796
100.0	101.33	0.0010437	1.6720	1.6730	419.06	2256.9	2676.0	1.3069	6.0485	7.3554
105.0	120.80	0.0010477	1.4182	1.4193	440.17	2243.6	2683.7	1.3630	5.9331	7.2962
110.0	143.27	0.0010519	1.2089	1.2099	461.32	2230.0	2691.3	1.4185	5.8203	7.2388
115.0	169.06	0.0010562	1.0352	1.0363	482.50	2216.2	2698.7	1.4733	5.7099	7.1832
120.0	198.54	0.0010606	0.8905	0.8915	503.72	2202.2	2706.0	1.5276	5.6017	7.1293

Table A-1-5 Saturated Steam: Temperature Table (SI) (cont.)

Temp C T	Abs. Press. kPa p	Specific Volume, m³/kg			Enthalpy, kJ/kg			Entropy, kJ/kg K		
		Sat. Liq. v_f	Evap. v_{fg}	Sat. Vapor v_g	Sat. Liquid h_f	Evap. h_{fg}	Sat. Vapor h_g	Sat. Liquid s_f	Evap. s_{fg}	Sat. Vapor s_g
125.0	232.1	0.0010652	0.7692	0.7702	524.99	2188.0	2713.0	1.5813	5.4957	7.0769
130.0	270.1	0.0010700	0.6671	0.6681	546.31	2173.6	2719.9	1.6344	5.3917	7.0261
134.0	313.1	0.0010750	0.5807	0.5818	567.68	2158.9	2726.6	1.6869	5.2897	6.9766
140.0	361.4	0.0010801	0.5074	0.5085	589.10	2144.0	2733.1	1.7390	5.1894	6.9284
145.0	415.5	0.0010853	0.4449	0.4460	610.59	2128.7	2739.3	1.7906	5.0910	6.8815
150.0	476.0	0.0010908	0.3914	0.3924	632.15	2113.2	2745.4	1.8416	4.9941	6.8358
155.0	543.3	0.0010964	0.3453	0.3464	653.77	2097.4	2751.2	1.8923	4.8989	6.7911
160.0	618.1	0.0011022	0.3057	0.3068	675.47	2081.3	2756.7	1.9425	4.8050	6.7475
165.0	700.8	0.0011082	0.2713	0.2724	697.25	2064.8	2762.0	1.9923	4.7126	6.7048
170.0	792.0	0.0011145	0.2414	0.2426	719.12	2047.9	2767.1	2.0416	4.6214	6.6630
175.0	892.4	0.0011209	0.21542	0.21654	741.07	2030.7	2771.8	2.0906	4.5314	6.6221
180.0	1002.7	0.0011275	0.19267	0.19380	763.12	2013.2	2776.3	2.1393	4.4426	6.5819
185.0	1123.3	0.0011344	0.17272	0.17386	785.26	1995.2	2780.4	2.1876	4.3548	6.5424
190.0	1255.1	0.0011415	0.15517	0.15632	807.52	1976.7	2784.3	2.2356	4.2680	6.5036
195.0	1398.7	0.0011489	0.13969	0.14084	829.88	1957.9	2787.8	2.2833	4.1821	6.4654
200.0	1554.9	0.0011565	0.12600	0.12716	852.37	1938.6	2790.9	2.3307	4.0971	6.4278
205.0	1724.3	0.0011644	0.11386	0.11503	874.99	1918.8	2793.8	2.3778	4.0128	6.3906
210.0	1907.7	0.0011726	0.10307	0.10424	897.73	1898.5	2796.2	2.4247	3.9293	6.3539
215.0	2106.0	0.0011811	0.09344	0.09463	920.63	1877.6	2798.3	2.4713	3.8463	6.3176
220.0	2319.8	0.0011900	0.08485	0.08604	943.67	1856.2	2799.9	2.5178	3.7639	6.2817
225.0	2550.	0.0011992	0.07715	0.07835	966.88	1834.3	2801.2	2.5641	3.6820	6.2461
230.0	2798.	0.0012087	0.07024	0.07145	990.27	1811.7	2802.0	2.6102	3.6006	6.2107
235.0	3063.	0.0012187	0.06403	0.06525	1013.83	1788.5	2802.3	2.6561	3.5194	6.1756
240.0	3348.	0.0012291	0.05843	0.05965	1037.60	1764.6	2802.2	2.7020	3.4386	6.1406
245.0	3652.	0.0012399	0.05337	0.05461	1061.58	1740.0	2801.6	2.7478	3.3579	6.1057
250.0	3978.	0.0012513	0.04879	0.05004	1085.78	1714.7	2800.4	2.7935	3.2773	6.0708
255.0	4325.	0.0012632	0.04463	0.04590	1110.23	1688.5	2798.7	2.8392	3.1968	6.0359

260.0	0.0012756	4694.	0.04086	0.04213	1134.94	1661.5	2796.4	2.8848	3.1161	6.0010
265.0	0.0012887	5088.	0.03742	0.03871	1159.93	1633.5	2793.5	2.9306	3.0353	5.9658
270.0	0.0013025	5506.	0.03429	0.03559	1185.23	1604.6	2789.9	2.9763	2.9541	5.9304
275.0	0.0013170	5950.	0.03142	0.03274	1210.86	1574.7	2785.5	3.0222	2.8725	5.8947
280.0	0.0013324	6420.	0.02879	0.03013	1236.84	1543.6	2780.4	3.0683	2.7903	5.8586
285.0	0.0013487	6919.	0.02638	0.02773	1263.21	1511.3	2774.5	3.1146	2.7074	5.8220
290.0	0.0013659	7446.	0.02417	0.02554	1290.01	1477.6	2767.6	3.1611	2.6237	5.7848
295.0	0.0013844	8004.	0.02213	0.02351	1317.27	1442.6	2759.8	3.2079	2.5389	5.7469
300.0	0.0014041	8593.	0.020245	0.021649	1345.05	1406.0	2751.0	3.2552	2.4529	5.7081
305.0	0.0014252	9214.	0.018502	0.019927	1373.40	1367.7	2741.1	3.3029	2.3656	5.6685
310.0	0.0014480	9870.	0.016886	0.018334	1402.39	1327.6	2730.0	3.3512	2.2766	5.6278
315.0	0.0014726	10561.	0.015383	0.016856	1432.09	1285.5	2717.6	3.4002	2.1856	5.5858
320.0	0.0014995	11289.	0.013980	0.015480	1462.60	1241.1	2703.7	3.4500	2.0923	5.5423
325.0	0.0015289	12056.	0.012666	0.014195	1494.03	1194.0	2688.0	3.5008	1.9961	5.4969
330.0	0.0015615	12863.	0.011428	0.012989	1526.52	1143.6	2670.2	3.5528	1.8962	5.4490
335.0	0.0015978	13712.	0.010256	0.011854	1560.25	1089.5	2649.7	3.6963	1.7916	5.3979
340.0	0.0016387	14605.	0.009142	0.010780	1595.47	1030.7	2626.2	3.6616	1.6811	5.3427
345.0	0.0016858	15545.	0.008077	0.009763	1632.52	966.4	2598.9	3.7193	1.5636	5.2828
350.0	0.0017411	16535.	0.007058	0.008799	1671.94	895.7	2567.7	3.7800	1.4376	5.2177
355.0	0.0018085	17577.	0.006051	0.007859	1716.63	813.8	2530.4	3.8489	1.2953	5.1442
360.0	0.0018959	18675.	0.005044	0.006940	1764.17	721.3	2485.4	3.9210	1.1390	5.0600
365.0	0.0020160	19833.	0.003996	0.006012	1817.96	610.0	2428.0	4.0021	0.9558	4.9479
370.0	0.0022136	21054.	0.002759	0.004973	1890.21	452.6	2342.8	4.1108	0.7036	4.8144
371.0	0.0022778	21306.	0.002446	0.004723	1910.50	407.4	2317.9	4.1414	0.6324	4.7738
372.0	0.0023636	21562.	0.002075	0.004439	1935.57	351.4	2287.0	4.1794	0.5446	4.7240
373.0	0.0024963	21820.	0.001588	0.004084	1970.50	273.5	2244.0	4.2326	0.4233	4.6559
374.0	0.0028427	22081.*	0.000623	0.003466	2046.72	109.5	2156.2	4.3493	0.1692	4.5185
374.15	0.00317	22120.*	0.0	0.00317	2107.37	0.0	2107.4	4.4429	0.0	4.4429

† Approximate Triple Point
* Approximate Critical Point
Source: Reprinted with permission from ASME.

Table A-1-6 Saturated Steam: Pressure Table (SI)

Abs. Press kPa p	Temp. °C T	Specific Volume, m³/kg			Enthalpy, kJ/kg			Entropy, kJ/kg K			Energy, kJ/kg	
		Sat. Liquid v_f	Evap. v_{fg}	Sat. Vapor v_g	Sat. Liquid h_f	Evap. h_{fg}	Sat. Vapor h_g	Sat. Liquid s_f	Evap. s_{fg}	Sat. Vapor s_g	Sat. Liquid u_f	Sat. Vapor u_g
1.0	6.983	0.0010001	129.21	129.21	29.34	2485.0	2514.4	0.1060	8.8706	8.9767	29.33	2385.2
1.1	8.380	0.0010001	118.04	118.04	35.20	2481.7	2516.9	0.1269	8.8149	8.9418	35.20	2387.1
1.2	9.668	0.0010002	108.70	108.70	40.60	2478.7	2519.3	0.1461	8.7640	8.9101	40.60	2388.9
1.3	10.866	0.0010003	100.76	100.76	45.62	2475.9	2521.5	0.1638	8.7171	8.8809	45.62	2390.5
1.4	11.985	0.0010004	93.92	93.92	50.31	2473.2	2523.5	0.1803	8.6737	8.8539	50.31	2392.0
1.5	13.036	0.0010006	87.98	87.98	54.71	2470.7	2525.5	0.1957	8.6332	8.8288	54.71	2393.5
1.6	14.026	0.0010007	82.76	82.77	58.86	2468.3	2527.4	0.2101	8.5952	8.8054	58.86	2394.8
1.8	15.855	0.0010010	74.03	74.03	66.52	2464.1	2530.6	0.2367	8.5260	8.7627	66.52	2397.4
2.0	17.513	0.0010012	67.01	67.01	73.46	2460.2	2533.6	0.2607	8.4639	8.7246	73.46	2399.6
2.2	19.031	0.0010015	61.23	61.23	79.81	2456.6	2536.4	0.2825	8.4077	8.6901	79.81	2401.7
2.4	20.433	0.0010018	56.39	56.39	85.67	2453.3	2539.0	0.3025	8.3563	8.6587	85.67	2403.6
2.6	21.737	0.0010021	52.28	52.28	91.12	2450.2	2541.3	0.3210	8.3089	8.6299	91.12	2405.4
2.8	22.955	0.0010024	48.74	48.74	96.22	2447.3	2543.6	0.3382	8.2650	8.6033	96.21	2407.1
3.0	24.100	0.0010027	45.67	45.67	101.00	2444.6	2545.6	0.3544	8.2241	8.5785	101.00	2408.6
3.5	26.694	0.0010033	39.48	39.48	111.85	2438.5	2550.4	0.3907	8.1325	8.5232	111.84	2412.2
4.0	28.983	0.0010040	34.80	34.80	121.40	2433.1	2554.5	0.4225	8.0530	8.4755	121.41	2415.3
4.5	31.035	0.0010046	31.14	31.14	129.99	2428.2	2558.2	0.4507	7.9827	8.4335	129.98	2418.1
5.0	32.898	0.0010052	28.19	28.19	137.77	2423.8	2561.6	0.4763	7.9197	8.3960	137.77	2420.6
5.5	34.605	0.0010058	25.77	25.77	144.91	2419.8	2564.7	0.4995	7.8626	8.3621	144.90	2422.9
6.0	36.183	0.0010064	23.74	23.74	151.50	2416.0	2567.5	0.5209	7.8104	8.3312	151.50	2425.1
6.5	37.651	0.0010069	22.015	22.016	157.64	2412.5	2570.2	0.5407	7.7622	8.3029	157.63	2427.0
7.0	39.025	0.0010074	20.530	20.531	163.38	2409.2	2572.6	0.5591	7.7176	8.2767	163.37	2428.9
7.5	40.316	0.0010079	19.238	19.239	168.77	2406.2	2574.9	0.5763	7.6760	8.2523	168.76	2430.6
8.0	41.534	0.0010084	18.104	18.105	173.86	2403.2	2577.1	0.5925	7.6370	8.2296	173.86	2432.3
9.0	43.787	0.0010094	16.203	16.204	183.28	2397.9	2581.1	0.6224	7.5657	8.1881	183.27	2435.3
10.	45.833	0.0010102	14.674	14.675	191.83	2392.9	2584.8	0.6493	7.5018	8.1511	191.82	2438.0
11.	47.710	0.0010111	13.415	13.416	199.68	2388.4	2588.1	0.6738	7.4439	8.1177	199.67	2440.5

12.	49.446	0.0010119	12.361	12.362	206.94	2384.3	2591.2	0.6963	7.3909	8.0872	206.93	2442.8
13.	51.062	0.0010126	11.465	11.466	213.70	2380.3	2594.0	0.7172	7.3420	8.0592	213.68	2445.0
14.	52.574	0.0010133	10.693	10.694	220.02	2376.7	2596.7	0.7367	7.2967	8.0334	220.01	2447.0
15.	53.997	0.0010140	10.022	10.023	225.97	2373.2	2599.2	0.7549	7.2544	8.0093	225.96	2448.9
16.	55.341	0.0010147	9.432	9.433	231.59	2370.0	2601.6	0.7721	7.2148	7.9869	231.58	2450.6
18.	57.826	0.0010160	8.444	8.445	241.99	2363.9	2605.9	0.8036	7.1424	7.9460	241.98	2453.9
20.	60.086	0.0010172	7.649	7.650	251.45	2358.4	2609.9	0.8321	7.0774	7.9094	251.43	2456.9
22.	62.162	0.0010183	6.994	6.995	260.14	2353.3	2613.5	0.8581	7.0184	7.8764	260.12	2459.6
24.	64.082	0.0010194	6.446	6.447	268.18	2348.6	2616.8	0.8820	6.9644	7.8464	268.16	2462.1
26.	65.871	0.0010204	5.979	5.980	275.67	2344.2	2619.9	0.9041	6.9147	7.8188	275.65	2464.4
28.	67.547	0.0010214	5.578	5.579	282.69	2340.0	2622.7	0.9248	6.8685	7.7933	282.66	2466.5
30.	69.124	0.0010223	5.228	5.229	289.30	2336.1	2625.4	0.9441	6.8254	7.7695	289.27	2468.6
35.	72.709	0.0010245	4.525	4.526	304.33	2327.2	2631.5	0.9878	6.7288	7.7166	304.29	2473.1
40.	75.886	0.0010265	3.992	3.993	317.65	2319.2	2636.9	1.0261	6.6448	7.6709	317.61	2477.1
45.	78.743	0.0010284	3.575	3.576	329.64	2312.0	2641.7	1.0603	6.5704	7.6307	329.59	2480.7
50.	81.345	0.0010301	3.239	3.240	340.56	2305.4	2646.0	1.0912	6.5035	7.5947	340.51	2484.0
55.	83.737	0.0010317	2.963	2.964	350.61	2299.3	2649.9	1.1194	6.4428	7.5623	350.56	2486.9
60.	85.954	0.0010333	2.731	2.732	359.93	2293.6	2653.6	1.1454	6.3873	7.5327	359.86	2489.7
65.	88.021	0.0010347	2.5335	2.5346	368.62	2288.3	2656.9	1.1696	6.3360	7.5055	368.55	2492.2
70.	89.959	0.0010361	2.3637	2.3647	376.77	2283.3	2660.1	1.1921	6.2883	7.4804	376.70	2494.5
75.	91.785	0.0010375	2.2158	2.2169	384.45	2278.6	2663.0	1.2131	6.2439	7.4570	384.37	2496.7
80.	93.512	0.0010387	2.0859	2.0870	391.72	2274.1	2665.8	1.2330	6.2022	7.4352	391.64	2498.8
90.	96.713	0.0010412	1.8682	1.8692	405.21	2265.0	2670.9	1.2696	6.1258	7.3954	405.11	2502.6
100.	99.632	0.0010434	1.6927	1.6937	417.51	2257.9	2675.4	1.3027	6.0571	7.3598	417.41	2506.1
110.	102.317	0.0010455	1.5482	1.5492	428.84	2250.8	2679.6	1.3330	5.9947	7.3277	428.73	2509.2
120.	104.808	0.0010476	1.4271	1.4281	439.36	2244.1	2683.4	1.3609	5.9375	7.2984	439.24	2512.1
130.	107.133	0.0010495	1.3240	1.3251	449.19	2237.8	2687.0	1.3868	5.8847	7.2715	449.05	2514.7
140.	109.315	0.0010513	1.2353	1.2363	458.42	2231.9	2690.3	1.4109	5.8356	7.2465	458.27	2517.2
150.	111.37	0.0010530	1.1580	1.1590	467.13	2226.2	2693.4	1.4336	5.7898	7.2234	466.97	2519.5
160.	113.32	0.0010547	1.0901	1.0911	475.38	2220.9	2696.2	1.4550	5.7467	7.2017	475.25	2521.7
180.	116.93	0.0010579	0.9762	0.9772	490.70	2210.8	2701.5	1.4944	5.6678	7.1622	490.51	2525.6
200.	120.23	0.0010608	0.8844	0.8854	504.70	2201.6	2706.3	1.5301	5.5967	7.1268	504.49	2529.2
220.	123.27	0.0010636	0.8088	0.8098	517.62	2193.0	2710.6	1.5627	5.5321	7.0949	517.39	2532.4

Table A-1-6 Saturated Steam: Pressure Table (SI) (cont.)

Abs. Press kPa p	Temp. C T	Specific Volume, m³/kg Sat. Liquid v_f	Evap. v_{fg}	Sat. Vapor v_g	Enthalpy, kJ/kg Sat. Liquid h_f	Evap. h_{fg}	Sat. Vapor h_g	Entropy, kJ/kg K Sat. Liquid s_f	Evap. s_{fg}	Sat. Vapor s_g	Energy, kJ/kg Sat. Liquid u_f	Sat. Vapor u_g
240.	126.09	0.0010663	0.7454	0.7465	529.6	2184.9	2714.5	1.5929	5.4728	7.0657	529.38	2535.4
260.	128.73	0.0010688	0.6914	0.6925	540.9	2177.3	2718.2	1.6209	5.4180	7.0389	540.60	2538.1
280.	131.20	0.0010712	0.6450	0.6460	551.4	2170.1	2721.5	1.6471	5.3670	7.0140	551.14	2540.6
300.	133.54	0.0010735	0.6045	0.6056	561.4	2163.2	2724.7	1.6716	5.3193	6.9909	561.11	2543.0
350.	138.89	0.0010789	0.5229	0.5240	584.3	2147.4	2731.6	1.7273	5.2119	6.9392	583.89	2548.2
400.	143.62	0.0010839	0.4611	0.4622	604.7	2133.0	2737.6	1.7764	5.1179	6.8943	604.24	2552.7
450.	147.92	0.0010885	0.4127	0.4138	623.2	2119.7	2742.9	1.8204	5.0343	6.8547	622.67	2556.7
500.	151.84	0.0010928	0.3736	0.3747	640.1	2107.4	2747.5	1.8604	4.9588	6.8192	639.57	2560.2
550.	155.47	0.0010969	0.3414	0.3425	655.8	2095.9	2751.7	1.8970	4.8900	6.7870	655.20	2563.3
600.	158.84	0.0011009	0.3144	0.3155	670.4	2085.0	2755.5	1.9308	4.8267	6.7575	669.76	2566.2
650.	161.99	0.0011046	0.29138	0.29249	684.1	2074.7	2758.9	1.9623	4.7681	6.7304	683.42	2568.7
700.	164.96	0.0011082	0.27157	0.27268	697.1	2064.9	2762.0	1.9918	4.7134	6.7052	696.29	2571.1
750.	167.76	0.0011116	0.25431	0.25543	709.3	2055.5	2764.8	2.0195	4.6139	6.6596	720.04	2575.3
800.	170.41	0.0011150	0.23914	0.24026	720.9	2046.5	2767.5	2.0457	4.6139	6.6596	720.04	2575.3
900.	175.36	0.0011213	0.21369	0.21481	742.6	2029.5	2772.1	2.0941	4.5250	6.6192	741.63	2478.8
1000.	179.88	0.0011274	0.19317	0.19429	762.6	2013.6	2776.2	2.1382	4.4446	6.5828	761.48	2581.9
1100.	184.07	0.0011331	0.17625	0.17738	781.1	1998.5	2779.7	2.1786	4.3711	6.5497	779.88	2584.5
1200.	187.96	0.0011386	0.16206	0.16320	798.4	1984.3	2782.7	2.2161	4.3033	6.5194	797.06	2586.9
1300.	191.61	0.0011438	0.14998	0.15113	814.7	1970.7	2785.4	2.2510	4.2403	6.4913	813.21	2589.0
1400.	195.04	0.0011489	0.13957	0.14072	830.1	1957.7	2787.8	2.2837	4.1814	6.4651	828.47	2590.8
1500.	198.29	0.0011539	0.13050	0.13166	844.7	1945.2	2789.9	2.3145	4.1261	6.4406	842.93	2592.4
1600.	201.37	0.0011586	0.12253	0.12369	858.6	1933.2	2791.7	2.3436	4.0739	6.4175	856.71	2593.8
1800.	207.11	0.0011678	0.10915	0.11032	884.6	1910.3	2794.8	2.3976	3.9775	6.3751	882.47	2596.3
2000.	212.37	0.0011766	0.09836	0.09954	908.6	1888.6	2797.2	2.4469	3.8898	6.3367	906.24	2598.2
2200.	217.24	0.0011850	0.08947	0.09065	931.0	1868.1	2799.1	2.4922	3.8093	6.3015	928.35	2599.6
2400.	221.78	0.0011932	0.08201	0.08320	951.9	1848.5	2800.4	2.5343	3.7347	6.2690	949.07	2600.7
2600.	226.04	0.0012011	0.07565	0.07686	971.7	1829.6	2801.4	2.5736	3.6651	6.2387	968.60	2601.5

2800.	230.05	0.0012088	0.07018	0.07139	990.5	1811.5	2802.0	2.6106	3.5998	6.2104	987.10	2602.1
3000.	233.84	0.0012163	0.06541	0.06663	1008.4	1793.9	2802.3	2.6455	3.5382	6.1837	1004.70	2602.1
3500.	242.54	0.0012345	0.05579	0.05703	1049.8	1752.2	2802.0	2.7253	3.3976	6.1228	1045.44	2602.4
4000.	250.33	0.0012521	0.04850	0.04975	1087.4	1712.9	2800.3	2.7965	3.2720	6.0685	1082.4	2601.3
4500.	257.41	0.0012691	0.04277	0.04404	1122.1	1675.6	2797.7	2.8612	3.1579	6.0191	1116.4	2599.5
5000.	263.91	0.0012858	0.03814	0.03943	1154.5	1639.7	2794.2	2.9206	3.0529	5.9735	1148.0	2597.0
5500.	269.93	0.0013023	0.03433	0.03563	1184.9	1605.0	2789.0	2.9757	2.9552	5.9309	1177.7	2594.0
6000.	275.55	0.0013187	0.03112	0.03244	1213.7	1571.3	2785.0	3.0273	2.8635	5.8908	1205.8	2590.4
6500.	280.82	0.0013350	0.028384	0.029719	1241.1	1538.4	2779.5	3.0759	2.7768	5.8527	1232.5	2586.3
7000.	285.79	0.0013513	0.026022	0.027373	1267.4	1506.0	2773.5	3.1219	2.6943	5.8162	1258.0	2581.0
7500.	290.50	0.0013677	0.023959	0.025327	1292.7	1474.2	2766.9	3.1657	2.6153	5.7811	1282.4	2577.0
8000.	294.97	0.0013842	0.022141	0.023525	1317.1	1442.8	2759.9	3.2076	2.5395	5.7471	1306.0	2571.7
9000.	303.31	0.0014179	0.019078	0.020495	1363.7	1380.9	2744.6	3.2867	2.3953	5.6820	1351.0	2560.1
10000.	310.96	0.0014526	0.016589	0.018041	1408.0	1319.7	2727.7	3.3605	2.2593	5.6198	1393.5	2547.3
11000.	318.05	0.0014887	0.014517	0.016006	1450.6	1258.7	2709.3	3.4304	2.1291	5.5595	1434.2	2533.2
12000.	324.65	0.0015268	0.012756	0.014283	1491.8	1197.4	2689.2	3.4972	2.0030	5.5002	1473.4	2517.8
13000.	330.83	0.0015672	0.011230	0.012797	1532.0	1135.0	2667.0	3.5616	1.8792	5.4408	1511.6	2500.6
14000.	336.64	0.0016106	0.009884	0.011495	1571.6	1070.7	2642.4	3.6242	1.7560	5.3803	1549.1	2481.4
15000.	342.13	0.0016579	0.008682	0.010340	1611.0	1004.0	2615.0	3.6859	1.6320	5.3178	1586.1	2459.9
16000.	347.33	0.0017103	0.007597	0.009308	1650.5	934.3	2584.9	3.7471	1.5060	5.2531	1623.2	2436.0
17000.	352.26	0.0017696	0.006601	0.008371	1691.7	859.9	2551.6	3.8107	1.3748	5.1855	1661.6	2409.3
18000.	356.96	0.0018399	0.005658	0.007498	1734.8	779.1	2513.9	3.8765	1.2362	5.1128	1701.7	2378.9
19000.	361.43	0.0019260	0.004751	0.006678	1778.7	692.0	2470.6	3.9429	1.0903	5.0332	1742.1	2343.8
20000.	365.70	0.0020370	0.003840	0.005877	1826.5	591.9	2418.4	4.0149	0.9263	4.9412	1785.7	2300.8
21000.	369.78	0.0022015	0.002822	0.005023	1866.3	461.3	2347.6	4.1048	0.7175	4.8223	1840.0	2242.1
22000.	373.69	0.0026714	0.001056	0.003728	2011.1	184.5	2195.6	4.2947	0.2852	4.5799	1952.4	2113.6
22120.*	374.15*	0.00317	0.0	0.00317	2107.4	0.0	2107.4	4.4429	0.0	4.4429	2037.3	2037.3

* Approximate Critical Point

Source: Reprinted with permission from ASME.

Table A-1-7 Superheated Steam: (No Liquid Present) (SI)

Abs. Press. kPa (Sat. Temp, C)		Sat. Liquid	Sat. Vapor	Temperature—C 20.	40.	60.	80.	100.	120.	140.	160.
1.0 (6.983)	v	0.0010	129.2	135.23	144.47	153.71	162.95	172.19	181.42	190.66	199.89
	h	29.34	2514.4	2538.6	2575.9	2613.3	2650.9	2688.6	2726.5	2764.6	2802.9
	s	0.1060	9.9767	9.0611	9.1842	9.3001	9.4096	9.5136	9.6125	9.7070	9.7975
2.0 (17.51)	v	0.0010	67.01	67.582	72.211	76.837	81.459	86.080	90.700	95.319	99.936
	h	73.46	2533.6	2538.3	2575.6	2613.1	2650.7	2688.5	2726.4	2764.5	2802.8
	s	0.2607	8.7246	8.7404	8.8637	8.9797	9.0894	9.1934	9.2924	9.3870	9.4775
3.0 (24.10)	v	0.0010	45.67		48.124	51.211	54.296	57.378	60.460	63.540	66.619
	h	101.00	2545.6		2575.4	2612.9	2650.6	2688.4	2726.3	2764.5	2802.8
	s	0.3544	8.5785		8.6760	8.7922	8.9019	9.0060	9.1051	9.1997	9.2902
4.0 (28.98)	v	0.0010	34.80		36.081	38.398	40.714	43.027	45.339	47.650	49.961
	h	121.41	2554.5		2575.2	2612.7	2650.4	2688.3	2726.2	2764.4	2802.7
	s	0.4225	8.4755		8.5426	8.6589	8.7688	8.8730	8.9721	9.0668	9.1573
5.0 (32.90)	v	0.0010	28.19		28.854	30.711	32.565	34.417	36.267	38.117	39.966
	h	137.77	2561.6		2574.9	2612.6	2650.3	2688.1	2726.1	2764.3	2802.6
	s	0.4763	8.3960		8.4390	8.5555	8.6655	8.7698	8.8690	8.9636	9.0542
6.0 (36.18)	v	0.0010	23.74		24.037	25.586	27.132	28.676	30.219	31.761	33.302
	h	151.50	2567.5		2574.7	2612.4	2650.1	2688.0	2726.0	2764.2	2802.6
	s	0.5209	8.3312		8.3543	8.4709	8.5810	8.6854	8.7846	8.8793	8.9700
8.0 (41.53)	v	0.0010	18.105			19.179	20.341	21.501	22.659	23.816	24.973
	h	173.86	2577.1			2612.0	2649.8	2687.8	2725.8	2764.1	2802.4
	s	0.5925	8.2296			8.3372	8.4476	8.5521	8.6515	8.7463	8.8370
10.0 (45.83)	v	0.0010	14.675			15.336	16.266	17.195	18.123	19.050	19.975
	h	191.85	2584.8			2611.6	2649.5	2687.5	2725.6	2763.9	2802.3
	s	0.6493	8.1511			8.2334	8.3439	8.4486	8.5481	8.6430	8.7338
15.0 (54.00)	v	0.0010	10.023			10.210	10.834	11.455	12.075	12.694	13.312
	h	225.97	2599.2			2610.6	2648.8	2686.9	2725.1	2763.5	2802.0
	s	0.7549	8.0093			8.0440	8.1551	8.2601	8.3599	8.4551	8.5460
	v	0.0010	7.650				8.1172	8.5847	9.0508	9.516	9.980

P, kPa (Tsat, °C)		Sat. liquid (f)	Sat. vapor (g)					
20.0	v							
(60.09)	h	251.45	2609.9	2648.0	2686.3	2724.6	2763.1	2801.6
	s	0.8321	7.9094	8.0206	8.1261	8.2262	8.3215	8.4127
30.0	v	0.0010	5.229	5.4007	5.7144	6.0267	6.3379	6.6483
(69.12)	h	289.30	2625.4	2646.5	2685.1	2723.6	2762.3	2801.0
	s	0.9441	7.7695	7.8300	7.9363	8.0370	8.1329	8.2243
40.0	v	0.0010	3.993	4.0424	4.2792	4.5146	4.7489	4.9825
(75.89)	h	317.65	2636.9	2644.9	2683.8	2722.6	2761.4	2800.3
	s	1.0261	7.6709	7.6937	7.8009	7.9023	7.9985	8.0903
50.0	v	0.0010	3.240		3.4181	3.6074	3.7955	3.9829
(81.35)	h	340.56	2646.0		2682.6	2721.6	2760.6	2799.6
	s	1.0912	7.5947		7.6953	7.7972	7.8940	7.9861
60.0	v	0.0010	2.732		2.8440	3.0025	3.1599	3.3165
(85.95)	h	359.93	2653.6		2681.3	2720.6	2759.8	2798.9
	s	1.1454	7.5327		7.6085	7.7111	7.8083	7.9008
80.0	v	0.0010	2.0870		2.1262	2.2464	2.3654	2.4836
(93.51)	h	391.72	2665.8		2678.8	2718.6	2758.1	2797.5
	s	1.2330	7.4352		7.4703	7.5742	7.6723	7.7655
100.0	v	0.0010	1.6937		1.6955	1.7927	1.8886	1.9838
(99.63)	h	417.51	2675.4		2676.2	2716.5	2756.4	2796.2
	s	1.3027	7.3598		7.3618	7.4670	7.5662	7.6601
150.0	v	0.0011	1.1590			1.1876	1.2529	1.3173
(111.4)	h	467.13	2693.4			2711.2	2752.2	2792.7
	s	1.4336	7.2234			7.2693	7.3709	7.4667
200.0	v	0.0011	0.8854				0.9349	0.9840
(120.2)	h	504.70	2706.3				2747.8	2789.1
	s	1.5301	7.1268				7.2298	7.3275
300.0	v	0.0011	0.6056				0.6167	0.6506
(133.5)	h	561.4	2724.7				2738.8	2781.8
	s	1.6716	6.9909				7.0254	7.1271
400.0	v	0.0011	0.4622					0.4837
(143.6)	h	604.7	2737.6					2774.2
	s	1.7764	6.8943					6.9805

Table A-1-7 Superheated Steam: (No Liquid Present) (SI) (cont.)

Abs. Press kPa (Sat. Temp, C)		180.	200.	220.	240.	260.	280.	300.	320.	340.
1.0 (6.983)	v	209.12	218.35	227.58	236.82	246.05	255.28	264.51	273.74	282.97
	h	2841.4	2880.1	2919.0	2958.1	2997.1	3037.0	3076.8	3116.9	3157.2
	s	9.8843	9.9679	10.0484	10.1262	10.2014	10.2743	10.3450	10.4137	10.4805
2.0 (17.51)	v	104.55	109.17	113.79	118.40	123.02	127.64	132.25	136.87	141.48
	h	2841.3	2880.0	2918.9	2958.0	2997.4	3037.0	3076.8	3116.9	3157.2
	s	9.5643	9.6479	9.7284	9.8062	9.8814	9.9543	10.0251	10.0938	10.1606
3.0 (24.10)	v	69.698	72.777	75.855	78.933	82.010	85.088	88.165	91.242	94.320
	h	2841.3	2880.0	2918.9	2958.0	2997.4	3037.0	3076.8	3116.9	3157.2
	s	9.3771	9.4607	9.5412	9.6190	9.6943	9.7672	9.8379	9.9066	9.9735
4.0 (28.98)	v	52.270	54.580	56.889	59.197	61.506	63.814	66.122	68.430	70.738
	h	2841.2	2879.9	2918.8	2958.0	2997.3	3036.9	3076.8	3116.8	3157.2
	s	9.2443	9.3279	9.4084	9.4862	9.5615	9.6344	9.7051	9.7738	9.8407
5.0 (32.90)	v	41.814	43.661	45.509	47.356	49.203	51.050	52.897	54.743	56.590
	h	2841.2	2879.9	2918.8	2957.9	2997.3	3036.9	3076.7	3116.8	3157.1
	s	9.1412	9.2248	9.3054	9.3832	9.4584	9.5313	9.6021	9.6708	9.7377
6.0 (36.18)	v	34.843	36.383	37.922	39.462	41.001	42.540	44.079	45.618	47.157
	h	2841.1	2879.8	2918.8	2957.9	2997.3	3036.9	3076.7	3116.8	3157.1
	s	9.0569	9.1406	9.2212	9.2990	9.3742	9.4472	9.5179	9.5866	9.6535
7.0 (41.53)	v	26.129	27.284	28.439	29.594	30.749	31.903	33.058	34.212	35.367
	h	2841.0	2879.7	2918.7	2957.8	2997.2	3036.8	3076.7	3116.8	3157.1
	s	8.9240	9.0077	9.0883	9.1661	9.2414	9.3143	9.3851	9.4538	9.5207
10.0 (45.83)	v	20.900	21.825	22.750	23.674	24.598	25.521	26.445	27.369	28.292
	h	2840.9	2879.6	2918.6	2957.8	2997.2	3036.8	3076.6	3116.7	3157.0
	s	8.8208	8.9045	8.9852	9.0630	9.1383	9.2113	9.2820	9.3508	9.4177
15.0 (54.00)	v	13.929	14.546	15.163	15.780	16.396	17.012	17.628	18.244	18.860
	h	2840.6	2879.4	2918.4	2957.6	2997.0	3036.6	3076.5	3116.6	3157.0
	s	8.6332	8.7170	8.7977	8.8757	8.9510	9.0240	9.0948	9.1635	9.2304
20.0 (60.09)	v	10.444	10.907	11.370	11.832	12.295	12.757	13.219	13.681	14.143
	h	2840.3	2879.2	2918.2	2957.4	2996.9	3036.5	3076.4	3116.5	3156.9
	s	8.5000	8.5839	8.6647	8.7426	8.8180	8.8910	8.9618	9.0306	9.0975

P (kPa) (Tsat °C)										
30.0 (69.12)	v	6.9582	7.2675	7.5766	7.8854	8.1940	8.5024	8.8108	9.1190	9.4272
	h	2839.8	2878.7	2917.8	2957.1	2996.6	3036.2	3076.1	3116.3	3156.7
	s	8.3119	8.3960	8.4769	8.5550	8.6305	8.7035	8.7744	8.8432	8.9102
40.0 (75.89)	v	5.2154	5.4478	5.6800	5.9118	6.1435	6.3751	6.6065	6.8378	7.0690
	h	2839.2	2878.2	2917.4	2956.7	2996.3	3036.0	3075.9	3116.1	3156.5
	s	8.1782	8.2625	8.3435	8.4217	8.4973	8.5704	8.6413	8.7102	8.7772
50.0 (81.35)	v	4.1697	4.3560	4.5420	4.7277	4.9133	5.0986	5.2839	5.4691	5.6542
	h	2838.6	2877.7	2917.0	2956.4	2995.9	3035.7	3075.7	3115.9	3156.3
	s	8.0742	8.1587	8.2399	8.3182	8.3939	8.4671	8.5380	8.6070	8.6740
60.0 (85.95)	v	3.4726	3.6281	3.7833	3.9383	4.0931	4.2477	4.4022	4.5566	4.7109
	h	2838.1	2877.3	2916.6	2956.0	2995.6	3035.4	3075.4	3115.6	3156.1
	s	7.9891	8.0738	8.1552	8.2336	8.3093	8.3826	8.4536	8.5226	8.5896
80.0 (93.51)	v	2.6011	2.7183	2.8350	2.9515	3.0678	3.1840	3.3000	3.4160	3.5319
	h	2836.9	2876.3	2915.8	2955.3	2995.0	3034.9	3075.0	3115.2	3155.7
	s	7.8544	7.9395	8.0212	8.0998	8.1757	8.2491	8.3202	8.3893	8.4564
100.0 (99.63)	v	2.0783	2.1723	2.2660	2.3595	2.4527	2.5458	2.6387	2.7316	2.8244
	h	2835.8	2875.4	2915.0	2954.6	2994.4	3034.4	3074.5	3114.8	3155.3
	s	7.7495	7.8349	7.9169	7.9958	8.0719	8.1454	8.2166	8.2857	8.3529
150.0 (111.4)	v	1.3811	1.4444	1.5073	1.5700	1.6325	1.6948	1.7570	1.8191	1.8812
	h	2832.9	2872.9	2912.9	2952.9	2992.9	3033.0	3073.3	3113.7	3154.3
	s	7.5574	7.6439	7.7266	7.8061	7.8826	7.9565	8.0280	8.0973	8.1646
200.0 (120.2)	v	1.0325	1.0804	1.1280	1.1753	1.2224	1.2693	1.3162	1.3629	1.4095
	h	2830.0	2870.5	2910.8	2951.1	2991.4	3031.7	3072.1	3112.6	3153.3
	s	7.4196	7.5072	7.5907	7.6707	7.7477	7.8219	7.8937	7.9632	8.0307
300.0 (133.5)	v	0.6837	0.7164	0.7486	0.7805	0.8123	0.8438	0.8753	0.9066	0.9379
	h	2824.0	2865.5	2906.6	2947.5	2988.2	3028.9	3069.7	3110.5	3151.4
	s	7.2222	7.3119	7.3971	7.4783	7.5562	7.6311	7.7034	7.7734	7.8412
400.0 (143.6)	v	0.5093	0.5343	0.5589	0.5831	0.6072	0.6311	0.6549	0.6785	0.7021
	h	2817.8	2860.4	2902.3	2943.9	2985.1	3026.2	3067.2	3108.3	3149.4
	s	7.0788	7.1708	7.2576	7.3402	7.4190	7.4947	7.5675	7.6379	7.061

v = specific volume, m³/kg
h = enthalpy, kJ/kg
s = entropy, kJ/kgK

Table A-1-7 Superheated Steam: (No Liquid Present) (SI) (cont.)

Abs. Press. kPa (Sat. Temp, C)		360.	380.	400.	420.	Temperature—C 440.	460.	480.	500.	520.
1.0 (6.983)	v	292.20	301.43	310.66	319.89	329.12	338.35	347.58	356.81	366.04
	h	3197.8	3238.6	3279.7	3321.1	3362.7	3404.6	3446.8	3489.2	3531.9
	s	10.5457	10.6091	10.6711	10.7317	10.7909	10.8488	10.9056	10.9612	11.0157
1.5 (13.04)	v	194.80	200.95	207.11	213.26	219.41	225.57	231.72	237.87	244.03
	h	3197.8	3238.6	3279.7	3321.1	3362.7	3404.6	3446.8	3489.2	3531.9
	s	10.3585	10.4220	10.4840	10.5445	10.6037	10.6617	10.7184	10.7741	10.8286
2.0 (17.51)	v	146.10	150.71	155.33	159.94	164.56	169.17	173.79	178.41	183.02
	h	3197.8	3238.6	3279.7	3321.1	3362.7	3404.6	3446.8	3489.2	3531.9
	s	10.2257	10.2892	10.3512	10.4118	10.4710	10.5289	10.5857	10.6413	10.6958
3.0 (24.10)	v	97.397	100.47	103.55	106.63	109.71	112.78	115.86	118.94	122.01
	h	3197.8	3238.6	3279.7	3321.1	3362.7	3404.6	3446.8	3489.2	3531.9
	s	10.0386	10.1021	10.1641	10.2246	10.2838	10.3418	10.3985	10.4541	10.5087
4.0 (28.98)	v	73.046	75.354	77.662	79.970	82.278	84.586	86.893	89.201	91.509
	h	3197.7	3238.6	3279.7	3321.0	3362.7	3404.6	3446.7	3489.2	3531.9
	s	9.9058	9.9693	10.0313	10.0918	10.1510	10.2090	10.2657	10.3214	10.3759
5.0 (32.90)	v	58.436	60.283	62.129	63.975	65.822	67.668	69.514	71.360	73.207
	h	3197.7	3238.6	3279.7	3321.0	3362.7	3404.6	3446.7	3489.2	3531.9
	s	9.8028	9.8663	9.9283	9.9888	10.0480	10.1060	10.1627	10.2184	10.2729
6.0 (36.18)	v	48.696	50.235	51.773	53.312	54.851	56.389	57.928	59.467	61.005
	h	3197.7	3238.5	3279.6	3321.0	3362.6	3404.5	3446.87	3489.2	3531.9
	s	9.7186	9.7821	9.8441	9.9047	9.9639	10.0218	10.0786	10.1342	10.1888
8.0 (41.53)	v	36.521	37.675	38.829	39.983	41.137	42.291	43.445	44.599	45.753
	h	3197.7	3238.5	3279.6	3321.0	3362.6	3404.5	3446.7	3489.1	3531.9
	s	9.5858	9.6493	9.7113	9.7719	9.8311	9.8890	9.9458	10.0014	10.0560
10.0 (45.83)	v	29.216	30.139	31.062	31.986	32.909	33.832	34.756	35.679	36.602
	h	3197.6	3238.5	3279.6	3321.0	3362.6	3404.5	3446.7	3489.1	3531.9
	s	9.4828	9.5463	9.6083	9.6689	9.7281	9.7860	9.8428	9.8984	9.9530
15.0	v	19.475	20.091	20.707	21.323	21.938	22.554	23.169	23.785	24.400
	h	3197.5	3238.4	3279.5	3320.9	3362.5	3404.4	3446.6	3489.1	3531.8

(54.00) s	9.2956	9.3591	9.4211	9.4817	9.5409	9.5988	9.6556	9.7112	9.7658
20.0 v	14.605	15.067	15.529	15.991	16.453	16.914	17.376	17.838	18.300
(60.09) h	3197.5	3238.3	3279.4	3320.8	3362.5	3404.4	3446.6	3489.0	3531.8
s	9.1627	9.2262	9.2882	9.3488	9.4081	9.4660	9.5228	9.5784	9.6330
30.0 v	9.7353	10.043	10.351	10.659	10.967	11.275	11.583	11.891	12.199
(69.12) h	3197.3	3238.2	3279.3	3320.7	3362.3	3404.2	3446.4	3488.9	3531.6
s	8.9754	9.0389	9.1010	9.1615	9.2208	9.2788	9.3355	9.3912	9.4458
40.0 v	7.3002	7.5314	7.7625	7.9935	8.2246	8.4556	8.6866	8.9176	9.1485
(75.89) h	3197.1	3238.0	3279.1	3320.5	3362.2	3404.1	3446.3	3488.8	3531.5
s	8.8424	8.9060	8.9680	9.0286	9.0879	9.1459	9.2027	9.2583	9.3129
50.0 v	5.8392	6.0242	6.2091	6.3941	6.5790	6.7638	6.9487	7.1335	7.3183
(81.35) h	3196.9	3237.8	3279.0	3320.4	3362.1	3404.0	4336.2	3488.7	3531.4
s	8.7392	8.8028	8.8649	8.9255	8.9848	9.0428	9.0996	9.1552	9.2098
60.0 v	4.8652	5.0194	5.1736	5.3277	5.4819	5.6360	5.7900	5.9441	5.0981
(85.95) h	3196.7	3237.7	3278.8	3320.2	3361.9	3403.9	3446.1	3488.6	3531.3
s	8.6549	8.7185	8.7806	8.8412	8.9005	8.9585	9.0153	9.0710	9.1256
80.0 v	3.6477	3.7634	3.8792	3.9948	4.1105	4.2261	4.3418	4.4574	4.5729
(93.51) h	3196.4	3237.3	3278.5	3320.0	3361.7	3403.6	3445.9	3488.4	3531.1
s	8.5217	8.5854	8.6475	8.7081	8.7675	8.8255	8.8823	8.9380	8.9926
100.0 v	2.9172	3.0098	3.1025	3.1951	3.2877	3.3803	3.4728	3.5653	3.6578
(99.63) h	3196.0	3237.0	3278.2	3319.7	3361.4	3403.4	3445.6	3488.1	3530.9
s	8.4183	8.4820	8.5442	8.6049	8.6642	8.7223	8.7791	8.8348	8.8894
150.0 v	1.9431	2.0051	2.0669	2.1288	2.1906	2.2524	2.3142	2.3759	2.4377
(111.4) h	3195.1	3236.2	3277.5	3319.0	3360.7	3402.8	3445.0	3487.6	3530.4
s	8.2301	8.2940	8.3562	8.4170	8.4764	8.5345	8.5914	8.6472	8.7018
200.0 v	1.4561	1.5027	1.5492	1.5956	1.6421	1.6885	1.7349	1.7812	1.8276
(120.2) h	3194.2	3235.4	3276.7	3318.3	3360.1	3402.1	3444.5	3487.0	3529.9
s	8.0964	8.1603	8.2226	8.2835	8.3429	8.4011	8.4581	8.5139	8.5686
300.0 v	0.9691	1.0003	1.0314	1.0625	1.0935	1.1245	1.1556	1.1865	1.2175
(133.5) h	3192.4	3233.7	3275.2	3316.8	3358.8	3400.9	3443.3	3486.0	3528.9
s	7.9072	7.9713	8.0338	8.0949	8.1545	8.2128	8.2698	8.3257	8.3805
400.0 v	0.7256	0.7491	0.7725	0.7959	0.8192	0.8426	0.8659	0.8892	0.9125
(143.6) h	3190.6	3232.1	3273.6	3315.4	3357.4	3399.7	3442.1	3484.9	3527.8
s	7.7723	7.8367	7.8994	7.9606	8.0203	8.0787	8.1359	8.1919	8.2468

Table A-1-7 Superheated Steam: (No Liquid Present) (SI) (cont.)

Abs. Press. kPa (Sat. Temp, C)		540.	560.	580.	600.	625.	650.	700.	750.	800.
1.0 (6.983)	v	375.27	384.50	393.74	402.97	414.50	426.04	449.12	472.19	495.27
	h	3574.9	3618.2	3661.8	3705.6	3760.8	3816.4	3928.9	4043.0	4158.7
	s	11.0693	11.218	11.1735	11.2243	11.2866	11.3476	11.4663	11.5807	11.6911
1.5 (13.04)	v	250.18	256.34	262.49	268.64	276.33	284.03	299.41	314.79	330.18
	h	3574.9	3618.2	3661.8	3705.6	3760.8	3816.4	3928.9	4043.0	4158.7
	s	10.8821	10.9347	10.9864	11.0372	11.0995	11.1605	11.2792	11.3935	11.5040
2.0 (17.51)	v	187.64	192.25	196.87	201.48	207.25	213.02	224.56	236.10	247.63
	h	3574.9	3618.2	3661.8	3705.6	3760.8	3816.4	3928.8	4043.0	4158.7
	s	10.7494	10.8019	10.8536	10.9044	10.9667	11.0277	11.1464	11.2608	11.3712
3.0 (24.10)	v	125.09	128.17	131.24	134.32	138.17	142.01	149.70	157.40	165.09
	h	3574.9	3618.2	3661.8	3705.6	3760.8	3816.4	3928.8	4043.0	4158.7
	s	10.5622	10.6148	10.6665	10.7173	10.7796	10.8406	10.9593	11.0736	11.1841
4.0 (28.98)	v	93.817	96.124	98.432	100.74	103.62	106.51	112.28	118.05	123.82
	h	3574.9	3618.2	3661.7	3705.6	3705.6	3816.4	3928.8	4043.0	4158.7
	s	10.4295	10.4820	10.5337	10.5845	10.6468	10.7078	10.8265	10.9409	11.0513
5.0 (32.90)	v	75.053	76.899	78.745	80.592	82.899	85.207	89.822	94.438	99.053
	h	3574.9	3618.2	3661.7	3705.6	3760.7	3816.3	3928.8	4043.0	4158.7
	s	10.3265	10.3790	10.4307	10.4815	10.5438	10.6049	10.7235	10.8379	10.9483
6.0 (36.18)	v	62.544	64.082	65.621	67.159	69.082	71.005	74.852	78.698	82.544
	h	3574.9	3618.2	3661.7	3705.6	3760.7	3816.3	3928.8	4043.0	4158.7
	s	10.2423	10.2949	10.3466	10.3973	10.4596	10.5207	10.6394	10.7537	10.8642
8.0 (41.53)	v	46.907	48.061	49.215	50.369	51.811	53.254	56.138	59.023	61.908
	h	3574.9	3618.2	3661.7	3705.5	3760.7	3816.3	3928.8	4043.0	4158.7
	s	10.1095	10.1621	10.2138	10.2646	10.3269	10.3879	10.5066	10.6210	10.7314
10.0 (45.83)	v	37.525	38.448	39.372	40.295	41.449	42.603	44.910	47.218	49.526
	h	3574.9	3618.1	3661.7	3705.5	3760.7	3816.3	3928.8	4042.9	4158.7
	s	10.0065	10.0591	10.1108	10.1616	10.2239	10.2849	10.4036	10.5180	10.6284
15.0 (54.00)	v	25.016	25.632	26.247	26.863	27.632	28.401	29.940	31.478	33.017
	h	3574.8	3618.1	3661.7	3705.5	3760.7	3816.3	3928.8	4042.9	4158.7
	s	9.8194	9.8719	9.9236	9.9744	10.0367	10.0978	10.2164	10.3308	10.4413

P, kPa (Tsat)		T₁	T₂	T₃	T₄	T₅	T₆	T₇	T₈	T₉
20.0 (60.09)	v	18.761	19.223	19.685	20.146	20.723	21.300	22.455	23.609	24.762
	h	3574.8	3618.0	3661.6	3705.4	3760.6	3816.2	3928.7	4042.9	4158.7
	s	9.6885	9.7391	9.7908	9.8416	9.9039	9.9650	10.0836	10.1980	10.3085
30.0 (69.12)	v	12.507	12.815	13.122	13.430	13.815	14.200	14.969	15.739	16.508
	h	3574.7	3618.0	3661.5	3705.4	3760.6	3816.2	3928.7	4042.8	4158.6
	s	9.4993	9.5519	9.6036	9.6544	9.7167	9.7778	9.8965	10.0109	10.1213
40.0 (75.89)	v	9.3795	9.6104	9.8413	10.072	10.361	10.649	11.227	11.804	12.381
	h	3574.6	3617.9	3661.4	3705.3	3760.5	3816.1	3928.6	4042.8	4158.6
	s	9.3665	9.4191	9.4708	9.5216	9.5839	9.6450	9.7636	9.8780	9.9885
50.0 (81.35)	v	7.5031	7.6878	7.8726	8.0574	8.2883	8.5192	8.9810	9.4427	9.9044
	h	3574.5	3617.8	3661.3	3705.2	3760.4	3816.0	3928.6	4042.7	4158.5
	s	9.2634	9.3160	9.3677	9.4185	9.4808	9.5419	9.6606	9.7750	9.8855
60.0 (85.95)	v	6.2521	6.4062	6.5602	6.7141	6.9066	7.0991	7.4839	7.8687	8.2535
	h	3574.4	3617.7	3661.3	3705.1	3760.3	3816.0	3928.5	4042.7	4158.5
	s	9.1792	9.2318	9.2835	9.3343	9.3966	9.4577	9.5764	9.6908	9.8013
80.0 (93.51)	v	4.6885	4.8040	4.9196	5.0351	5.1795	5.3239	5.6126	5.9013	6.1899
	h	3574.2	3617.5	3661.1	3705.0	3760.2	3815.8	3928.4	4042.6	4158.4
	s	9.0462	9.0988	9.1506	9.2014	9.2637	9.3248	9.4436	9.5580	9.6685
100.0 (99.63)	v	3.7503	3.8428	3.9352	4.0277	4.1432	4.2588	4.4898	4.7208	4.9517
	h	3574.0	3617.3	3660.9	3704.8	3760.0	3815.7	3928.2	4042.5	4158.3
	s	8.9431	8.9957	9.0474	9.0982	9.1606	9.2217	9.3405	9.4549	9.5654
150.0 (111.4)	v	2.4994	2.5611	2.6628	2.6845	2.7616	2.8386	2.9927	3.1468	3.3008
	h	3574.5	3616.9	3660.5	3704.4	3759.6	3815.3	3927.9	4042.2	4158.0
	s	8.7555	8.8082	8.8599	8.9108	8.9732	9.0343	9.1531	9.2676	9.3781
200.0 (120.2)	v	1.8739	1.9202	1.9666	2.0129	2.0707	2.1286	2.2442	2.3598	2.4754
	h	3573.0	3616.4	3660.0	3704.0	3759.3	3815.0	3927.6	4041.9	4157.8
	s	8.6223	8.6750	8.7268	8.7776	8.8401	8.9012	9.0201	9.1346	9.2452
300.0 (133.5)	v	1.2485	1.2794	1.3103	1.3412	1.3799	1.4185	1.4957	1.5728	1.6499
	h	3572.0	3615.5	3659.2	3703.2	3758.5	3814.2	3927.0	4041.4	4157.3
	s	8.4343	8.4870	8.5389	8.5898	8.6523	8.7135	8.8325	8.9471	9.0577
400.0 (143.6)	v	0.9357	0.9590	0.9822	1.0054	1.0344	1.0634	1.1214	1.1793	1.2372
	h	3571.1	3614.6	3658.3	3702.3	3757.7	3813.5	3926.4	4040.8	4156.9
	s	8.3006	8.3534	8.4053	8.4563	8.5189	8.5802	8.6992	8.8139	8.9246

Table A-1-7 Superheated Steam: (No Liquid Present) (SI) (cont.)

Abs. Press. kPa (Sat. Temp. C)		Sat. Liquid	Sat. Vapor	Temperature—C 160.
500.0 (151.8)	v	0.0011	0.3747	0.38347
	h	640.1	2747.5	2766.4
	s	1.8604	6.8192	6.8631
600.0 (158.8)	v	0.0011	0.3155	0.31655
	h	670.4	2755.5	2758.2
	s	1.9308	6.7575	6.7640
800.0 (170.4)	v	0.0011	0.2403	
	h	720.9	2767.5	
	s	2.0457	6.6596	
1000.0 (179.9)	v	0.0011	0.1943	
	h	762.6	2776.3	
	s	2.1382	6.5828	
1500.0 (198.3)	v	0.0012	0.1317	
	h	844.7	2789.9	
	s	2.3145	6.4406	
2000.0 (212.4)	v	0.0012	0.09954	
	h	908.6	2797.2	
	s	2.4469	6.3367	
3000.0 (233.8)	v	0.0012	0.0666	
	h	1008.4	2802.3	
	s	2.6455	6.1837	
4000.0 (250.3)	v	0.0013	0.0498	
	h	1087.4	2800.3	
	s	2.7965	6.0685	
5000.0 (263.9)	v	0.0013	0.03943	
	h	1154.5	2794.2	
	s	2.9206	5.9735	
	v	0.0013	0.0324	

Pressure (Temp)			
6000.0 (275.5)	v		
	h	1213.7	2785.0
	s	3.0273	5.8908
8000.0 (295.0)	v	0.0014	0.0235
	h	1317.1	2759.9
	s	3.2076	5.7471
10000.0 (311.0)	v	0.0015	0.01804
	h	1408.8	2727.7
	s	3.3605	5.6198
15000.0 (342.1)	v	0.0017	0.0134
	h	1611.0	2615.0
	s	3.6859	5.3178
20000.0 (365.7)	v	0.0020	0.0059
	h	1826.5	2418.4
	s	4.0149	4.9412
30000.0	v		
	h		
	s		
40000.0	v		
	h		
	s		
50000.0	v		
	h		
	s		
60000.0	v		
	h		
	s		
80000.0	v		
	h		
	s		
100000.0	v		
	h		
	s		

Table A-1-7 continued

Abs. Press. kPa (Sat. Temp, C)		180.	200.	220.	240.	260.	280.	300.	320.	340.
500.0 (151.8)	v	0.4045	0.4250	0.4450	0.4647	0.4841	0.5034	0.5226	0.5416	0.5606
	h	2811.4	2855.1	2898.0	2940.1	2981.9	3023.4	3064.8	3106.1	3147.4
	s	6.9647	7.0592	7.1478	7.2317	7.3115	7.3879	7.4614	7.5322	7.6008
600.0 (158.8)	v	0.3346	0.3520	0.3690	0.3857	0.4021	0.4183	0.4344	0.4504	0.4663
	h	2804.8	2849.7	2893.5	2936.4	2978.7	3020.6	3062.3	3103.9	3145.4
	s	6.8691	6.9662	7.0567	7.1419	7.2228	7.3000	7.3740	7.4454	7.5143
800.0 (170.4)	v	0.2471	0.2608	0.2740	0.2869	0.2995	0.3119	0.3241	0.3363	0.3483
	h	2791.1	2838.6	2884.2	2928.6	2972.1	3014.9	3057.3	3099.4	3141.4
	s	6.7122	6.8148	6.9094	6.9976	7.0807	7.1595	7.2348	7.3070	7.3767
1000.0 (179.9)	v	0.1944	0.2059	0.2169	0.2276	0.2379	0.2480	0.2580	0.2678	0.2776
	h	2776.5	2826.8	2874.6	2920.6	2965.2	3009.0	3052.1	3094.9	3137.4
	s	6.5835	6.6922	6.7911	6.8825	6.9680	7.0485	7.1251	7.1984	7.2689
1500.0 (198.3)	v		0.1324	0.1406	0.1483	0.1556	0.1628	0.1697	0.1765	0.1832
	h		2794.7	2848.6	2899.2	2947.3	2993.7	3038.9	3083.3	3127.0
	s		6.4508	6.5624	6.6630	6.7550	6.8405	6.9207	6.9967	7.0693
2000.0 (212.4)	v			0.1021	0.1084	0.1144	0.1200	0.1255	0.1308	0.1360
	h			2819.9	2875.0	2928.1	2977.5	3025.0	3071.2	3116.3
	s			6.3829	6.4943	6.5941	6.6852	6.7696	6.8487	6.9235

(Handwritten annotations appear near the 250 °C column: for 600.0 kPa — 0.3959, 2957.6, 7.1824; for 800.0 kPa — 0.2932, 2950.3, 7.0392.)

3000.0 (233.8)	v	0.06816	0.07283	0.07712	0.08166	0.08500	0.08871
	h	2822.9	2885.1	2942.0	2995.1	3045.4	3093.9
	s	6.2241	6.3432	6.4479	6.5422	6.6285	6.7088
4000.0 (250.3)	v		0.05172	0.05544	0.05883	0.06200	0.06499
	h		2835.6	2902.0	2962.0	3017.5	3069.8
	s		6.1353	6.2576	6.3642	6.4593	6.5461
5000.0 (263.9)	v			0.04222	0.04530	0.04810	0.05070
	h			2856.9	2925.5	2987.2	3044.1
	s			6.0886	6.2105	6.3163	6.4106
6000.0 (275.5)	v			0.03317	0.03614	0.03874	0.04111
	h			2804.9	2885.0	2954.2	3016.5
	s			5.9270	6.0692	6.1880	6.2913
8000.0 (295.0)	v				0.02426	0.02681	0.02896
	h				2786.8	2878.7	2955.3
	s				5.7942	5.9519	6.0790
10000.0 (311.0)	v					0.01926	0.02147
	h					2783.5	2883.4
	s					5.7145	5.8803

Table A-1-7   Superheated Steam: (No Liquid Present) (SI) (cont.)

Abs. Press. kPa (Sat. Temp. C)		Temperature—C								
		360.	380.	400.	420.	440.	460.	480.	500.	520.
500.0 (151.8)	v	0.5795	0.5984	0.6172	0.6359	0.6547	0.6734	0.6921	0.7108	0.7294
	h	3188.8	3230.4	3272.1	3314.0	3356.1	3398.4	3441.0	3483.8	3526.8
	s	7.6673	7.7319	7.7948	7.8561	7.9160	7.9745	8.0318	8.0879	8.1428
600.0 (158.8)	v	0.4821	0.4979	0.5136	0.5293	0.5450	0.5606	0.5762	0.5918	0.6074
	h	3187.0	3228.7	3270.6	3312.6	3354.8	3397.2	3439.8	3482.7	3525.8
	s	7.5810	7.6459	7.7090	7.7705	7.8305	7.8891	7.9465	8.0027	8.0577
800.0 (170.4)	v	0.3603	0.3723	0.3842	0.3960	0.4078	0.4196	0.4314	0.4432	0.4549
	h	3183.4	3225.4	3267.5	3309.7	3352.1	3394.7	3437.5	3480.5	3523.7
	s	7.4441	7.5094	7.5729	7.6347	7.6950	7.7539	7.8115	7.8678	7.9230
1000.0 (179.9)	v	0.2873	0.2969	0.3065	0.3160	0.3256	0.3350	0.3445	0.3540	0.3634
	h	3179.7	3222.0	3264.4	3306.9	3349.5	3392.2	3435.1	3478.3	3521.6
	s	7.3368	7.4027	7.4665	7.5287	7.5893	7.6484	7.7062	7.7627	7.8181
1500.0 (198.3)	v	0.1898	0.1964	0.2029	0.2094	0.2158	0.2223	0.2287	0.2350	0.2414
	h	3170.4	3213.5	3256.6	3299.7	3342.8	3386.0	3429.3	3472.8	3516.5
	s	7.1389	7.2060	7.2709	7.3340	7.3953	7.4550	7.5133	7.5703	7.6261
2000.0 (212.4)	v	0.1411	0.1461	0.1511	0.1561	0.1610	0.1659	0.1707	0.1756	0.1804
	h	3160.8	3204.9	3248.7	3292.4	3336.0	3379.7	3423.4	3467.3	3511.3
	s	6.9950	7.0635	7.1296	7.1935	7.2555	7.3159	7.3748	7.4323	7.4885
3000.0 (233.8)	v	0.09232	0.09584	0.09931	0.1027	0.1061	0.1095	0.1128	0.1161	0.1194
	h	3140.9	3187.0	3232.5	3277.5	3322.3	3367.0	3411.6	3456.2	3500.9
	s	6.7844	6.8561	6.9246	6.9906	7.0543	7.1160	7.1760	7.2345	7.2916
4000.0 (250.3)	v	0.06787	0.07066	0.07338	0.07604	0.07866	0.08125	0.08381	0.08634	0.08886
	h	3119.9	3168.4	3215.7	3262.3	3308.3	3354.0	3399.6	3445.0	3490.4
	s	6.6265	6.7019	6.7733	6.8414	6.9069	6.9702	7.0314	7.0909	7.1489
5000.0 (263.9)	v	0.05316	0.05551	0.05779	0.06001	0.06218	0.06431	0.06642	0.06849	0.07055
	h	3097.6	3148.8	3198.3	3246.5	3294.0	3340.9	3387.4	3433.7	3479.8
	s	6.4966	6.5762	6.6508	6.7215	6.7890	6.8538	6.9164	6.9770	7.0360

P (kPa) (Tsat)										
6000.0 (275.5)	v	0.04330	0.04539	0.04738	0.04931	0.05118	0.05302	0.05482	0.05659	0.05834
	h	3074.0	3128.3	3180.1	3230.3	3279.3	3327.4	3375.0	3422.2	3469.1
	s	6.3836	6.4680	6.5462	6.6196	6.6893	6.7559	6.8199	6.8818	6.9417
8000.0 (295.0)	v	0.03088	0.03265	0.03431	0.03589	0.03740	0.03887	0.04030	0.04170	0.04308
	h	3022.7	3084.2	3141.6	3196.2	3248.7	3299.7	3349.6	3398.8	3447.4
	s	6.1872	6.2828	6.3694	6.4493	6.5240	6.5945	6.6617	6.7262	6.7883
10000.0 (311.0)	v	0.02331	0.02493	0.02641	0.02779	0.02911	0.03036	0.03158	0.03276	0.03391
	h	2964.8	3035.7	3099.9	3159.7	3216.2	3270.5	3323.2	3374.6	3425.1
	s	6.0110	6.1213	6.2182	6.3057	6.3861	6.4612	6.5321	6.5994	6.6640
15000.0 (342.1)	v	0.01256	0.01428	0.01566	0.01686	0.01794	0.01895	0.01989	0.02080	0.02166
	h	2770.8	2887.7	2979.1	3057.0	3126.9	3191.5	3252.4	3310.6	3366.8
	s	5.5677	5.7497	5.8876	6.0016	6.1010	6.1904	6.2724	6.3487	6.4204
20000.0 (365.7)	v		0.008246	0.009947	0.01120	0.01224	0.01315	0.01399	0.01477	0.01551
	h		2660.2	2820.5	2932.9	3023.7	3102.7	3174.4	3241.1	3304.2
	s		5.3165	5.5585	5.7232	5.8523	5.9616	6.0581	6.1456	6.2262
30000.0	v		0.001874	0.002831	0.004921	0.006227	0.007189	0.007985	0.008681	0.009310
	h		1837.7	2161.8	2558.0	2754.0	2887.7	2993.9	3085.0	3166.6
	s		4.0021	4.4896	5.0706	5.3499	5.5349	5.6779	5.7972	5.9014
40000.0	v		0.001682	0.001909	0.002371	0.003200	0.004137	0.004941	0.005616	0.006205
	h		1776.4	1934.1	2145.7	2399.4	2617.1	2779.8	2906.8	3013.7
	s		3.8814	4.1190	4.4285	4.7893	5.0906	5.3097	5.4762	5.6128
50000.0	v		0.001589	0.001729	0.001938	0.002269	0.002747	0.003308	0.003882	0.004408
	h		1746.8	1877.7	2026.6	2199.7	2387.2	2564.9	2723.0	2854.9
	s		3.8110	4.0083	4.2262	4.4723	4.7316	4.9709	5.1782	5.3466
60000.0	v		0.001528	0.001632	0.001771	0.001962	0.002226	0.002565	0.002952	0.003358
	h		1728.4	1847.3	1975.0	2113.5	2263.2	2418.8	2570.6	2712.6
	s		3.7589	3.9383	4.1252	4.3221	4.5291	4.7385	4.9374	5.1189
80000.0	v		0.001445	0.001518	0.001605	0.001710	0.001841	0.001999	0.002188	0.002405
	h		1707.0	1814.2	1924.1	2036.6	2152.5	2272.8	2397.4	2524.0
	s		3.6807	3.8425	4.0033	4.1633	4.3237	4.4855	4.6488	4.8104

Temperature—C

Abs. Press. kPa (Sat. Temp, C)		360.	380.	400.	420.	440.	460.	480.	500.	520.
100000.0	v	0.001390		0.001446	0.001511	0.001587	0.001675	0.001777	0.001893	0.002024
	h	1696.3		1797.6	1899.0	2000.3	2102.7	2207.7	2316.1	2427.2
	s	3.6211		3.7738	3.9223	4.0664	4.2079	4.3492	4.4913	4.6331

Abs. Press. kPa (Sat. Temp, C)		540.	560.	580.	600.	625.	650.	700.	750.	800.
500.0 (151.8)	v	0.7481	0.7667	0.7853	0.8039	0.8272	0.8504	0.8968	0.9432	0.9896
	h	3570.1	3613.6	3657.4	3701.5	3757.0	3812.8	3925.8	4040.3	4156.4
	s	8.1967	8.2496	8.3016	8.3526	8.4152	8.4766	8.5957	8.7105	8.8213
600.0 (158.8)	v	0.6230	0.6386	0.6541	0.6696	0.6890	0.7084	0.7471	0.7858	0.8245
	h	3569.1	3612.7	3656.6	3700.7	3756.2	3812.1	3925.1	4039.8	4155.9
	s	8.1117	8.1647	8.2167	8.2678	8.3305	8.3919	8.5111	8.6259	8.7368
800.0 (170.4)	v	0.4666	0.4783	0.4900	0.5017	0.5163	0.5309	0.5600	0.5891	0.6181
	h	3567.2	3610.9	3654.8	3699.1	3754.7	3810.7	3923.9	4038.7	4155.0
	s	7.9771	8.0302	8.0824	8.1336	8.1964	8.2579	8.3773	8.4923	8.6033
1000.0 (179.9)	v	0.3728	0.3822	0.3916	0.4010	0.4127	0.4244	0.4477	0.4710	0.4943
	h	3565.2	3609.0	3653.1	3697.4	3753.1	3809.3	3922.7	4037.6	4154.1
	s	7.8724	7.9256	7.9779	8.0292	8.0921	8.1537	8.2734	8.3885	8.4997
1500.0 (198.3)	v	0.2477	0.2540	0.2605	0.2667	0.2745	0.2824	0.2980	0.3136	0.3292
	h	3560.4	3604.5	3648.8	3693.3	3749.3	3805.7	3919.6	4034.9	4151.7
	s	7.6808	7.7343	7.7869	7.8385	7.9017	7.9636	8.0838	8.1993	8.3108
2000.0 (212.4)	v	0.1852	0.1900	0.1947	0.1995	0.2054	0.2114	0.2232	0.2349	0.2467
	h	3555.5	3599.9	3644.4	3689.2	3745.5	3802.1	3916.5	4032.2	4149.4
	s	7.5435	7.5974	7.6503	7.7022	7.7657	7.8279	7.9485	8.0645	8.1763
3000.0 (233.8)	v	0.1226	0.1259	0.1291	0.1323	0.1364	0.1404	0.1483	0.1562	0.1641
	h	3545.7	3590.6	3635.7	3681.0	3737.8	3795.0	3910.3	4026.8	4144.7
	s	7.3474	7.4020	7.4554	7.5079	7.5721	7.6349	7.7564	7.8733	7.9857

P (kPa) (Tsat °C)										
4000.0 (250.3)	v	0.1229	0.1169	0.1109	0.1049	0.1018	0.09876	0.09631	0.09384	0.09135
	h	4140.0	4021.4	3904.1	3787.9	3730.2	3672.8	3627.0	3581.4	3535.8
	s	7.8495	7.7363	7.6187	7.4961	7.4328	7.3680	7.3149	7.2608	7.2055
5000.0 (263.9)	v	0.09809	0.09329	0.08845	0.08356	0.08109	0.07862	0.07662	0.07461	0.07259
	h	4135.3	4016.1	3897.9	3780.7	3722.5	3664.5	3618.2	3572.0	3525.9
	s	7.7431	7.6292	7.5108	7.3872	7.3233	7.2578	7.2042	7.1494	7.0934
6000.0 (275.5)	v	0.08159	0.07755	0.07348	0.06936	0.06728	0.06518	0.06349	0.06179	0.06008
	h	4130.7	4010.7	3891.7	3773.5	3714.8	3656.8	3609.4	3562.7	3515.9
	s	7.6554	7.5409	7.4217	7.2971	7.2326	7.1664	7.1122	7.0568	7.0000
8000.0 (295.0)	v	0.06096	0.05788	0.05477	0.05161	0.05001	0.04839	0.04709	0.04577	0.04443
	h	4121.3	3999.9	3879.2	3759.2	3699.3	3639.5	3591.7	3543.8	3495.7
	s	7.5158	7.3999	7.2790	7.1523	7.0866	7.0191	6.9636	6.9068	6.8484
10000.0 (311.0)	v	0.04858	0.04608	0.04355	0.04096	0.03965	0.03832	0.03724	0.03615	0.03504
	h	4112.0	3989.1	3866.8	3744.7	3683.8	3622.7	3573.7	3524.5	3475.1
	s	7.4058	7.2886	7.1660	7.0373	6.9703	6.9013	6.8446	6.7863	6.7261
15000.0 (342.1)	v	0.03209	0.03036	0.02859	0.02677	0.02584	0.02488	0.02411	0.02331	0.02250
	h	4088.6	3962.1	3835.4	3708.3	3644.3	3579.8	3527.7	3475.0	3421.4
	s	7.2013	7.0806	6.9536	6.8195	6.7492	6.6764	6.6160	6.5535	6.4885
20000.0 (365.7)	v	0.02385	0.02250	0.02111	0.01967	0.01893	0.01816	0.01753	0.01688	0.01621
	h	4065.3	3935.0	3803.8	3671.1	3603.8	3535.5	3479.9	3423.0	3364.7
	s	7.0511	6.9267	6.7953	6.6554	6.5814	6.5043	6.4398	6.3724	6.3015
30000.0	v	0.01562	0.01465	0.01365	0.01258	0.01202	0.01144	0.01095	0.01043	0.009890
	h	4018.5	3880.3	3739.7	3595.0	3520.2	3443.0	3378.9	3312.1	3241.7
	s	6.8288	6.6970	6.5560	6.4033	6.3212	6.2340	6.1597	6.0805	5.9949
40000.0	v	0.01152	0.01075	0.009930	0.009053	0.008584	0.008088	0.007667	0.007219	0.006735
	h	3971.7	3825.5	3674.8	3517.0	3433.8	3346.4	3272.4	3193.4	3108.0
	s	6.6606	6.5210	6.3701	6.2035	6.1122	6.0135	5.9276	5.8340	5.7302
50000.0	v	0.009076	0.008420	0.007720	0.006960	0.006550	0.006111	0.005734	0.005328	0.004888
	h	3925.3	3770.9	3610.2	3438.9	3346.8	3248.3	3163.2	3070.7	2968.9
	s	6.5222	6.3749	6.2138	6.0331	5.9320	5.8207	5.7221	5.6124	5.4886

Table A-1-7 Superheated Steam: (No Liquid Present) (SI) (cont.)

		540.	560.	580.	600.	625.	650.	700.	750.	800.
60000.0	v	0.003755	0.004135	0.004496	0.004835	0.005229	0.005596	0.006269	0.006885	0.007460
	h	2838.3	2951.7	3055.8	3151.6	3261.4	3362.4	3547.0	3717.4	3879.6
	s	5.2755	5.4132	5.5367	5.6477	5.7717	5.8827	6.0775	6.2483	6.4031
80000.0	v	00.02641	0.002886	0.003132	0.003379	0.03682	0.003974	0.004519	0.005017	0.005481
	h	2648.2	2765.1	2874.9	2980.3	3104.6	3220.3	3428.7	3616.7	3792.8
	s	4.9650	5.1072	5.2374	5.3595	5.4999	5.6270	5.8470	6.0354	6.2034
100000.0	v	0.002168	0.002326	0.002493	0.002668	0.002891	0.003106	0.003536	0.003952	0.004341
	h	2538.6	2648.2	2754.5	2857.5	2985.8	3105.3	3324.4	3526.1	3714.3
	s	4.7719	4.9050	5.0311	5.1505	5.2954	5.4267	5.6579	5.8600	6.0397

Source: Reprinted with permission from ASME.

Appendix A-2: Properties of Refrigerant-12

Table A-2-1 Saturation Properties—Temperature Table

Temp. °F	Pressure		Volume ft³/lbm		Density lbm/ft³		Enthalpy Btu/lbm			Entropy Btu/lbm R		Temp. °F
	Psia	Psig	Liquid v_f	Vapor v_g	Liquid $1/v_f$	Vapor $1/v_g$	Liquid h_f	Latent h_{fg}	Vapor h_g	Liquid s_f	Vapor s_g	
−152	0.13799	29.64024*	0.0095673	197.58	104.52	0.0050614	−23.106	83.734	60.628	−0.063944	0.20818	−152
−151	0.14561	29.62473*	0.0095747	187.84	104.44	0.0053238	−22.901	83.634	60.733	−0.063280	0.20764	−151
−150	0.15359	29.60849*	0.0095822	178.65	104.36	0.0055976	−22.697	83.534	60.837	−0.062619	0.20711	−150
−149	0.16194	29.59150*	0.0095897	169.97	104.28	0.0058832	−22.492	83.435	60.943	−0.061959	0.20658	−149
−148	0.17067	29.57372*	0.0095971	161.78	104.20	0.0061811	−22.288	83.336	61.048	−0.061302	0.20606	−148
−147	0.17980	29.55512*	0.0096047	154.05	104.12	0.0064915	−22.083	83.237	61.154	−0.060647	0.20554	−147
−146	0.18935	29.53568*	0.0096122	146.74	104.03	0.0068149	−21.879	83.138	61.259	−0.059995	0.20503	−146
−145	0.19933	29.51537*	0.0096198	139.83	103.95	0.0071517	−21.674	83.039	61.365	−0.059344	0.20452	−145
−144	0.20975	29.49414*	0.0096274	133.29	103.87	0.0075025	−21.470	82.941	61.471	−0.058696	0.20402	−144
−143	0.22064	29.47198*	0.0096350	127.10	103.79	0.0078675	−21.265	82.842	61.577	−0.058049	0.20353	−143
−142	0.23200	29.44884*	0.0096426	121.25	103.71	0.0082474	−21.061	82.744	61.683	−0.057405	0.20304	−142
−141	0.24386	29.42470*	0.0096502	115.71	103.62	0.0086425	−20.857	82.646	61.789	−0.056763	0.20256	−141
−140	0.25623	29.39951*	0.0096579	110.46	103.54	0.0090533	−20.652	82.548	61.896	−0.056123	0.20208	−140
−139	0.26913	29.37324*	0.0096656	105.48	103.46	0.0094803	−20.448	82.451	62.003	−0.055484	0.20161	−139
−138	0.28258	29.34586*	0.0096733	100.77	103.38	0.0099241	−20.244	82.353	62.109	−0.054848	0.20114	−138

* Inches of Mercury below 1 atm.

Temp. °F	Pressure Psia	Pressure Psig	Volume ft³/lbm Liquid v_f	Volume ft³/lbm Vapor v_g	Density lbm/ft³ Liquid $1/v_f$	Density lbm/ft³ Vapor $1/v_g$	Enthalpy Btu/lbm Liquid h_f	Enthalpy Btu/lbm Latent h_{fg}	Enthalpy Btu/lbm Vapor h_g	Entropy Btu/lbm R Liquid s_f	Entropy Btu/lbm R Vapor s_g	Temp. °F
-137	0.29660	29.31732*	0.0096811	96.292	103.29	0.010385	-20.039	82.255	62.216	-0.054214	0.20068	-137
-136	0.31120	29.28759*	0.0096888	92.050	103.21	0.010864	-19.835	82.158	62.323	-0.053582	0.20023	-136
-135	0.32641	29.25663*	0.0096966	88.023	103.13	0.011361	-19.631	82.061	62.430	-0.052952	0.19978	-135
-134	0.34224	29.22439*	0.0097044	84.201	103.05	0.011876	-19.426	81.964	62.538	-0.052324	0.19933	-134
-133	0.35872	29.19084*	0.0097123	80.571	102.96	0.012411	-19.222	81.867	62.645	-0.051697	0.19889	-133
-132	0.37587	29.15593*	0.0097201	77.123	102.88	0.012966	-19.018	81.770	62.752	-0.051073	0.19845	-132
-131	0.39370	29.11962*	0.0097280	73.846	102.80	0.013542	-18.813	81.673	62.860	-0.050450	0.19802	-131
-130	0.41224	29.08187*	0.0097359	70.730	102.71	0.014138	-18.609	81.577	62.968	-0.049830	0.19760	-130
-129	0.43152	29.04262*	0.0097438	67.768	102.63	0.014756	-18.404	81.480	63.076	-0.049211	0.19718	-129
-128	0.45155	29.00184*	0.0097518	64.949	102.55	0.015397	-18.200	81.384	63.184	-0.048594	0.19676	-128
-127	0.47236	28.95947*	0.0097598	62.267	102.46	0.016060	-17.996	81.288	63.292	-0.047979	0.19635	-127
-126	0.49397	28.91546*	0.0097678	59.714	102.38	0.016746	-17.791	81.191	63.400	-0.047366	0.19594	-126
-125	0.51641	28.86978*	0.0097758	57.283	102.29	0.017457	-17.587	81.096	63.509	-0.046754	0.19554	-125
-124	0.53970	28.82236*	0.0097838	54.968	102.21	0.018192	-17.383	81.000	63.617	-0.046144	0.19514	-124
-123	0.56387	28.77315*	0.0097919	52.761	102.13	0.018953	-17.178	80.904	63.726	-0.045537	0.19475	-123
-122	0.58894	28.72211*	0.0098000	50.658	102.04	0.019740	-16.974	80.808	63.834	-0.044930	0.19436	-122
-121	0.61494	28.66917*	0.0098081	48.653	101.96	0.020554	-16.769	80.712	63.943	-0.044326	0.19397	-121
-120	0.64190	28.61429*	0.0098163	46.741	101.87	0.021395	-16.565	80.617	64.052	-0.043723	0.19359	-120
-119	0.66984	28.55741*	0.0098245	44.916	101.79	0.022264	-16.360	80.521	64.161	-0.043122	0.19322	-119
-118	0.69879	28.49846*	0.0098327	43.175	101.70	0.023162	-16.155	80.425	64.270	-0.042522	0.19285	-118
-117	0.72878	28.43740*	0.0098409	41.513	101.62	0.024089	-15.951	80.330	64.379	-0.041925	0.19248	-117
-116	0.75984	28.37416*	0.0098491	39.926	101.53	0.025047	-15.746	80.234	64.488	-0.041329	0.19212	-116
-115	0.79200	28.30869*	0.0098574	38.410	101.45	0.026035	-15.541	80.139	64.598	-0.040734	0.19176	-115
-114	0.82528	28.24092*	0.0098657	36.961	101.36	0.027056	-15.337	80.044	64.707	-0.040141	0.19140	-114
-113	0.85973	28.17078*	0.0098740	35.577	101.28	0.028108	-15.132	79.949	64.817	-0.039550	0.19105	-113
-112	0.89537	28.09823*	0.0098824	34.253	101.19	0.029195	-14.927	79.853	64.926	-0.038960	0.19070	-112

Temp											
−111	0.19036	−0.038372	65.036	79.759	−14.723	0.030315	101.10	32.987	0.0098908	28.02318*	0.93223
−110	0.19002	−0.037786	65.145	79.663	−14.518	0.031470	101.02	31.777	0.0098992	27.94558*	0.97034
−109	0.18968	−0.037201	65.255	79.568	−14.313	0.032660	100.93	30.618	0.0099076	27.8654*	1.0097
−108	0.18935	−0.036618	65.365	79.473	−14.108	0.033888	100.85	29.509	0.0099161	27.7824*	1.0505
−107	0.18902	−0.036036	65.475	79.378	−13.903	0.035152	100.76	28.448	0.0099246	27.6968*	1.0925
−106	0.18870	−0.035455	65.585	79.283	−13.698	0.036455	100.67	27.431	0.0099331	27.6083*	1.1360
−105	0.18838	−0.034877	65.696	79.188	−13.492	0.037796	100.59	26.458	0.0099416	27.5169*	1.1809
−104	0.18806	−0.034299	65.806	79.094	−13.288	0.039177	100.50	25.525	0.0099502	27.4225*	1.2273
−103	0.18775	−0.033723	65.916	78.998	−13.082	0.040600	100.41	24.631	0.0099588	27.3251*	1.2751
−102	0.18744	−0.033149	66.027	78.904	−12.877	0.042063	100.33	23.774	0.0099674	27.2245*	1.3245
−101	0.18713	−0.032576	66.137	78.809	−12.672	0.043569	100.24	22.952	0.0099761	27.1208*	1.3754
−100	0.18683	−0.032005	66.248	78.714	−12.466	0.045119	100.15	22.164	0.0099847	27.0138*	1.4280
−99	0.18653	−0.031435	66.358	78.619	−12.261	0.046713	100.07	21.407	0.0099935	26.9034*	1.4822
−98	0.18623	−0.030866	66.469	78.524	−12.055	0.048352	99.978	20.682	0.010002	26.7896*	1.5381
−97	0.18594	−0.030299	66.579	78.429	−11.850	0.050037	99.891	19.985	0.010011	26.6723*	1.5957
−96	0.18565	−0.029733	66.690	78.334	−11.644	0.051769	99.803	19.316	0.010020	26.5514*	1.6551
−95	0.18536	−0.029169	66.801	78.239	−11.438	0.053550	99.715	18.674	0.010029	26.4268*	1.7163
−94	0.18508	−0.028606	66.911	78.144	−11.233	0.055379	99.627	18.057	0.010037	26.2984*	1.7794
−93	0.18480	−0.028044	67.022	78.049	−11.027	0.057259	99.539	17.465	0.010046	26.1662*	1.8443
−92	0.18452	−0.027484	67.133	77.954	−10.821	0.059189	99.451	16.895	0.010055	26.0301*	1.9112
−91	0.18425	−0.026925	67.244	77.859	−10.615	0.061172	99.363	16.347	0.010064	25.8899*	1.9800
−90	0.18398	−0.026367	67.355	77.764	−10.409	0.063207	99.274	15.821	0.010073	25.7456*	2.0509
−89	0.18371	−0.025811	67.466	77.669	−10.203	0.065297	99.185	15.315	0.010082	25.5971*	2.1238
−88	0.18345	−0.025256	67.577	77.574	−9.9971	0.067441	99.097	14.828	0.010091	25.4443*	2.1988
−87	0.18318	−0.024702	67.688	77.479	−9.7908	0.069642	99.008	14.359	0.010100	25.2872*	2.2760
−86	0.18293	−0.024150	67.799	77.384	−9.5845	0.071900	98.919	13.908	0.010109	25.1255*	2.3554
−85	0.18267	−0.023599	67.911	77.289	−9.3782	0.074216	98.830	13.474	0.010118	24.9593*	2.4371
−84	0.18242	−0.023049	68.022	77.194	−9.1717	0.076591	98.740	13.056	0.010128	24.7884*	2.5210
−83	0.18217	−0.022500	68.133	77.098	−8.9652	0.079027	98.651	12.654	0.010137	24.6127*	2.6073
−82	0.18192	−0.021953	68.244	77.003	−8.7586	0.081525	98.561	12.266	0.010146	24.4321*	2.6960
−81	0.18168	−0.021407	68.355	76.907	−8.5519	0.084085	98.472	11.893	0.010155	24.2466*	2.7871

Table A-2-1 Saturation Properties—Temperature Table (cont.)

Temp. °F	Pressure Psia	Pressure Psig	Volume ft³/lbm Liquid v_f	Volume ft³/lbm Vapor v_g	Density lbm/ft³ Liquid $1/v_f$	Density lbm/ft³ Vapor $1/v_g$	Enthalpy Btu/lbm Liquid h_f	Enthalpy Btu/lbm Latent h_{fg}	Enthalpy Btu/lbm Vapor h_g	Entropy Btu/lbm R Liquid s_f	Entropy Btu/lbm R Vapor s_g	Temp. °F
−80	2.8807	24.0560*	0.010164	11.533	98.382	0.086708	−8.3451	76.812	68.467	−0.020862	0.18143	−80
−79	2.9769	23.8603*	0.010174	11.186	98.292	0.089397	−8.1383	76.716	68.578	−0.020319	0.18120	−79
−78	3.0756	23.6592*	0.010183	10.852	98.201	0.092151	−7.9314	76.620	68.689	−0.019776	0.18096	−78
−77	3.1770	23.4528*	0.010193	10.529	98.111	0.094973	−7.7244	76.525	68.801	−0.019235	0.18073	−77
−76	3.2811	23.2409*	0.010202	10.218	98.021	0.097863	−7.5173	76.429	68.912	−0.018695	0.18050	−76
−75	3.3879	23.0234*	0.010211	9.9184	97.930	0.10082	−7.3101	76.333	69.023	−0.018156	0.18027	−75
−74	3.4975	22.8002*	0.010221	9.6290	97.839	0.10385	−7.1029	76.238	69.135	−0.017619	0.18004	−74
−73	3.6100	22.5712*	0.010230	9.3497	97.748	0.10695	−6.8955	76.142	69.246	−0.017083	0.17982	−73
−72	3.7254	22.3362*	0.010240	9.0802	97.657	0.11013	−6.6881	76.046	69.358	−0.016547	0.17960	−72
−71	3.8438	22.0953*	0.010249	8.8199	97.566	0.11338	−6.4806	75.949	69.468	−0.016013	0.17938	−71
−70	3.9651	21.8482*	0.010259	8.5687	97.475	0.11670	−6.2730	75.853	69.580	−0.015481	0.17916	−70
−69	4.0896	21.5948*	0.010269	8.3260	97.383	0.12011	−6.0653	75.757	69.692	−0.014949	0.17895	−69
−68	4.2172	21.3350*	0.010278	8.0916	97.292	0.12359	−5.8574	75.660	69.803	−0.014418	0.17874	−68
−67	4.3479	21.0688*	0.010288	7.8651	97.200	0.12714	−5.6496	75.564	69.914	−0.013889	0.17853	−67
−66	4.4819	20.7959*	0.010298	7.6462	97.108	0.13078	−5.4416	75.467	70.025	−0.013361	0.17833	−66
−65	4.6193	20.5164*	0.010308	7.4347	97.016	0.13451	−5.2336	75.371	70.137	−0.012834	0.17812	−65
−64	4.7599	20.2299*	0.010317	7.2302	96.924	0.13831	−5.0254	75.273	70.248	−0.012308	0.17792	−64
−63	4.9040	19.9365*	0.010327	7.0325	96.831	0.14220	−4.8172	75.177	70.360	−0.011783	0.17772	−63
−62	5.0516	19.6360*	0.010337	6.8412	96.739	0.14617	−4.6088	75.080	70.471	−0.011259	0.17753	−62
−61	5.2027	19.3284*	0.010347	6.6563	96.646	0.15023	−4.4004	74.982	70.582	−0.010736	0.17733	−61
−60	5.3575	19.0133*	0.010357	6.4774	96.553	0.15438	−4.1919	74.885	70.693	−0.010214	0.17714	−60
−59	5.5159	18.6908*	0.010367	6.3043	96.460	0.15862	−3.9832	74.788	70.805	−0.009694	0.17695	−59
−58	5.6780	18.3607*	0.010377	6.1367	96.367	0.16295	−3.7745	74.691	70.916	−0.009174	0.17676	−58
−57	5.8439	18.0229*	0.010387	5.9746	96.274	0.16738	−3.5657	74.593	71.027	−0.008656	0.17658	−57
−56	6.0137	17.6773*	0.010397	5.8176	96.180	0.17189	−3.3567	74.495	71.138	−0.008139	0.17639	−56

Temp												Temp
-55	0.17621	-0.007622	71.249	74.397	-3.1477	0.17650	96.086	5.6656	0.010407	17.3237*	6.1874	-55
-54	0.17603	-0.007107	71.360	74.299	-2.9386	0.18121	95.993	5.5184	0.010417	16.9619*	6.3650	-54
-53	0.17585	-0.006593	71.472	74.201	-2.7294	0.18602	95.899	5.3758	0.010428	16.5920*	6.5467	-53
-52	0.17568	-0.006080	71.583	74.103	-2.5200	0.19092	95.804	5.2377	0.010438	16.2136*	6.7326	-52
-51	0.17551	-0.005568	71.694	74.005	-2.3106	0.19593	95.710	5.1039	0.010448	15.8268*	6.9226	-51
-50	0.17533	-0.005056	71.805	73.906	-2.1011	0.20104	95.616	4.9742	0.010459	15.4313*	7.1168	-50
-49	0.17516	-0.004546	71.916	73.807	-1.8914	0.20625	95.521	4.8485	0.010469	15.0271*	7.3153	-49
-48	0.17500	-0.004037	72.027	73.709	-1.6817	0.21157	95.426	4.7267	0.010479	14.6139*	7.5183	-48
-47	0.17483	-0.003529	72.138	73.610	-1.4719	0.21699	95.331	4.6085	0.010490	14.1917*	7.7257	-47
-46	0.17467	-0.003022	72.249	73.511	-1.2619	0.22252	95.236	4.4940	0.010500	13.7603*	7.9375	-46
-45	0.17451	-0.002516	72.359	73.411	-1.0519	0.22816	95.141	4.3828	0.010511	13.3196*	8.1540	-45
-44	0.17435	-0.002011	72.470	73.312	-0.8417	0.23391	95.045	4.2751	0.010521	12.8693*	8.3751	-44
-43	0.17419	-0.001507	72.581	73.212	-0.6314	0.23978	94.949	4.1705	0.010532	12.4095*	8.6010	-43
-42	0.17403	-0.001003	72.691	73.112	-0.4211	0.24576	94.854	4.0691	0.010543	11.9399*	8.8316	-42
-41	0.17388	-0.000501	72.802	73.013	-0.2106	0.25185	94.758	3.9706	0.010553	11.4605*	9.0671	-41
-40	0.17373	0	72.913	72.913	0	0.25806	94.661	3.8750	0.010564	10.9709*	9.3076	-40
-39	0.17357	0.000500	73.023	72.812	0.2107	0.26439	94.565	3.7823	0.010575	10.4712*	9.5530	-39
-38	0.17343	0.001000	73.134	72.712	0.4215	0.27084	94.469	3.6922	0.010586	9.9611*	9.8035	-38
-37	0.17328	0.001498	73.243	72.611	0.6324	0.27741	94.372	3.6047	0.010596	9.441*	10.059	-37
-36	0.17313	0.001995	73.354	72.511	0.8434	0.28411	94.275	3.5198	0.010607	8.909*	10.320	-36
-35	0.17299	0.002492	73.464	72.409	1.0546	0.29093	94.178	3.4373	0.010618	8.367*	10.586	-35
-34	0.17285	0.002988	73.575	72.309	1.2659	0.29788	94.081	3.3571	0.010629	7.814*	10.858	-34
-33	0.17271	0.003482	73.685	72.208	1.4772	0.30495	93.983	3.2792	0.010640	7.250*	11.135	-33
-32	0.17257	0.003976	73.795	72.106	1.6887	0.31216	93.886	3.2035	0.010651	6.675*	11.417	-32
-31	0.17243	0.004469	73.904	72.004	1.9003	0.31949	93.788	3.1300	0.010662	6.088*	11.706	-31
-30	0.17229	0.004961	74.015	71.903	2.1120	0.32696	93.690	3.0585	0.010674	5.490*	11.999	-30
-29	0.17216	0.005452	74.125	71.801	2.3239	0.33457	93.592	2.9890	0.010685	4.880*	12.299	-29
-28	0.17203	0.005942	74.234	71.698	2.5358	0.34231	93.493	2.9214	0.010696	4.259*	12.604	-28
-27	0.17189	0.006431	74.344	71.596	2.7479	0.35018	93.395	2.8556	0.010707	3.625*	12.916	-27
-26	0.17177	0.006919	74.454	71.494	2.9601	0.35820	93.296	2.7917	0.010719	2.979*	13.233	-26
-25	0.17164	0.007407	74.563	71.391	3.1724	0.36636	93.197	2.7295	0.010730	2.320*	13.556	-25
-24	0.17151	0.007894	74.673	71.288	3.3848	0.37466	93.098	2.6691	0.010741	1.649*	13.886	-24
-23	0.17139	0.008379	74.782	71.185	3.5973	0.38311	92.999	2.6102	0.010753	0.966*	14.222	-23

Table A-2-1 Saturation Properties—Temperature Table (cont.)

Temp. °F	Pressure		Volume ft³/lbm		Density lbm/ft³		Enthalpy Btu/lbm			Entropy Btu/lbm R		Temp. °F
	Psia	Psig	Liquid v_f	Vapor v_g	Liquid $1/v_f$	Vapor $1/v_g$	Liquid h_f	Latent h_{fg}	Vapor h_g	Liquid s_f	Vapor s_g	
−22	14.564	0.270*	0.010764	2.5529	92.899	0.39171	3.8100	71.081	74.891	0.008864	0.17126	−22
−21	14.912	0.216	0.010776	2.4972	92.799	0.40045	4.0228	70.978	75.001	0.009348	0.17114	−21
−20	15.267	0.571	0.010788	2.4429	92.699	0.40934	4.2357	70.874	75.110	0.009831	0.17102	−20
−19	15.628	0.932	0.010799	2.3901	92.599	0.41839	4.4487	70.770	75.219	0.010314	0.17090	−19
−18	15.996	1.300	0.010811	2.3387	92.499	0.42758	4.6618	70.666	75.328	0.010795	0.17078	−18
−17	16.371	1.675	0.010823	2.2886	92.399	0.43694	4.8751	70.561	75.436	0.011276	0.17066	−17
−16	16.753	2.057	0.010834	2.2399	92.298	0.44645	5.0885	70.456	75.545	0.011755	0.17055	−16
−15	17.141	2.445	0.010846	2.1924	92.197	0.45612	5.3020	70.352	75.654	0.012234	0.17043	−15
−14	17.536	2.840	0.010858	2.1461	92.096	0.46595	5.5157	70.246	75.762	0.012712	0.17032	−14
−13	17.939	3.243	0.010870	2.1011	91.995	0.47595	5.7295	70.141	75.871	0.013190	0.17021	−13
−12	18.348	3.652	0.010882	2.0572	91.893	0.48611	5.9434	70.036	75.979	0.013666	0.17010	−12
−11	18.765	4.069	0.010894	2.0144	91.791	0.49643	6.1574	69.930	76.087	0.014142	0.16999	−11
−10	19.189	4.493	0.010906	1.9727	91.689	0.50693	6.3716	69.824	76.196	0.014617	0.16989	−10
−9	19.621	4.925	0.010919	1.9320	91.587	0.51759	6.5859	69.718	76.304	0.015091	0.16978	−9
−8	20.059	5.363	0.010931	1.8924	91.485	0.52843	6.8003	69.611	76.411	0.015564	0.16967	−8
−7	20.506	5.810	0.010943	1.8538	91.382	0.53944	7.0149	69.505	76.520	0.016037	0.16957	−7
−6	20.960	6.264	0.010955	1.8161	91.280	0.55063	7.2296	69.397	76.627	0.016508	0.16947	−6
−5	21.422	6.726	0.010968	1.7794	91.177	0.56199	7.4444	69.291	76.735	0.016979	0.16937	−5
−4	21.891	7.195	0.010980	1.7436	91.074	0.57354	7.6594	69.183	76.842	0.017449	0.16927	−4
−3	22.369	7.673	0.010993	1.7086	90.970	0.58526	7.8745	69.075	76.950	0.017919	0.16917	−3
−2	22.854	8.158	0.011005	1.6745	90.867	0.59718	8.0898	68.967	77.057	0.018388	0.16907	−2
−1	23.348	8.652	0.011018	1.6413	90.763	0.60927	8.3052	68.859	77.164	0.018855	0.16897	−1
0	23.849	9.153	0.011030	1.6089	90.659	0.62156	8.5207	68.750	77.271	0.019323	0.16888	0
1	24.359	9.663	0.011043	1.5772	90.554	0.63404	8.7364	68.642	77.378	0.019789	0.16878	1
2	24.878	10.182	0.011056	1.5463	90.450	0.64670	8.9522	68.533	77.485	0.020255	0.16869	2

3	0.16860	0.020719	77.592	68.424	9.1682	0.65957	90.345	1.5161	0.011069	10.708	25.404	3
4	0.16851	0.021184	77.698	68.314	9.3843	0.67263	90.240	1.4867	0.011082	11.243	25.939	4
5	0.16842	0.021647	77.805	68.204	9.6005	0.68588	90.135	1.4580	0.011094	11.787	26.483	**5**
6	0.16833	0.022110	77.911	68.094	9.8169	0.69934	90.030	1.4299	0.011107	12.340	27.036	6
7	0.16824	0.022572	78.017	67.984	10.033	0.71300	89.924	1.4025	0.011121	12.901	27.597	7
8	0.16815	0.023033	78.123	67.873	10.250	0.72687	89.818	1.3758	0.011134	13.471	28.167	8
9	0.16807	0.023494	78.229	67.762	10.467	0.74094	89.712	1.3496	0.011147	14.051	28.747	9
10	0.16798	0.023954	78.335	67.651	10.684	0.75523	89.606	1.3241	0.011160	14.639	29.335	10
11	0.16790	0.024413	78.440	67.539	10.901	0.76972	89.499	1.2992	0.011173	15.236	29.932	11
12	0.16782	0.024871	78.546	67.428	11.118	0.78443	89.392	1.2748	0.011187	15.843	30.539	12
13	0.16774	0.025329	78.651	67.315	11.336	0.79935	89.285	1.2510	0.011200	16.459	31.155	13
14	0.16765	0.025786	78.757	67.203	11.554	0.81449	89.178	1.2278	0.011214	17.084	31.780	14
15	0.16758	0.026243	78.861	67.090	11.771	0.82986	89.070	1.2050	0.011227	17.719	32.415	15
16	0.16750	0.026699	78.966	66.977	11.989	0.84544	88.962	1.1828	0.011241	18.364	33.060	16
17	0.16742	0.027154	79.071	66.864	12.207	0.86125	88.854	1.1611	0.011254	19.018	33.714	17
18	0.16734	0.027608	79.176	66.750	12.426	0.87729	88.746	1.1399	0.011268	19.682	34.378	18
19	0.16727	0.028062	79.280	66.636	12.644	0.89356	88.637	1.1191	0.011282	20.356	35.052	19
20	0.16719	0.028515	79.385	66.522	12.863	0.91006	88.529	1.0988	0.011296	21.040	35.736	20
21	0.16712	0.028968	79.488	66.407	13.081	0.92679	88.419	1.0790	0.011310	21.734	36.430	21
22	0.16704	0.029420	79.593	66.293	13.300	0.94377	88.310	1.0596	0.011324	22.439	37.135	22
23	0.16697	0.029871	79.697	66.177	13.520	0.96098	88.201	1.0406	0.011338	23.153	37.849	23
24	0.16690	0.030322	79.800	66.061	13.739	0.97843	88.091	1.0220	0.011352	23.878	38.574	24
25	0.16683	0.030772	79.904	65.946	13.958	0.99613	87.981	1.0039	0.011366	24.614	39.310	25
26	0.16676	0.031221	80.007	65.829	14.178	1.0141	87.870	0.98612	0.011380	25.360	40.056	26
27	0.16669	0.031670	80.111	65.713	14.398	1.0323	87.760	0.96874	0.011395	26.117	40.813	27
28	0.16662	0.032118	80.214	65.596	14.618	1.0507	87.649	0.95173	0.011409	26.884	41.580	28
29	0.16655	0.032566	80.316	65.478	14.838	1.0694	87.537	0.93509	0.011424	27.663	42.359	29
30	0.16648	0.033013	80.419	65.361	15.058	1.0884	87.426	0.91880	0.011438	28.452	43.148	30
31	0.16642	0.033460	80.522	65.243	15.279	1.1076	87.314	0.90286	0.011453	29.252	43.948	31
32	0.16635	0.033905	80.624	65.124	15.500	1.1271	87.202	0.88725	0.011468	30.064	44.760	32
33	0.16629	0.034351	80.726	65.006	15.720	1.1468	87.090	0.87197	0.011482	30.887	45.583	33
34	0.16622	0.034796	80.828	64.886	15.942	1.1668	86.977	0.85702	0.011497	31.721	46.417	34

Table A-2-1 Saturation Properties—Temperature Table (cont.)

Temp. °F	Pressure Psia	Pressure Psig	Volume ft³/lbm Liquid v_f	Volume ft³/lbm Vapor v_g	Density lbm/ft³ Liquid $1/v_f$	Density lbm/ft³ Vapor $1/v_g$	Enthalpy Btu/lbm Liquid h_f	Enthalpy Btu/lbm Latent h_{fg}	Enthalpy Btu/lbm Vapor h_g	Entropy Btu/lbm R Liquid s_f	Entropy Btu/lbm R Vapor s_g	Temp. °F
35	47.263	32.567	0.011512	0.84237	86.865	1.1871	16.163	64.767	80.930	0.035240	0.16616	35
36	48.120	33.424	0.011527	0.82803	86.751	1.2077	16.384	64.647	81.031	0.035683	0.16610	36
37	48.989	34.293	0.011542	0.81399	86.638	1.2285	16.606	64.527	81.133	0.036126	0.16604	37
38	49.870	35.174	0.011557	0.80023	86.524	1.2496	16.828	64.406	81.234	0.036569	0.16598	38
39	50.763	36.067	0.011573	0.78676	86.410	1.2710	17.050	64.285	81.335	0.037011	0.16592	39
40	51.667	36.971	0.011588	0.77357	86.296	1.2927	17.273	64.163	81.436	0.037453	0.16586	40
41	52.584	37.888	0.011603	0.76064	86.181	1.3147	17.495	64.042	81.537	0.037893	0.16580	41
42	53.513	38.817	0.011619	0.74798	86.066	1.3369	17.718	63.919	81.637	0.038334	0.16574	42
43	54.454	39.758	0.011635	0.73557	85.951	1.3595	17.941	63.796	81.737	0.038774	0.16568	43
44	55.407	40.711	0.011650	0.72341	85.836	1.3823	18.164	63.673	81.837	0.039213	0.16562	44
45	56.373	41.677	0.011666	0.71149	85.720	1.4055	18.387	63.550	81.937	0.039652	0.16557	45
46	57.352	42.656	0.011682	0.69982	85.604	1.4289	18.611	63.426	82.037	0.040091	0.16551	46
47	58.343	43.647	0.011698	0.68837	85.487	1.4527	18.835	63.301	82.136	0.040529	0.16546	47
48	59.347	44.651	0.011714	0.67715	85.371	1.4768	19.059	63.177	82.236	0.040966	0.16540	48
49	60.364	45.668	0.011730	0.66616	85.254	1.5012	19.283	63.051	82.334	0.041403	0.16535	49
50	61.394	46.698	0.011746	0.65537	85.136	1.5258	19.507	62.926	82.433	0.041839	0.16530	50
51	62.437	47.741	0.011762	0.64480	85.018	1.5509	19.732	62.800	82.532	0.042276	0.16524	51
52	63.494	48.798	0.011779	0.63444	84.900	1.5762	19.957	62.673	82.630	0.042711	0.16519	52
53	64.563	49.867	0.011795	0.62428	84.782	1.6019	20.182	62.546	82.728	0.043146	0.16514	53
54	65.646	50.950	0.011811	0.61431	84.663	1.6278	20.408	62.418	82.826	0.043581	0.16509	54
55	66.743	52.047	0.011828	0.60453	84.544	1.6542	20.634	62.290	82.924	0.044015	0.16504	55
56	67.853	53.157	0.011845	0.59495	84.425	1.6808	20.859	62.162	83.021	0.044449	0.16499	56
57	68.977	54.281	0.011862	0.58554	84.305	1.7078	21.086	62.033	83.119	0.044883	0.16494	57
58	70.115	55.419	0.011879	0.57632	84.185	1.7352	21.312	61.903	83.215	0.045316	0.16489	58
59	71.267	56.571	0.011896	0.56727	84.065	1.7628	21.539	61.773	83.312	0.045748	0.16484	59

60	0.16479	0.046180	83.409	61.643	21.766	1.7909	83.944	0.55839	0.011913	57.737	72.433
61	0.16474	0.046612	83.505	61.512	21.993	1.8193	83.823	0.54967	0.011930	58.917	73.613
62	0.16470	0.047044	83.601	61.380	22.221	1.8480	83.701	0.54112	0.011947	60.111	74.807
63	0.16465	0.047475	83.696	61.248	22.448	1.8771	83.580	0.53273	0.011965	61.320	76.016
64	0.16460	0.047905	83.792	61.116	22.676	1.9066	83.457	0.52450	0.011982	62.543	77.239
65	0.16456	0.048336	83.887	60.982	22.905	1.9364	83.335	0.51642	0.012000	63.781	78.477
66	0.16451	0.048765	83.982	60.849	23.133	1.9666	83.212	0.50848	0.012017	65.033	79.729
67	0.16447	0.049195	84.077	60.715	23.362	1.9972	83.089	0.50070	0.012035	66.300	80.996
68	0.16442	0.049624	84.171	60.580	23.591	2.0282	82.965	0.49305	0.012053	67.583	82.279
69	0.16438	0.050053	84.266	60.445	23.821	2.0595	82.841	0.48555	0.012071	68.880	83.576
70	0.16434	0.050482	84.359	60.309	24.050	2.0913	82.717	0.47818	0.012089	70.192	84.888
71	0.16429	0.050910	84.453	60.172	24.281	2.1234	82.592	0.47094	0.012108	71.520	86.216
72	0.16425	0.051338	84.546	60.035	24.511	2.1559	82.467	0.46383	0.012126	72.863	87.559
73	0.16421	0.051766	84.639	59.898	24.741	2.1889	82.341	0.45686	0.012145	74.222	88.918
74	0.16417	0.052193	84.732	59.759	24.973	2.2222	82.215	0.45000	0.012163	75.596	90.292
75	0.16412	0.052620	84.825	59.621	25.204	2.2560	82.089	0.44327	0.012182	76.986	91.682
76	0.16408	0.053047	84.916	59.481	25.435	2.2901	81.962	0.43666	0.012201	78.391	93.087
77	0.16404	0.053473	85.008	59.341	25.667	2.3247	81.835	0.43016	0.012220	79.813	94.509
78	0.16400	0.053900	85.100	59.201	25.899	2.3597	81.707	0.42378	0.012239	81.250	95.946
79	0.16396	0.054326	85.191	59.059	26.132	2.3951	81.579	0.41751	0.012258	82.704	97.400
80	0.16392	0.054751	85.282	58.917	26.365	2.4310	81.450	0.41135	0.012277	84.174	98.870
81	0.16388	0.055177	85.373	58.775	26.598	2.4673	81.322	0.40530	0.012297	85.66	100.36
82	0.16384	0.055602	85.463	58.631	26.832	2.5041	81.192	0.39935	0.012316	87.16	101.86
83	0.16380	0.056027	85.553	58.488	27.065	2.5413	81.063	0.39351	0.012336	88.68	103.38
84	0.16376	0.056452	85.643	58.343	27.300	2.5789	80.932	0.38776	0.012356	90.22	104.92
85	0.16372	0.056877	85.732	58.198	27.534	2.6170	80.802	0.38212	0.012376	91.77	106.47
86	0.16368	0.057301	85.821	58.052	27.769	2.6556	80.671	0.37657	0.012396	93.34	108.04
87	0.16364	0.057725	85.910	57.905	28.005	2.6946	80.539	0.37111	0.012416	94.93	109.63
88	0.16360	0.058149	85.998	57.757	28.241	2.7341	80.407	0.36575	0.012437	96.53	111.23
89	0.16357	0.058573	86.086	57.609	28.477	2.7741	80.275	0.36047	0.012457	98.15	112.85
90	0.16353	0.058997	86.174	57.461	28.713	2.8146	80.142	0.35529	0.012478	99.79	114.49
91	0.16349	0.059420	86.261	57.311	28.950	2.8556	80.008	0.35019	0.012499	101.45	116.15

Table A-2-1 Saturation Properties—Temperature Table (cont.)

Temp. °F	Pressure Psia	Pressure Psig	Volume Liquid v_f	Volume Vapor v_g	Density Liquid $1/v_f$	Density Vapor $1/v_g$	Enthalpy Liquid h_f	Enthalpy Latent h_{fg}	Enthalpy Vapor h_g	Entropy Liquid s_f	Entropy Vapor s_g	Temp. °F
92	117.82	103.12	0.012520	0.34518	79.874	2.8970	29.187	57.161	86.348	0.059844	0.16345	92
93	119.51	104.81	0.012541	0.34025	79.740	2.9390	29.425	57.009	86.434	0.060267	0.16341	93
94	121.22	106.52	0.012562	0.33540	79.605	2.9815	29.663	56.858	86.521	0.060690	0.16338	94
95	122.95	108.25	0.012583	0.33063	79.470	3.0245	29.901	56.705	86.606	0.061113	0.16334	95
96	124.70	110.00	0.012605	0.32594	79.334	3.0680	30.140	56.551	86.691	0.061536	0.16330	96
97	126.46	111.76	0.012627	0.32133	79.198	3.1120	30.380	56.397	86.777	0.061959	0.16326	97
98	128.24	113.54	0.012649	0.31679	79.061	3.1566	30.619	56.242	86.861	0.062381	0.16323	98
99	130.04	115.34	0.012671	0.31233	78.923	3.2017	30.859	56.086	86.945	0.062804	0.16319	99
100	131.86	117.16	0.012693	0.30794	78.785	3.2474	31.100	55.929	87.029	0.063227	0.16315	100
101	133.70	119.00	0.012715	0.30362	78.647	3.2936	31.341	55.772	87.113	0.063649	0.16312	101
102	135.56	120.86	0.012738	0.29937	78.508	3.3404	31.583	55.613	87.196	0.064072	0.16308	102
103	137.44	122.74	0.012760	0.29518	78.368	3.3877	31.824	55.454	87.278	0.064494	0.16304	103
104	139.33	124.63	0.012783	0.29106	78.228	3.4357	32.067	55.293	87.360	0.064916	0.16301	104
105	141.25	126.55	0.012806	0.28701	78.088	3.4842	32.310	55.132	87.442	0.065339	0.16297	105
106	143.18	128.48	0.012829	0.28303	77.946	3.5333	32.553	54.970	87.523	0.065761	0.16293	106
107	145.13	130.43	0.012853	0.27910	77.804	3.5829	32.797	54.807	87.604	0.066184	0.16290	107
108	147.11	132.41	0.012876	0.27524	77.662	3.6332	33.041	54.643	87.684	0.066606	0.16286	108
109	149.10	134.40	0.012900	0.27143	77.519	3.6841	33.286	54.478	87.764	0.067028	0.16282	109
110	151.11	136.41	0.012924	0.26769	77.376	3.7357	33.531	54.313	87.844	0.067451	0.16279	110
111	153.14	138.44	0.012948	0.26400	77.231	3.7878	33.777	54.146	87.923	0.067873	0.16275	111
112	155.19	140.49	0.012972	0.26037	77.087	3.8406	34.023	53.978	88.001	0.068296	0.16271	112
113	157.27	142.57	0.012997	0.25680	76.941	3.8941	34.270	53.809	88.079	0.068719	0.16268	113
114	159.36	144.66	0.013022	0.25328	76.795	3.9482	34.517	53.639	88.156	0.069141	0.16264	114
115	161.47	146.77	0.013047	0.24982	76.649	4.0029	34.765	53.468	88.233	0.069564	0.16260	115
116	163.61	148.91	0.013072	0.24641	76.501	4.0584	35.014	53.296	88.310	0.069987	0.16256	116

117	0.16253	0.070410	88.386	53.123	35.263	4.1145	76.353	0.24304	0.013097	151.06	165.76	117
118	0.16249	0.070833	88.461	52.949	35.512	4.1713	76.205	0.23974	0.013123	153.24	167.94	118
119	0.16245	0.071257	88.536	52.774	35.762	4.2288	76.056	0.23647	0.013148	155.43	170.13	119
120	0.16241	0.071680	88.610	52.597	36.013	4.2870	75.906	0.23326	0.013174	157.65	172.35	120
121	0.16237	0.072104	88.684	52.420	36.264	4.3459	75.755	0.23010	0.013200	159.89	174.59	121
122	0.16234	0.072528	88.757	52.241	36.516	4.4056	75.604	0.22698	0.013227	162.15	176.85	122
123	0.16230	0.072952	88.830	52.062	36.768	4.4660	75.452	0.22391	0.013254	164.43	179.13	123
124	0.16226	0.073376	88.902	51.881	37.021	4.5272	75.299	0.22089	0.013280	166.73	181.43	124
125	0.16222	0.073800	88.973	51.698	37.275	4.5891	75.145	0.21791	0.013308	169.06	183.76	125
126	0.16218	0.074225	89.044	51.515	37.529	4.6518	74.991	0.21497	0.013335	171.40	186.10	126
127	0.16214	0.074650	89.115	51.330	37.785	4.7153	74.836	0.21207	0.013363	173.77	188.47	127
128	0.16210	0.075075	89.184	51.144	38.040	4.7796	74.680	0.20922	0.013390	176.16	190.86	128
129	0.16206	0.075501	89.253	50.957	38.296	4.8448	74.524	0.20641	0.013419	178.57	193.27	129
130	0.16202	0.075927	89.321	50.768	38.553	4.9107	74.367	0.20364	0.013447	181.01	195.71	130
131	0.16198	0.076353	89.389	50.578	38.811	4.9775	74.209	0.20091	0.013476	183.46	198.16	131
132	0.16194	0.076779	89.456	50.387	39.069	5.0451	74.050	0.19821	0.013504	185.94	200.64	132
133	0.16189	0.077206	89.522	50.194	39.328	5.1136	73.890	0.19556	0.013534	188.45	203.15	133
134	0.16185	0.077633	89.588	50.000	39.588	5.1829	73.729	0.19294	0.013563	190.97	205.67	134
135	0.16181	0.078061	89.653	49.805	39.848	5.2532	73.568	0.19036	0.013593	193.52	208.22	135
136	0.16177	0.078489	89.718	49.608	40.110	5.3244	73.406	0.18782	0.013623	196.09	210.79	136
137	0.16172	0.078917	89.781	49.409	40.372	5.3965	73.243	0.18531	0.013653	198.69	213.39	137
138	0.16168	0.079346	89.844	49.210	40.634	5.4695	73.079	0.18283	0.013684	201.31	216.01	138
139	0.16163	0.079775	89.906	49.008	40.898	5.5435	72.914	0.18039	0.013715	203.95	218.65	139
140	0.16159	0.080205	89.967	48.805	41.162	5.6184	72.748	0.17799	0.013746	206.62	221.32	140
141	0.16154	0.080635	90.028	48.601	41.427	5.6944	72.581	0.17561	0.013778	209.30	224.00	141
142	0.16150	0.081065	90.087	48.394	41.693	5.7713	72.413	0.17327	0.013810	212.02	226.72	142
143	0.16145	0.081497	90.146	48.187	41.959	5.8493	72.244	0.17096	0.013842	214.76	229.46	143
144	0.16140	0.081928	90.204	47.977	42.227	5.9283	72.075	0.16868	0.013874	217.52	232.22	144
145	0.16135	0.082361	90.261	47.766	42.495	6.0083	71.904	0.16644	0.013907	220.30	235.00	145
146	0.16130	0.082794	90.318	47.553	42.765	6.0895	71.732	0.16422	0.013941	223.12	237.82	146
147	0.16125	0.083227	90.373	47.338	43.035	6.1717	71.559	0.16203	0.013974	225.95	240.65	147
148	0.16120	0.083661	90.428	47.122	43.306	6.2551	71.386	0.15987	0.014008	228.81	243.51	148
149	0.16115	0.084096	90.482	46.904	43.578	6.3395	71.211	0.15774	0.014043	231.70	246.40	149

Table A-2-1 Saturation Properties—Temperature Table (cont.)

Temp. °F	Pressure Psia	Pressure Psig	Volume ft³/lbm Liquid v_f	Volume ft³/lbm Vapor v_g	Density lbm/ft³ Liquid $1/v_f$	Density lbm/ft³ Vapor $1/v_g$	Enthalpy Btu/lbm Liquid h_f	Enthalpy Btu/lbm Latent h_{fg}	Enthalpy Btu/lbm Vapor h_g	Entropy Btu/lbm R Liquid s_f	Entropy Btu/lbm R Vapor s_g	Temp. °F
150	249.31	234.61	0.014078	0.15564	71.035	6.4252	43.850	46.684	90.534	0.084531	0.16110	150
151	252.24	237.54	0.014113	0.15356	70.857	6.5120	44.124	46.462	90.586	0.084967	0.16105	151
152	255.20	240.50	0.014148	0.15151	70.679	6.6001	44.399	46.238	90.637	0.085404	0.16099	152
153	258.19	243.49	0.014184	0.14949	70.500	6.6893	44.675	46.012	90.687	0.085842	0.16094	153
154	261.20	246.50	0.014221	0.14750	70.319	6.7799	44.951	45.784	90.735	0.086280	0.16088	154
155	264.24	249.54	0.014258	0.14552	70.137	6.8717	45.229	45.554	90.783	0.086719	0.16083	155
156	267.30	252.60	0.014295	0.14358	69.954	6.9648	45.508	45.322	90.830	0.087159	0.16077	156
157	270.39	255.69	0.014333	0.14166	69.770	7.0592	45.787	45.088	90.875	0.087600	0.16071	157
158	273.51	258.81	0.014371	0.13976	69.584	7.1551	46.068	44.852	90.920	0.088041	0.16065	158
159	276.65	261.95	0.014410	0.13789	69.397	7.2523	46.350	44.614	90.964	0.088484	0.16059	159
160	279.82	265.12	0.014449	0.13604	69.209	7.3509	46.633	44.373	91.006	0.088927	0.16053	160
161	283.02	268.32	0.014489	0.13421	69.019	7.4510	46.917	44.130	91.047	0.089371	0.16047	161
162	286.24	271.54	0.014529	0.13241	68.828	7.5525	47.202	43.885	91.087	0.089817	0.16040	162
163	289.49	274.79	0.014570	0.13062	68.635	7.6556	47.489	43.637	91.126	0.090263	0.16034	163
164	292.77	278.07	0.014611	0.12886	68.441	7.7602	47.777	43.386	91.163	0.090710	0.16027	164
165	296.07	281.37	0.014653	0.12712	68.245	7.8665	48.065	43.134	91.199	0.091159	0.16021	165
166	299.40	284.70	0.014695	0.12540	68.048	7.9743	48.355	42.879	91.234	0.091608	0.16014	166
167	302.76	288.06	0.014738	0.12370	67.850	8.0838	48.647	42.620	91.267	0.092059	0.16007	167
168	306.15	291.45	0.014782	0.12202	67.649	8.1950	48.939	42.360	91.299	0.092511	0.16000	168
169	309.56	294.86	0.014826	0.12037	67.447	8.3080	49.233	42.097	91.330	0.092964	0.15992	169
170	313.00	298.30	0.014871	0.11873	67.244	8.4228	49.529	41.830	91.359	0.093418	0.15985	170
171	316.47	301.77	0.014917	0.11710	67.038	8.5394	49.825	41.562	91.387	0.093874	0.15977	171
172	319.97	305.27	0.014963	0.11550	66.831	8.6579	50.123	41.290	91.413	0.094330	0.15969	172
173	323.50	308.80	0.015010	0.11392	66.622	8.7783	50.423	41.015	91.438	0.094789	0.15961	173
174	327.06	312.36	0.015058	0.11235	66.411	8.9007	50.724	40.736	91.460	0.095248	0.15953	174

175	0.15945	0.095709	91.481	40.455	51.026	9.0252	66.198	0.11080	0.015106	315.94	330.64	175
176	0.15936	0.096172	91.501	40.171	51.330	9.1518	65.983	0.10927	0.015155	319.55	334.25	176
177	0.15928	0.096636	91.519	39.883	51.636	9.2805	65.766	0.10775	0.015205	323.20	337.90	177
178	0.15919	0.097102	91.535	39.592	51.943	9.4114	65.547	0.10625	0.015256	326.87	341.57	178
179	0.15910	0.097569	91.549	39.297	52.252	9.5446	65.326	0.10477	0.015308	330.57	345.27	179
180	0.15900	0.098039	91.561	38.999	52.562	9.6802	65.102	0.10330	0.015360	334.30	349.00	180
181	0.15891	0.098509	91.571	38.697	52.874	9.8182	64.877	0.10185	0.015414	338.06	352.76	181
182	0.15881	0.098982	91.579	38.391	53.188	9.9587	64.649	0.10041	0.015468	341.85	356.55	182
183	0.15871	0.099457	91.585	38.081	53.504	10.102	64.418	0.098992	0.015524	345.68	360.38	183
184	0.15861	0.099933	91.589	37.767	53.822	10.248	64.185	0.097584	0.015580	349.53	364.23	184
185	0.15850	0.10041	91.590	37.449	54.141	10.396	63.949	0.096190	0.015637	353.41	368.11	185
186	0.15839	0.10089	91.590	37.127	54.463	10.547	63.711	0.094810	0.015696	357.32	372.02	186
187	0.15828	0.10138	91.579	36.800	54.786	10.702	63.470	0.093443	0.015756	361.26	375.96	187
188	0.15817	0.10186	91.580	36.469	55.111	10.859	63.225	0.092089	0.015816	365.24	379.94	188
189	0.15805	0.10235	91.572	36.133	55.439	11.020	62.978	0.090747	0.015878	369.24	383.94	189
190	0.15793	0.10284	91.561	35.792	55.769	11.183	62.728	0.089418	0.015942	373.28	387.98	190
191	0.15780	0.10333	91.548	35.447	56.101	11.351	62.475	0.088101	0.016006	377.35	392.05	191
192	0.15768	0.10382	91.531	35.096	56.435	11.521	62.218	0.086796	0.016073	381.44	396.14	192
193	0.15755	0.10432	91.511	34.739	56.772	11.696	61.958	0.085502	0.016140	385.57	400.27	193
194	0.15741	0.10482	91.488	34.377	57.111	11.874	61.694	0.084218	0.016209	389.74	404.44	194
195	0.15727	0.10532	91.462	34.009	57.453	12.056	61.426	0.082946	0.016280	393.93	408.63	195
196	0.15713	0.10583	91.433	33.636	57.797	12.242	61.155	0.081683	0.016352	398.16	412.86	196
197	0.15698	0.10634	91.400	33.256	58.144	12.433	60.879	0.080431	0.016426	402.42	417.12	197
198	0.15683	0.10685	91.363	32.869	58.494	12.628	60.599	0.079188	0.016502	406.71	421.41	198
199	0.15667	0.10737	91.323	32.476	58.847	12.828	60.315	0.077953	0.016580	411.03	425.73	199
200	0.15651	0.10789	91.278	32.075	59.203	13.033	60.026	0.076728	0.016659	415.39	430.09	200
201	0.15634	0.10841	91.230	31.668	59.562	13.243	59.732	0.075511	0.016741	419.78	434.48	201
202	0.15617	0.10894	91.176	31.252	59.924	13.459	59.433	0.074301	0.016826	424.21	438.91	202
203	0.15599	0.10947	91.118	30.828	60.290	13.680	59.128	0.073099	0.016912	428.66	443.36	203
204	0.15580	0.11001	91.055	30.396	60.659	13.908	58.818	0.071903	0.017002	433.15	447.85	204
205	0.15561	0.11055	90.987	29.955	61.032	14.141	58.502	0.070714	0.017094	437.68	452.38	205

Table A-2-1 Saturation Properties—Temperature Table (cont.)

Temp. °F	Pressure Psia	Pressure Psig	Volume ft³/lbm Liquid v_f	Volume ft³/lbm Vapor v_g	Density lbm/ft³ Liquid $1/v_f$	Density lbm/ft³ Vapor $1/v_g$	Enthalpy Btu/lbm Liquid h_f	Enthalpy Btu/lbm Latent h_{fg}	Enthalpy Btu/lbm Vapor h_g	Entropy Btu/lbm R Liquid s_f	Entropy Btu/lbm R Vapor s_g	Temp. °F
206	456.94	442.24	0.017188	0.069531	58.179	14.382	61.409	29.505	90.914	0.11109	0.15541	206
207	461.53	446.83	0.017286	0.068353	57.849	14.630	61.790	29.045	90.835	0.11164	0.15521	207
208	466.16	451.46	0.017387	0.067179	57.513	14.886	62.175	28.574	90.749	0.11220	0.15499	208
209	470.82	456.12	0.017492	0.066009	57.168	15.149	62.565	28.092	90.657	0.11276	0.15477	209
210	475.52	460.82	0.017601	0.064843	56.816	15.422	62.959	27.599	90.558	0.11332	0.15453	210
211	480.25	465.55	0.017713	0.063679	56.455	15.704	63.359	27.093	90.452	0.11390	0.15429	211
212	485.01	470.31	0.017830	0.062517	56.084	15.996	63.764	26.573	90.337	0.11448	0.15404	212
213	489.82	475.12	0.017952	0.061355	55.703	16.299	64.174	26.040	90.214	0.11506	0.15377	213
214	494.65	479.95	0.018079	0.060193	55.312	16.613	64.591	25.490	90.081	0.11566	0.15349	214
215	499.53	484.83	0.018212	0.059030	54.908	16.941	65.014	24.925	89.939	0.11626	0.15320	215
216	504.44	489.74	0.018351	0.057864	54.492	17.282	65.444	24.341	89.785	0.11687	0.15290	216
217	509.38	494.68	0.018497	0.056694	54.062	17.639	65.881	23.738	89.619	0.11749	0.15257	217

218	0.15223	0.11813	89.440	23.113	66.327	18.012	53.616	0.055518	0.018651	499.66	514.36	218
219	0.15187	0.11877	89.247	22.465	66.782	18.405	53.153	0.054334	0.018814	504.68	519.38	219
220	0.15149	0.11943	89.036	21.790	67.246	18.818	52.670	0.053140	0.018986	509.73	524.43	220
221	0.15108	0.12010	88.808	21.086	67.722	19.255	52.167	0.051934	0.019169	514.82	529.52	221
222	0.15064	0.12079	88.559	20.350	68.209	19.720	51.638	0.050711	0.019365	519.95	534.65	222
223	0.15017	0.12150	88.286	19.575	68.711	20.215	51.082	0.049468	0.019576	525.12	539.82	223
224	0.14966	0.12223	87.985	18.757	69.228	20.747	50.494	0.048200	0.019804	530.32	545.02	224
225	0.14911	0.12298	87.651	17.888	69.763	21.322	49.868	0.046900	0.020053	535.56	550.26	225
226	0.14850	0.12377	87.278	16.958	70.320	21.949	49.196	0.045559	0.020327	540.84	555.54	226
227	0.14782	0.12459	86.857	15.953	70.904	22.642	48.468	0.044166	0.020632	546.15	560.85	227
228	0.14705	0.12545	86.373	14.854	71.519	23.418	47.669	0.042702	0.020978	551.50	566.20	228
229	0.14617	0.12638	85.806	13.629	72.177	24.307	46.778	0.041140	0.021378	556.90	571.60	229
230	0.14512	0.12739	85.122	12.229	72.893	25.358	45.758	0.039435	0.021854	562.33	577.03	230
231	0.14380	0.12852	84.249	10.553	73.696	26.672	44.544	0.037492	0.022450	567.80	582.50	231
232	0.14191	0.12987	82.986	8.335	74.651	28.538	42.988	0.035041	0.023262	573.31	588.01	232
233.6 (Critical)	0.1359	0.1359	78.86	0	78.86	34.84	34.84	0.02870	0.02870	582.2	596.9	233.6 (Critical)

Source: Copyright 1955–56 by DuPont Company. Used by permission.,

Table A-2-2 Superheated Freon-12 Table

Temp., F	5 lbf/in.^{2}			10 lbf/in.^{2}			15 lbf/in.^{2}		
	v	h	s	v	h	s	v	h	s
0	8.0611	78.582	0.19663	3.9809	78.246	0.18471	2.6201	77.902	0.17751
20	8.4265	81.309	0.20244	4.1691	81.014	0.19061	2.7494	80.712	0.18319
40	8.7903	84.090	0.20812	4.3556	83.828	0.19635	2.8770	83.561	0.18931
60	9.1528	86.922	0.21367	4.5408	86.689	0.20197	3.0031	86.451	0.19498
80	9.5142	89.806	0.21912	4.7248	89.596	0.20746	3.1281	89.393	0.20051
100	9.8747	92.738	0.22445	4.9079	92.548	0.21283	3.2521	92.357	0.20593
120	10.234	95.717	0.22968	5.0903	95.546	0.21809	3.3754	95.373	0.21122
140	10.594	98.743	0.23481	5.2720	98.586	0.22325	3.4981	98.429	0.21640
160	10.952	101.812	0.23985	5.4533	101.669	0.22830	3.6202	101.525	0.22148
180	11.311	104.925	0.24479	5.6341	104.793	0.23326	3.7419	104.661	0.22646
200	11.668	108.079	0.24964	5.8145	107.957	0.23813	3.8632	107.835	0.23135
220	12.026	111.272	0.25441	5.9946	111.159	0.24291	3.9841	111.046	0.23614

Temp., F	20 lbf/in.^{2}			25 lbf/in.^{2}			30 lbf/in.^{2}		
	v	h	s	v	h	s	v	h	s
20	2.0391	80.403	0.17829	1.6125	80.088	0.17414	1.3278	79.765	0.17065
40	2.1373	83.289	0.18419	1.6932	83.012	0.18012	1.3969	82.730	0.17671
60	2.2340	86.210	0.18992	1.7723	85.965	0.18591	1.4644	85.716	0.18257
80	2.3295	89.169	0.19550	1.8502	88.950	0.19155	1.5306	88.729	0.18826
100	2.4241	92.164	0.20095	1.9271	91.068	0.19704	1.5957	91.770	0.19379
120	2.5179	95.198	0.20628	2.0032	95.021	0.20240	1.6600	94.843	0.19918
140	2.6110	98.270	0.21149	2.0786	98.110	0.20763	1.7237	97.948	0.20445
160	2.7036	101.380	0.21659	2.1535	101.234	0.21276	1.7868	101.086	0.20960
180	2.7957	104.528	0.22159	2.2279	104.393	0.21778	1.8494	104.258	0.21463
200	2.8874	107.712	0.22649	2.3019	107.588	0.22269	1.9116	107.464	0.21957

T	v	h	s	v	h	s	v	h	s
220	2.9789	110.932	0.23130	2.3756	110.817	0.22752	1.9735	110.702	0.22440
240	3.0700	114.186	0.23602	2.4491	114.080	0.23225	2.0351	113.973	0.22915

	35 lbf/in.²			40 lbf/in.²			50 lbf/in.²		
T	v	h	s	v	h	s	v	h	s
40	1.1850	82.442	0.17375	1.0258	82.148	0.17112	0.80248	81.540	0.16655
60	1.2442	85.463	0.17968	1.0789	85.206	0.17712	0.84713	84.676	0.17271
80	1.3021	88.504	0.18542	1.1306	88.277	0.18292	0.89025	87.811	0.17862
100	1.3589	91.570	0.19100	1.1812	91.367	0.18854	0.93216	90.953	0.18431
120	1.4148	94.663	0.19643	1.2309	94.480	0.19401	0.97313	94.110	0.18988
140	1.4701	97.785	0.20172	1.2798	97.620	0.19933	1.0133	97.286	0.19527
160	1.5248	100.938	0.20689	1.3282	100.788	0.20453	1.0529	100.485	0.20051
180	1.5789	104.122	0.21195	1.3761	103.985	0.20961	1.0920	103.708	0.20563
200	1.6327	107.338	0.21690	1.4236	107.212	0.21457	1.1307	106.958	0.21064
220	1.6862	110.586	0.22175	1.4707	110.469	0.21944	1.1690	110.235	0.21553
240	1.7394	113.865	0.22651	1.5176	113.757	0.22420	1.2070	113.539	0.22032
260	1.7923	117.175	0.23117	1.5642	117.074	0.22888	1.2447	116.871	0.22502

	60 lbf/in.²			70 lbf/in.²			80 lbf/in.²		
T	v	h	s	v	h	s	v	h	s
60	0.69210	84.126	0.16892	0.58088	83.552	0.16556	⋮	⋮	⋮
80	0.72964	87.330	0.17497	0.61458	86.832	0.17175	0.52795	86.316	0.16885
100	0.76588	90.528	0.18079	0.64685	90.091	0.17768	0.55734	89.640	0.17489
120	0.80110	93.731	0.18641	0.67803	93.343	0.18339	0.58556	92.945	0.18070
140	0.83551	96.945	0.19186	0.70836	96.597	0.18891	0.61286	96.242	0.18629
160	0.86928	100.776	0.19716	0.73800	99.862	0.19427	0.63943	99.512	0.19170
180	0.90252	103.427	0.20233	0.76708	103.141	0.19918	0.66543	102.851	0.19696
200	0.93531	106.700	0.20736	0.79571	106.439	0.20455	0.69095	106.174	0.20207
220	0.96775	109.997	0.21229	0.82397	109.756	0.20951	0.71609	109.513	0.20706
240	0.99988	113.319	0.21710	0.85191	113.096	0.21435	0.74090	112.872	0.21193
260	1.0318	116.666	0.22182	0.87959	116.459	0.21909	0.76544	116.251	0.21669
280	1.0634	120.039	0.22644	0.90705	119.846	0.22373	0.78975	119.652	0.22135

$v\ [=]\ \text{ft}^3/\text{lbm}$ $h\ [=]\ \text{Btu/lbm}$ $s\ [=]\ \text{Btu/lbmR}$

Table A-2-2 Superheated Freon-12 Table (cont.)

Temp., F	90 lbf/in.² v	h	s	100 lbf/in.² v	h	s	125 lbf/in.² v	h	s
100	0.48749	89.175	0.17234	0.43138	88.694	0.16996	0.32943	87.407	0.16455
120	0.51346	92.536	0.17824	0.45562	92.116	0.17597	0.35086	91.008	0.17087
140	0.53845	95.879	0.18391	0.47881	95.507	0.18172	0.37098	94.023	0.17686
160	0.56268	99.216	0.18938	0.50118	98.884	0.18726	0.39015	98.023	0.18258
180	0.58629	102.557	0.19469	0.52291	102.257	0.19262	0.40857	101.484	0.18807
200	0.60941	105.905	0.19984	0.54413	105.633	0.19782	0.42642	104.934	0.19338
220	0.63213	109.267	0.20486	0.56492	109.018	0.20287	0.44380	108.380	0.19853
240	0.65451	112.644	0.20976	0.58538	112.415	0.20780	0.46081	111.829	0.20353
260	0.67662	116.040	0.21455	0.60554	115.828	0.21261	0.47750	115.287	0.20840
280	0.69849	119.456	0.21923	0.62546	119.259	0.21731	0.49394	118.756	0.21316
300	0.72016	122.892	0.22381	0.64518	122.707	0.22191	0.51016	122.238	0.21780
320	0.74166	126.349	0.22830	0.66472	126.176	0.22641	0.52619	125.737	0.22235

Temp., F	150 lbf/in.² v	h	s	175 lbf/in.² v	h	s	200 lbf/in.² v	h	s
120	0.28007	89.800	0.16629	...	...	...	...	...	...
140	0.29815	93.498	0.17256	0.24595	92.373	0.16859	0.20579	91.137	0.1480
160	0.31566	97.112	0.17819	0.26198	96.142	0.17478	0.22121	95.100	0.17130
180	0.33200	100.675	0.18415	0.27697	99.823	0.18062	0.23535	98.921	0.17737
200	0.34769	104.206	0.18958	0.29120	103.447	0.18620	0.24860	102.652	0.18311
220	0.36285	107.720	0.19483	0.30485	107.036	0.19156	0.26117	106.325	0.18860
240	0.3761	111.226	0.19992	0.31804	110.605	0.19674	0.27323	109.962	0.19387
260	0.39203	114.732	0.20485	0.33087	114.162	0.20175	0.28189	113.576	0.19896
280	0.40617	118.242	0.20967	0.34339	117.717	0.20662	0.29623	117.178	0.20390
300	0.42008	121.761	0.21436	0.35567	121.273	0.21137	0.30730	120.775	0.20870
320	0.43379	125.290	0.21894	0.36773	124.835	0.21599	0.31815	124.373	0.21337
340	0.44733	128.833	0.22343	0.37963	128.407	0.22052	0.32881	127.974	0.21793

Temp	250 lbf/in.² v	250 h	250 s	300 lbf/in.² v	300 h	300 s	400 lbf/in.² v	400 h	400 s
160	0.16249	92.717	0.16462	. . .	. . .	. . .	. . .	. . .	. . .
180	0.17605	96.925	0.17130	0.13482	94.556	0.16537	. . .	. . .	. . .
200	0.18824	100.930	0.17747	0.14697	98.975	0.17217	0.091005	93.718	0.16092
220	0.19952	104.809	0.18326	0.15774	103.136	0.17838	0.10316	99.046	0.16888
240	0.21014	108.607	0.18877	0.16761	107.140	0.18419	0.11300	103.735	0.17568
260	0.22027	112.351	0.19404	0.17685	111.043	0.18969	0.12163	108.105	0.18183
280	0.23001	116.060	0.19913	0.18562	114.879	0.19495	0.12949	112.286	0.18756
300	0.23944	119.747	0.20405	0.19402	118.670	0.20000	0.13680	116.343	0.19298
320	0.24862	123.420	0.20882	0.20214	122.430	0.20489	0.14372	120.318	0.19814
340	0.25759	127.088	0.21346	0.21002	126.171	0.20963	0.15032	124.235	0.20310
360	0.26639	130.754	0.21799	0.21770	129.900	0.21423	0.15668	128.112	0.20789
380	0.27504	134.423	0.22241	0.22522	133.624	0.21872	0.16285	131.961	0.21250

Temp	500 lbf/in.² v	500 h	500 s	600 lbf/in.² v	600 h	600 s
220	0.064207	92.397	0.15683	. . .	. . .	. . .
240	0.077620	99.218	0.16672	0.047488	91.024	0.15335
260	0.087054	104.526	0.17421	0.061922	99.741	0.16566
280	0.094923	109.277	0.18072	0.070859	105.637	0.17374
300	0.10190	113.729	0.18666	0.078059	110.729	0.18053
320	0.10829	117.997	0.19221	0.084333	115.420	0.18603
340	0.11426	122.143	0.19746	0.090017	119.871	0.19227
360	0.11992	126.205	0.20217	0.095289	124.167	0.19757
380	0.12533	130.207	0.20730	0.10025	128.355	0.20262
400	0.13054	134.166	0.21196	0.10498	132.466	0.20746
420	0.13559	138.096	0.21648	0.10952	136.523	0.21213
440	0.14051	142.004	0.22087	0.11391	140.539	0.21664

Appendix A-3: Properties

Table A-3-1 Properties of Solids

(Values are for room temperature unless otherwise noted in brackets)

Material Description	Specific Heat Btu/(lb)(F)	Density lb/ft³	Thermal Conductivity Btuh/(ft²)(F/ft)	Emissivity Ratio	Surface Condition
Aluminum (alloy 1100)	0.214	171	128	0.09	commercial sheet
				0.20	heavily oxidized
Aluminum Bronze (76%Cu, 22%Zn, 2% Al)	0.09	517	58		
Alundum (aluminum oxide)	0.186				
Asbestos: fiber	0.25	150	0.097		
insulation	0.20	36	0.092		
Ashes, wood	0.20	40	0.041 [122]	0.93	"paper"
Asphalt	0.22	132	0.43		
Bakelite	0.35	81	9.7		
Bell Metal	0.086 [122]				
Bismuth Tin	0.040		37.6		
Brick, building	0.2	123	0.4	0.93	
Brass:					
red (85% Cu, 15% Zn)	0.09	548	87	0.030	highly polished
yellow (65% Cu, 35% Zn)	0.09	519	69	0.033	highly polished
Bronze	0.104	530	17 [32]		
Cadmium	0.055	540	53.7	0.02	
Carbon (gas retort)	0.17		0.20 [2]	0.81	
Cardboard			0.04		
Cellulose	0.32	3.4	0.033		
Cement (Portland clinker)	0.16	120	0.017		
Chalk	0.215	143	0.48	0.34	about 250 F
Charcoal (wood)	0.20	15	0.03		
Chrome Brick	0.17	200	0.03 [392]		
Clay	0.22	63	0.67		

Material					Remarks
Coal	0.3	90			
Coal Tars	0.35 [104]	75	0.098 [32]		
Coke (petroleum, powdered)	0.36 [752]	62	0.07		
Concrete (stone)	0.156 [392]	144	0.55 [752], 0.54		
Copper (electrolytic)	0.092	556	227	0.072	commercial, shiny
Cork (granulated)	0.485	5.4	0.028 [23]		
Cotton (fiber)	0.319	95	0.024		
Cryolite (AlF₃ · 3NaF)	0.253	181			
Diamond	0.147	151	27		
Earth (dry and packed)		95	0.037	0.41	
Felt		20.6	0.03		
Fireclay Brick	0.198 [212]	112	0.58 [392]	0.75	at 1832 F
Flourspar (CaF₂)	0.21	199	0.63		
German Silver (nickel silver)	0.09	545	19	0.135	polished
Glass:					
crown (soda-lime)	0.18	154	0.59 [200]	0.94	smooth
flint (lead)	0.117	267	0.79		
pyrex	0.20	139	0.59 [200]		
"wool"	0.157	3.25	0.022		
Gold	0.0312	1208	172	0.02	highly polished
Graphite:					
powder	0.165	117	0.106	0.75	
"Karbate" (impervious)	0.16	78	75	0.903	on a smooth plate
Gypsum	0.259	93	0.25	0.95	
Hemp (fiber)	0.323	57.5			
Ice: [32 F]	0.487		1.3		
[−4 F]	0.465		1.41		
Iron:					
cast	0.12 [212]	450	27.6 [129]	0.435	freshly turned
wrought		485	34.9	0.94	dull, oxidized
Lead	0.0309	707	20.1	0.28	gray, oxidized
Leather (sole)		62.4	0.092		
Limestone	0.217	103	0.54	0.36 to 0.90	at 145 to 380 F
Linen	0.055		0.05		
Litharge (lead monoxide)		490			
Magnesia:					
powdered	0.234 [212]	49.7	0.35 [117]		
light carbonate		13	0.34		
Magnesite Brick	0.222 [212]	158	2.2 [400]		
Magnesium	0.241	108	91	0.55	oxidized

Table A-3-1 Properties of Solids (cont.)

(Values are for room temperature unless otherwise noted in brackets)

Material Description	Specific Heat Btu/(lb)(F)	Density lb/ft³	Thermal Conductivity Btuh/(ft²)(F/ft)	Emissivity Ratio	Emissivity Surface Condition
Marble	0.21	162^b	1.5	0.931	light gray, polished
Nickel	0.105	555	34.4^u	0.045	electroplated, polished
Paints:					
White lacquer				0.80^n	on rough plate
White enamel				0.91^n	
Black lacquer				0.80^n	
Black shellac		63	0.15^u	0.91^n	"matte" finish
Flat black lacquer				0.96	
Aluminum lacquer				0.39	
Paper	0.32	58	0.075^b	0.92	on rough plate pasted on tinned plate
Paraffin	0.69	56	0.14 [32]		
Plaster		132	0.43 [167]	0.91	rough
Platinum	0.032	1340	39.9^u	0.054^b	polished
Porcelain	0.18	162	1.3^u	0.92	glazed
Pyrites (Copper)	0.131	262			
Pyrites (Iron)	0.136 [156]	310			
Rock Salt	0.219	136			
Rubber:					
Vulcanized (soft)	0.48	68.6	0.08^t	0.86	rough
(hard)		74.3	0.092^t	0.95	glossy
Sand	0.191	94.6	0.19^b		
Sawdust		12	0.03^b		
Silica	0.316	140	0.83 [200]		
Silver	0.0560	654	245	0.02	polished and at 440 F

Substance					
Snow (freshly fallen)		7	0.34		
(at 32 F)		31	1.3		
Steel (mild)	0.12	489	26.2		
Stone (quarried)	0.2	95		0.12	cleaned
Tar:					
pitch	0.59	67	0.51		
bituminous		75	0.41		
Tin	0.0556	455	37.5	0.06	bright and at 122 F
Tungsten	0.032	1210	116	0.032	filament at 80 F
Wood:					
Hardwoods:	0.45/0.65	23/70	0.065/0.148		
Ash, white		43	0.0992		
Elm, American		36	0.0884		
Hickory		50			
Mahogany		34			
Maple, sugar		45	0.075		
Oak, white	0.570	47	0.108	0.90	planed
Walnut, black		39	0.102		
Softwoods:		22/46	0.061/0.093		
Fir, white		27	0.068		
Pine, white		27	0.063		
Spruce		26	0.065		
Wool:					
Fiber	0.325	82			
Fabric		6.9/20.6	0.021/0.037		
Zinc:					
Cast	0.092	445	65	0.05	polished
Hot-rolled	0.094	445	62		
Galvanizing				0.23	fairly bright

Source: Reprinted with permission from the 1977 Fundamentals Volume, ASHRAE Handbook and Product Directory.

Table A-3-2 Thermal Properties of Gases (Atmospheric Pressure)

T, F	ρ, lbm/ft^3	c_p, Btu/lbm-F	$\mu \times 10^5$ lbm/ ft-sec	$\nu \times 10^3$, ft^2/sec	k, Btu/hr-ft-F	Pr	$\beta \times 10^3$, 1/R
Air							
0	0.086	0.239	0.110	0.130	0.0133	0.73	2.18
32	0.081	0.240	1.165	0.145	0.0140	0.72	2.03
100	0.071	0.240	1.285	0.180	0.0154	0.72	1.79
200	0.060	0.241	1.440	0.239	0.0174	0.72	1.52
300	0.052	0.243	1.610	0.306	0.0193	0.71	1.32
400	0.046	0.245	1.750	0.378	0.0212	0.689	1.16
500	0.0412	0.247	1.890	0.455	0.0231	0.683	1.04
600	0.0373	0.250	2.000	0.540	0.0250	0.685	0.943
700	0.0341	0.253	2.14	0.625	0.0268	0.690	0.862
800	0.0314	0.256	2.24	0.717	0.0286	0.697	0.794
900	0.0291	0.259	2.36	0.815	0.0303	0.705	0.735
1000	0.0271	0.262	2.47	0.917	0.0319	0.713	0.685
1500	0.0202	0.276	3.00	1.47	0.0400	0.739	0.510
2000	0.0161	0.286	3.45	2.14	0.0471	0.753	0.406

2500	0.0133	0.292	3.69	2.80	0.051	0.763	0.338
3000	0.0114	0.297	3.86	3.39	0.054	0.765	0.289
Steam							
212	0.0372	0.451	0.870	0.234	0.0145	0.96	1.49
300	0.0328	0.456	1.000	0.303	0.0171	0.95	1.32
400	0.0288	0.462	1.130	0.395	0.0200	0.94	1.16
500	0.0258	0.470	1.265	0.490	0.0228	0.94	1.04
600	0.0233	0.477	1.420	0.610	0.0257	0.94	0.943
700	0.0213	0.485	1.555	0.725	0.0288	0.93	0.862
800	0.0196	0.494	1.700	0.855	0.0321	0.92	0.794
900	0.0181	0.50	1.810	0.987	0.0355	0.91	0.735
1000	0.0169	0.51	1.920	1.13	0.0388	0.91	0.685
1200	0.0149	0.53	2.14	1.44	0.0457	0.88	0.603
1400	0.0133	0.55	2.36	1.78	0.053	0.87	0.537
1600	0.0120	0.56	2.58	2.14	0.061	0.87	0.485
1800	0.0109	0.58	2.81	2.58	0.068	0.88	0.442
2000	0.0100	0.60	3.03	3.03	0.076	0.86	0.406
2500	0.0083	0.64	3.58	4.30	0.096	0.86	0.338
3000	0.0071	0.67	4.00	5.75	0.114	0.86	0.289

Table A-3-3 Thermal Properties of Liquids

T, F	ρ lbm/ft^3	c_p, Btu/lbm-R	$\mu \times 10^3$ lbm/ ft-sec	k, Btu/hr-ft-R	Pr	$\beta \times 10^4$, 1/R
Water						
32	62.4	1.01	1.20	0.319	13.7	-0.37
40	62.4	1.00	1.04	0.325	11.6	0.20
50	62.4	1.00	0.88	0.332	9.55	0.49
60	62.3	0.999	0.76	0.340	8.03	0.85
70	62.3	0.998	0.658	0.347	6.82	1.2
80	62.2	0.998	0.578	0.353	5.89	1.5
90	62.1	0.997	0.514	0.359	5.13	1.8
100	62.0	0.998	0.458	0.364	4.52	2.0
150	61.2	1.00	0.292	0.384	2.74	3.1
200	60.1	1.00	0.205	0.394	1.88	4.0
250	58.8	1.01	0.158	0.396	1.45	4.8
300	57.3	1.03	0.126	0.395	1.18	6.0
350	55.6	1.05	0.105	0.391	1.02	6.9
400	53.6	1.08	0.091	0.381	0.927	8.0
450	51.6	1.12	0.080	0.367	0.876	9.0
500	49.0	1.19	0.071	0.349	0.87	10.0
550	45.9	1.31	0.064	0.325	0.93	11.0
600	42.4	1.51	0.058	0.292	1.09	12.0

Sodium

200	58.0	0.33	0.47	49.8	0.011	1.50
400	56.3	0.32	0.29	46.4	0.007	2.00
700	53.7	0.31	0.19	41.8	0.005	
1000	51.2	0.30	0.14	37.8	0.004	
1300	48.6	0.30	0.12	34.5	0.004	

Freon-12 (CCl_2F_2)

−40	94.8	0.211	0.284	0.040	5.4	
−20	93.0	0.214	0.250	0.040	4.8	
0	91.2	0.217	0.231	0.041	4.4	10.3
20	89.2	0.220	0.210	0.042	4.0	10.5
32	87.2	0.223	0.200	0.042	3.8	13.4
60	83.0	0.231	0.180	0.042	3.5	17.2
100	78.5	0.240	0.160	0.040	3.5	21.0
120	75.9	0.244	0.155	0.039	3.5	25.0

Light Oil

60	57.0	0.43	58.20	0.077	1170	3.8
80	56.8	0.44	27.80	0.077	570	3.8
100	56.0	0.46	15.30	0.076	340	3.9
150	54.3	0.48	5.30	0.075	122	4.0
200	54.0	0.51	2.50	0.074	62	4.2
250	53.0	0.52	1.39	0.074	35	4.4
300	51.8	0.54	.83	0.073	22	4.5

Table A-3-4 Approximate Values of c_p, c_v, and R

Gas*	c_p		c_v		R	
	$\dfrac{Btu}{lbm\text{-}R}$	$\dfrac{kJ}{kg \cdot K}$	$\dfrac{Btu}{lbm\text{-}R}$	$\dfrac{kJ}{kg \cdot K}$	$\dfrac{Btu}{lbm\text{-}R}$	$\dfrac{ft\text{-}lbf}{lbm\text{-}R}$
Air	0.240	1	0.171	0.716	0.0685	53.3
CO	0.250	1.04	0.179	0.746	0.0709	55.2
CO_2	0.204	0.85	0.159	0.665	0.0451	35.1
N_2	0.247	1.04	0.176	0.787	0.0708	55.1
O_2	0.220	0.917	0.158	0.661	0.0621	48.3

Source: Reprinted with permission from the 1977 *Fundamentals Volume, ASHRAE Handbook and Product Directory.*

*Assumed to be an ideal gas at low pressure.

Table A-3-5 Convective Surface Resistances

		Surface Emissivity					
		Non-reflective $\epsilon = 0.90$		Reflective $\epsilon = 0.20$		Reflective $\epsilon = 0.05$	
Position of Surface	Direction of Heat Flow	h_i	R	h_i	R	h_i	R
Still Air							
Horizontal	Upward	1.63	0.61	0.91	1.10	0.76	1.32
Sloping 45°	Upward	1.60	0.62	0.88	1.14	0.73	1.37
Vertical	Horizontal	1.46	0.68	0.74	1.35	0.59	1.70
Sloping 45°	Downward	1.32	0.76	0.60	1.67	0.45	2.22
Horizontal	Downward	1.08	0.92	0.37	2.70	0.22	4.55
		h_o	R	h_o	R	h_o	R
Moving Air (any position)							
15-mph wind (for winter)	Any	6.00	0.17				
$7\frac{1}{2}$-mph wind (for summer)	Any	4.00	0.25				

Source: Reprinted with permission from the 1977 *Fundamentals Volume, ASHRAE Handbook and Product Directory.*

Note: All conductance values expressed in Btu/hr-ft²-F.

Table A-3-6 Thermal Resistance of Air Spaces*

Position of Air Space	Direction of Heat Flow	Air Space Mean Temp., F	Air Space Temp. Diff., F	Thermal Resistance R ($\frac{3}{4}$-in. Air Space)					Thermal Resistance R (4-in. Air Space)				
				E = 0.03	E = 0.05	E = 0.2	E = 0.5	E = 0.82	E = 0.03	E = 0.05	E = 0.2	E = 0.5	E = 0.82
Horizontal	Up	90	10	2.39	2.26	1.63	1.05	0.76	2.94	2.75	1.87	1.14	0.80
		50	30	1.72	1.67	1.37	1.0	0.78	2.14	2.06	1.62	1.13	0.85
		50	10	2.33	2.23	1.71	1.16	0.87	2.88	2.73	1.99	1.29	0.94
		0	20	1.83	1.79	1.52	1.16	0.93	2.28	2.22	1.81	1.33	1.03
		0	10	2.23	2.16	1.78	1.31	1.02	2.77	2.67	2.11	1.48	1.12
		-50	20	1.73	1.71	1.52	1.25	1.05	2.17	2.12	1.84	1.46	1.20
		-50	10	2.11	2.07	1.81	1.44	1.18	2.64	2.57	2.18	1.66	1.33
Sloping 45°	Up	90	10	3.01	2.81	1.90	1.15	0.81	3.22	3.00	1.98	1.18	0.82
		50	30	2.02	1.95	1.54	1.09	0.83	2.32	2.22	1.71	1.17	0.88
		50	10	2.94	2.78	2.02	1.30	0.94	3.18	3.00	2.13	1.34	0.96
		0	20	2.16	2.27	1.74	1.29	1.01	2.50	2.42	1.95	1.40	1.08
		0	10	2.81	2.71	2.13	1.49	1.13	3.09	2.97	2.49	1.57	1.17
		-50	20	1.99	1.96	1.72	1.38	1.14	2.39	2.33	2.00	1.56	1.26
		-50	10	2.66	2.59	2.19	1.67	1.33	2.97	2.89	2.40	1.79	1.41
Vertical	Horizontal	90	10	3.54	3.28	2.10	1.22	0.84	3.73	3.44	2.16	1.24	0.91
		50	30	2.95	2.80	2.04	1.32	0.96	2.74	2.62	1.94	1.28	0.94
		50	10	3.72	3.48	2.36	1.43	1.01	3.69	3.45	2.34	1.43	1.01
		0	20	3.23	3.10	2.36	1.60	1.19	2.98	2.86	2.22	1.53	1.16
		0	10	3.96	3.76	2.73	1.76	1.28	3.59	3.42	2.55	1.69	1.24
		-50	20	3.15	3.06	2.51	1.85	1.44	2.89	2.82	2.35	1.76	1.39
		-50	10	4.23	4.07	3.16	2.18	1.64	3.45	3.34	2.70	1.95	1.50
Sloping 45°	Down	90	10	3.51	3.24	2.09	1.21	0.84	4.84	4.36	2.50	1.34	0.90
		50	30	3.47	3.27	2.27	1.41	1.01	3.61	3.39	2.33	1.44	1.02
		50	10	3.82	3.57	2.40	1.45	1.02	4.79	4.41	2.75	1.57	1.08
		0	20	3.83	3.65	2.67	1.74	1.27	3.92	3.73	2.71	1.75	1.27
		0	10	4.23	4.04	2.88	1.82	1.31	4.67	4.39	3.05	1.89	1.35
		-50	20	4.05	3.90	3.05	2.13	1.61	3.82	3.68	2.92	2.06	1.57
		-50	10	4.84	4.63	3.48	2.33	1.72	4.47	4.29	3.29	2.24	1.67

Source: Reprinted with permission from the 1977 *Fundamentals Volume, ASHRAE Handbook and Product Directory.*

*See Table A-3-7 for E values.

Table A-3-7 Effective Emissivities of Air Spaces

Surface	Reflectivity in Percent	Average Emissivity ϵ	Effective Emissivity E of Air Space	
			With One Surface Having Emissivity ϵ and Other 0.90	With Both Surfaces of Emissivity ϵ
Aluminum foil, bright	92 to 97	0.05	0.05	0.03
Aluminum sheet	80 to 95	0.12	0.12	0.06
Aluminum coated paper, polished	75 to 84	0.20	0.20	0.11
Steel, galvanized, bright	70 to 80	0.25	0.24	0.15
Aluminum paint	30 to 70	0.50	0.47	0.35
Building materials: wood, paper, glass, masonry, nonmetallic paints	5 to 15	0.90	0.82	0.82

Source: Reprinted with permission from the 1977 Fundamentals Volume, ASHRAE Handbook and Product Directory.

Appendix A-4: Thermal Properties of Typical Building and Insulating Materials—(Design Values)

Table A-4 Thermal Properties of Typical Building and Insulating Materials Per Inch Thickness

Description	Density (lb/ft³)	Conductivity (k)	Conductance (C)	Customary Unit Resistance[b] (R) Per inch thickness (1/k)	Customary Unit Resistance[b] (R) For thickness listed (1/C)	Specific Heat, Btu/(lb)(deg F)	SI Unit Resistance[b] (R) (m·K)/W	SI Unit Resistance[b] (R) (m²·K)/W
BUILDING BOARD								
Boards, Panels, Subflooring, Sheathing Woodboard Panel Products								
Asbestos-cement board . . . . :	120	4.0	—	0.25	—	0.24	1.73	
Asbestos-cement board. . . . . . 0.125 in.	120	—	33.00	—	0.03			0.005
Asbestos-cement board. . . . . . 0.25 in.	120	—	16.50	—	0.06			0.01
Gypsum or plaster board . . . . 0.375 in.	50	—	3.10	—	0.32	0.26		0.06
Gypsum or plaster board . . . . . 0.5 in.	50	—	2.22	—	0.45			0.08
Gypsum or plaster board . . . . 0.625 in.	50	—	1.78	—	0.56			0.10
Plywood (Douglas Fir) . . . . :	34	0.80	—	1.25	—	0.29	8.66	
Plywood (Douglas Fir) . . . . . . 0.25 in.	34	—	3.20	—	0.31			0.05
Plywood (Douglas Fir) . . . . . 0.375 in.	34	—	2.13	—	0.47			0.08
Plywood (Douglas Fir) . . . . 0.5 in.	34	—	1.60	—	0.62			0.11
Plywood (Douglas Fir) . . . . . 0.625 in.	34	—	1.29	—	0.77			0.19

Table A-4 Thermal Properties of Typical Building and Insulating Materials Per Inch Thickness (cont.)

Description	Density (lb/ft³)	Conductivity (k)	Conductance (C)	Resistance^b (R) Per inch thickness (1/k)	Resistance^b (R) For thickness listed (1/C)	Specific Heat, Btu/(lb)(deg F)	SI Unit Resistance^b (R) (m·K)/W	SI Unit Resistance^b (R) (m²·K)/W
Plywood or wood panels. .... 0.75 in.	34	—	1.07	—	0.93	0.29		0.16
Vegetable Fiber Board								
Sheathing, regular density .... 0.5 in.	18	—	0.76	—	1.32	0.31		0.23
.. 0.78125 in.	18	—	0.49	—	2.06			0.36
Sheathing intermediate density ... 0.5 in.	22	—	0.82	—	1.22	0.31		0.21
Nail-base sheathing. .... 0.5 in.	25	—	0.88	—	1.14	0.31		0.20
Shingle backer. .... 0.375 in.	18	—	1.06	—	0.94	0.31		0.17
Shingle backer. .... 0.3125 in.	18	—	1.28	—	0.78			0.14
Sound deadening board. .... 0.5 in.	15	—	0.74	—	1.35	0.30		0.24
Tile and lay-in panels, plain or acoustic	18	0.40	—	2.50	—	0.14	17.33	
.... 0.5 in.	18	—	0.80	—	1.25			0.22
.... 0.75 in.	18	—	0.53	—	1.89			0.33
Laminated paperboard..	30	0.50	—	2.00	—	0.33	13.86	
Homogeneous board from repulped paper	30	0.50	—	2.00	—	0.28	13.86	
Hardboard								
Medium density	50	0.73	—	1.37	—	0.31	9.49	
High density, service temp. service underlay ...	55	0.82	—	1.22	—	0.32	8.46	
High density, std. tempered	63	1.00	—	1.00	—	0.32	6.93	
Particleboard								
Low density ...	37	0.54	—	1.85	—	0.31	12.82	
Medium density ...	50	0.94	—	1.06	—	0.31	7.35	
High density.. ...	62.5	1.18	—	0.85	—	0.31	5.89	
Underlayment.... .... 0.625 in.	40	—	1.22	—	0.82	0.29		0.14
Wood subfloor. .... 0.75 in.		—	1.06	—	0.94	0.33		0.17

BUILDING MEMBRANE								
Vapor—permeable felt	—	—	16.70	—	*0.06*			*0.01*
Vapor—seal, 2 layers of mopped 15-lb felt	—	—	8.35	—	*0.12*			*0.02*
Vapor—seal, plastic film	—	—	—	—	Negl.			
FINISH FLOORING MATERIALS								
Carpet and fibrous pad	—	—	0.48	—	*2.08*	0.34		*0.37*
Carpet and rubber pad	—	—	0.81	—	*1.23*	0.33		*0.22*
Cork tile 0.125 in.	—	—	3.60	—	*0.28*	0.48		*0.05*
Terrazzo 1 in.	—	—	12.50	—	*0.08*	0.19		*0.01*
Tile—asphalt, linoleum, vinyl, rubber	—	—	20.00	—	*0.05*	0.30		*0.01*
vinyl asbestos						0.24		
ceramic						0.19		
Wood, hardwood finish 0.75 in.			1.47		*0.68*			0.12
INSULATING MATERIALS								
Blanket and Batt								
Mineral Fiber, fibrous form processed from rock, slag, or glass								
approx.[e] 2–2.75 in.	0.3–2.0	—	0.143	—	*7*[d]	0.17–0.23		*1.23*
approx.[e] 3–3.5 in.	0.3–2.0	—	0.091	—	*11*[d]			*1.94*
approx.[e] 5.50–6.5	0.3–2.0	—	0.053	—	*19*[d]			*3.35*
approx.[e] 6–7 in.	0.3–2.0		0.045		*22*[d]			*3.87*
approx.[d] 8.5 in.	0.3–2.0		0.033		*30*[d]			*5.28*
Board and Slabs								
Cellular glass	8.5	0.38	—	*2.63*	—	0.24	*18.23*	
Glass fiber, organic bonded	4–9	0.25	—	*4.00*	—	0.23	*27.72*	
Expanded rubber (rigid)	4.5	0.22	—	*4.55*	—	0.40	*31.53*	
Expanded polystyrene extruded Cut cell surface	1.8	0.25	—	*4.00*	—	0.29	*27.72*	
Expanded polystyrene extruded Smooth skin surface	2.2	0.20	—	*5.00*	—	0.29	*34.65*	
Expanded polystyrene extruded Smooth skin surface	3.5	0.19	—	*5.26*	—		*36.45*	
Expanded polystyrene, molded beads	1.0	0.28	—	*3.57*	—	0.29	*24.74*	
Expanded polyurethane[f] (R-11 exp.)	1.5	0.16	—	*6.25*	—	0.38	*43.82*	
(Thickness 1 in. or greater)	2.5							

Table A-4 Thermal Properties of Typical Building and Insulating Materials Per Inch Thickness (cont.)

Description	Customary Unit						SI Unit	
	Density (lb/ft³)	Conductivity (k)	Conductance (C)	Resistance^b (R) Per inch thickness (1/k)	Resistance^b (R) For thickness listed (1/C)	Specific Heat, Btu/(lb)(deg F)	Resistance^b (R) (m·K)/W	Resistance^b (R) (m²·K)/W
Mineral fiber with resin binder	15	0.29	—	3.45	—	0.17	23.91	
Mineral fiberboard, wet felted								
Core or roof insulation................	16–17	0.34	—	2.94	—	0.19	20.38	
Acoustical tile......................	18	0.35	—	2.86	—		19.82	
Acoustical tile......................	21	0.37	—	2.70	—		18.71	
Mineral fiberboard, wet molded								
Acoustical tile^g	23	0.42	—	2.38	—	0.14	16.49	
Wood or cane fiberboard								
Acoustical tile^g ....0.5 in.	—	—	0.80	—	1.25	0.31		0.22
Acoustical tile^g ....0.75 in.	—	—	0.53	—	1.89			0.33
Interior finish (plank, tile).............	15	0.35	—	2.86	—	0.32	19.82	
Wood shredded (cemented in preformed slabs).............	22	0.60	—	1.67	—	0.31	11.57	
LOOSE FILL								
Cellulosic insulation (milled paper or wood pulp).............	2.3–3.2	0.27–0.32	—	3.13–3.70	—	0.33	21.69–25.64	
Sawdust or shavings	8.0–15.0	0.45	—	2.22	—	0.33	15.39	
Wood fiber, softwoods	2.0–3.5	0.30	—	3.33	—	0.33	23.08	
Perlite, expanded	5.0–8.0	0.37	—	2.70	—	0.26	18.71	
Mineral fiber (rock, slag or glass)								
approx.^e 3.75–5 in.	0.6–2.0	—	—	—	11	0.17		1.94
approx.^e 6.5–8.75 in.	0.6–2.0	—	—	—	19			3.35
approx.^e 7.5–10 in.	0.6–2.0	—	—	—	22			3.87
approx.^e 10.25–13.75 in.	0.6–2.0	—	—	—	30			5.28
Vermiculite, exfoliated.............	7.0–8.2	0.47	—	2.13	—	3.20	14.76	
	4.0–6.0	0.44	—	2.27	—		15.73	

(Table rotated 90° on page; column headers appear on the preceding page. Columns reconstructed below: Density (lb/ft³) · Conductivity k · Conductance C · Resistance 1/k (per inch) · Resistance 1/C (for thickness) · Specific Heat · and two additional resistance columns.)

Material	Density	k	C	R, 1/k	R, 1/C	Sp. Heat		
ROOF INSULATION[h] Preformed, for use above deck. Different roof insulations are available in different thicknesses to provide the design C values listed.[h] Consult individual manufacturers for actual thickness of their material ...	—	—	0.72 to 0.12	—	1.39 to 8.33	—	— —	0.24 to 1.47
MASONRY MATERIALS								
CONCRETES								
Cement mortar ...	116	5.0	—	0.20	—	0.21	1.39	
Gypsum-fiber concrete 87.5% gypsum, 12.5% wood chips ...	51	1.66	—	0.60	—		4.16	
Lightweight aggregates including expanded shale, clay or slate; expanded slags; cinders; pumice; vermiculite; also cellular concretes	120	5.2	—	0.19	—		1.32	
	100	3.6	—	0.28	—		1.94	
	80	2.5	—	0.40	—		2.77	
	60	1.7	—	0.59	—		4.09	
	40	1.15	—	0.86	—		5.96	
	30	0.90	—	1.11	—		7.69	
	20	0.70	—	1.43	—		9.91	
Perlite, expanded ...	40	0.93	—	1.08	—	0.32	7.48	
	30	0.71	—	1.41	—		9.77	
	20	0.50	—	2.00	—		13.86	
Sand and gravel or stone aggregate (oven dried) ...	140	9.0	—	0.11	—	0.22	0.76	
Sand and gravel or stone aggregate (not dried) ...	140	12.0	—	0.08	—		0.55	
Stucco ...	116	5.0	—	0.20	—		1.39	
MASONRY UNITS								
Brick, common[i] ...	120	5.0	—	0.20	—	0.19	1.39	
Brick, face[i] ...	130	9.0	—	0.11	—		0.76	
Clay tile, hollow:								
1 cell deep ... 3 in.	—	—	1.25	—	0.80	0.21		0.14
1 cell deep ... 4 in.	—	—	0.90	—	1.11			0.20
2 cells deep ... 6 in.	—	—	0.66	—	1.52			0.27
2 cells deep ... 8 in.	—	—	0.54	—	1.85			0.33
2 cells deep ... 10 in.	—	—	0.45	—	2.22			0.39
3 cells deep ... 12 in.	—	—	0.40	—	2.50			0.44

Table A-4 Thermal Properties of Typical Building and Insulating Materials Per Inch Thickness (cont.)

Description	Density (lb/ft³)	Conductivity (k)	Conductance (C)	Resistanceᵇ (R) Per inch thickness (1/k)	Resistanceᵇ (R) For thickness listed (1/C)	Specific Heat, Btu/(lb)(deg F)	SI Resistanceᵇ (R) (m·K)/W	SI Resistanceᵇ (R) (m²·K)/W
Concrete blocks, three oval core:								
Sand and gravel aggregate .... 4 in.	—	—	1.40	—	0.71	0.22		0.13
.... 8 in.	—	—	0.90	—	1.11			0.20
.... 12 in.	—	—	0.78	—	1.28			0.23
Cinder aggregate .... 3 in.	—	—	1.16	—	0.86	0.21		0.15
.... 4 in.	—	—	0.90	—	1.11			0.20
.... 8 in.	—	—	0.58	—	1.72			0.30
.... 12 in.	—	—	0.53	—	1.89			0.33
Lightweight aggregate .... 3 in.	—	—	0.79	—	1.27	0.21		0.22
(expanded shale, clay, slate .... 4 in.	—	—	0.67	—	1.50			0.26
or slag; pumice) .... 8 in.	—	—	0.50	—	2.00			0.35
.... 12 in.	—	—	0.44	—	2.27			0.40
Concrete blocks, rectangular core.*j								
Sand and gravel aggregate								
2 core, 8 in. 36 lb. k*	—	—	0.96	—	1.04	0.22		0.18
Same with filled coresj*	—	—	0.52	—	1.93	0.22		0.34
Lightweight aggregate (expanded shale, clay, slate or slag, pumice):								
3 core, 6 in. 19 lb. k*	—	—	0.61	—	1.65	0.21		0.29
Same with filled coresl*	—	—	0.33	—	2.99			0.53
2 core, 8 in. 24 lb. k*	—	—	0.46	—	2.18			0.38
Same with filled coresl*	—	—	0.20	—	5.03			0.89
3 core, 12 in. 38 lb.k*	—	—	0.40	—	2.48			0.44
Same with filled coresl*	—	—	0.17	—	5.82			1.02
Stone, lime or sand.	—	12.50	—	0.08	—	0.19	0.55	
Gypsum partition tile:								
3 × 12 × 30 in. solid	—	—	0.79	—	1.26	0.19		0.22
3 × 12 × 30 in. 4-cell	—	—	0.74	—	1.35			0.24
4 × 12 × 30 in. 3-cell	—	—	0.60	—	1.67			0.29

METALS
(See Chapter 37, Table 3)

PLASTERING MATERIALS

Description	Density	Conductivity (k)	Conductance (C)	Resistance (1/k)	Resistance (1/C)	Specific Heat	Resistance (1/k)	Resistance (1/C)
Cement plaster, sand aggregate	116	5.0	—	0.20	—	0.20	1.39	—
Sand aggregate .... 0.375 in.	—	—	13.3	—	*0.08*	0.20	—	*0.01*
Sand aggregate .... 0.75 in.	—	—	6.66	—	*0.15*	—	—	*0.03*
Gypsum plaster:								
Lightweight aggregate .... 0.5 in.	45	—	3.12	—	*0.32*	—	—	*0.06*
Lightweight aggregate .... 0.625 in.	45	—	2.67	—	*0.39*	—	—	*0.07*
Lightweight agg. on metal lath .... 0.75 in.	—	—	2.13	—	*0.47*	—	—	*0.08*
Perlite aggregate	45	1.5	—	0.67	—	0.20	4.64	—
Sand aggregate	105	5.6	—	0.18	—	0.20	1.25	—
Sand aggregate .... 0.5 in.	105	—	11.10	—	*0.09*	—	—	*0.02*
Sand aggregate .... 0.625 in.	105	—	9.10	—	*0.11*	—	—	*0.02*
Sand aggregate on metal lath .... 0.75 in.	—	—	7.70	—	*0.13*	—	—	*0.02*
Vermiculite aggregate	45	1.7	—	0.59	—	0.32	4.09	—

ROOFING

Description	Density	Conductivity (k)	Conductance (C)	Resistance (1/k)	Resistance (1/C)	Specific Heat	Resistance (1/k)	Resistance (1/C)
Asbestos-cement shingles	120	—	4.76	—	*0.21*	0.20	—	*0.04*
Asphalt roll roofing	70	—	6.50	—	*0.15*	0.24	—	*0.03*
Asphalt shingles	70	—	2.27	—	*0.44*	0.36	—	*0.08*
Built-up roofing .... 0.375 in.	70	—	3.00	—	*0.33*	0.30	—	*0.06*
Slate .... 0.5 in.	—	—	20.00	—	*0.05*	0.35	—	*0.01*
Wood shingles, plain and plastic film faced	—	—	1.06	—	*0.94*	0.30	—	*0.17*

SIDING MATERIALS (On Flat Surface)

Shingles

Description	Density	Conductivity (k)	Conductance (C)	Resistance (1/k)	Resistance (1/C)	Specific Heat	Resistance (1/k)	Resistance (1/C)
Asbestos-cement	120	—	4.75	—	*0.21*	0.31	—	*0.04*
Wood, 16 in., 7.5 exposure	—	—	1.15	—	*0.87*	0.31	—	*0.15*
Wood, double, 16-in., 12-in. exposure	—	—	0.84	—	*1.19*	0.28	—	*0.21*
Wood, plus insul. backer board, 0.3125 in.	—	—	0.71	—	*1.40*	0.31	—	*0.25*

Siding

Description	Density	Conductivity (k)	Conductance (C)	Resistance (1/k)	Resistance (1/C)	Specific Heat	Resistance (1/k)	Resistance (1/C)
Asbestos-cement, 0.25 in., lapped	—	—	4.76	—	*0.21*	0.24	—	*0.04*
Asphalt roll siding	—	—	6.50	—	*0.15*	0.35	—	*0.03*
Asphalt insulating siding (0.5 in. bed.)	—	—	0.69	—	*1.46*	0.35	—	*0.26*
Wood, drop, 1 × 8 in.	—	—	1.27	—	*0.79*	0.28	—	*0.14*
Wood, bevel, 0.5 × 8 in., lapped	—	—	1.23	—	*0.81*	0.28	—	*0.14*
Wood, bevel, 0.75 × 10 in., lapped	—	—	0.95	—	*1.05*	0.28	—	*0.18*
Wood, plywood, 0.375 in., lapped	—	—	1.59	—	*0.59*	0.29	—	*0.10*
Wood, medium density siding, 0.4375 in.	40	1.49	—	0.67	—	0.28	4.65	—

Table A-4 Thermal Properties of Typical Building and Insulating Materials Per Inch Thickness (cont.)

Description	Customary Unit						SI Unit	
	Density (lb/ft³)	Conductivity (k)	Conductance (C)	Resistance^b (R)		Specific Heat, Btu/(lb)(deg F)	Resistance^b (R)	
				Per inch thickness (1/k)	For thickness listed (1/C)		(m·K)/W	(m²·K)/W
Aluminum or Steel^m, over sheathing								
Hollow-backed	—	—	1.61	—	0.61	0.29		0.11
Insulating-board backed nominal 0.375 in.	—	—	0.55	—	1.82	0.32		0.32
Insulating-board backed nominal 0.375 in., foil backed	—	—	0.34	—	2.96			0.52
Architectural glass	—	—	10.00	—	0.10	0.20		0.02
WOODS								
Maple, oak, and similar hardwoods	45	1.10	—	0.91	—	0.30	6.31	
Fir, pine, and similar softwoods	32	0.80	—	1.25	—	0.33	8.66	
Fir, pine, and similar softwoods . . . . 0.75 in.	32	—	1.06	—	0.94	0.33		0.17
. . . . 1.5 in.		—	0.53	—	1.89			0.33
. . . . 2.5 in.		—	0.32	—	3.12			0.60
. . . . 3.5 in.		—	0.23	—	4.35			0.75

Notes for Table A-4

[a] Representative values for dry materials were selected by ASHRAE TC4.4, Insulation and Moisture Barriers. They are intended as design (not specification) values for materials in normal use. For properties of a particular product, use the value supplied by the manufacturer or by unbiased tests

[b] Resistance values are the reciprocals of *C* before rounding off *C* to two decimal places.

[c] Also see Insulating Materials, Board.

[d] Does not include paper backing and facing, if any.

[e] Conductivity varies with fiber diameter. Insulation is produced by different densities; therefore, there is a wide variation in thickness for the same *R*-value among manufacturers. No effort should be made to relate any specific *R*-value to any specific thickness. Commercial thicknesses generally available range from 2 to 8.5.

[f] Values are for aged board stock.

[g] Insulating values of acoustical tile vary, depending on density of the board and on type, size, and depth of perforations.

[h] The U. S. Department of Commerce, *Simplified Practice Recommendation for Thermal Conductance Factors for Preformed Above-Deck Roof Insulation*, No. R 257-55, recognizes the specification of roof insulation on the basis of the *C*-values shown. Roof insulation is made in thicknesses to meet these values.

[i] Face brick and common brick do not always have these specific densities. When density is different from that shown, there will be a change in thermal conductivity.

[j] Data on rectangular core concrete blocks differ from the above data on oval core blocks, due to core configuration, different mean temperatures, and possibly differences in unit weights. Weight data on the oval core blocks tested are not available.

[k] Weights of units approximately 7.625 in. high and 15.75 in. long. These weights are given as a means of describing the blocks tested, but conductance values are all for 1 ft² of area.

[l] Vermiculite, perlite, or mineral wool insulation. Where insulation is used, vapor barriers or other precautions must be considered to keep insulation dry.

[m] Values for metal siding applied over flat surfaces vary widely, depending on amount of ventilation of air space beneath the siding; whether air space is reflective or nonreflective; and on thickness, type, and application of insulating backing-board used. Values given are averages for use as design guides, and were obtained from several guarded hotbox tests (ASTM C236) or calibrated hotbox (BSS 77) on hollow-backed types and types made using backing-boards of wood fiber, foamed plastic, and glass fiber. Departures of ±50% or more from the values given may occur.

Source: Reprinted with permission from the 1977 *Fundamentals Volume, ASHRAE Handbook and Product Directory.*

Appendix A-5: Blackbody Emissive Power within a Wavelength Band or Interval

It is often desirable to know how much emission occurs in a specific portion of the total wavelength spectrum. This emission is expressed most conveniently as a fraction of the total emissive power. The fraction between wavelength λ_1 and λ_2 is designated $f_{\lambda_1 - \lambda_2}$ and is given by

$$f_{\lambda_1 - \lambda_2} = \frac{\displaystyle\int_{\lambda_1}^{\lambda_2} E_{b\lambda}(\lambda)d\lambda}{\displaystyle\int_0^{\infty} E_{b\lambda}(\lambda)d\lambda} \qquad \text{(A-5-1)}$$

$$= \frac{1}{\sigma T^4} \int_{\lambda_1}^{\lambda_2} E_{b\lambda}(\lambda)d\lambda$$

Equation (A-5-1) is conveniently broken down into two integrals as follows:

$$f_{\lambda_1 - \lambda_2} = \frac{1}{\sigma T^4}\left[\int_0^{\lambda_2} E_{b\lambda}(\lambda)d\lambda - \int_0^{\lambda_1} E_{b\lambda}(\lambda)d\lambda\right] \qquad \text{(A-5-2)}$$

$$= f_{0-\lambda_2} - f_{0-\lambda_1}$$

These values may thus be calculated at a given T and, between any two wavelengths, the fraction of total emission may be determined by subtraction. Table A-5-1 will aid in this calculation.

Because values of $f_{0-\lambda}$ as expressed above vary with temperature, an additional manipulation, eliminating T as a separate variable, is helpful. We may define f in terms of the product λT and modify Equation (A-5-2) to read

$$f_{\lambda_1 T - \lambda_2 T} = \frac{1}{\sigma}\left[\int_0^{\lambda_2 T} \frac{E_{b\lambda}(\lambda)}{T^5}d(\lambda T) = \int_0^{\lambda_1 T} \frac{E_{b\lambda}(\lambda)}{T^5}d(\lambda T)\right] \qquad \text{(A-5-3)}$$

$$= f_{0-\lambda_2 T} - f_{0-\lambda_1 T}$$

Note that $E_{b\lambda}/T^5$ is a function of λT; thus $f_{0-\lambda T}$ may be evaluated and tabulated.

Table A-5-1 Blackbody Radiation Functions

λT	$\dfrac{E_{b\lambda} \times 10^5}{\sigma T^5}$	$\dfrac{E_{b(0-\lambda T)}}{\sigma T^4}$	λT	$\dfrac{E_{b\lambda} \times 10^5}{\sigma T^5}$	$\dfrac{E_{b(0-\lambda T)}}{\sigma T^4}$
1000	0.0000394	0	10200	5.378	0.7076
1200	0.001184	0	10400	5.146	0.7181
1400	0.01194	0	10600	4.925	0.7282
1600	0.0618	0.0001	10800	4.714	0.7378
1800	0.2070	0.0003	11000	4.512	0.7474
2000	0.5151	0.0009	11200	4.320	0.7559
2200	1.0384	0.0025	11400	4.137	0.7643
2400	1.791	0.0053	11600	3.962	0.7724
2600	2.753	0.0098	11800	3.795	0.7802
2800	3.872	0.0164	12000	3.673	0.7876
3000	5.081	0.0254	12200	3.485	0.7947
3200	6.312	0.0368	12400	3.341	0.8015
3400	7.506	0.0506	12600	3.203	0.8081
3600	8.613	0.0667	12800	3.071	0.8144
3800	9.601	0.0850	13000	2.947	0.8204
			13200	2.827	0.8262
4000	10.450	0.1051			
4200	11.151	0.1267	13400	2.714	0.8317
4400	11.704	0.1496	13600	2.605	0.8370
4600	12.114	0.1734	13800	2.052	0.8421
4800	12.392	0.1979	14000	2.416	0.8470
			14200	2.309	0.8517
5000	12.556	0.2229			
5200	12.607	0.2481	14400	2.219	0.8563
5400	12.571	0.2733	14600	2.134	0.8606
5600	12.458	0.2983	14800	2.052	0.8648
5800	12.282	0.3230	15000	1.972	0.8688
			16000	1.633	0.8868
6000	12.053	0.3474			
6200	11.783	0.3712	17000	1.360	0.9017
6400	11.480	0.3945	18000	1.140	0.9142
6600	11.152	0.4171	19000	0.962	0.9247
6800	10.808	0.4391	20000	0.817	0.9335
7000	10.451	0.4604	21000	0.702	0.9411
7200	10.089	0.4809	22000	0.599	0.9475
7400	9.723	0.5007	23000	0.516	0.9531
7600	9.357	0.5199	24000	0.448	0.9589
7800	8.997	0.5381	25000	0.390	0.9621
8000	8.642	0.5558	26000	0.341	0.9657
8200	8.293	0.5727	27000	0.300	0.9689
8400	7.954	0.5890	28000	0.265	0.9718
8600	7.624	0.6045	29000	0.234	0.9742
8800	7.304	0.6195	30000	0.208	0.9765
9000	6.995	0.6337	40000	0.0741	0.9881
9200	6.697	0.6474	50000	0.0326	0.9941
9400	6.411	0.6606	60000	0.0165	0.9963
9600	6.136	0.6731	70000	0.0092	0.9981
9800	5.872	0.6851	80000	0.0055	0.9987
10000	5.619	0.6966	90000	0.0035	0.9990
			100000	0.0023	0.9992
			0		1.0000

Note: λ [=] microns; T [=] R; $E_{b\lambda}$ [=] Btu/hr-ft^2-μ. For SI conversion, multiply column 1 (4) by 0.555556 (the resulting units are μK) and column 2 (5) by 0.0059554467 (the resulting units are W/m$^2 \cdot$ K$^5 \cdot \mu$). Column 3 (6) does not change.

Table A-6-1 Dimensions and Properties of Steel Pipe

Nominal[a] Size	ASTM[b] Schedule	Diameter OD In.	Diameter ID In.	Wall Thickness In.	Surface Area Sq Ft/Lin Ft OD	Surface Area Sq Ft/Lin Ft ID	Section Area Sq In. OD	Section Area Sq In. ID	Area of Metal Sq In.	Volume Gal/Lin Ft	Weight[c] (plain end) Lb/Lin Ft	Working Pressure Psia
⅛	40 (s)	0.405	0.269	0.068	0.106	0.0704	0.129	0.0568	0.0720	0.00295	0.244	314[d(1)]
	80 (x)	0.405	0.215	0.095	0.106	0.0563	0.129	0.0363	0.0925	0.00189	0.314	1084
¼	40 (s)	0.540	0.364	0.088	0.141	0.0953	0.229	0.104	0.125	0.00541	0.424	649
	80 (x)	0.540	0.302	0.119	0.141	0.0791	0.229	0.0716	0.157	0.00372	0.535	1353
⅜	40 (s)	0.675	0.493	0.091	0.177	0.129	0.358	0.191	0.167	0.00992	0.567	574
	80 (x)	0.675	0.423	0.126	0.177	0.111	0.358	0.140	0.217	0.00730	0.738	1191
½	40 (s)	0.840	0.622	0.109	0.220	0.163	0.554	0.304	0.250	0.0158	0.850	697
	80 (x)	0.840	0.546	0.147	0.220	0.143	0.554	0.234	0.320	0.0122	1.09	1266
	XX	0.840	0.252	0.294	0.220	0.0660	0.554	0.0499	0.504	0.00259	1.71	3824
¾	40 (s)	1.050	0.824	0.113	0.275	0.216	0.886	0.533	0.333	0.0277	1.13	604
	80 (x)	1.050	0.742	0.154	0.275	0.194	0.866	0.432	0.434	0.0225	1.47	1078
	XX	1.050	0.434	0.308	0.275	0.114	0.866	0.148	0.718	0.00768	2.44	3134
1	40 (s)	1.315	1.049	0.133	0.344	0.275	1.36	0.864	0.494	0.0449	1.68	651
	80 (x)	1.315	0.957	0.179	0.344	0.251	1.36	0.719	0.639	0.0374	2.17	1083
	XX	1.315	0.599	0.358	0.344	0.157	1.36	0.282	1.08	0.0146	3.66	2963
1¼	40 (s)	1.660	1.380	0.140	0.435	0.361	2.16	1.50	0.669	0.0777	2.27	440
	80 (x)	1.660	1.278	0.191	0.435	0.335	2.16	1.28	0.881	0.0666	3.00	805
	XX	1.660	0.896	0.382	0.435	0.235	2.16	0.630	1.53	0.0328	5.21	2318
1½	40 (s)	1.900	1.610	0.145	0.497	0.421	2.84	2.04	0.800	0.1058	2.72	417
	80 (x)	1.900	1.500	0.200	0.497	0.393	2.84	1.77	1.07	0.0918	3.65	756
	XX	1.900	1.100	0.400	0.497	0.288	2.84	0.950	1.89	0.0494	6.41	2122
2	40 (s)	2.375	2.067	0.154	0.622	0.541	4.43	3.36	1.07	0.174	3.65	376
	80 (x)	2.375	1.939	0.218	0.622	0.508	4.43	2.95	1.48	0.153	5.02	690
	XX	2.375	1.503	0.436	0.622	0.393	4.43	1.77	2.66	0.0922	9.03	1861

Nominal Size	Schedule											
2½	40 (s)	2.875	2.469	0.203	0.753	0.646	6.49	4.79	1.70	0.249	5.79	505
	80 (x)	2.875	2.323	0.276	0.753	0.608	6.49	4.24	2.25	0.220	7.66	806
	XX	2.875	1.771	0.552	0.753	0.364	6.49	2.46	4.03	0.128	13.7	2048
3	40 (s)	3.500	3.068	0.216	0.916	0.803	9.62	7.39	2.23	0.384	7.57	454
	80 (x)	3.500	2.900	0.300	0.916	0.759	9.62	6.61	3.02	0.343	10.3	734
	XX	3.500	2.300	0.600	0.916	0.602	9.62	4.15	5.47	0.216	18.5	1829
3½	40 (s)	4.000	3.548	0.226	1.05	0.929	12.6	9.89	2.68	0.514	9.11	425
	80 (x)	4.000	3.364	0.318	1.05	0.881	12.6	8.89	3.68	0.462	12.5	692
	XX*	4.000	2.728	0.636	1.05	0.714	12.6	5.85	6.72	0.304	22.9	1699
4	40 (s)	4.500	4.026	0.237	1.18	1.05	15.9	12.7	3.17	0.661	10.8	403
	80 (x)	4.500	3.826	0.337	1.18	1.00	15.9	11.5	4.41	0.597	14.9	663
	XX	4.500	3.152	0.674	1.18	0.825	15.9	7.80	8.10	0.405	27.5	1602
5	40 (s)	5.563	5.047	0.258	1.46	1.32	24.3	20.0	4.30	1.04	14.6	498[d(2)]
	80 (x)	5.563	4.813	0.375	1.46	1.26	24.3	18.2	6.11	0.945	20.8	825
	XX	5.563	4.063	0.750	1.46	1.06	24.3	13.0	11.3	0.673	38.6	1951
6	40 (s)	6.625	6.065	0.280	1.73	1.59	34.5	28.9	5.58	1.50	18.0	467
	80 (x)	6.625	5.761	0.432	1.73	1.51	34.5	26.1	8.40	1.35	28.6	825
	XX	6.625	4.897	0.864	1.73	1.28	34.5	18.8	15.6	0.978	53.1	1912
8	30 (s)	8.625	8.071	0.277	2.26	2.11	58.4	51.2	7.26	2.66	24.7	351
	40 (s)	8.625	7.981	0.322	2.26	2.09	58.4	50.0	8.40	2.60	28.6	431
	80 (x)	8.625	7.625	0.500	2.26	2.00	58.4	45.7	12.8	2.37	43.4	753
	XX	8.625	6.875	0.875	2.26	1.80	58.4	37.1	21.3	1.93	72.4	1460
10	(s)	10.750	10.192	0.279	2.81	2.67	90.8	81.6	9.18	4.24	31.2	285
	30	10.750	10.136	0.307	2.81	2.65	90.8	80.7	10.1	4.19	34.2	324
	40 (x)	10.750	10.020	0.365	2.81	2.62	90.8	78.9	11.9	4.10	40.5	405
	60	10.750	9.750	0.500	2.81	2.55	90.8	74.7	16.1	3.88	54.7	600
12	30 (s)	12.750	12.090	0.330	3.34	3.17	128.	115.	12.9	5.96	43.8	299
	(s)	12.750	12.000	0.375	3.34	3.14	128.	113.	14.6	5.88	49.6	352
	(x)	12.750	11.750	0.500	3.34	3.08	128.	108.	19.2	5.63	65.4	503
14	30 (s)	14.000	13.250	0.375	3.67	3.46	154.	138.	16.0	7.17	54.6	458[d(3)]
	(x)	14.000	13.000	0.500	3.67	3.15	154.	133.	21.2	6.70	72.1	653

Table A-6-1 Dimensions and Properties of Steel Pipe (cont.)

Nominal Size[a]	ASTM[b] Schedule	Diameter OD In.	Diameter ID In.	Wall Thickness In.	Surface Area Sq Ft/Lin Ft OD	Surface Area Sq Ft/Lin Ft ID	Section Area Sq In. OD	Section Area Sq In. ID	Area of Metal Sq In.	Volume Gal/Lin Ft	Weight[c] (plain end) Lb/Lin Ft	Working Pressure Psia
16	30 (s)	16.000	15.250	0.375	4.18	3.99	201.	183.	18.4	9.48	62.4	400[d(3)]
	40 (x)	16.000	15.000	0.500	4.18	3.93	201.	177.	24.3	9.18	82.8	570
18	(s)	18.000	17.250	0.375	4.71	4.52	254.	234.	20.7	12.1	70.6	355
	(x)	18.000	17.000	0.500	4.71	4.45	254.	227.	27.4	11.8	93.5	506
20	20 (s)	20.000	19.250	0.375	5.23	4.51	314.	291.	23.2	15.2	78.6	319
	30 (s)	20.000	19.000	0.500	5.23	4.97	314.	284.	30.6	14.7	104.2	454
24	20	24.000	23.250	0.375	6.29	6.08	452.	426.	26.8	22.1	94.6	265
	(s)	24.000	23.000	0.500	6.29	6.03	452.	415.	36.9	21.5	125.5	378

* 3½ double extra strong is no longer considered in ASTM specification but some pipe of this size is still manufactured.

[a] The sizes for wrought iron are approximately the same except wall thickness is slightly heavier. See *ASTM A-72.*

[b] *American Society for Testing and Materials* Schedule. The numbers 30, 40, etc., refer to the *ASTM* Schedule; the letter (s) refers to the former designation *Standard Weight*; the letter (x) refers to the former designation *Extra Strong*; the letters XX refer to the former designation *Double Extra Strong.* Threaded and coupled (T and C) pipe is slightly heavier.

[c] Weight per foot is based on plain end pipe.

[d] Working pressure for welded joints—see formula in Table 1. Refer to Table 1 for type of weld.

(1) Working pressure based on an allowable fiber stress of 6225 psi (for 250 F).
(2) Working pressure based on an allowable fiber stress of 8400 psi (for 250 F).
(3) Working pressure based on an allowable fiber stress of 12000 psi (for 250 F).

Note: Standard-weight pipe is generally furnished with threaded ends in random lengths of 16 to 22 ft, although when ordered with plain ends, 5 percent may be in lengths of 12 to 16 ft. Five percent of the total number of lengths ordered may be jointers which are two pieces coupled together. Extra-strong pipe is generally furnished with plain ends in random lengths of 12 to 22 ft, although 5 percent may be in lengths of 6 to 12 ft.

Source: Reprinted with permission from the 1977 Fundamentals Volume, ASHRAE Handbook and Product Directory.

Table A-6-2 Dimensions and Properties of Copper Tube

Nominal Size	Type	Diameter OD In.	Diameter ID In.	Wall Thickness In.	Surface Area Sq Ft/Lin Ft OD	Surface Area Sq Ft/Lin Ft ID	Section Area Sq In. OD	Section Area Sq In. ID	Area of Metal Sq In.	Volume Gal/Lin Ft	Weight[gs] Lb/Lin Ft	Working Pressure[b] Psia
1/4	K°	0.375	0.305	0.035	0.0982	0.0798	0.110	0.0730	0.0374	0.00379	0.145	918
	L°	0.375	0.315	0.030	0.0982	0.0825	0.110	0.0779	0.0324	0.00404	0.126	764
3/8	K	0.500	0.402	0.049	0.131	0.105	0.196	0.127	0.0695	0.00660	0.269	988
	L	0.500	0.430	0.035	0.131	0.113	0.196	0.145	0.0512	0.00753	0.198	677
1/2	K	0.625	0.527	0.049	0.164	0.138	0.306	0.218	0.0887	0.0113	0.344	779
	L	0.625	0.545	0.040	0.164	0.143	0.306	0.233	0.0735	0.0121	0.285	625
5/8	K	0.750	0.652	0.049	0.193	0.171	0.441	0.334	0.108	0.0174	0.418	643
	L	0.750	0.666	0.042	0.193	0.174	0.441	0.348	0.0934	0.0181	0.362	547
3/4	K	0.875	0.745	0.065	0.229	0.195	0.601	0.436	0.165	0.0227	0.641	747
	L	0.875	0.785	0.045	0.229	0.206	0.601	0.484	0.117	0.0250	0.455	497
1	K	1.125	0.995	0.065	0.295	0.260	0.994	0.778	0.216	0.0405	0.839	574
	L	1.125	1.025	0.050	0.295	0.268	0.994	0.825	0.169	0.0442	0.655	432
1 1/4	K	1.375	1.245	0.065	0.360	0.326	1.48	1.22	0.268	0.0634	1.04	466
	L	1.375	1.265	0.055	0.360	0.331	1.48	1.26	0.228	0.0655	0.884	387
	M	1.375	1.291	0.042	0.360	0.338	1.48	1.31	0.176	0.0681	0.682	293
	DWV	1.375	1.295	0.040	0.360	0.339	1.48	1.32	0.163	0.0684	0.650	—
1 1/2	K	1.625	1.481	0.072	0.425	0.388	2.07	1.72	0.351	0.0894	1.36	421
	L	1.625	1.505	0.060	0.425	0.394	2.07	1.78	0.295	0.0925	1.14	359
	M	1.625	1.527	0.049	0.425	0.400	2.07	1.83	0.243	0.0950	0.940	289
	DWV	1.625	1.541	0.042	0.425	0.403	2.07	1.86	0.205	0.0969	0.809	—
2	K	2.125	1.959	0.083	0.556	0.513	3.56	3.01	0.532	0.157	2.06	376
	L	2.125	1.985	0.070	0.556	0.520	3.56	3.10	0.452	0.161	1.75	316
	M	2.125	2.009	0.058	0.556	0.526	3.56	3.17	0.377	0.164	1.46	255
	DWV	2.125	2.041	0.042	0.556	0.534	3.56	3.27	0.288	0.142	1.07	—

Table A-6-2 Dimensions and Properties of Copper Tube (cont.)

Nominal Size	Type	Diameter OD In.	Diameter ID In.	Wall Thickness In.	Surface Area Sq Ft/Lin Ft OD	Surface Area Sq Ft/Lin Ft ID	Section Area Sq In. OD	Section Area Sq In. ID	Area of Metal Sq In.	Volume Gal/Lin Ft	Weight[a] Lb/Lin Ft	Working Pressure[b] Psia
2½	K	2.625	2.435	0.095	0.687	0.638	5.41	4.66	0.755	0.242	2.93	352
	L	2.625	2.465	0.080	0.687	0.645	5.41	4.77	0.640	0.247	2.48	295
	M	2.625	2.495	0.065	0.687	0.653	5.41	4.89	0.523	0.254	2.03	234
3	K	3.125	2.907	0.109	0.818	0.761	7.67	6.64	1.03	0.345	4.00	343
	L	3.125	2.945	0.090	0.818	0.771	7.67	6.81	0.858	0.354	3.33	278
	M	3.125	2.981	0.072	0.818	0.780	7.67	6.98	0.691	0.362	2.68	220
	DWV	3.125	3.035	0.045	0.818	0.796	7.67	7.23	0.435	0.376	1.69	—
3½	K	3.625	3.385	0.120	0.949	0.886	10.3	9.00	1.32	0.468	5.12	324
	L	3.625	3.425	0.100	0.949	0.897	10.3	9.21	1.11	0.478	4.29	268
	M	3.625	3.459	0.083	0.949	0.906	10.3	9.40	0.924	0.489	3.58	218
4	K	4.125	3.857	0.134	1.08	1.01	13.3	11.7	1.68	0.607	6.51	315
	L	4.125	3.905	0.110	1.08	1.02	13.3	12.0	1.39	0.623	5.38	256

Size	Type											
	M	4.125	3.935	0.095	1.08	1.03	13.3	12.2	1.20	0.634	4.66	217
	DWV	4.125	4.009	0.058	1.08	1.05	13.3	12.6	0.67	0.656	2.87	—
5	K	5.125	4.805	0.160	1.34	1.26	20.7	18.1	2.50	0.940	9.67	307
	L	5.125	4.875	0.125	1.34	1.28	20.7	18.7	1.96	0.971	7.61	234
	M	5.125	4.907	0.109	1.34	1.29	20.7	18.9	1.72	0.981	6.66	203
6	K	6.225	5.741	0.192	1.60	1.50	29.4	25.9	3.50	1.35	13.9	308
	L	6.125	5.845	0.140	1.60	1.53	29.4	26.8	2.63	1.39	10.2	221
	M	6.125	5.881	0.122	1.60	1.54	29.4	27.2	2.30	1.42	8.92	190
	DWV	6.125	5.959	0.083	1.60	1.56	29.4	27.9	1.58	1.55	6.10	—
8	K	8.125	7.583	0.271	2.13	1.99	51.8	45.2	6.69	2.34	25.9	330
	L	8.125	7.725	0.200	2.13	2.02	51.8	46.9	4.98	2.43	19.3	239
	M	8.125	7.785	0.170	2.13	2.04	51.8	47.6	4.25	2.47	16.5	200
10	K	10.125	9.449	0.338	2.65	2.47	70.5	70.1	10.4	3.65	40.3	332
	L	10.125	9.625	0.250	2.65	2.52	80.5	72.8	7.76	3.79	30.1	241
	M	10.125	9.701	0.212	2.65	2.54	80.5	73.9	6.60	3.84	25.6	202
12	K	12.125	11.315	0.405	3.17	2.96	115.	101.	14.9	5.24	57.8	334
	L	12.125	11.565	0.280	3.17	3.03	115.	105.	10.4	5.45	40.4	225
	M	12.125	11.617	0.254	3.17	3.04	115.	106.	9.47	5.50	36.7	204

Source: Reprinted with permission from the 1977 Fundamentals Volume, ASHRAE Handbook and Product Directory.

Appendix B-1:
Proof *u*=*u*(T) Only for an Ideal Gas

Making the assumption that a gas is ideal implies much more than just dictating an equation of state. Of particular import, it implies that the internal energy u, and therefore the enthalpy, are functions of temperature only.

As a first step in this proof, we must acquire some mathematical background. Consider a state function, say $G = G(T, v)$. Then

$$dG = \left(\frac{\partial G}{\partial T}\right)_v dT + \left(\frac{\partial G}{\partial v}\right)_T dv \tag{B-1-1}$$

Now if we have an equation of state $T = T(p, v)$, then

$$dT = \left(\frac{\partial T}{\partial p}\right)_v dp + \left(\frac{\partial T}{\partial v}\right)_p dv \tag{B-1-2}$$

Eliminating dT from Equations (B-1-1) and (B-1-2) yields

$$dG = \left(\frac{\partial G}{\partial T}\right)_v \left(\frac{\partial T}{\partial p}\right)_v dp + \left[\left(\frac{\partial G}{\partial T}\right)_v \left(\frac{\partial T}{\partial v}\right)_p + \left(\frac{\partial G}{\partial v}\right)_T\right] dv \tag{B-1-3}$$

Note that Equation (B-1-3) implies $G = G(p, v)$; thus

$$dG = \left(\frac{\partial G}{\partial p}\right)_v dp + \left(\frac{\partial G}{\partial v}\right)_p dv \tag{B-1-4}$$

Comparing Equations (B-1-3) and (B-1-4) yields

$$\left(\frac{\partial G}{\partial T}\right)_v \left(\frac{\partial T}{\partial p}\right)_v = \left(\frac{\partial G}{\partial p}\right)_v$$

$$\left(\frac{\partial G}{\partial v}\right)_p = \left(\frac{\partial G}{\partial T}\right)_v \left(\frac{\partial T}{\partial v}\right)_p + \left(\frac{\partial G}{\partial v}\right)_T \tag{B-1-5}$$

It is easily seen that G could be a function of any two of p, v, or T and that the result would be similar, but with the p, v, and T cyclically interchanged. In general, then, all results could be written as

$$\left(\frac{\partial G}{\partial x}\right)_y = \left(\frac{\partial G}{\partial z}\right)_y \left(\frac{\partial z}{\partial x}\right)_y \tag{B-1-6}$$

and

566

$$\left(\frac{\partial G}{\partial x}\right)_y = \left(\frac{\partial G}{\partial z}\right)_x \left(\frac{\partial z}{\partial x}\right)_y + \left(\frac{\partial G}{\partial z}\right)_z \qquad \text{(B-1-7)}$$

where x, y, and z may be p, v, and T in any order.

Now let us look, in particular, at the internal energy $u = u(T, v)$ and the first law of thermodynamics:

$$\delta q = du + p\,dv$$

Let

$$dG = \frac{\delta q}{T} = \frac{1}{T}\left[\left(\frac{\partial u}{\partial T}\right)_v dt + \left(\frac{\partial u}{\partial v}\right)_T dv\right] + \frac{p}{T}dv$$

$$= \frac{1}{T}\left[p + \left(\frac{\partial u}{\partial v}\right)_T\right]dv + \frac{1}{T}\left(\frac{\partial u}{\partial T}\right)_v dT \qquad \text{(B-1-8)}$$

If we assert that G is a point function, the cross derivatives of Equation (B-1-8) are equal. The result is

$$\left(\frac{\partial u}{\partial v}\right)_T = -p + T\left(\frac{\partial p}{\partial T}\right)_v \qquad \text{(B-1-9)}$$

Note that this expression is valid for any frictionless process.

By restricting our attention to an ideal gas where $pv = RT$

$$\left(\frac{\partial p}{\partial T}\right)_v = \frac{R}{v}$$

and then

$$\left(\frac{\partial u}{\partial v}\right)_T = \frac{RT}{v} - p = 0 \qquad \text{(B-1-10)}$$

Equation (B-1-10) indicates that the internal energy is not a function of v.

Now let us use Equation (B-1-6) with $G = u$, $x = p$, $y = T$, and $z = v$:

$$\left(\frac{\partial u}{\partial p}\right)_T = \left(\frac{\partial u}{\partial v}\right)_T \left(\frac{\partial v}{\partial p}\right)_T = 0 \qquad \text{(B-1-11)}$$

Equation (B-1-11) indicates that the internal energy is not a function of p either. The only property remaining is T.

Therefore, $u = u(T)$ only for an ideal gas. Recall that the definition of enthalpy is

$$h = u + pv$$

$$= u + RT \qquad \text{for an ideal gas}$$

Therefore, $h = h(T)$ only for an ideal gas.

Appendix B-2:
Using the First Law of Thermodynamics

Process Equation for a Reversible Adiabatic Process Involving an Ideal Gas

The first law for an open system, $d(KE) = d(PE) = 0$, in the steady-state–steady-flow process is

$$\delta q = dh + \delta w \tag{B-2-1}$$

If, in addition, the process is adiabatic ($\delta q = 0$) and frictionless ($\delta w = -v\, dp$), this first law becomes

$$0 = dh - v\, dp \tag{B-2-2}$$

Recall that for an idea gas $dh = c_p\, dT$ and $pv = RT$. To eliminate the dT, the equation of state is used (i.e., $dT = (p\, dv + v\, dp)/R$). So the first law becomes

$$0 = \frac{c_p}{R}(p\, dv + v\, dp) - v\, dp \tag{B-2-3}$$

This may be rearranged to the following form

$$c_p p\, dv = -(c_p - R)v\, dp \tag{B-2-4}$$

$$= -c_v v\, dp \quad \text{(that is, } c_p - c_v = R\text{)}$$

or

$$k\frac{dv}{v} + \frac{dp}{p} = 0 \quad \left(k = \frac{c_p}{c_v}\right) \tag{B-2-5}$$

At this point we must impose the restriction of constant specific heat ratio in order to integrate Equation (B-2-5) (that is, k is constant). Integration yields

$$pv^k = \text{constant} \tag{B-2-6}$$

Equation (B-2-6) presents the process equation for a reversible adiabatic process involving an ideal gas with constant specific heat ratio. Note that the same expression could be obtained several other ways—by

568

starting (1) with the first law for a closed system; (2) with the $T \, ds$ relation, Equations (6-16) and (6-17); or (3) with Equation (6-22).

c_v and c_p from the Definition

In order to relate the specific heat definitions of c_v and c_p and the general definition of heat capacity, let us consider the first for a closed system in which there is no change in kinetic or potential energy. Thus for a reversible process

$$\delta q = du + p \, dv \tag{B-2-7}$$

If, in addition, we assume that $u = u\,(T,\,v)$ then

$$du = \left(\frac{\partial u}{\partial T}\right)_v dT + \left(\frac{\partial u}{\partial v}\right)_T dv \tag{B-2-8}$$

Eliminating du of Equation (B-2-7) by using Equation (B-2-8) yields, after rearranging,

$$\delta q = \left[p + \left(\frac{\partial u}{\partial v}\right)_T\right] dv + \left(\frac{\partial u}{\partial T}\right)_v dT \tag{B-2-9}$$

Now we divide Equation (B-2-9) by dT:

$$\frac{\delta q}{dT} = \left[p + \left(\frac{\partial u}{\partial v}\right)_T\right]\frac{dv}{dT} + \left(\frac{\partial u}{\partial T}\right)_v$$

Note that $\delta q/dT$ is a heat capacity. Now consider a constant-volume process

$$c_v = \left(\frac{\delta q}{dT}\right)_v = \left(\frac{\partial u}{\partial T}\right)_v \tag{B-2-10}$$

To determine the relationship for c_p we begin with the definition of enthalpy and assume that $h = h\,(p,\,T)$. Thus, since $h = u + pv$,

$$dh = du + p \, dv + v \, dp \tag{B-2-11}$$

Using Equation (B-2-7) in Equation (B-2-11) yields

$$\delta q = dh - v \, dp \tag{B-2-12}$$

Now

$$dh = \left(\frac{\partial h}{\partial p}\right)_T dp + \left(\frac{\partial h}{\partial T}\right)_p dT \tag{B-2-13}$$

so

$$\delta q = \left[\left(\frac{\partial h}{\partial p}\right)_T - v\right] dp + \left(\frac{\partial h}{\partial T}\right)_p dT \tag{B-2-14}$$

Dividing Equation (B-2-14) by dT yields

$$\frac{\delta q}{dT} = \left[\left(\frac{\partial h}{\partial p}\right)_T - v\right]\frac{dp}{dT} + \left(\frac{\partial h}{\partial T}\right)_p \qquad \text{(B-2-15)}$$

If one limits the process to a constant-pressure process, $dp = 0$. Thus

$$c_p = \left(\frac{\delta q}{dT}\right)_p = \left(\frac{\partial h}{\partial T}\right)_p \qquad \text{(B-2-16)}$$

Appendix B-3:
Proof of the Equivalence of the Clausius and Kelvin–Planck Statements of the Second Law

Although the Clausius and the Kelvin–Planck statements of the second law of thermodynamics sound very different, they are entirely equivalent. To demonstrate this, we will describe situations such that if one statement is violated, the other is also violated. Figure B-3-1 describes a system, composed of a normal engine and a refrigerator that violates the Clausius statement (inside the dashed rectangle). Let us determine the heat exchanges at the reservoirs, calculate the net work done, and, using the first law of thermodynamics, decide if the Kelvin–Planck statement is violated. The net heat from the low-temperature reservoir is zero $(Q_2 - Q_2 = 0)$ whereas the heat transferred into the high-temperature reservoir is $Q_1 - Q_2$. The net work is also $Q_1 - Q_2$. Thus we have violated the Kelvin–Planck statement.

Now let us set up a situation such that the Kelvin–Planck statement is violated (see Figure B-3-2). The dashed rectangle outlines the system composed of a perfectly good refrigerator and an engine violating the Kelvin–Planck statement. As before, let us add the heats and so forth.

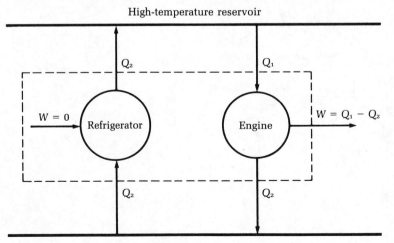

Figure B-3-1. A combination that violates the Clausius statement.

The net heat transferred from the low-temperature reservoir is Q_2; the net transfer into the high-temperature reservoir is also Q_2. Thus we have violated the Clausius statement.

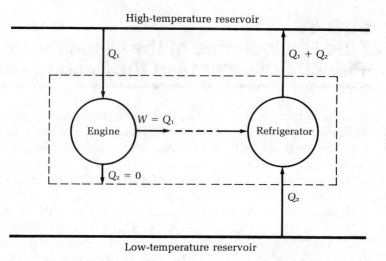

Figure B-3-2. A combination that violates the Kelvin–Planck statement.

Appendix B-4:
Proof of Corollary *A* of the Second Law

Corollary *A* of the second law states: No engine operating between two reservoirs can be more efficient than a Carnot engine operating between the same two reservoirs. To demonstrate this idea, consider two engines. Let *C* represent a Carnot engine, and let *I* represent an engine not operating in a Carnot cycle. Let us further assume that both of these engines operate between the same two reservoirs and deliver the same work, *W*. The Carnot engine receives *Q* from the high-temperature reservoir, while the other one receives *Q'*. Therefore, the Carnot engine rejects $Q - W$ to the low-temperature reservoir while the other engine rejects $Q' - W$. (See Figure B-4-1.) Let us assume the efficiencies of these two engines are related in the following fashion

$$\eta_I > \eta_C \tag{B-4-1}$$

From the definition of efficiency

$$\frac{W}{Q'} > \frac{W}{Q} \tag{B-4-2}$$

which implies

$$Q' < Q \tag{B-4-3}$$

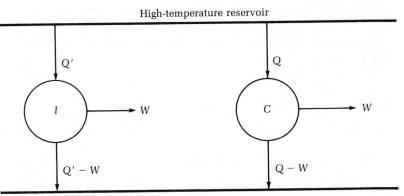

High-temperature reservoir

Low-temperature reservoir

Figure B-4-1. Schematic of two thermodynamic engines.

Now let the Carnot engine be set up to run as a refrigerator. Note that since it is reversible, this action may be accomplished with the magnitudes of Q and W remaining unchanged. Let us now define our system to be the combination of the non-Carnot engine still producing work W—and the reversed Carnot engine being driven by the other engine. (See Figure B-4-2.)

We now can determine the heat transferred from the low- to the high-temperature reservoir. At the low-temperature reservoir, the net heat leaving is $(Q - Q')$. Notice that heat $(Q - Q')$ is entering the high-temperature reservoir. This arrangement contradicts the Clausius statement of the second law.

Going back over the work we have done here, we see that the error must be in the assumption that $\eta_I > \eta_C$. We can only conclude that Equation (B-4-1) is in error and that

$$\eta_I \le \eta_C \tag{B-4-4}$$

which is corollary A.

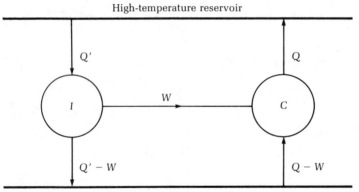

Figure B-4-2. Schematic of a thermodynamic engine and refrigerator (a reversed engine).

Appendix B-5: Evidence of the Truth of the Inequality of Clausius

Consider the heat engine system indicated in Figure B-5-1. We know that the efficiency of this heat engine, if it is reversible, is greater than it would be if there are any irreversible processes in the cycle. Thus

$$\eta_{rev} > \eta_{irrev}$$

and

$$\frac{W}{Q_{in}}\bigg|_{rev} > \frac{W}{Q_{in}}\bigg|_{irrev}$$

or

$$\frac{Q_{in} - Q_{out}}{Q_{in}}\bigg|_{rev} > \frac{Q_{in} - Q_{out}}{Q_{in}}\bigg|_{irrev}$$

For the reversible (Carnot) cycle

$$\frac{T_{in} - T_{out}}{T_{in}}\bigg|_{rev} > \frac{Q_{in} - Q_{out}}{Q_{in}}\bigg|_{irrev}$$

Since T_{in} and T_{out} represent temperatures of the reservoirs ($c \rightarrow$ infinity), they will not be different for the reversible and the irreversible engines. Thus, for the irreversible case,

$$\frac{T_{in} - T_{out}}{T_{in}} > \frac{Q_{in} - Q_{out}}{Q_{in}}$$

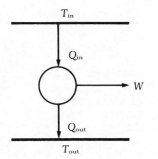

Figure B-5-1. A heat engine system.

This will reduce to

$$\frac{T_{out}}{T_{in}} < \frac{Q_{out}}{Q_{in}}$$

or

$$\frac{Q_{out}}{T_{out}} > \frac{Q_{in}}{T_{in}} \qquad \textbf{(B-5-1)}$$

At this point we will approximate a reversible cycle by a series of Carnot cycles (Figure B-5-2). In this figure:

$$\left.\begin{array}{c} \overline{ab} \\ \overline{cd} \\ \overline{ef} \\ \overline{gh} \\ . \\ . \\ . \end{array}\right\} \quad \text{is an isotherm of} \quad \left\{\begin{array}{c} T_1 \\ T_2 \\ T_3 \\ T_4 \\ . \\ . \\ . \end{array}\right.$$

Note that this approximation will come into better agreement with the cycle as the number of Carnot cycles increases. That is, for cycles $\overline{abcd}$ and $\overline{efgh}$, the effects of their common sides cancel each other ($\overline{bc}$ is down and $\overline{he}$ is up). Thus the cycle is approximated by a series of alternating adiabatic and isothermal lines. As the number of cycles increases, the sawtooth approximation more nearly represents the cycle.

Therefore, for Carnot cycles

$$\frac{Q_1}{T_1} = \frac{Q_2}{T_2} \qquad \textbf{(B-5-2)}$$

$$\frac{Q_3}{T_3} = \frac{Q_4}{T_4}$$

and so on. We may, of course, add these quantities to get

$$\frac{Q_1}{T_1} + \frac{Q_3}{T_3} + \cdots = \frac{Q_2}{T_2} + \frac{Q_4}{T_4} + \cdots$$

or

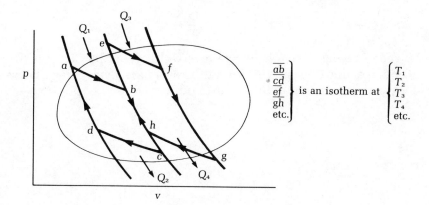

Figure B-5-2. Approximation of a reversible cycle.

$$\sum_{i=1}^{n} \frac{Q_i}{T_i} = \sum_{j=1}^{n} \frac{Q_j}{T_j} \tag{B-5-3}$$

The subscript i represents the portions of the cycle with "heat in" while the subscript j represents the portions of the cycle in which heat is rejected. Hence we may rewrite Equation (B-5-2), on the case of a very large number of little Carnot cycles (that is, infinite), as

$$\int_{in} \frac{\delta Q}{T} = \int_{out} \frac{\delta Q}{T} \tag{B-5-4}$$

or

$$\oint \frac{\delta Q}{T} = 0 \tag{B-5-5}$$

Recall Equation (B-5-1). If we were to apply the same arguments leading to Equation (B-5-4), but this time with an irreversible portion of the cycle, we would get

$$\int_{in} \frac{\delta Q}{T} > \int_{out} \frac{\delta Q}{T}$$

or

$$\oint \frac{\delta Q}{T} < 0 \tag{B-5-6}$$

Thus we may combine Equations (B-5-5) and (B-5-6) to get

$$\oint \frac{\delta Q}{T} \le 0$$

where the equal sign is for the reversible cycle (the Clausius theorem) and the unequal sign is for the irreversible cycle (the Clausius inequality). Notice that this is just corollary F.

Appendix B-6:
Derivation of the General Conduction Equation

To treat more than one-dimensional heat transfer, we need only to apply the first law of thermodynamics (conservation of energy) and Fourier's law. Consider the heat conducted into and out of a unit volume in all three coordinate directions. Figure B-6-1 details only the x direction. The first law, written for this geometric situation, yields

$$\begin{pmatrix} \text{Rate of} \\ \text{heat inflow} \end{pmatrix} + \begin{pmatrix} \text{rate of} \\ \text{heat generation by} \\ \text{internal sources} \end{pmatrix}$$

$$= \begin{pmatrix} \text{rate of} \\ \text{heat outflow} \end{pmatrix} + \begin{pmatrix} \text{rate of} \\ \text{change in} \\ \text{internal energy} \end{pmatrix} \quad \textbf{(B-6-1)}$$

This energy balance yields algebraically

$$q_x + q_y + q_z + q_{gen} = q_{x+dx} + q_{y+dy} + q_{z+dz} + \frac{dE}{dt} \quad \textbf{(B-6-2)}$$

and the energy quantities are given by

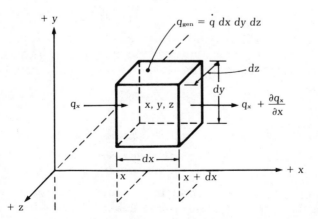

Figure B-6-1. Nomenclature for derivation of the general heat conduction equation in Cartesian coordinates.

$$q_x = -k \, dy \, dz \, \frac{\partial T}{\partial x} \qquad \begin{array}{l}\text{(rate of heat transfer into} \\ \text{left face in x direction)}\end{array} \qquad \textbf{(B-6-3)}$$

$$q_{x+dx} \doteq -\left[k\frac{\partial T}{\partial x} + \frac{\partial}{\partial x}\left(k\frac{\partial T}{\partial x}\right) dx \right] dy \, dz \qquad \begin{array}{l}\text{(rate of heat transfer} \\ \text{out of right face in x} \\ \text{direction)}\end{array}$$

$$\textbf{(B-6-4)}$$

$$q_y = -k \, dx \, dz \, \frac{\partial T}{\partial y} \qquad \begin{array}{l}\text{(rate of heat transfer near} \\ \text{face in y direction)}\end{array} \qquad \textbf{(B-6-5)}$$

$$q_{y+dy} \doteq -\left[k\frac{\partial T}{\partial y} + \frac{\partial}{\partial y}\left(k\frac{\partial T}{\partial y}\right) dy \right] dx \, dz \qquad \begin{array}{l}\text{(rate of heat transfer out} \\ \text{of face in y direction)}\end{array}$$

$$\textbf{(B-6-6)}$$

$$q_z = -k \, dx \, dy \, \frac{\partial T}{\partial z} \qquad \begin{array}{l}\text{(rate of heat transfer into} \\ \text{lower face in z direction)}\end{array} \qquad \textbf{(B-6-7)}$$

$$q_{z+dz} \doteq -\left[k\frac{\partial T}{\partial z} + \frac{\partial}{\partial z}\left(k\frac{\partial T}{\partial z}\right) dz \right] dx \, dy \qquad \begin{array}{l}\text{(rate of heat transfer} \\ \text{into upper face} \quad \textbf{(B-6-8)} \\ \text{in z direction)}\end{array}$$

Notice that Equations (B-6-3) through (B-6-8) represent only the conduction terms; moreover, Equations (B-6-4), (B-6-6), and (B-6-8) are only linear approximations to q_{x+dx}, q_{y+dy}, and q_{z+dz} respectively. The net heat transfer by conduction (only) is

$$-(q_{x+dx} - q_x) - (q_{y+dy} - q_y) - (q_{z+dz} - q_z)$$

$$= \frac{\partial}{\partial x}\left(k\frac{\partial T}{\partial x}\right) + \frac{\partial}{\partial y}\left(k\frac{\partial T}{\partial y}\right) + \frac{\partial}{\partial z}\left(k\frac{\partial T}{\partial z}\right) \qquad \textbf{(B-6-9)}$$

Note that q_{gen} is an approximation that is supposed to account for all other forms of energy which appear in the unit volume by means other than conduction (for example, heating due to electrical activity). Thus

$$q_{gen} = \dot{q} \, dx \, dy \, dz \qquad \textbf{(B-6-10)}$$

Notice that

$$\frac{dE}{dt} = \rho c \, dx \, dy \, dz \, \frac{\partial T}{\partial t} \qquad \textbf{(B-6-11)}$$

This term represents the accumulation of energy within the unit volume. Thus the general three-dimensional heat conduction equation reduces to

$$\frac{\partial}{\partial x}\left(k\frac{\partial T}{\partial x}\right) + \frac{\partial}{\partial y}\left(k\frac{\partial T}{\partial y}\right) + \frac{\partial}{\partial z}\left(k\frac{\partial T}{\partial z}\right) + \dot{q} = \rho c\frac{\partial T}{\partial t} \qquad \textbf{(B-6-12)}$$

For constant thermal conductivity, Equation (B-6-12) may be further reduced to

$$\frac{\partial^2 T}{\partial x^2} + \frac{\partial^2 T}{\partial y^2} + \frac{\partial^2 T}{\partial z^2} + \frac{\dot{q}}{k} = \frac{1}{\alpha}\frac{\partial T}{\partial t} \qquad \textbf{(B-6-13)}$$

where $\alpha = k/\rho c$. This combination of physical properties $(k/\rho c)$ is therefore a property (α) and is called the **thermal diffusivity.** The magnitude of this variable indicates how fast the heat will disperse (or diffuse) from a point in the material. The larger α is, the faster the heat will disperse. One can see how large magnitudes of α occur. If the thermal conductivity is high or the thermal heat capacity is small, the heat at a point is diffused rapidly. Note that usually α is assumed to be a constant and is associated with the transient part of the heat conduction equation.

Recalling that heat is a vectorlike quantity, we can rewrite Equation (B-6-12) in a very general form by appealing to the definitions of vector operation. Thus the vector form of the three-dimensional heat conduction equation is

$$\nabla \cdot (k\nabla T) + \dot{q} = \rho c\frac{\partial T}{\partial t} \qquad \textbf{(B-6-14)}$$

while Equation (B-6-13) is

$$\nabla^2 T + \frac{\dot{q}}{k} = \rho c\frac{\partial T}{\partial t} \qquad \textbf{(B-6-15)}$$

This is a very convenient form, since to change coordinate systems one needs only to recall the definition of $\nabla \cdot (k\nabla T)$ in that system. For example, in a cylindrical situation

$$\frac{1}{r}\frac{\partial}{\partial r}\left(kr\frac{\partial T}{\partial r}\right) + \frac{1}{r^2}\frac{\partial}{\partial \phi}\left(k\frac{\partial T}{\partial \phi}\right) + \frac{\partial}{\partial z}\left(k\frac{\partial T}{\partial z}\right) + \dot{q} = \rho c\frac{\partial T}{\partial t}$$

One final note: In this appendix k, in general, has been allowed to be a function of temperature. But temperature is a function of position (and time), so k is really a function of position (and time).

Appendix B-7:
Blackbody Radiation

The quantum theory resulted from the failure of classic physics to explain experimental facts of thermal radiation. In particular, classic physics could not explain why the intensity of the radiant energy emitted by a blackbody depends on the wavelength. To obtain a true answer, it was necessary to develop a theory for the emission of radiation that was based on a concept entirely different from those of classic physics. The new concept, that of **quanta,** or discrete corpuscles of energy, must be applied to all problems involving the emission and absorption of electromagnetic radiation.

A hot body emits radiation in the form of heat; and thermal radiation differs from x-rays, for example, by having longer wavelengths (λ):

$$\lambda > 7000 \text{ Å} \qquad \text{infrared}$$

$$7000 \text{ Å} > \lambda > 4000 \text{ Å} \qquad \text{visible light}$$

$$4000 \text{ Å} > \lambda > \quad 50 \text{ Å} \qquad \text{ultraviolet radiation}$$

$$\lambda < \quad 50 \text{ Å} \qquad \text{x rays}$$

$$\lambda(\text{x rays}) > \lambda(\gamma \text{ rays})$$

At any temperature, the emitted heat energy is distributed over a continuous spectrum of wavelengths, and this spectral distribution changes with temperature. At low temperatures, relatively long wavelengths are emitted. As the temperature increases, the distribution of energy among the different wavelengths shifts to shorter wavelengths.

In 1879, after considering measurements by Tyndall on the radiation of hot platinum wire, Stefan suggested the empirical rule

$$W = \epsilon\sigma T^4 \qquad\qquad \textbf{(B-7-1)}$$

where W is the rate of emission of radiant energy per unit area (called **total emissive power**), ϵ is the emissivity of the surface ($0 \leq \epsilon \leq 1$), and σ is the Stefan–Boltzmann constant. In 1884 Boltzmann derived Equation (B-7-1) from thermodynamic principles. The W of Equation (B-7-1) is used here only for the sake of history. It is, of course, exactly equivalent to ϵE_b used in the text.

582

To explain the discrepancy between theory and experiment, physicists chose to study the thermal radiation from a "blackbody." The absorptivity of a body is the fraction of the radiant energy incident on its surface that is absorbed. A blackbody is, by definition, a body with absorptivity 1. **Kirchhoff's law** states that the ratio of the monochromatic emissivity to the monochromatic absorptivity is the same for all bodies at the same temperature and is equal to the emissive power of a blackbody at this temperature ($\epsilon_\lambda/\alpha_\lambda = \epsilon_{b\lambda}$). Thus no body emits radiant energy at a greater rate than a blackbody; a blackbody is the most efficient emitter as well as the most efficient absorber.

When the monochromatic emissive power of a blackbody is measured as a function of λ, the curve shown in Figure B-7-1 is obtained. The monochromatic emissive power of spectral emittance, W_λ, is the emissive power per unit range of λ and thus

$$W = \int_0^\infty W_\lambda \, d\lambda$$

Note that W, total emissive power, depends only on temperature and not on the nature of the body. As T increases, W_λ (λ_{max}) increases and the position of this maximum is shifted in the direction of smaller wavelengths. The total radiant energy for a given temperature is represented by the area under the curve (the Stefan–Boltzmann law) and increases as the fourth power of the absolute temperature.

In 1893 Wien showed that it was possible to predict certain features of the spectral distribution of thermal radiation from the laws of classic physics. In his theoretical treatment of the problem Wien used a quantity called the **monochromatic energy density** (U_λ) rather than the monochromatic emissive power (W_λ). The energy density represents the energy contained in a cubic centimeter of volume. The energy den-

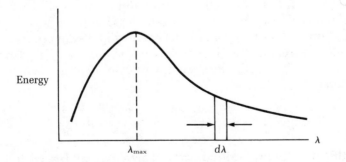

Figure B-7-1. Emissive power of blackbody.

sity and the emissive power are directly related; the proportionality factor contains only geometrical factors and the velocity of light, neither of which depends on λ or temperature. Therefore, it is permissible to use U_λ and W_λ interchangeably as long as absolute values are not required.

In 1896 Wien attempted to predict the shape of the blackbody spectral distribution by assuming that:

1. The radiation was produced by emitters or oscillators of molecular size.

2. The frequency of the radiation was proportional to the kinetic energy of the oscillators.

3. The intensity in any particular wavelength range was proportional to the number of oscillators with the requisite energy.

From the thermodynamic considerations, he obtained

$$U_\lambda = \frac{C_1}{\lambda^5} e^{-C_2/\lambda T} \qquad (C_1 \text{ and } C_2 \text{ are constants}) \qquad \text{(B-7-2)}$$

This formula fit the experimental data well at short wavelengths, but for longer λ it predicted values of U_λ that were too small. It also failed at high temperatures. Even so, Wien's work did imply that the true expression for the distribution should have the form of Equation (B-7-2) for short wavelengths.

Another attempt used classic dynamics and statistics and, in a much more general way, derived an expression. It was about the best that classic physics could do. The result was

$$U_\lambda = \frac{8\pi\kappa T}{\lambda^4} \qquad \text{(B-7-3)}$$

This is called the **Rayleigh–Jeans formula** and it agrees well with the experimental distribution for long wavelengths. (For small wavelengths it does not.) Note that as λ gets small, the energy radiated increases. This increase is in disagreement with experiment (for small λ, U_λ is also very small). And, in addition, the total energy density diverges. Thus for a finite temperature

$$U = \int_0^\infty U_\lambda \, d\lambda = \int_0^\infty \frac{8\pi\kappa T}{\lambda^4} \, d\lambda = \frac{8\pi\kappa T}{5} \lim_{\epsilon \to 0} \frac{1}{\epsilon^5} \qquad \text{(B-7-4)}$$

Therefore the total energy emitted per unit area per unit time is infinite. Couldn't we use this infinite energy to cope with the energy crisis?

Planck in 1901 finally solved the problem by postulating

$$U_\lambda = \frac{8\pi hc}{\lambda^5} \frac{1}{e^{hc/\lambda \kappa T} - 1} \tag{B-7-5}$$

This distribution fits the real-life spectral distribution. It should be emphasized that Planck's assumptions were revolutionary. They were not based on the ordinary lines of reasoning of classic physics but represent an empirical modification of classic physics that brought the theoretical deductions into harmony with experiment. In summary, then, consider the following points:

1. Planck's radiation law leads to the Stefan–Boltzmann law:

$$U = \int_0^\infty U(\lambda)\, d\lambda = 8\pi ch \int \frac{1}{e^{hc/\lambda \kappa T} - 1} \frac{d\lambda}{\lambda^5} = \sigma' T^4$$

$$\sigma' = \frac{8}{15} \frac{\pi^5 \kappa^4}{(ch)^3} = 7.061\,(10^{15})\ \text{erg/cm}^3 \cdot \text{K}^4$$

2. We can compare Planck's law to the Wien and Rayleigh–Jean equations. In the case of λT large, we may expand the term in the denominator:

$$e^{hc/\lambda \kappa T} - 1 = 1 + \frac{hc}{\lambda \kappa T} + \cdots - 1$$

$$\doteq \frac{hc}{\lambda \kappa T}$$

So

$$U_\lambda = \frac{8\pi \kappa T}{\lambda^4} \qquad \text{(Rayleigh–Jeans formula)}$$

In the case of λT small, $e^{hc/\lambda \kappa T} > 1$ so we neglect 1 in the denominator:

$$U_\lambda = \frac{8\pi hc}{\lambda^5} e^{-hc/\lambda \kappa T} \qquad \text{(Wien's formula)}$$

3. Planck's radiation law leads to Wien's displacement law. The position of the maximum value of energy density is obtained by taking

$$\frac{dU_\lambda}{d\lambda} = 0 \quad \text{yields} \quad \lambda_{max} T = \frac{ch}{\kappa} \frac{1}{4.965}$$

Appendix B-8:
Configuration Factor Development

Consider two incremental areas of two different surfaces at constant temperatures T_1 and T_2, respectively (Figure B-8-1). Temperature is not a function of position. The two surfaces are also assumed to be black-body emitters.

The flux (q/A) leaving a point on dA_1 and arriving at dA_2 is

$$\frac{E_1}{\pi} \cos \phi_1 d\omega \qquad \text{(B-8-1)}$$

where E_1 is σT_1^4 and $d\omega$ represents the solid angle (in steradians); that is,

$$d\omega = \frac{dA_2}{r^2} \cos \phi_2 \qquad \text{(B-8-2)}$$

So the expression for the flux is

$$\frac{E_1}{\pi} \cos \phi_1 \cos \phi_2 \frac{dA_2}{r^2} \qquad \text{(B-8-3)}$$

The total energy leaving all of dA_1 and being incident on dA_2 is

$$dE_{1 \to 2} = \frac{E_1}{\pi} \cos \phi_1 \cos \phi_2 \frac{dA_1 dA_2}{r^2} \qquad \text{(B-8-4)}$$

A similar expression for the total energy leaving dA_2 and being incident on dA_1 is

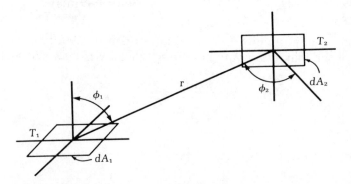

Figure B-8-1. Incremental areas for radiant energy interchange.

$$dE_{2\to1} = \frac{E_2}{\pi} \cos \phi_2 \cos \phi_1 \frac{dA_2 dA_1}{r^2} \qquad \text{(B-8-5)}$$

Of importance to engineers is the net energy transfer from dA_1 to dA_2. Thus

$$d(q/A) \text{ net} = dE_{1\to2} - dE_{2\to1}$$

$$= (E_1 - E_2) \cos \phi_1 \cos \phi_2 \frac{dA_1 dA_2}{\pi r^2} \qquad \text{(B-8-6)}$$

Recalling that the temperature T (and thus T^4) is not a function of position, we see that the total net energy is

$$(q/A) \text{ net} = \int_{A_1} \int_{A_2} dE$$

$$= (E_1 - E_2) \int_{A_1} \int_{A_2} \cos \phi_1 \cos \phi_2 \frac{dA_1 dA_2}{\pi r^2}$$

$$= \sigma(T_1^4 - T_2^4) A_1 F_{12} \qquad \text{(B-8-7)}$$

where

$$A_1 F_{12} = \int_{A_1} \int_{A_2} \cos \phi_1 \cos \phi_2 \frac{dA_1 dA_2}{\pi r^2} \qquad \text{(B-8-8)}$$

By definition F_{ij} is the configuration factor of A_i with respect to A_j and is equal to the fraction of energy leaving A_i that is *directly* incident on A_j. It may be easily seen from Equation (B-8-8) that

$$A_i F_{ij} = A_j F_{ji}$$

Do not be confused by the apparent difference between the verbal definition and the mathematical definition of Equation (B-8-8). Even though the equation is strictly geometric in nature, it is consistent with the verbal definition.

As is easily seen from Equation (B-8-8), evaluating a configuration factor from the basic operational definition is a big task indeed. Thus configuration factor algrebra must be used to solve problems with only a minimum of information (areas, charts, and the like). The basic rules are as follows:

1. Basic reciprocal relation:

$$A_i F_{ij} = A_j F_{ji}$$

2. Summation relation (a conservation of energy statement):

$$\sum_{j=1}^{n} F_{ij} = 1$$

3. Decomposition: The energy leaving an area and being incident on another area that may be considered to be composed of two or more subdivisions is equal to the sum of the energy incident on the individual subdivisions:

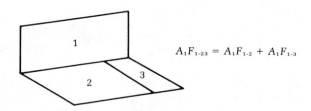

$$A_1 F_{1\text{-}23} = A_1 F_{1\text{-}2} + A_1 F_{1\text{-}3}$$

4. Corresponding corners: If two finite rectangular areas, either adjoint or directly opposed and parallel, are divided into small areas by lines parallel to the outer boundaries, simple configuration-factor relations may be deduced from the definition. These relationships are presented in Figure B-8-2.

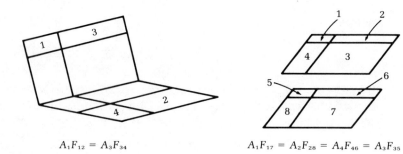

$$A_1 F_{12} = A_3 F_{34} \qquad\qquad A_1 F_{17} = A_2 F_{28} = A_4 F_{46} = A_3 F_{35}$$

(a) Adjoint (no two areas need be equal). (b) Parallel and directly opposed (opposite areas must be equal, $A_1 = A_5$, $A_3 = A_7$, etc.).

Figure B-8-2. Composite surfaces used in radiant interchange.

Appendix B-9: Heat Transfer Dimensionless Groups

Name	Symbol	Significance	Application
Biot number	Bi	$\dfrac{hL}{k}$	transient conduction
Colburn j factor	j	$\left(\dfrac{h}{\rho Vc}\right) Pr^{2/3}$ or $St Pr^{2/3}$	forced convection
Fourier modulus	Fo	$\dfrac{\alpha t}{L^2}$	transient conduction
Friction factor	f	$\dfrac{2g_c}{\rho V^2}\dfrac{D}{L}\Delta P$	forced convection
Graetz number	Gz	$\dfrac{mc_p}{kL}$ or $RePr\dfrac{D}{L}$	laminar convection
Grashof number	Gr	$\dfrac{L^3\rho^2\beta g\Delta T}{\mu^2}$	free convection
Mach number	M	$\dfrac{V}{a}$	high-speed flow
Nusselt number	Nu	$\dfrac{hL}{k}$	convection
Peclet number	Pe	$\dfrac{DV\rho c_p}{k}$ or $RePr$	forced convection
Prandtl number	Pr	$\dfrac{c_p\mu}{k}$	convection
Rayleigh number	Ra	$\dfrac{L^3\rho^2\beta g\Delta Tc_p}{\mu k}$ or $GrPr$	free convection
Reynolds number	Re	$\dfrac{LV\rho}{\mu}$	forced convection
Stanton number	St	$\dfrac{h}{\rho Vc_p}$ or $\dfrac{Nu}{RePr}$	forced convection

Appendix B-10:   More History

Galileo Galilei (1564–1642, Italian) invented a thermometer in 1592, but it did not have a well-founded scale.

Otto von Guericke (1602–1686, German) brilliantly demonstrated, before the emperor at Regensburg in 1654, that he could produce a vacuum with his experiment of the Magdeburg hemispheres.

Evangelista Torricelli (1608–1647, Italian) was primarily a mathematician. But his pioneering work, which correctly distinguished between weight and pressure and demonstrated that air has weight, was fundamental in the beginnings of thermodynamics.

Edme Mariotte (1620–1684, French) was a scientist who independently discovered Boyle's law.

Robert Boyle (1627–1691, English) was the most famous scientist of his day (like Newton, 20 years later). He developed a vacuum pump and deduced his ideal-gas law (p/T = constant) in 1662. Besides his scientific endeavors he led the life of a courtier and public figure. He was also a devout Christian and biblical scholar who endeavored to show that religion and science were not only reconcilable but integrally related.

Sir Isaac Newton (1642–1727, English), often credited with being the greatest scientist of all time, invented a thermometer in 1720 long before the discovery of the first law of thermodynamics. He proposed a scale with zero for the ice point and 12 for the normal human body temperature. He credited his scientific successes to hard work and patient thought.

G. Amontons (1663–1705, French) made major contributions to thermometry in two papers in 1702 and 1703.

Gabriel Daniel Fahrenheit (1686–1736, German) was the first to use mercury-in-glass thermometers indicating the temperature in degrees. Fahrenheit's scale was a modification of one proposed by Sir Isaac Newton. Fahrenheit lowered the zero of Newton's scale to the temperature of a salt–ice mixture and made the degree smaller so that body temperature was 96. Measurements showed the ice and steam points to be at 32 and 212, respectively, based on the reference points at 0 and 96. Subsequently, 32 and 212 were adopted as reference points. Refinements in thermometers since that time have revealed that the minimum temperature of the salt–ice mixture and the normal body temperature are not exactly 0 and 96 on the present Fahrenheit scale.

Anders Celsius (1701–1744, Swedish) was an astronomer from a distinguished scientific family. In 1742 he devised the thermometric scale used today and bears his name (formerly the Centigrade scale).

Joseph Black (1728–1799, Scottish) is best known for his enunciation of the concept of latent heat of transformation and his rediscovery of what turned out to be carbon dioxide.

James Watt (1736–1819, Scottish) was trained to be an instrument maker. While serving in this capacity at the University of Glasgow, he was called upon to repair a model of a Newcomen steam engine. He improved the engine and in fact held patents on most of the basic features of the modern reciprocating steam engine. He also carried on extended research on the properties of steam, about which practically nothing was known at the time. After many attempts at building a commercially viable steam engine and many scrapes with financial disaster, Watt eventually manufactured a successful engine. Engines built by Boulton & Watt, Birmingham, played an important part in the industrial growth of Great Britain during the nineteenth century.

Jacques Alexandre César Charles (1746–1823, French) was an experimentalist and codiscoverer of the ideal-gas law relating volume and temperature (v/T = constant). Actually, it was not Charles who published a description of this gas behavior but Gay-Lussac.

Benjamin Thompson (1753–1814, American) was born in Woburn, Massachusetts, but was made a count of the Holy Roman Empire for the cannon-boring experiments he made while in Bavaria. In these experiments he discovered the equivalence of work and heat (1797) while boring solid metal submerged in water. He convinced himself, but not the world, that the caloric theory of heat (a theory that supposed heat to be a substance without mass) did not explain all known phenomena of heat and that work and heat were in some manner related.

John Dalton (1766–1844, English) was a self-taught scientist. Though color-blind (red), Dalton made most of his contributions in atomic theory and meteorology. His law of partial pressures was a major discovery.

Thomas J. Seebeck (1770–1831, Estonian) discovered the thermocouple in 1821. Although he was educated as a doctor of medicine at Göttingen University in Germany, he chose to lecture and experiment in the physical sciences.

Joseph Louis Gay-Lussac (1778–1850, French) was a chemist who presented the ideas of Charles regarding thermal expansion of an ideal gas. Also to his credit, Gay-Lussac announced a "law" that gases combine chemically in simple proportions by volume. This idea was further extended by Dalton and Avogadro.

Jean-Charles-Athanase Peltier (1785–1845, French) discovered that the junction of two dissimilar metals will absorb or reject heat dependent upon the direction of electrical current. Today this effect is the basis of a thermometer called the thermocouple.

The Reverend **Robert Stirling** (1790–1878, Scottish) was the first person to propose the use of regeneration in heat-engine cycles. He spent 29 years with his brother James designing, improving, and selling his "air engine." The Stirling brothers did not understand the thermodynamic reasons for regeneration (thermodynamics itself had not yet evolved). It was James who had the idea of closing and pressurizing the system, using gas at all times with pressure greater than atmospheric.

Nicholas Leonard Sadi Carnot (1796–1832, French), who lived during the turbulent Napoleonic period, was an officer in the French army engineers. In the only paper he published during his lifetime, "Reflections on the Motive Power of Heat," he devised and analyzed the Carnot cycle. In this paper, written when he was only 23 or 24, he originated the use of cycles in thermodynamic analysis and laid the foundations for the second law by describing and analyzing the Carnot cycle and stating the Carnot principle. Even though he employed the caloric theory in his reasoning, his conclusions are correct because the second law is a principle that is independent of the first law. Carnot's cycle is independent of the theory of heat as well as the working substance.

John Ericsson (1803–1889, Swedish) built a steam locomotive, the Novelty. Among his other inventions were the revolving naval gun turret, the marine screw propeller, and the steam fire engine. He is well known as the designer and builder of the ironclad *Monitor* used by the United States in answer to the Confederate *Merrimac* during the Civil War. Under Ericsson's supervision the *Monitor* was built in only 126 days. During the last few years of his life, he was a recluse in New York City and a disbeliever in the telephone.

Julius Robert von Mayer (1814–1878, German) independently deduced the first law of thermodynamics and properly applied this conservation law. Domestic grief and lack of appreciation of his work prompted Mayer to attempt suicide. Although he was treated in a mental institute and released, his mind never completely recovered.

Alphonse-Eugene Beau de Rochas (1815–1893, French) was an engineer who originated the principle of the four-stroke internal combustion engine. He patented this idea in 1862, but did not develop the engine.

James Prescott Joule (1818–1889, English) inherited a large brewery in Manchester, England. His financial independence made it possible for him to devote his life to scientific research, chiefly in the fields of electricity and thermodynamics. His research was significant in the budding science of thermodynamics, in which he established two fundamental principles: one was the equivalence of heat and work; the other was the dependence of the internal energy change of a perfect gas upon temperature change. As a result of this work, the modern kinetic theory of heat superseded the caloric theory of heat. Joule once remarked, "I believe I have done two or three little things, but nothing to make a fuss about."

William John MacQuorn Rankine (1820–1872, Scottish), while a professor of civil engineering at the University of Glasgow, made several outstanding contributions to the development of thermodynamics—he was the first to write formally on the subject—and its engineering applications. He was a versatile genius and a prolific contributor to the engineering and scientific literature of his day.

Hermann Ludwig Ferdinand Helmholtz (1821–1894, German) is best known for his statement of the law of the conservation of energy (the first law of thermodynamics).

Rudolph Julius Emmanuel Clausius (1822–1888, German), a mathematical physicist, was a genius in mathematical investigations of natural phenomena. After a study of the work of Sadi Carnot he presented in 1850 a clear general statement of the second law. He applied the second law and showed the value of the property he called "entropy" in an exhaustive treatise on steam engines. In addition, his work in the kinetic theory of gases prompted J. C. Maxwell to credit him with being its founder.

William Thomson (1824–1907, English), knighted Lord Kelvin, was professor of natural philosophy at the University of Glasgow for 53 years. He is said by some to be the greatest English physicist. Before his graduation from Cambridge, he had already established a reputation in scientific circles by his original thinking. He contributed most to the science of thermodynamics—having established a thermometric scale of absolute temperatures that is independent of the properties of any gas, having helped establish the first law of thermodynamics on a firm foundation, and having stated significantly the second law. In 1851, he presented a paper in which the first and second laws were combined for the first time.

Pierre Eugene Marcelin Berthelot (1827–1907, French) was a founder of thermochemistry and coined the terms *exothermic* and *endothermic* to describe whether heat leaves or is absorbed by a reaction.

George B. Brayton (1830–1892, American) invented a breech-loading gun, a riveting machine, and a sectional steam generator in addition to the internal combustion engine for which he is best remembered. The Brayton engine, developed around 1870, was a reciprocating oil-burning engine with fuel injection directly into the cylinder and a compressor that was separate from the power cylinder. Although his cycle was first used with reciprocating engines, it is now used only for gas turbines.

James Clerk Maxwell (1831–1879, Scottish), at the age of fifteen, presented a paper to the Royal Society of Edinburgh on the calculation of the refractive index of a material. By the time he was 29, he was a professor of natural philosophy at Kings College, London. He wrote on many scientific matters, but his greatest contributions were in electromagnetic theory. In thermodynamics, he contributed the Maxwell relations.

Nikolaus A. Otto (1832–1891, German), with his partner Eugen Langen, built a gas engine in 1867 in Dertz, Germany, and began marketing it. In 1876, Otto produced a successful four-stroke cycle engine that was far superior to any internal combustion engine previously built. The principle of the four-stroke cycle, however, had been worked out in 1862 by a Frenchman, Alphonse Beau de Rochas.

Gottlieb Wilhelm Daimler (1834–1900, German) patented, in 1885, the first high-speed internal combustion, vertical single-cylinder engine. In 1889 a twin-cylinder V-type engine was patented and used in French cars.

Johannes Dederik van der Waals (1837–1923, Dutch) worked in the area of thermodynamics that deals with the behavior of liquids and gases. Using the work of Clausius, van der Waals postulated the equation of state that bears his name. In 1910, he won a Nobel prize for this work.

Josiah Willard Gibbs (1839–1903, American) received from Yale University in 1863 the first Ph.D. in engineering awarded in America. He undoubtedly contributed more to the science of thermodynamics than any other American, not the least of which is the Gibbs phase rule, but the name of this man, one of the outstanding scientists of all time, is virtually unknown to the general public.

Sir James Dewar (1842–1923, Scottish), educated at Edinburgh University, was elected a professor at Cambridge and later at the Royal Institute in London. His major contributions were his studies of low-temperature phenomena and his invention of the vacuum flask (Dewar flask).

Ludwig Boltzmann (1844–1906, Austrian) had several great achievements—especially the development of statistical mechanics and the statistical explanation of the second law of thermodynamics. During his lifetime he made extensive calculations in the kinetic theory of gases and derived Stefan's law of blackbody radiation using thermodynamics. His statistical approach to the entropy concept was monumental. In fact, the mathematical relation linking entropy and probability is carved on a monument at Boltzmann's grave.

Jacobus Hendricus Van't Hoff (1852–1911, Dutch) was the first Nobel laureate in chemistry (1901). From his first publication in 1874 until his death, he was an active researcher in areas of chemistry that overlapped with thermodynamics.

Herke Karnerlingh Onnes (1853–1926, Dutch) won the 1913 Nobel prize for his successful experiment to produce liquid helium in 1908. His efforts to solidify helium failed, but his student, Willem Henduk Keesom, did succeed in 1926.

Sir Dugald Clerk (1854–1932, Scottish) invented the two-stroke Clerk cycle internal combustion engine used on light motorcycles. He built the engine in 1876 and patented the two-stroke engine in 1881.

Rudolf Diesel (1858–1913, German) obtained in 1893 a patent on the type of engine that now bears his name. One of his engines blew up at the first injection of fuel and Diesel narrowly escaped being killed. Years of tedious and costly experiment elapsed before he produced a successful engine in 1899. He disappeared mysteriously in 1913 while crossing the English Channel in a storm.

Max Karl Ernst Ludwig Planck (1858–1947, German) introduced the quantum theory in 1900 for which he won the Nobel prize in 1918. Planck worked on the writings of Clausius in the area of thermodynamics and clarified the concept of entropy. The roots of his Nobel prize quantum theory are in his mastery of thermodynamics.

Hugh Lougbourne Callendar (1863–1930, English) authored papers on internal combustion engines, thermometric scales, radiation, vapor pressure, and the boiling point of substances. While serving as a professor at several educational institutions (McGill University, University College in London, Imperial College), he directed his primary efforts toward experimentation.

Walther Hermann Nernst (1864–1941, German) was one of the founders of modern physical chemistry, and he also made fundamental contributions to thermodynamics. From 1887 until he retired in 1933, Nernst conducted important research. In 1906 he announced his "third law of thermodynamics" and received the 1920 Nobel prize for it.

Wilhelm Wien (1864–1928, German) was an assistant to Helmholtz. Using thermodynamic principles, Wien studied thermal radiation (paralleling the work of Planck). For his efforts Wien won the 1911 Nobel prize in physics.

Constantin Caratheodory (1873–1950, Greek) was a great mathematician who presented an alternative logical structure of the second law without using the word *heat*.

Percy Williams Bridgman (1882–1961, American) conducted research on materials at extremely high pressures (10^5 atm) and studied their thermodynamic behavior. For this work he received a Nobel prize.

Bibliography

Thermodynamics

Coad, W. J. "Energy Effectiveness Factor." *Heating/Piping/Air Conditioning* (August 1976).

――――. "Second Law Concepts: I." *Heating/Piping/Air Conditioning* (February 1979).

Eastop, T. D. and McConkey, A. *Applied Thermodynamics for Engineering Technologists*. 3rd ed. London: Longmans, 1978.

Faires, V. M. and Simmang, C. M. *Thermodynamics*. 6th ed. New York: Macmillan, 1978.

Gaggioli, R. A. "The Concept of Available Energy." *Chemical Engineering Science* 16(1961):87–96.

――――. "The Concepts of Thermodynamic Friction, Thermal Available Energy, Chemical Available Energy and Thermal Energy." *Chemical Engineering Science* 17(1962):523–530.

Gaggioli, R. A. and Petit, P. J. "Second Law Analysis for Pinpointing the True Inefficiencies in Fuel Conversion Systems." *ACS Symposium Series* 21(2)(1976):56–75. This article also appeared in *Chemical Technology* 1(8)(1977):496–506.

Gaggioli, R. A. and Wepfer, W. J. "Available-Energy Costing—A Cogeneration Case Study." Paper presented at A.I.Ch.E. Meeting 1978.

Gibbs, J. W. *Collected Works*. Vol. 1. New Haven: Yale University Press, 1948.

Gyftopoulos, E. P., Keenan, J. H., and Hatsopoulos, G. N. "Thermodynamics." *Encyclopedia Brittanica*, 1975.

Hatsopoulos, G. N. and Keenan, J. H. *Principles of General Thermodynamics*. New York: Wiley, 1965.

Holman, J. P. *Thermodynamics*. 3rd ed. New York: McGraw-Hill, 1980.

Jones, B. J. and Hawkins, G. A. *Engineering Thermodynamics*. New York: Wiley, 1960.

Keenan, J. H. "A Steam Chart for Second Law Analysis." *Transactions ASME* 54(1932):195.

――――. *Thermodynamics*. New York: Wiley, 1941.

Lay, J. E. *Thermodynamics*. Columbus: Merrill, 1963.

Obert, E. F. *Thermodynamics*. New York: McGraw-Hill, 1948.

――――. *Concepts of Thermodynamics*. New York: McGraw-Hill, 1960.

Tien, C. L. and Lienhard, J. H. *Statistical Thermodynamics*. New York: Holt, Rinehart & Winston, 1971.

Tribus, M. and Evans, R. *Thermoeconomics*. UCLA Report 52-63. Los Angeles: UCLA, 1962.

Tribus, M. and McIrvine, E. "Energy and Information." *Scientific American* (September 1971):121–128.

Van Wylen, G. J. and Sonntag, R. E. *Fundamentals of Classical Thermodynamics*. New York: Wiley, 1965.

Wark, K. *Thermodynamics*. 3rd ed. New York: McGraw-Hill, 1977.

Heat Transfer

ASHRAE. *Handbook of Fundamentals*. New York: ASHRAE, 1981.

Chapman, A. J. *Principles of Heat Transfer*. 3rd ed. New York: Intext, 1974.

Gebhart, B. *Heat Transfer*. 2nd ed. New York: McGraw-Hill, 1971.

Holman, J. P. *Heat Transfer*. 5th ed. New York: McGraw-Hill, 1981.

Kreith, F. *Principles of Heat Transfer*. 3rd ed. New York: Intext, 1973.

Schenck, H. Jr. *Heat Transfer Engineering*. Englewood Cliffs, N.J.: Prentice-Hall, 1959.

Thomas, L. C. *Fundamentals of Heat Transfer*. Englewood Cliffs, N.J.: Prentice-Hall, 1980.

Answers to Selected Exercises

Chapter 1

1-1. 35.0 lbf   **1-3.** 45.9 psia; 6608 psfa or 3.12 std atm   **1-5** 1.88 psig;
16.48 psia   **1-7** 0.03953 lbf/in.^{2} **1-9** 3.6×10^{-9} lbm   **1-13.** $T_R = -218$
1-15. x inexact, y exact, z inexact   **1-17.** $_1W_2 = 429.03$ N · m
1-21. 344,737.7 Pa   **1-23.** exact   **1-25.** 152.37 kg
1-27. m $= 2.97$ (10^2) kg/min $= 3.9293$ lbm/hr   **1-29.** $V_2 = 12.71$ m/s
1-31. (a) $E = 10$ mV, $T = 35.9$ C; (b) $E = 20$ mV, $T = 68.0$ C;
(c) $E = 50$ mV, $T = 149.4$ C   **1-33.** -50.851 kJ/kg (on air);
compressor work is 50.9 kJ/kg

Chapter 2

2-1.

Substance	T. F	p, psia	v, ft^3/lbm	u, Btu/lbm
	20	35.7	0.174	21.68
Freon-12	38	50	0.5	59.52
	114	50	0.960	84.3
	100	1000	0.01608	67.7
Water	80	0.5073	20	79.33
	544.75	1000	0.40	1048.6

Substance	h, Btu/lbm	s, Btu/lbm-R	Condition
Freon-12	22.83	0.049	x = 0.15
	64.07	0.133	x = 0.744
	93.2	0.1882	75° SH
Water	70.68	0.12901	544.75 − 100 = 444.75 SC
	81.22	0.1547	x = 0.0316
	1123	1.321	x = 0.893

2-3.

(a)		0.016293	120.4	0.2141
(b)		0.0163	118	0.2149
(c)	29.8 psi	8.75	816.4	
(d)	250.3 F		314	0.502
(e)		3.06	1322	1.677

2-5. 52.87 Btu/lbm

2-7.

Inlet	Outlet
v = 0.016580	v = 1.2146
h = 170.32	h = 1845.0
u = 167.26	u = 1620.2
s = 0.29281	s = 1.8474
Condition: 344.75° SC	Condition: 1055.25° SH

2-9. $T = 327.81$ F, $v = 4.3415$ **2-11.** $T_F = 303$, $V_i = 0.01773$ lbm, $\Delta h = 23$ Btu/lbm, $\Delta u = 21.1$ Btu/lbm, $\Delta U = 0.0.3874$ Btu

2-13. 0.09474 ft = 1.1369 in.   **2-15.** $T_2 = 120.6$ F

2-17. $v_2 = 0.01608$ ft³/lbm; -458.63 ft³/lbm

2-19. 57.53 Btu/lbm   **2-21.** 324.79 kJ/kg = 139.65 Btu/lbm

2-23. (a) $p = 50$ psia, $v = 0.96$ ft³/lbm, $T \simeq 115$;
(b) $p = 40$ psia, $V = 0.96$ ft³/lbm, $x = 0.973$;
(c) $p = 70$ psia, $V = 0.77$ ft³/lbm, $T = 182$;
(d) $p = 70$ psia, $V = 0.55$ ft³/lbm, $x = 0.953$;
(e) $T = 102$ F, $V = 0.20$ ft³/lbm, $x = 0.653$

2.25. (a) 1.494 lbm; (b) 4.506 lbm, 19.9796 ft³

2-27. (a) $_1W_2 = 536.19$ Btu/lbm; (b) $_1W_2 = 264.3$ Btu/lbm

2-29. $T \triangleq 5T_2 - 4T_1$ **2-31.** (a) $T = 20$ C, $u = 2000$ kJ/kg, $x = 0.826$;
(b) $p = 2$ MPa, $v = 0.1$ m³/kg, S_1H_1 1.1 C, $T = 213.5$ C;
(c) $T = 140$ C, $v = 0.5089$ m³/kg, $x = 1$ (sat. vap.);
(d) $p = 4$ MPa, $v = 0.04$ m³/kg, $x = 0.798$;
(e) $p = 2$ MPa, $v = 0.111$ m³/kg, SH 36.1 C, $T = 248.5$ C

2-33. 1564.92 kJ/kg   **2-35.** 0.31 kJ/(kg · k)

~1.0

2-37. $w = -416.6$ kJ/kg

2-39. $m_g = 5.794$ kg, $m_f = 4.21$ kg, $V_g = 84.998$ m³, $V_f = 0.002$m³

2-41. c_p(average) $= 0.9275$ kJ/(kgm K)

Chapter 3

3-1. $T = 312.03$ F, $v = 3.33$ ft³/lbm   **3-3.** (a) 252.7 F; (b) 20.84 ft³/lbm;
(c) 63.048 Btu/lbm   **3-5.** $du = -34,884$ Btu/hr

3-7. (a) -0.178 Btu/lbm-R; (b) 5833.3 Btu/hr   **3-9.** 48.96 Btu/lbm, 38.16
Btu/lbm, 0.0774 Btu/lbm-R   **3-11.** 277.6 ft/sec

3-13. (a) 23.058 ft³/lbm; (b) $\Delta v = 21.87$ ft³/lbm, 3016.03 ft³/min,
$\Delta h = c_p(T_2 - T_1) = 145.2$ Btu/lbm, $\Delta u = c_v(T_2 - T_1) = 103.45$ Btu/lbm,
$\Delta s = 0.3174$ Btu/lbm R, $\dot{m} = 7848$ lbm/hr   **3-15.** -7187.4 ft-lbf

3-17. 589 N·m   **3-19.** $m = (100 - x) = 8.4\%$ **3-21.** 0.35 ft³

3-23. process ab: $p_1V_1 = RT$; process adb: $3/2$ RT

3-27. $s_2 - s_1 = -0.008735$ Btu/lbm-F   **3-29.** $du = [1/(k - 1)] d(pv)$,
$dh = [k/(k - 1)] d(pv)$ **3-31.** -180.07 kJ/kg   **3-33.** $P_r = 0.452$,
$T_r = 1.04$; from compressibility chart $Z = 0.85$

3-35. $q = -3309.25$ kJ/kg   **3-37.** 33.96%

3-43. (a) $\dot{m} = 5.22(10^4)$ kg/hr; (b) $\dot{V} = 5.49(10^4)$ L/hr $= 0.915(10^6)$ L/hr;
(c) $T = 1100$ C $\rightarrow T_2 - T_1 = 950$ C;
(d) $\Delta h = 4865.1 - 632.2 = 4232.9$ kJ/kg $= \Delta h$

Chapter 4

4-1. 101.74, 0.8294, 1505.1, 1718.1, 85.1, 1.2, 1175.1, 1111.8, 57.8, 0.027,
0.054, 81.74, 5.18, 120, 1749.6, 544.6, 0   **4-3.** (a) 684 Btu; $\Delta H = 957.95$
Btu, $Q = MC_p\Delta T = 684$ Btu, $W = 0$; (b) 960 Btu; $\Delta u = 685.57$ Btu; 274 Btu;
$Q = 960$ Btu   **4-5.** wattage $= 74.766$ kW, current $= 339.85$ A

4-7. 3564.3 hp   **4-9.** 3290 ft/sec   **4-11.** (a) -1015.57 Btu/lbm;
(b) -976.47 Btu/lbm   **4-13.** 279.226 ft³/min   **4-15.** $-\dot{w} = 3.73$ hp

4-17. 34.9%   **4-19.**(a) 7.078×10^{12} Btu/hr; (b) 6.43×10^5 lbm/hr;
(c) 5.37×10^{12} Btu/hr   **4-21.** 2.55×10^{11} Btu/hr   **4-23.** $\Delta u = 10$ kJ

4-25 (a) $T_1 = 110$ F; (b) $T_2 = -30$ F; (c) $x_2 = 0.563$;
(d) $\Delta v = 1.7043$ ft³/lbm   **4-27.** $\dot{m} = 78.5$ lbm/hr

4-29. $\Delta u = 2785$ Btu   **4-31.** $\Delta h = 316$ Btu/lbm

4-33. $m = 0.4$ lbm/m, $Q = 200$ Btu/hr   **4-35.** $_aQ_b = 9$ RT_1, $17/2$ RT_1

4-37. $h_2 = 1148.5$ Btu/lbm   **4-39.** (a) $V = 19.76$ ft³; (b) $V_2 = 7.236$ ft³;
(c) $T_2 = 388.1$ R; (d) $W = 2.28762$ (10^4) ft-lbm $= 29.4$ Btu

4-41 (a) 1736.26 kg/hr; (b) 0.613 m³/hr; (c) 0.747 cm

4-43. $T_2 = T_1 = 260$ C, $1.069(10^{12})$ m³/kg

4-45. $D = 0.389$ m   **4-47.** (a) $3.0794(10^8)$ kJ/hr; (b) $4.678(10^8)$ kJ/hr

Chapter 5

5-1. (a) cop = 1.83; (b) 0.4   **5.3** 1.99 kW   **5.5** 21.1 kW
5-7. x = 92.4 ft^2 **5-9.** 0.08 hp, 311.4 F   **5-11.** 166.67 kJ
5-13. 151.89 kJ, 143.73 Btu   **5-15.** 291 Btu, 200 Btu, η_{HE} = 0.6875,
η_R = 1.5   **5-17.** 521.4 Btu/min   **5-19.** (a) T = 482 K, 209 C; (b) 1664 kJ;
(c) W_1 = 264 kJ, W_2 = 159 kJ; (d) η_1 = 0.3975, η_2 = 0.3975
5-21. W = 4.9 kJ   **5-25.** T_b = 298.5075, T_f = 218.5075

Chapter 6

6-1. (a) $\geq$, $\geq$, =, $\leq$; (b) $\geq$, $>$, $\geq$, $\leq$. $\leq$, $\geq$ **6-3.** (a) $S_2 - S_1 =$
$c_p \ln(T_2/T_1)$; (b) like (a) if ideal path is made; (c) $S_2 - S_1 > 0$;
(d) like (b) if ideal path is made
6-5. (Ia) 4.843(10^5) Btu/hr; (Ib) 0.3956 Btu/lbm > 0.031294 Btu/lbm-F;
(II) claim is all right   **6-7.** w = -93.35 Btu/lbm, -0.052446 Btu/lbm-R,
-34.34 Btu/lbm   **6-9.** (a) 1204.1 R, 325.11 lbf/in.^{2}; (b) 90.7 hp
6-11. (a) 290 C; (b) 1.14318 m^3/kg; (c) -373.8 kJ/kg; (d) -523.3 kJ/kg;
(e) 522 kJ/kg   **6-13.** 68.83 F   **6-15.** 7.723 in.^{2}
6-17. 3225.27 hp   **6-19.** (a) -359.3 Btu; (b) -359.3 Btu out
6-21. (a) 0.92; (b) 460 kW-hr; (c) -0.5 kW-hr/K, 0.5 kW-hr/K;
(d) -0.333 kW-hr/K, 0.5 kW-hr/K; (e) 0, 0.1667 kW-hr/K   **6-23.** no
6-25. no   **6-27.** -0.024514 Btu/lbm-R   **6-29.** 0.304
6-35. w = -10.36 Btu/lbm   **6-37** 0.28715 kJ/kg · K
6-39. 0.29423(10^6) kJ/kg, 0.4119(10^6) kJ/kg   **6-41.** no work is done—it is a
free expansion   **6-47.** (a) η = 0.1711, w = 130($s_2 - s_1$);
(b) η = 0.27892, w = 201.7($s_2 - s_1$); (c) η = 0.206, w = 130($s_2 - s_1$);
(d) η = 0.3421, w = 260($s_2 - s_1$)

Chapter 7

7-1. 9661.7 Btu/lbm   **7-3.** (a) m = 2805.6 lbm/hr;
(b) w = mw = 0.2624 hp   **7-5.** $-w$ = 1.776 hp   **7-7.** 0.547 Btu/lbm
7-9. \$303.92/hr   **7-11.** 5.11 hp   **7-13.** 364.37 hp
7-15. 8102.54 R   **7-17.** T^2/T^1 = 14.97, w = 243.6 Btu/lbm
7-19. (a) 8343 Btu/lbm; (b) 1260.6 Btu/lbm; (c) 40%; (d) 3034.34 kW;
(e) 204.54/hr   **7-21.** w_T = 190.89 kW, η = 79%
7-23. (a) 95.22 mw; (c) 34.5%   **7-25.** (a) 1.0793 × 10^9 Btu/hr;
(b) 5.3728 × 10^9 Btu/hr; (c) 24.1%; (d) 34.8%; (e) 69.2%
7-27. (a) 22,750 Btu; (b) 293.64 Btu/R
7-29. (a) -18, 112 = 32 Btu/hr;   **7-31.** (a) 13.42 hp; (b) 2.5;
(c) 100,156 Btu/hr   **7-33.** (a) V_2 = 307 m/sec; (b) η = 99.89%

7-35. (b) 5.34   **7-37.** 2.77
7-39. (a)

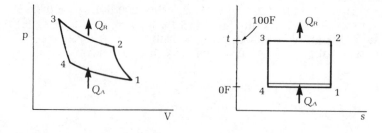

(b) x_1 = 96.2%; x_4 = 29.36%; (c) 3.714   **7-41.** (a) η = 0.6627,
W = 397.6 Btu; (b) Δs_{high} = −0.3614 Btu/K, Δs_{low} = 0.3614 Btu/K
7-43. (a) − 92.4 Btu/lbm; (b) 244.6 Btu/lbm; (c) 197.5 Btu/lbm; (d) 0.43
7-45. (a) D = 0.228 ft = 2.74 in.; (b) 2.01(10^4) kW = 131.43 hp;
(c) 2.01(10^4) kW; (d) 2.203 × 10^8 Btu/hr; (e) 2.89 × 10^8 Btu/hr; (f) 23.7%;
(g) 27.3%; (h) $458.51/hr   **7-47.** w = 64.3 hp, m_f = 0.551 lbm/min
7-49. 324 C   **7-51.** 0.1029 kJ/(kg · K)   **7-53.** Δs = 0.179 kJ/(kg · K),
w = − 176.4 kJ/kg, Q = − 22.16 kJ/kg   **7-55.** η = 25.7%,
$-w_p$ = 4.01, η = 36.6%   **7-57.** A_T = 6.19(10^{-4}) m^2,
A_2 = 0.070 m^2 **7-59.** 40.07%   **7-61.** 56.7%   **7-63.** 1.73
7-65. m = 2.928(10^5) lbm/min   **7-67.** $V_{7\%}$ = 2.089(10^9) ft^3
7-69. 21.47 hp   **7-71.** 616.89 ft^2 **7-73.** $-w$ = 0.16187 in.-lbf/sec
7-75. 0.349   **7-81.** (a) 232.39 K; (b) 21.13 K

Chapter 8

8-1.

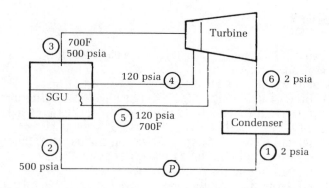

(b) 29%; (c) 7.9   **8-3.** $W_{p/1} = 159.28$ hp, $W_{p/2} = 2700$ hp,
$d = 12.48$ ft   **8-5.** $\eta = 46.6\%$
8-7. 720.5 kJ/kg, 23%   **8-9.** (a) 35%; (b) 0.099
8-11. 50.9%   **8-13.** 28%   **8-15.** $\Delta\eta = 31.8$

Chapter 9

9-1. (a) 103.2 Btu/lbm; (b) 11.0 Btu/lbm; (c) 134.8 Btu/lbm;
(d) 92.2 Btu/lbm; (e) 92.2 Btu/lbm; (f) 92.1 Btu/lbm; (g) 0
9-3. (a) $m = 1126(10^6)$ lbm/hr; (b) 3.28 Btu/lbm; (c) 98 Btu/lbm;
(d) 12.58 Btu/lbm; (e) 1.11 Btu/lbm; (f) 86.53 Btu/lbm; (g) 128.9 Btu/lbm;
(h) 2.54%; (i) 3.79%; (j) 83.25 Btu/lbm   **9-5.** (a) 190.5 Btu/lbm;
(b) -52 Btu/lbm; (c) -27 Btu/lbm; (d) -31 Btu/lbm
9-7. worse—more auxiliary energy would be added in A since it will encounter heat exchange loss.   **9-9.** 45.01 kJ/kg

Chapter 10

10-1. $P_{O_2} = 52.59$ kN/m^2, $P_{N_2} = 196.8$ kPa, $P_{CO} = 171$ kPa
10-3. 325 lbm, 32.6 lbm/lb-mole   **10-5.** 0.14833 Btu/lbm-R
10-7. (a) 39.4%; (b) 0.0086 lbm/lbm; (c) 53.15 F; (d) 1.19 lbm
10-9. (a) 65.2 F; (b) 30.2 Btu/lbm; (c) 0.0112 lbm/lbm

10-11.

Dry Bulb	Wet Bulb	Dew Point	Humidity Ratio	ϕ	h	v
80	63.5	53.7	0.0083	40	28.8	13.8
70	55	43	0.0053	38	23.1	13.47
100	78	70	0.016	39	42	14.47
97	77	68	0.0151	40	40	14.3
79	65	57	0.01	46	30	13.8
86	60.0	40	0.0052	20	26.4	13.85
40	29	11	0.001	20	10.3	12.6
74	65	60	0.011	60	30	13.7
85	70	62	0.012	47	33.8	14.0
80	30	80	0.0239	100	43.8	14.1

10-13.

Dry-Bulb Temperature, C	Wet-Bulb Temperature, C	Dew-Point Temperature, C	Humidity Ratio, kg/kg	Relative Humidity, %	Enthalpy, kJ/kg	Specific Volume, m³/kg
26.5	17.3	12.2	0.0087	41	49	0.86
21	13	7.6	0.006	40	36.7	0.84
38	25.4	21	0.0155	38	78.2	0.905
41.7	29.2	25.5	0.0207	40	95	0.92
22.2	17	14.2	0.01	60	48	0.85
32	16	4	0.005	18	44.5	0.87
4	− 2	− 10	0.001	20	6.5	0.78
39.8	23.4	16	0.0115	26	70	0.904
30	21	17	0.012	45	61	0.875
27	27	27	0.0227	100	85.5	0.88

10-15. 4.02×10^{-4} psi, 4.02×10^{-3} psi   **10-17.** (a) 0.0088 kg/kg;
(b) 11.8 C; (c) 0.584 kg   **10-19.** (a) 3.16 lbm/min; (b) − 5860 Btu/min
10-21. (a) 49%; (b) 41 F; (c) 0.0054 lbm/lbm; (d) 20.3 Btu/lbm;
(e) h_F = 25.2 Btu/lbm; (f) 4.9 Btu/lbm; (g) 25%   **10-23.** 96 F
10-25. 0.01044 lbm/lbm air, 0.467, 30.65 Btu/lbm-hp, 13.842 ft³/lbm
10-27. 214 m/s   **10-29.** (a) 257.8 tons removed; (b) 1287 lbm/hr;
(c) 140.5 tons, 143.3 tons, 143.2 tons; (d) 120.0 tons, 115.4 tons;
(e) 54.2 F, 115.0 tons   **10-31.** 114.5 F   **10-33.** 22.23 Btu/lbm,
0.0070 lbm/lbm   **10-35.** (a) t_{DB} = 81.5 F;
(b) w = 0.0112 lbm/lbm; (c) ϕ = 48%; (d) h = 32.0 Btu/lbm;
(e) T_{AP} = 60.5 F   **10-37.** T_2 = 28.2 C, Q = − 2.77 kJ
10-39. T_2 = 74 F, T_{WB} = 64 F   **10-41.** (a) 81.967 lbm/hr;
(b) T = 65 F, T = 53 F; (c) 0.50   **10-43.** (a) 76.6 F, 54.5 F;
(b) 25.9 Tons   **10-45.** W_m = 0.00524 lbm/hr, T_m = 54.5 F, ϕ_m = 53%
10-47. T_2 = 63.2 F, 0.00552 lbm   **10-49.** ϕ = 45.8%, T_{dp} = 75.3 F,
m = 0.463   **10-51.** (a) 0.0177 lb/kg; (b) T_{dp} = 52 F; (c) ϕ = 32%
10-53. normally within 1 F for a shielded thermometer in air–water vapor
mixtures (because the Lewis relation is approximately equal to 1.0)
10-55. m = 69.5 lb/min, 15,250 Btu/hr, 7430 Btu/hr, RH = 70%

Chapter 11

11-1. 8.0 lb air/lb c   **11-3.** (a) A/F = 14.55 lb air/lb fuel,
m_{H_2} = 1.34 lb H_2O/lb fuel, P_w = 2.47 psia, DP = 133.9 F;
(b) 5.94%, 11.04%, 83.02%   **11-5.** 32.5%   **11-7** (a) 1.286 lbm/lb fuel;
(b) 1.924 psia; (c) 15.1%; (d) 379.5 ft³/lb   **11-9.** 17.49 lb air/lb fuel
11-11. (a) wet: CO_2: 11.50%,: 11.65%, O_2: 2.49%, N_2: 74.40%; dry: CO_2:
12.95%, O_2: 2.82%, N_2: 24.20%; (b) 120 F **11-13.** (a) 19,385 Btu/lbm;

(b) 103.6 lbm air/gal fuel   **11-15.** (a) 1672.5 Btu/lbm$_f$; (b) 29.4 Btu/lbm$_f$
11-17. (a) $23,417; (b) $2,629.85   **11-19.** See Problem 11-7

Chapter 12

12-1 (a) k(air) = 0.02 Btu/hr-ft-F = 0.0346 W/(m $\cdot$ C);
(b) k(water) = 0.35 Btu/hr-ft-F = 0.60 W/(m $\cdot$ C);
(c) k(concrete) = 0.55 Btu/hr-ft-F = 0.95 W/(m $\cdot$ C);
(d) k(steel) = 26 Btu/hr-ft-F = 45 W/(m $\cdot$ C);
(e) c(steel) = 0.12 Btu/hr-ft-F = 502.416 J/(kg $\cdot$ K);
(f) c(water) = 0.24 Btu/lbm-F = 1004.8 J/(kg $\cdot$ K);
(g) c_p(air) = 1 Btu/lbm-F = 4186.8 J/(kg $\cdot$ K)
12-3. (a) C(3/4-in. plywood) = 1.07 Btu/hr-ft^2 = 3.3746 W/m^2;
(b) C(2-in. insulated roof deck) = 0.022 Btu/hr-ft^2 = 0.074 W/m^2;
(c) C(1/2-in. sand aggregate gypsum plaster) = 2.25 Btu/hr-ft^2 = 7.1 W/m^2
12-5 ΔX_c = 23.48 in.   **12-7.** T_1 = 1786 F; (a) 7.04 in.;
(b) 0.774 F/ft; (c) 2698.1 F/ft   **12-9.** 258 Btu/hr-ft
12-11. q_A = $U(T_1 - T_2)$ = 10.9 Btu/hr-ft^2-F   **12-13.** (a) $T(X = 1)$ = 525;
(b) T = 600 − 300/4 = 525°   **12-15.** 87.36 F, − 2.7K in. at X = 0.75
12-17. 300.22°   **12-19.** 7.61175(10^2) amp
12-21. t = 0.31223 hr = 18.7358 min
12-23. (a) T_1 = 341.9 F, T_2 = 343.9 F; (b) 106 F; (c) 54.19 Btu/hr
12-25. 438 F, 6.75 W   **12-27.** 0.909
12-31. (a) 10.8238 Btu/hr; (b) 125.32 F   **12-35.** 66.7 F
12-37. U = 1/R = 104.724 per foot   **12-39.**(a) 336.1 F;
(b) 3.2 Btu/hr-ft^2-F   **12-41.** 853.2 lbm/hr, V = 189.6 ft^3/min
12-43. 955.6 Btu/hr   **12-45.** 3784 Btu/hr = 1.109 kW
12-47. L = 0.129 ft   **12-49.** 4.608 F   **12-51.** 32.8 Btu/hr-ft^2-F
12-53. T_0 = 80.6 F   **12-55.** h = 953.7 Btu/hr-ft^2-F   **12-57.** 14.47 W/m
12-59. 2/15   **12-61.** 0.027 W/cm$^2\mu$ **12-63.** 37$\frac{1}{2}$%
12-65. 4.625(10^{-10}), 5.39(10^{-6})   **12-67.** 289.8 Btu/hr
12-69. 4920.4 Btu/hr   **12-71.** 168.9 Btu/hr-ft   **12-73.** 397.347 Btu/hr-ft^2
12-75. (a) 2103 Btu/hr, − 2120 Btu/hr, 0; (b) 3¢/hr; (c) 94 F
12-77. −424 F   **12-79.** 6.6908(10^7) Btu/hr   **12-81.** − 64 F
12-83. 114.53 lbm/hr   **12-85.** 2246.4 Btu/hr

Index

Nomenclature

a = acceleration (ft/sec^2) (m/s^2)

A = area (ft^2) (m^2)

c = specific heat (Btu/lbm-F) (kj/(kg·K))

D = diameter (ft) (m)

E = energy (Btu) (kJ)

E_b = Blackbody emissive power (Btu/hr-ft^2) (W/m^2)

F = force (lbf) (N)

F_{ij} = configuration factor (angle factor)

g = local acceleration of gravity (ft/sec^2) (m/s^2)

g_c = mass to force conversion factor (32.174 lbm-ft/lbf-s^2) (kg·m/N·s^2)

Gr = $g\beta L^3(T_w-T_\infty)/\nu^2$ (Grashof number)

h = specific enthalpy (Btu/lbm) (kJ/kg)

= convection coefficient (Btu/hr-ft^2-F) (W/m^2·c)

H = enthalpy (Btu) (kJ)

I = intensity (Btu/hr-ft^2-sr) (W/m^2·sr)

J = mechanical equivalent of heat (778.2 ft-lbf/Btu)

k = c_p/c_v, ratio of specific heats

= thermal conductivity (Btu/hr-ft-F) (W/m·C)

m = mass (lbm) (kg)

$\dot{m}$ = mass rate (lbm/sec) (kg/s)

M = molecular weight (lbm/lbm-mole) (kg/kg-mole)

n = m/M, number of moles

= polytropic index

Nu = hL/k, (Nusselt number)

Q = heat (Btu) (kJ)

$\dot{Q}$ = heat rate (Btu/hr) (W)

q = heat per unit mass (Btu/lbm or kJ/kg) in thermodynamics

= heat (Btu/hr or W) in heat transfer

$\dot{q}$ = volume generation coefficient (Btu/hr-ft^3) (W/m^3)

p = pressure (lbf/in.^{2}) (Pa=N/m^2)

P = perimeter (ft) (m)

Pr = $\mu c_p/k$ (Prandtl number)

$\bar{R}$ = universal gas constant (1545 ft-lbf/mole-R) (8.3132 J/mole-K)

R = $\bar{R}/M$, gas constant (lbf-ft/F) (kN·M/kg·K)

= thermal resistance (F/Btu) (K/kJ)

= radiosity (Btu/hr-ft^2) (W/m^2)

Re = $\rho VL/\mu$ (Reynolds number)

St = $Nu/RePr$ (Stanton number)

s = specific entropy (Btu/F-lbm) (kJ/kg·K)

S = entropy (Btu/F) (kJ/K)

t = time (sec)

T = temperature (F, C, R, K)

u = specific internal energy (Btu/lbm) (kJ/kg)

U = internal energy (Btu) (kJ)

= overall heat transfer coefficient (Btu/hr-ft^2-F) (W/m^2·k)

v = specific volume (ft^3/lbm) (m^3/kg)

$\bar{v}$ = volume per mole (ft^3/mole) (m^3/mole)

V = volume (ft^3) (m^3)

V = velocity (ft/sec) (m/s)

w = specific work (ft-lbf/lbm) (kJ/kg)

W = work (ft-lbf) (kJ)

= humidity ratio

W_b = blackbody emissive power (Btu/hr-ft^2) (W/m^2)

x = quality

Z = compressibility factor

α = $k/\rho c$, thermal diffusivity (ft^2/sec) (m^2/s)

= absorptance

β = coefficient of performance

= volumetric coefficient of thermal expansion $[(1/v)/(\partial v/\partial T)_p]$

ϵ = emittance

η = efficiency

κ = Boltzmann constant

λ = wavelength (ft) (m)

μ = viscosity (lbm/ft-sec) (kg/m·s)

ν = μ/ρ, kinematic viscosity (ft^2/sec) (m^2/s)

ρ = density (lbm/ft^3) (kg/m^3)

= reflectance

σ = Stefan–Boltzmann constant (0.1714×10^{-8} Btu/hr-ft^2-R^4) (5.6697×10^{-8} W/m^2·K^4)

τ = transmittance

ϕ = angle (degrees)

ω = solid angle (steradian)